# Molecular Crystals
# and Molecules

This is Volume 29 of
PHYSICAL CHEMISTRY
A series of monographs
Edited by ERNEST M. LOEBL, *Polytechnic Institute of Brooklyn*

A complete list of the books in this series appears at the end of the volume.

# Molecular Crystals and Molecules

A. I. Kitaigorodsky

*Academy of Sciences*
*Moscow, USSR*

ACADEMIC PRESS   New York and London      1973

*A Subsidiary of Harcourt Brace Jovanovich, Publishers*

ACADEMIC PRESS, INC.
111 Fifth Avenue, New York, New York 10003

*United Kingdom Edition published by*
ACADEMIC PRESS, INC. (LONDON) LTD.
24/28 Oval Road, London NW1

LIBRARY OF CONGRESS CATALOG CARD NUMBER: 78-182616

PRINTED IN THE UNITED STATES OF AMERICA

Sci
R

# Contents

PREFACE                                                                                          *ix*

## Chapter I   Structure of Crystals

### A. Close-Packing Principle

1. Geometrical Model of a Molecular Crystal                                                      1
2. Determination of Intermolecular Radii                                                        10
3. Packing Coefficient                                                                          18
4. Close Packing and Crystal Symmetry                                                           21
5. Closest-Packed Plane Groups of Symmetry                                                      24
6. Space Groups Suitable for Close Packing of Molecules                                         33

### B. Typical Structures

7. Linear Aromatic Systems                                                                      38
8. Nonlinear Condensed Aromatic Molecules of Symmetry *mm* and *mmm*                             43
9. Structure of Normal Paraffins                                                                48
10. Organo-Iron Compounds                                                                       62
11. Tetraaryl Compounds                                                                         67
12. Polymorphic Modifications                                                                   71
13. Hydrogen Bonds in Crystals                                                                  74

### C. Crystals with Elements of Disorder

14. Rigid Disorder                                                                              85
15. Rotational Crystalline State                                                                89

### D. Binary Systems

16. Conditions for Formation of Solid Solutions                                                 94
17. Determination of Phase Diagrams                                                            105
18. X-Ray Diffraction of Solid Solution Crystals                                               108
19. Geometrical Analysis and Energy Calculations                                               115
20. Molecular Compounds                                                                        121
    References                                                                                 130

# Chapter II   Lattice Energy

**A. Interactions of Molecules**

1. van der Waals Forces 134
2. Calculation of the Lattice Energy of Molecular Crystals 140

**B. Electrostatic Energy**

3. The Dipole–Dipole Interaction in a Molecular Crystal 144
4. The Quadrupole Energy 154
5. Concluding Remarks 160

**C. The Device of Atom–Atom Potentials**

6. The van der Waals Interactions in a Molecular Crystal 161
7. Potential Curves 163
8. Energy as a Function of Lattice Parameters 167
9. Calculation of the Structure and of the Energy Surfaces for Benzene, Naphthalene, and Anthracene Crystals 170
10. The Condition for the Structural Stability of an Organic Crystal and the Principle of Close Packing 184
11. The Effect of the Crystalline Field on the Shape of a Molecule 186
    References 190

# Chapter III   Lattice Dynamics

1. The Equations of Motion 193
2. Selection of the Coordinate System 196
3. The Coupling Coefficients 197
4. The Limiting Frequencies and Their Eigenvectors 202
5. The Dynamic Problem for a Naphthalene Crystal 213
6. Calculation of Crystal Dynamics by the Method of Atom–Atom Potentials 219
7. The Mean Vibration Amplitude 227
8. Reorientation of Molecules 230
   References 232

# Chapter IV   Methods of Investigating Structure and Molecular Movement

**A. Diffraction Methods**

1. Methods of Structure Determination, Their Accuracy and Objectivity 233
2. Principles of the Diffraction Method of Studying Crystal Structure 238
3. Sphericity of Atoms 244
4. Accuracy of Structural Determinations 247
5. Comparison of X-Ray, Electron, and Neutron Diffraction Analyses 249

6. Finding and Elucidating the Structures of Molecular Crystals                         256
7. Heat Wave Scattering                                                                  262

**B. Nuclear Magnetic Resonance**

1. Theoretical Fundamentals of Nuclear Magnetic Resonance in a Solid                     269
2. Investigation of Molecular Movement in a Crystal by the NMR Method                    274
3. Determining Proton Coordinates in Organic Crystals                                    277
4. Theory of Nuclear Quadrupole Resonance                                                282
5. Use of NQR in Studying the Structure of Molecular Crystals                            288
   References                                                                            293

## Chapter V   **Thermodynamic Experiments**

1. Measuring Thermal Expansion                                                           294
2. Measuring the Elasticity Tensor of a Single Crystal                                   307
3. Calculating Elastic Constants of Single Crystals from Experimentally
   Measured Elastic Wave Velocities                                                      310
4. Elasticity Tensors of Naphthalene, Stilbene, Tolan, and Dibenzyl Single
   Crystals at Room Temperature and Normal Pressure                                      313
5. Investigation of the Elastic Properties of Polycrystalline Samples                    321
6. Measuring and Calculating Elastic Properties of Polycrystals                          325
7. Calorimetry                                                                           328
8. Isothermal Compressibility                                                            330
9. Measuring the Heat of Sublimation                                                     331
   References                                                                            336

## Chapter VI   **The Theory of Thermodynamics**

1. General Relationships                                                                 337
2. Specific Features of the Thermodynamics of Molecular Crystals.
   Introduction of the Characteristic Temperature                                        343
3. Experimental Characteristic Temperature                                               351
4. Thermodynamic Functions of a Naphthalene Crystal                                      357
5. Choice of an Optimal Quasi-Harmonic Model                                             369
6. Calculation of the Quasi-Harmonic Model by the Atom–Atom
   Potential Method                                                                      372
   References                                                                            380

## Chapter VII   **Conformations of Organic Molecules**

1. The Mechanical Model of a Molecule                                                    381
2. Parameters for Conformational Calculations                                            386
3. Internal Rotation in Molecules                                                        394
4. Conformations of Aliphatic Molecules                                                  398
5. Ethylenic, Conjugated, and Aromatic Systems                                           408
6. Geometry of Molecules and Thermochemical Properties of Substances                     426
7. Consistent Force Field                                                                441
   References                                                                            446

Chapter VIII  **Conformations of Macromolecules and Biopolymers**

1. The Structure of Stereoregular Macromolecules in Crystals 451
2. Conformations of Peptides and Proteins 479
3. Conformations of Polynucleotides and Nucleic Acids 510
   References 531

AUTHOR INDEX 537
SUBJECT INDEX 548

# Preface

This book deals with some of the problems of molecular crystallography, and certain aspects of molecular structure. At first sight the selection of material may appear somewhat unsystematic. True, the treatment of molecular crystals is restricted to the problems of lattice structure, dynamics, and thermodynamics, and molecular structure itself is discussed only from the viewpoint of molecular conformation.

However, the material in the book has by no means been selected at random; if anything, it is somewhat one-sided. This book is neither a textbook nor a manual. It contains mainly results obtained by the author and this makes unavoidable some degree of imbalance in the treatment of the subject.

The general idea that brings together the entire range of problems discussed in the book is the atom–atom potential model. An analysis of the electronic structure of molecules is not fundamental to the purposes of this book. On the contrary, a much more rational approach is to consider the atoms composing a molecule as the basic building blocks of our model. The energy of the relationships involved are thus confined to atomic interactions only.

Thus, we deal with a certain type of model which is undoubtedly approximate but which, nevertheless, rather successfully deals with the problems of the structures and thermodynamic properties of molecular crystals, and also those concerning the conformations of small and large molecules. The model in question is that of noncovalent interactions, and it is precisely these interactions that underlie the problems treated in the book.

The author hopes that the two parts of the book which deal, respectively, with crystals and molecules are sufficiently self-consistent. Both parts deal with organic substances; inorganic molecular crystals occur but rarely, and, conversely, organic crystals, with the exception of organic salts, always belong to the class of molecular crystals. The physics of molecular crystals is essentially organic crystal physics.

Until now textbooks on solid-state physics have devoted but a few pages to molecular crystals and, as a rule, have not gone beyond the properties of nitrogen, oxygen, carbon oxide, and other similar crystals. This approach distorts the real situation. Crystals composed of small molecules are by no means representative of the class. The "common salt" of molecular crystal physics is the naphthalene crystal.

For a long time solid organic substances have been ignored as materials. Therefore, several major problems of molecular crystal physics are still awaiting their investigator. The thermal and mechanical properties of crystals, their molecular composition, and other problems treated from the point of view that organic matter is a solid have, until recently, attracted the attention of very few physicists.

As to chemists, they usually ask merely for information on the structure of a molecule, since this information is relevant to chemical reactions. This kind of pressure on the part of chemists has resulted in an unbalanced development of our subject: organic crystal structures have been studied mostly because the only way to establish the structures of complex molecules is to examine them in the crystalline state.

The lack of active interest in the structures and properties of organic crystals as such, and the very few publications which seek to correlate the arrangement of molecules in a crystal with the crystal's properties, may be excused as long as organic substances are regarded as dissolved chemical reagents. In recent years, however, researchers have begun to take a closer look at problems associated with solid organic substances. Synthetic polymeric compounds have come to the fore, peculiar features of solid-state reactions have been observed and, finally, and this is perhaps the most important development, it is becoming obvious that numerous biological processes are most intimately linked with the mutual arrangement of organic molecules and their parts in protein and nucleic acid crystals. Molecular crystal physics will undoubtedly become the basis for research along these lines.

The author began studying the regularities in the mutual orientations of molecules in crystals 25 years ago. At the first stage of these studies, a simple geometrical model was proposed for the interpretation of crystal structures. It was found that if a molecule is bounded by van der Waals' radii and is thus "shaped," a crystal can be represented as a close packing of solid molecules. This geometrical model—a first approximation model—and its potential use for predicting the structures of crystals are dealt with in the author's book "Organic Chemical Crystallography" (Akademizdat, 1955; Consultants Bureau, New York, 1961).

The principle of close packing of molecules in crystals naturally suggests the idea of describing the energy of molecular interactions as the sum of the interactions of the component atoms. This idea gave rise to a new improved model which can be used to predict a crystal's structure to a greater degree of accuracy, and also to estimate quantitatively the thermodynamic properties of a crystalline compound.

The use of the atom–atom potential scheme does not, however, fully define the model of a crystal. Lattice dynamics can be described in different ways. We have chosen a quasi-harmonic model. Simple models are more approximate, but then they are more general. For the purpose of our research, it is probably more important to be able to predict the structure and properties (even allowing the predictions to be not quite accurate), than to seek ideal agreement between theory and experiment by treating the subject in terms of special-purpose models. In my opinion it is more advantageous to have a rough theory applicable to most molecular crystals than a fine theory useful only for crystals of benzene or urotropin.

Our studies in the field of intermolecular interactions, i.e., in the sphere of the structures and properties of organic crystals, have been developing concurrently with our research on molecular conformations since we treated them as different applications of the same model.

Historically, studies of conformations have received considerably greater attention than have studies of the structures and properties of organic crystals. Therefore, the part of the book devoted to conformations of molecules is essentially a review of the results published in the literature. It is easy to understand the reasons for the spectacular progress of conformational research which has used the atom–atom potential method as a reliable starting point. While experimental structural investigations of small molecules are relatively easy to perform, the situation is just the opposite with macromolecules. It is often impossible to determine the conformation of a molecule experimentally; in other cases, such as, for example, protein molecules, the experiment becomes extremely complicated. Therefore, attempts to determine a priori the molecular conformations should be encouraged. It is particularly important to apply conformational calculations to biological substances since the results that may be expected justify the effort. Science takes an interest in all biopolymers since each such molecules has a part in the life cycle.

This book has been written with the assistance of my younger colleagues. I am especially grateful to R. M. Myasnikova for her valuable cooperation in writing the first chapter. The part of the book dealing with molecular conformations has been written by V. G. Dashevsky. In other chapters I have used material from the publications, reviews, and theses of K. V. Mirskaya, A. P. Ryzhenkov, B. V. Koreshkov, V. F. Teslenko, Yu. T. Struchkov, R. L. Avoyan, E. Mukhtarov, Yu. V. Mnyukh, E. I. Fedin, and G. K. Semin. I. E. Kozlova has been a great help with the calculations and in preparing the manuscript for publication.

Very special thanks are due to David Harker of the Roswell Park Memorial Institute, Buffalo. His generous and able contributions in reading and editing both manuscript and proof are greatly appreciated.

I am greatly indebted to all my colleagues, without whose help this book would not have been possible.

# Molecular Crystals
# and Molecules

# Chapter I
## Structure of Crystals

### A. Close-Packing Principle†

1. GEOMETRICAL MODEL OF A MOLECULAR CRYSTAL

The distinguishing geometrical peculiarity of a molecular crystal is self-evident. If it is possible to single out groups of atoms in a crystal in which the distance from each atom of the group to at least one other atom of the same group is significantly smaller than the distances to the adjacent atoms of other groups, the crystal is said to be molecular. Thus, for example, in hydrocarbon crystals spacings between H atoms in different molecules are not less than 2.2–2.4 Å, whereas distances from these atoms to chemically bonded C or N atoms are about 1 Å. Intramolecular spacings between bonded C atoms are 1.2–1.5 Å, while the atoms of different molecules are at least 3.3 Å apart. The order of difference between intramolecular and intermolecular distances in crystals composed of diatomic O and N molecules is about the same.

Organic molecules containing hydroxyl (OH) and amine (NH$_2$) groups

---

† For a detailed description of the principle of close packing of molecules and illustration of this principle on a large body of material, see Ref. [1].

1

display a tendency to form so-called hydrogen bonds. In a hydrogen-bonded structure, two O atoms, or one O atom and one N atom, or two N atoms of adjacent molecules are often so arranged that the proton is situated approximately on a straight line drawn between these two atoms. In this case, the difference between the intramolecular and intermolecular distances is somewhat smaller: The proton positioned between two O atoms lies 1 Å from the O atom of the same molecule, and 1.6–1.8 Å from that of a different molecule. But in this case, too, the molecule may be quite unambiguously defined as a group of closely bonded atoms.

When speaking about molecular crystals, we shall mean mainly crystals of organic substances. Only relatively few inorganic compounds, such as nitrogen or oxygen crystals, or molecules built from nitrogen and phosphorous atoms, or carbonyls or complex compounds of certain metals, form molecular crystals. There are inorganic substances which may be classified as molecular crystals only formally, since their intramolecular and intermolecular distances do not differ by more than 5 10%. As to organic substances, they are all molecular crystals, with the obvious exception of organic salts. The class of organic compounds is infinite, and, therefore, the physics of molecular crystals is primarily the physics of organic crystals.

The mutual orientation of molecules in a crystal is conditioned by the shortest distances between the atoms of adjacent molecules. In most instances the packing of the molecules is determined by the interactions between H atoms, or the interaction of H atoms with atoms of other elements.

In early X-ray analyses, the coordinates of the H atoms were not determined, and the summary tables which presented the results listed only the distances between the C atoms of adjacent molecules, which is completely insufficient for determining the type of packing.

Recent papers report the coordinates of H atoms obtained by computing the smallest $R$-factor by the least-squares method (see Chapter IV). These data should be treated with caution. As a rule, the values of the valence bond lengths of a H atom derived by X-ray diffraction analysis are considerably underestimated. Apart from such systematic errors, X-ray analysis is also often plagued with large random errors. In determining proton coordinates, particular attention should be given to neutron diffraction studies. The nuclear magnetic resonance (NMR) technique also offers many potentialities so far largely unexplored. However, one need not necessarily resort to experimental analysis, because the a priori localization of H atoms does not involve any particular difficulties.

Studies of gas molecules have furnished a great body of information on the lengths of C–H, O–H, etc. bonds; this enables the correct prediction of the bond lengths in a majority of the compounds that have not yet been studied. Data obtained from gas molecule experiments can, with a high degree of certainty, be applied to the crystalline state in molecular crystal studies. The crystalline field has no effect whatsoever on bond lengths and influences but slightly the valence angles. The only material effect of the crystalline field is that it changes the conformation of the molecule whenever the possibility of rotation about single bonds presents itself. The effect of the crystalline field will be discussed in Section II.11, therefore, I will restrict myself to pointing out that this effect does not handicap the a priori determination of the coordinates of H atoms.

After the positions of all the atoms in a molecule have been determined by appropriate methods, we can pass on to the analysis of the molecular packing. The fundamental result of such an analysis is that the shortest distances between the atoms of the same chemical elements vary quite insignificantly; this fact graphically demonstrates that a crystal can be, to a fair approximation, described by a model, which may be called a geometrical model or a first-approximation model. The geometrical model is constructed as follows: We start by analyzing spacings between the atoms of neighboring molecules to find the shortest interatomic distances for all molecular pairs (e.g., for molecules related by translations along the $a$ axis, the $c$ axis, and by the symmetry operations of the twofold screw axis for the group $P2_1/a$). These are "determining contacts." Then, we select the values of the intermolecular atomic radii whose sums show the closest agreement with experimental data. These intermolecular radii are the mean radii for a given structure. With the aid of these radii, a volumetric model of the molecule can be constructed. It is obvious that since the sums of the mean intermolecular radii differ somewhat from the actual distances, the model will constitute a packing in which some atom pairs overlap slightly, while other pairs do not touch each other. It will always be possible to produce ideal packing (i.e., a packing in which no molecules are suspended in empty space and none overlap) by minor translational and rotational shifts of the molecules.

The difference between the ideal and actual packings is usually very small. For example, for a naphthalene crystal, the ideal packing with coordination number 12 in an experimentally observed cell at room temperature is obtained if we take a radius of 1.76 Å for carbon, and a radius of 1.18 Å for hydrogen. The orientations of the molecules in the ideal and the experimentally observed structures do not differ by more than 2–3°. A fact essential for the technique of X-ray diffraction analysis is that the ideal packing derived from strictly geometrical considerations furnishes a structural model that can be employed for the initial calculation of structure amplitudes.

The mean values of the intermolecular radii vary, though within narrow limits, from structure to structure.

Consider several typical examples illustrating the validity of the geometrical model, in other words, the possibility of representing an organic crystal structure by close packing of molecules connected by intermolecular radii.

(a) The structure seeker apparatus.

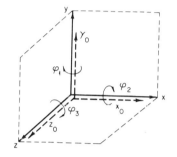

(b) The designation of the angles.

*Fig. 1.*

(The projections of one molecule are inserted into the hollows of others.) To make such a representation more descriptive, it is desirable to have both volumetric models of molecules and a structure seeker, which is an instrument enabling the determination of the various types of molecular stacking in a unit cell of a given size. Figure 1 illustrates this operational technique. In certain cases the nature of the molecular packing can be clearly seen from appropriate figures. For example, Fig. 2 shows the $xyO$ projection of the structure of 1,5-dinitronaphthalene [2]. The structure is characterized by a rather planar unit cell: $a = 7.76$, $b = 16.32$, $c = 3.70$ Å, $\beta = 101° 48'$, space group $P2_1/a$, $Z = 2$. The plane of the naphthalene nucleus of the molecule practically coincides with the $ab$ plane of the unit cell. The N atoms lie in the same plane, while the O atoms project from the plane alternately upward and downward by about 0.79 Å. The planes of the nitro groups make an angle of 49° with the plane of the aromatic ring, and the molecule retains its centrosymmetry. If we position the hydrogen atoms in each molecule and connect

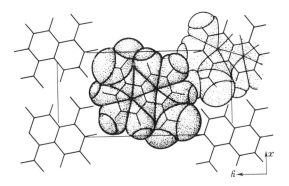

**Fig. 2.** Projection $xyO$ of the 1,5-dinitronaphthalene structure.

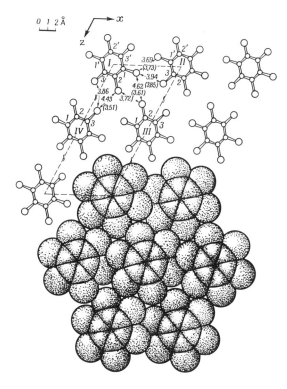

**Fig. 3.** Projection $xOz$ of hexachlorobenzene structure.

all the molecules by appropriate intermolecular radii, a peculiar feature of the molecular packing becomes obvious: The "projections" of one molecule get into the "hollows" of adjacent molecules, so that the molecules are closely packed with the minimum voids between them possible for the given cell.

Another simple example demonstrating the projection-to-hollow principle is the structure of hexachlorobenzene [3] (unit cell: $a = 8.08$, $b = 3.87$, $c = 16.65$ Å, $\beta = 117.0°$, space group $P2_1/c$, $Z = 2$). The plane of the hexachlorobenzene molecule is inclined to the plane $xOz$ at a small angle (approximately 22°), therefore, molecular packing can be seen well in the projection on the $ac$ face (Fig. 3).

It should be emphasized once again that, as is evidenced by the analysis of a large number of molecular structures, molecular packing on the dovetail principle is a general rule of organic chemical crystallography. Because of such packing, organic structures usually have a high coordination number, i.e., a large number of contacting molecules. Due to the irregular shape of the molecules, this circumstance is not evident and must be proved (see Section I.9). One can also refer to the experiment which shows that the coordination number 12 is the most common for organic structures (the same as for the closest spherical packing). Structures with 10 or 14 contacting molecules are less frequent.

This "model" approach to the principles of structure of organic crystals permitted the author to develop 25 years ago the geometrical-analysis technique that made possible (at any rate for a cell of known dimensions) the a priori determination of the structure of a molecular crystal, i.e., before making direct X-ray diffraction study.

The essence of the geometrical analysis will be made clear from the description of the structure of 2,6-dimethylnaphthalene, which was obtained using this method [4]. 2,6-Dimethylnaphthalene crystals are orthorhombic; the parameters of the units cell are: $a = 7.54$, $b = 6.07$, $c = 20.20$ Å, space group $V_h^{15} = Pbca$, $Z = 4$. Thus, the 2,6-dimethylnaphthalene molecule (I) takes a

(I)

centrosymmetric position in a crystal. Four models of these molecules constructed by means of approximate intermolecular radii $R_C = 1.72$ Å and $R_H = 1.2$ Å were arranged on the structure-seeker apparatus so that the distances between their centers corresponded to the unit-cell dimensions. It was then found that only one version of initial molecule orientation is possible (the orientations of the remaining three molecules are automatically derived from that of the first one by means of the symmetry elements of the given space group): $\phi_1 = 5°$, $\phi_2 = 2°$, $\phi_3 = 23°$ (for the notation of the angles see Fig. 1). The low accuracy of the operation on the structure seeker was compensated for by subsequent geometrical calculations. Figure 4 shows the *ab* cross section of the crystal cell drawn through C atoms 1 and 4 of molecules related by the translation *b*, i.e., through the most closely packed place. The cross section is constructed for angles $\phi_1 = \phi_2 = 0$. From this figure it can be seen that contacts determining molecule orientation exist, first, between the C4 atom of one molecule and the H1 atom of the second molecule separated from the first one by the translation *b*; second, the smallest rotation through angle $\phi_1$ will bring the same H1 atom into contact with the carbon atom of the molecule related to the previous two molecules by the *b* glide plane which is perpendicular to *a*.

The first contact is defined by the angles $\phi_2$ and $\phi_3$, the second contact by angles $\phi_1$ and $\phi_3$. The following notation will be used: $r_{CC}$ the length of the C–C bond in the naphthalene nucleus, $r'_{CC}$, the internuclear length of the bond nucleus carbon–methyl group carbon; $r_{CH}$ the length of the C–H bond; $\varepsilon$, HCC angle of the methyl group; $R_H$ and $R_C$, intermolecular radii of hydrogen and carbon atoms. Now we write the conditions for the contacts:

a. for molecules related by the translation *b*:

$$(R_H + R_C)^2 = b^2 + (2r_{CC} + r_{CH})^2 - 2b(r_{CC} + r_{CH})\cos\phi_3 - \cos\phi_2$$

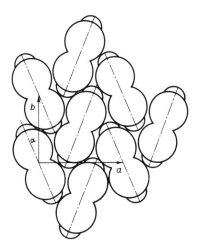

**Fig. 4.** Cross section *ab* of 2,6-dimethylnaphthalene unit cell with maximum number of contacts.

b. for molecules related by a glide plane:

$$(R_H + R_C)^2 = [\tfrac{1}{2}b - r_{CH}\cos\phi_3]^2 + [(a/b) - (2r_{CC} + r_{CH})\sin\phi_3$$
$$- 2r_{CC}\cos 30° \sin\phi_1]$$

Consideration of the cross section drawn through the methyl group centers gives the condition of contact for the case of free rotation of the $-CH_3$ group. Two more conditions of the contact will be derived from consideration of the contacts between molecules belonging to different layers (we shall not give all the formulas here, but refer the reader to the original paper).

Putting the bond lengths equal to $r_{CC} = 1.405 \pm 0.005$, $r_{CH} = 1.085 \pm 0.005$, and $r'_{CC} = 1.54 \pm 0.005$, we obtain for the angles which determine molecular orientation in the cell the following values: $\phi_1 = -3°55'$, $\phi_2 = 4°10'$, $\phi_3 = 22°45'$, and for intermolecular radii values, $R_H = 1.19$ and $R_C = 1.72$ Å.

Knowing the values of $\phi_1, \phi_2, \phi_3$, one can readily calculate the coordinates of all the atoms in the unit cell. The values of structure amplitudes $F_{hkl}$ calculated from these coordinates showed satisfactory agreement with the experimental data, and thereby confirmed the correctness of the structure so obtained.

In this way, geometrical analysis makes it possible (of course, in not too complicated cases) to determine a molecular packing which can be called ideal. It should not be very different from the real packing found by direct X-ray diffraction studies.

It must be understood that differences between the ideal and real molecule packing patterns are not accidental. A rough model (a first-approximation model) quite naturally fails to satisfy the minimum molecular interaction energy requirement. However, in certain cases, as, e.g., for an iodoform crystal, the reasons for the departure of the real packing from the ideal one can be interpreted within the framework of the geometrical model and without energy calculations.

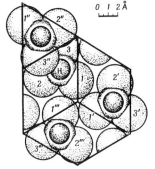

*Fig. 5.* Projection of iodoform structure on face *ab*.

The crystals of iodoform $CHI_3$ belong to space group $C_6^6 = C6_3$ with two molecules in a cell. The cell parameters are $a = 6.83$, $c = 7.53$ Å. The C and H atoms are positioned on the 3 axis, the I atoms are in general positions [5]. Consideration of Fig. 5, which presents the packing of molecules in the $ab$ layer, suggests the following question: Why does compression of packing not occur through a small rotation of the molecule around the 3 axis, so that each iodine atom touches not two iodine atoms of two molecules, but four iodine atoms of three molecules, as shown in Fig. 6? The explanation seems to lie in the fact that such an arrangement in the layer would involve a considerable reduction in the layer stacking density.

Let us now compare the parameters of the cells of the real structure and the ideal structure, which has "incompressible" packing in the $ab$ layer: actual structure: $a = 6.83$, $c = 7.53$ Å; $V = 303$ Å$^3$; ideal structure: $a = 6.67$, $c = 7.96$ Å, $V = 307$ Å$^3$.

The cell volumes are practically the same; in other words, closer packing in the layer does not result in an increase of the density of packing in the cell as the layers move apart. The shortest distances $I \cdots I$ between the molecules in different layers increase as follows: $2 \cdots 3''$ from 4.41 to 4.60 Å; $3 \cdots 3''$ from 4.43 to 4.50 Å; the distance $H \cdots 3''$ remains exactly equal to the sum of the intermolecular radii.

Thus, the real structure of iodoform whose molecular packing coefficient (see below) $k = 0.76$ is "better" than the ideal structure.

It should be pointed out that the ideal packing, i.e., packing with equal intermolecular distances, may not necessarily be optimum from the viewpoint of our second-approximation model (atom–atom potential scheme) which treats interaction between molecules as a sum of generally valid atom–atom potentials (see Chapter II) and considers forces between the atoms to be central forces. A pronounced minimum of potential energy for the calculated structure of a naphthalene crystal (see Chapter II, 180) need not correspond to a structure in which the determining contacts between the atoms of the same element are equal to each other.

Consequently, minor differences between the real and ideal packing patterns of molecules in this crystal are adequately interpreted by the second-approximation model. In particular, the "short" $C \cdots H$ distance between molecules

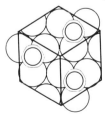

*Fig. 6.* Hypothetical iodoform structure with incompressible packing in layer $ab$.

related by axis $2_1$, and a somewhat "increased" $H \cdots H$ distance between the molecules of adjacent layers (see distances in Fig. 35) are obtained automatically through minimization-of-energy calculated with universal equilibrium distances on the atom–atom potential curves.

Thus, a departure from the additivity law in the shortest intermolecular contacts should certainly not be taken as an indication of "specific" forces. From the point of view of our more sophisticated model it may turn out more advantageous to compress contacts in certain places and to expand them elsewhere than to retain equal contact distances between all pairs of the molecules.† The approximation of the geometrical model consists merely in the fact that it is concerned only with atomic contacts, though interaction between the atoms does not diminish so quickly that this interaction can be neglected for atomic pairs separated by distances exceeding the sum of the radii.

It is obvious that the nearest atoms of the adjacent molecules play a key role, but differences between the ideal and real packings depend primarily on the interaction of atoms lying farther apart. In Chapter II it will be shown that the structure is determined by adjacent atoms located at distances up to 10 Å from the initial molecule.

Now let us turn to the geometrical model of a crystal.

## 2. Determination of Intermolecular Radii

It is required to find the shape of a molecule proceeding from the concept of "contact" between each pair of adjacent molecules. Contacts may be effected by one pair of atoms or several pairs of atoms, atoms of the same species or different species. One should take into account the shortest distances between the atoms of each pair of adjacent molecules, which are what we call "determining" contacts.

To avoid ambiguity in the selection of the magnitudes of the intermolecular radii from the interatomic distance, investigations must be started using structures where the determining contacts are definitely known to involve atoms of one chemical element. Proceeding then to structures with two, three, etc., types of contacts, one may use data on the intermolecular radii of the atoms of one species derived in the previous study for determining new radii.

A sufficient number of molecular crystals is known in which the surface of the molecules is composed of the atoms of one species only. The structures of

---

†In particular, I strongly doubt that the introduction of "specific forces" is required in the interpretation of a structure of the alloxan type (the appropriate reference is [9]).

such crystals may serve as standards for determining intermolecular radii. The description of some of these structures follows.

### a. Intermolecular Radius of the Hydrogen Atom

Adamantane (sym-tricyclodecane $(CH)_4(CH_2)_6$) is crystallized in a cubic face-centered cell with $a = 9.426 \pm 0.008$ Å, space group: $T_d{}^2 = F\bar{4}3m$, $Z = 4$; Molecular symmetry: point group, $\bar{4}3m$; in crystal $\bar{4}3m$. Figure 7 illustrates the crystalline structure of adamantane (hydrogen atoms are not shown).

X-ray diffraction studies [6] have given only the coordinates of the C atoms. The coordinates of H atoms can be easily found geometrically under the assumption that the lengths of C–H bonds in the CH and $CH_2$ groups are equal (1.08 Å) and that all the valence angles C–C–H and H–C–H are tetrahedral.

The adamantane molecule bounded by hydrogen atoms is approximately spherical. Adjacent molecules make contact only through hydrogen atoms: The H atom in each CH group touches H atoms in three CH groups in adjacent molecules; each H atom in any $CH_2$ group touches one H atom of a $CH_2$ group in an adjacent molecule. There are no CH–$CH_2$ contacts via H atoms.

Of these 24 contacts only two are symmetrically independent, and in both cases the H $\cdots$ H distance is found to equal 2.34 Å. It follows that the intermolecular radius of a hydrogen atom $R_H = 1.17$ Å.

No subsequent X-ray diffraction studies have required corrections of this value $R_H$.

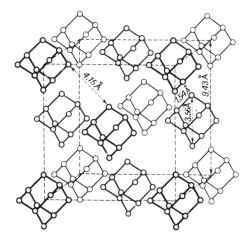

*Fig. 7.* Crystalline structure of adamantane. H atoms are not shown.

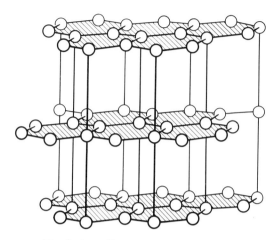

**Fig. 8.** Crystalline structure of graphite.

*b. Intermolecular Radius of the Carbon Atom*

It would be logical to estimate the intermolecular carbon radius from the structure of graphite. The structure of graphite displays hexagonal symmetry: space group $D_{6h}^4 = C6/mmc$, $Z = 4$; $a = 2.46$, $c = 6.70$ Å. The structure is laminar (Fig. 8). Atoms in each layer are arranged so as to form a net of regular hexagons with sides of 1.42 Å. In other words, each layer is one infinite aromatic molecule. Atoms have no valence bonds across the layer, and, therefore, C atoms are located at a distance of one intermolecular diameter. The layers are stacked upon each other with a displacement, so that translation $c$ equals twice the distance between adjacent layers. The shortest distance between the atoms of adjacent layers is 3.35 Å.

Net C⋯C distances are found from consideration of the structures of crystals formed by large aromatic molecules, which may be regarded as "fragments" of the graphite layer. In all these cases (a number of structures will be discussed in Part B of this chapter) the molecules are packed with maximum density so that a largest possible number of atoms of one molecule are accommodated between the atoms of the adjacent molecules. But the

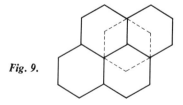

**Fig. 9.**

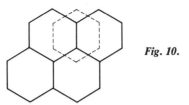

*Fig. 10.*

cellular structure renders unfeasible 100% projection-to-hollow packing. It would seem optimum to locate an atom of an adjacent molecule in the center of a hexagon. This is just the case of graphite, but then another atom happens to be located precisely over the underlying atom (Fig. 9). This means that one-half of all the superpositions are optimum; six adjacent atoms are placed at an equal distance $(3.35^2 + 1.4^2)^{1/2} = 3.64$ Å, and the other half of the superpositions are less advantageous (atom-to-atom) with one distance of 3.35 Å and three distances of 3.64 Å.

The most logical approach (although logic here is always a relative concept) would be to set the mean radius equal to $\frac{1}{2}(0.9 \times 3.64 + 0.1 \times 3.35)$, that is, $R_C = 1.80$ Å.

It is not surprising that aromatic molecules present various solutions to the stacking problem. In most cases, all atoms assume the same position. It would be easy to understand that all C atoms of two adjacent aromatic molecules will take up the same position with respect to their nearest neighbors in the case of the stacking shown in Fig. 10. The nearest neighbor will now be located at the distance $(d^2 + 0.7^2)^{1/2}$, two at distances $(d^2 + 1.2^2)^{1/2}$, another two at distance of $(d^2 + 1.2^2 + 1.4^2)^{1/2}$, and one at a distance of $(d^2 + 2.1^2)^{1/2}$. With $d = 3.35$ Å as in graphite, this would give, respectively, values of 1.72, 1.91, and 1.98 Å. It may be thought that the latter two figures are exaggerated and assumed that the corresponding "contacts" involve three neighbors (see the figure). But a distance of 3.35 Å between the planes is never realized in aromatic molecules. Usually it ranges within 3.4–3.5 Å (see Part B of this chapter), confirming a carbon radius $R_C = 1.80$ Å. The aliphatic C atom (i.e., an atom bonded to four neighbors) is "hidden" and is not involved in such contacts. Its radius is consequently not shown in the geometrical model.

There is a very small number of structures with $-C\equiv C-C\equiv C$ bonds in which $C \cdots C$ contact between adjacent molecules occurs. In these structures also the same figure, 1.80 Å, seems to be most reasonable for the carbon atom.

### c. Intermolecular Radius of the Nitrogen Atom

The intermolecular radius of a nitrogen atom can readily be determined from the symmetric and simple crystal structure of cyanuric triazide (II) [7].

$$
\begin{array}{c}
\text{N}{=}\text{N}{\equiv}\text{N} \\
|\\
\text{N}\overset{\displaystyle \text{C}}{\diagdown}\text{N}\\
\parallel\\
\text{N}{\equiv}\text{N}{=}\text{N}\overset{\displaystyle \text{C}}{\diagup}\overset{}{\underset{\text{N}}{\diagdown}}\overset{\displaystyle \text{C}}{\diagup}\text{N}{=}\text{N}{\equiv}\text{N}
\end{array}
$$

(II)

The molecular array makes the crystal structure of this compound completely unique. The space group is $C_{6h}^2 = C6_3/m$, $Z = 2$, the molecules fully retain their $3/m$ symmetry in the crystal. This may be classified as the rare case of a planar molecule lying in a symmetry plane in which a closely packed layer is formed with molecular coordination 6 (Fig. 11). The next layer is laid upon the first one with a displacement which ensures a high stacking density (Fig. 12). The contacts determining the packing pattern are between nitrogen atoms. The number of independent contacts of this kind is four; the shortest inter-molecular distances $N \cdots N$ give the radius $R_N = 1.58$ Å. This value is reasonably well confirmed in other structures as well.

### d. Intermolecular Radius of the Oxygen Atom

Contacts of the type $O \cdots O$ are determining, e.g., in such structures as hexanitrobenzene. The crystalline structure of hexanitrobenzene [8] is characterized by a monoclinic unit cell with parameters $a = 13.32$, $b = 9.13$, $c = 9.68$ Å, $\beta = 95.5°$, space group $I2/c$, $Z = 4$. The molecules occupy the

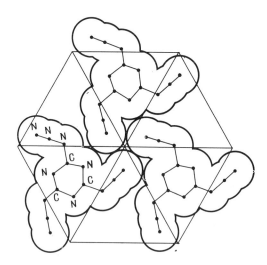

**Fig. 11.** Cyanuric triazide. Close packing in layer $ab$.

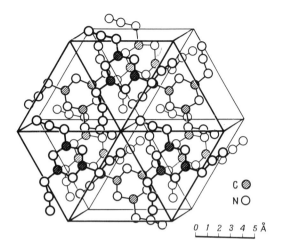

*Fig. 12.* Arrangement of cyanuric triazide molecules in unit cell.

special position on the twofold symmetry axis. The benzene ring of the molecule is planar and coplanar with the C–N bonds. Due to steric hindrances, the planes of the nitro groups are turned about the C–N bond through an angle of 53°. The molecules are loosely packed in the crystal (packing coefficient $k = 0.64$). The molecular array is given in Fig. 13. The structure is built up from molecular layers which are parallel to the ($10\bar{1}$) plane. The arrangement of molecules in one such layer can be seen in Fig. 14. The packing of the molecules in this layer is practically hexagonal with coordination number 6. This highly regular method of packing is dictated by the specific shape of the molecule resembling a round tablet. These layers of molecules are stacked with maximum density due to the centers of symmetry in the crystal.

As a result, an initial molecule I lying on the twofold axis $[Oy\frac{1}{2}]$ is surrounded by 14 adjacent molecules; i.e., it acquires the maximum coordination number for organic crystals. There are no intermolecular contacts C $\cdots$ C in the structure since the benzene ring is screened by the nitro groups. Accordingly, the distance between the closest-packed layers is large: $d_{10\bar{1}}/2 = 4.00$ Å (usually in crystals of aromatic compounds the molecular planes are spaced at 3.4–3.6 Å). A large number of intermolecular distances of the O $\cdots$ O type lie within the range 3.02–3.42 Å. However, an analysis of the molecular packing reveals that the contacts of the O $\cdots$ O type involve atoms spaced at 3.02–3.12 Å. The mean of these distances is 3.07 Å, which means that the intermolecular radius of an oxygen atom $R_O = 1.53$ Å.

In the iron pentacarbonyl structure (for the description of the structure see p. 65) the nearest O atoms of adjacent molecules lie 3.15 Å apart.

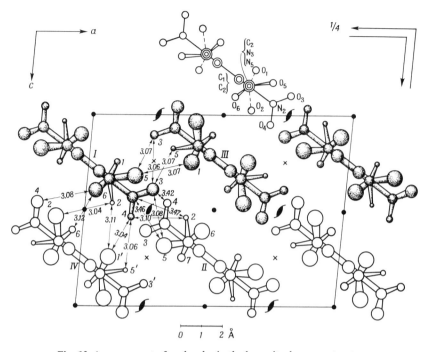

**Fig. 13.** Arrangement of molecules in the hexanitrobenzene structure.

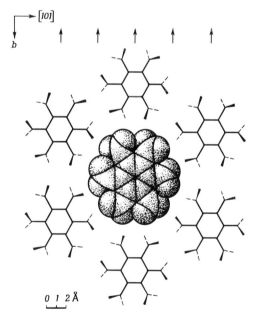

**Fig. 14.** Hexanitrobenzene. Closest-packed molecular layer.

For determining the oxygen atom radius, we may use the alloxan structure (III) [9]. The crystals of this compound are tetragonal; the dimensions of the

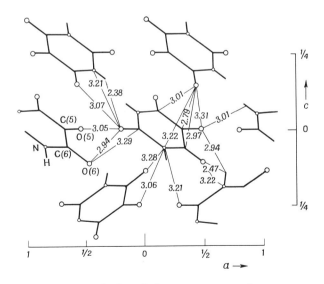

(III)

unit cell are $a = 5.886$, $c = 14.100$, space group $P4_12_12, Z = 4$. The molecule is in a special position on the twofold axis. Figure 15 shows the projection of the structure on the $ac$ face; contacts are between oxygen atoms of adjacent molecules separated by distances of 2.94, 3.05, and 3.07 Å. Hence, the mean radius $R_0 = 1.51$ Å. Bondi [10] gives in his work $R_0 = 1.52$ Å.

The value $R_0 = 1.36$ Å given in the author's book [1] is somewhat too small. Recent energy calculations carried out for the $CO_2$ structure [11] have confirmed the value $R_0 = 1.52$ Å.

The same procedure can be applied for estimating the intermolecular radii of other atoms as well.

It should be emphasized once again that the significance and potentialities of the geometrical model should not be overestimated. It is clear that identical intermolecular contacts may vary slightly in length both in the same structure

*Fig. 15.* Projection of alloxan structure on face $ac$.

and in different structures. It is not these variations that should seem surprising, but the fact that they are very small. The intermolecular contact distances are expressed by the same figures to within only a few percent, and the additivity law is also fulfilled with fair accuracy.

As to the differences themselves, their study is of limited interest, and is warranted only when similar structures are dealt with, in particular, structures that form homologous series. Generally, these differences can be attributed to the inconsistency of the geometrical model with the molecular interaction energy formula as well as to the various roles of molecular vibrations in different crystals. Variations in the characteristic temperatures, and the differences between the melting points and room temperature—at which most experiments are done—motivate a somewhat cautious approach to geometrical conclusions about molecular interaction.

Discrepancies between the ideal and real packing patterns and variations in the mean intermolecular radii from structure to structure can be interpreted on the basis of the second-approximation model to be described in the following chapters.

### 3. PACKING COEFFICIENT

Within the framework of the geometrical model it is worthwhile to investigate not so much the variations in the distances, as to use average figures for describing the packing density. The same intermolecular radii can be used for a large group of compounds, and the volume of a molecule calculated from these figures. X-ray diffraction experiments give the volume per molecule in a crystal. The ratio of the former volume to the latter will be termed the molecular packing coefficient.

It is useful to know the following regularities: The packing coefficients for the overwhelming majority of crystals are between 0.65–0.77, i.e., of the same order as the close-packing coefficients of spheres and ellipsoids. If the shape of a molecule is such that no packing can be effected with a packing coefficient above 0.6, it can be predicted that a drop in temperature will cause vitrification of the substance. Another interesting regularity is the following: Morphotropic changes associated with a loss of symmetry are accompanied by an increase in the packing density. Here is an illustrative example: The crystals of $Sn(C_6H_5)_4$, $Sn(C_6H_4CH_3)_4$, and $Sn(C_6H_4OCH_3)_4$ are tetragonal. Their packing coefficients are, respectively, 0.70, 0.68, and 0.62. The $Sn(C_6H_4OC_2H_5)_4$ crystal is monoclinic and has a packing coefficient of 0.67. Loss of symmetry is balanced by an increase in packing density.

Experimental studies of benzene, naphthalene, and anthracene have shown [12] that when the packing coefficient $k$ is above 0.68 these substances are in the solid state. When changing from a solid phase to a liquid phase, $k$ instantly drops to 0.58. A further increase of the liquid temperature up to the boiling point causes a decrease of $k$, e.g., for naphthalene, down to 0.51. At $k \leqslant 0.5$ these substances become gaseous.

Another example illustrating the physical meaning of the packing coefficient is as follows: The moduli of compressibility of tolane, dibenzyl, and stilbene crystals are, respectively, 4.07, 4.54, 6.50 (in $10^{10}$ dynes/cm²), and their packing coefficients are 0.69, 0.71, 0.72. Thus, for analogous structures, compressibility is inversely related to the packing coefficient.

For calculating the molecular packing coefficient in a crystal, it is necessary to know the volume of the molecule. Calculation of the volume of the molecule is feasible if one knows bond lengths, valence angles, and intermolecular contact distances.

After bounding the molecule by intermolecular contact radii, the total volume of the molecule is broken up into portions (increments) contributed to the total volume by the individual atoms. Figure 16 illustrates the construction of the model of a 4-chlorodiphenyl molecule. First, a part of a sphere with radius $R_C = 1.80$ Å is constructed about the center of each C atom, and a sphere with radius $R_H = 1.17$ Å about the center of each H atom (a convenient scale is 1 Å = 1 cm). The spheres intersect in circles; a plane drawn through the

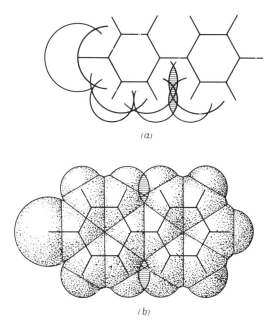

*Fig. 16.* Model of the 4-chlorodiphenyl molecule: (a) initial stage of construction; (b) finished model.

*Table 1*

VOLUME INCREMENTS OF CERTAIN COMMON GROUPS OF ATOMS

| Atom or atom group | ΔV Å³ | | ΔV Å³ |
|---|---|---|---|
| Aromatic carbon atom | | 8.4 | 19.9 |
| C–H group (aromatic) | | 14.7 | 9.6 |
| C–H group (aliphatic) | | 11.1 | 26.0 |
| CH₂ group | | 17.1 | 13.1 |
| CH₃ group | | 23.5 | 5.9 |
| Quaternary carbon atom | | 5.0 | 10.0 |
| Nitro group | | 23.0 | 23.1 |
| Amino group | | 19.7 | 2.0 |

circumference will "cut off" the portions of the spheres which are not involved in the external "shaping" of the molecule. The uppermost portion of Fig. 16 shows such a model of a 4-chlorodiphenyl molecule. It is partitioned into separate parts: nine increments $\diagdown$C–H, two increments $\diagdown$C, and one $\diagdown$C–Cl. Such volume increments can be calculated by the formula

$$\Delta V = \tfrac{4}{3}\pi R^3 - \sum_i \tfrac{1}{3}\pi h_i{}^2 (3R - h_i)$$

where $R$ is the intermolecular radius of the atom concerned; the $R_i$ are inter-molecular radii of the atoms that are valence-bonded with this atom and are positioned at distance $d_i$ from this atom; the height of the cut-off segment is

$$h_i = R - \frac{R^2 + d_i{}^2 - R_i{}^2}{2d_i}$$

The volume increments of some atomic combinations most commonly occurring in molecules are listed in Table 1. They can be used for calculating the packing coefficient. When estimating the volume of a molecule with the aid of increments, it should always be remembered that this scheme disregards steric hindrances which often arise between the atoms of one molecule not connected by valence bonds (e.g., the cross-hatched portions in the diphenyl molecule in Fig. 16b correspond to the overlapping spheres of hydrogen atoms which occupy positions called by chemists the ortho positions).

Appropriate corrections may be introduced if necessary. It has been found possible to write a general computer program for calculating the volume of any complex molecule.

## 4. CLOSE PACKING AND CRYSTAL SYMMETRY

The analysis of molecular packing in an organic crystal suggests the following conclusion: The mutual arrangement of the molecules in a crystal is always such that the "projections" of one molecule fit into the "hollows" of adjacent molecules. It would not, of course, be feasible to carry out one-by-one examin-ation of all possible packings of molecules for unit cells of different symmetries and different dimensions. Nonetheless, the statement that the real structure is one of the most closely packed of all conceivable patterns seems realistic enough.

Zorky and Poraj-Koshits† have made an attempt to verify this hypothesis

---

†See, for example, the collection of articles "Modern Problems of Physical Chemistry," Volume I, [13].

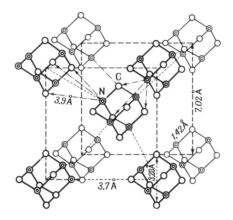

*Fig. 17.* Crystalline structure of hexamethylenetetramine. H atoms are not shown.

within the framework of the geometrical model for a two-dimensional case. They have calculated all possible packings for several models of molecules. It has been found that among them there is a small number of packing patterns with high density. The experimentally observed structures have been found to be among the closest-packed structures. It can be asserted that the real packing has the highest density with an accuracy of about 0.01 of the packing coefficient.

Close packing can exist if the molecular coordination number is sufficiently high. The experimental evidence is that the coordination number is usually 12; sometimes the specific shape of a molecule provides packing with coordination numbers of 10 and, naturally, 14.

An example of a crystal structure with coordination number 14 is that of hexamethylenetetramine $N_4(CH_2)_6$. The crystals of this substance have a body-centered cubic unit cell with symmetry $T_d^3 = I\bar{4}3m$ and $a = 7.02$ Å, $Z = 2$. Figure 17 shows the arrangement of molecules in this structure (H atoms are not shown). The molecule positioned in the center of the cube has $N \cdots H$ contacts with the eight molecules located at the vertices of the cube; besides, the $C \cdots C$ contacts of the $CH_2-$ groups relate this molecule with six other molecules placed in the centers of the adjacent cells. At coordination number 14 the packing coefficient of the hexamethylenetetramine molecules has the common value of 0.72. However, the distances between atoms in contact are somewhat larger than generally accepted: $N \cdots H = 2.86$ Å instead of $1.57 + 1.17 = 2.74$ Å; $C \cdots C = 3.72$ Å. The most probable explanation is that exceptionally large surfaces of the molecules are in contact.

Coordination number 10 is to be found, for example, in the structure of urea. The crystals of this compound have a tetragonal cell [14] with space group

$$O=C\overset{NH_2}{\underset{NH_2}{<}}$$

(IV)

$V_d^3 = P\bar{4}2_1m$, $Z = 2$. The molecule fully retains its symmetry mm in the crystal. The parameters of the cell are $a = 5.67$, $c = 4.73$ Å. The molecular array in the unit cell appears in Fig. 18. The initial molecule touches two other molecules related by translation along $c$, four molecules related to it by the $\bar{4}$ axis (or $2_1$ or a glide plane), and another four molecules related to the latter by translation along $c$.

High coordination numbers may appear if molecules are arranged in a crystal with appropriate symmetry. In the publication cited above the author made an attempt to predict all possible space groups in organic crystals on the basis of the close-packing principle. The space groups that were most highly suitable for the purpose were selected as follows. First, we determined the symmetry of two-dimensional layers, which provided conditions for packing molecules into a layer with coordination number 6, with the molecules tilted arbitrarily relative to the axes of the layer cell. In the general case, there are only two such layers of molecules of an arbitrary shape: an oblique net with inversion centers, and a layer with a rectangular net cell built up by

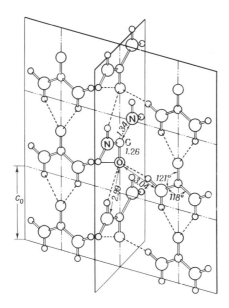

*Fig. 18.* Crystalline structure of urea.

translations and with a twofold screw axis parallel to a translation axis. The next operation was to select space groups in which these layers are possible and are stacked so as to yield high coordination numbers. It is obvious, for example, that the stacking of layers with the aid of a mirror symmetry plane is excluded here.

Detailed consideration will be given to only the planar case. This consideration will bring to light the main principles underlying the approach to the selection of space groups convenient for a molecular crystal.

### 5. Closest-Packed Plane Groups of Symmetry

Let us consider the possibilities for molecular packing in each of the 17 plane groups of symmetry.

A plane layer of molecules is said to be *close-packed* if it is packed with coordination number 6, i.e., if each molecule touches six neighbors.

A layer of molecules is *closest-packed* if coordination number 6 is feasible for any orientation of the molecules with respect to the axes of the unit cell.

A layer of molecules will be termed a layer of *maximum density* for a given symmetry of the figure in the layer if coordination 6 is possible at any tilt (at any orientation with respect to the axes of the unit cell) compatible with the figure's retaining its symmetry in the layer.

It is obvious that for generalizing such a consideration, the shape of the molecule must be arbitrary.

As is apparent from Fig. 19, translation $t_1$ can always be selected such that molecules related through this translation make closest contacts. With the help of a second translation $t_2$ that is noncolinear with $t_1$, a third molecule can be adjusted to the first two molecules so that it fits into a hollow between them,

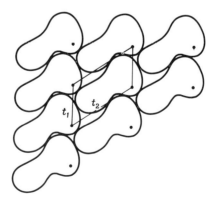

*Fig. 19.* Dense layer with symmetry $p1$.

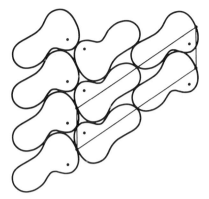

*Fig. 20.* Dense layer with symmetry *p*2.

i.e., touches both these molecules. Since no restrictions were imposed on translations $t_1$ and $t_2$, no obstacles exist for selecting a unit mesh with the minimum area for a molecule of a given shape. Thus, in plane group $C_1^1 = p1$ the closest-packed layer can be produced by selecting values $t_1$ and $t_2$ and an angle between them, whatever the shape of the molecule.

The formation of a closest-packed layer is similarly easy in centrosymmetric group $C_2^1 = p2$. Indeed, (see Fig. 20) translation $t_1$ which relates two closely touching molecules remains the same as in the previous case. By varying $t_2$ and the angle between $t_1$ and $t_2$ one can always arrange the third molecule so that it touches both initial molecues and is related to them by an inversion center. The resulting layer will be closest packed, with coordination 6 and a centrosymmetric oblique unit mesh containing two molecules.

Plane groups $C_s^1 = pm$ and $C_{2v}^1 = pmm$ are not suitable for the formation of any layer with coordination 6. Indeed, molecules related by a glide line (the

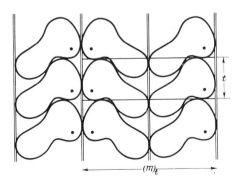

*Fig. 21.* Layer with symmetry *pm*.

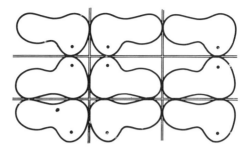

*Fig. 22.* Layer with symmetry *pmm.*

trace of a perpendicular glide plane or $2_1$ axis lying in the layer plane) are packed "projection to projection" yielding thereby a layer with coordination 4 (Figs. 21 and 22).

Consider plane group $C_s^{II} = pg$. By means of translation $t_1$ we form a row of closely touching molecules.

By varying the origin and the magnitude of translation $t_2$ normal to $t_1$, it is possible to arrange the glide line aligned with $t_2$ so that a molecule derived by its action from the initial molecule falls into the hollow between two molecules of the first row (Fig. 23). A layer with coordination 6 is obtained that will have a closest-packed structure, because no restrictions are imposed on the tilts of the molecule to the cell axes and no special requirements are imposed on the shape of the molecule. Similar reasoning can be used for the centrosymmetric group $C_{2v}^{II} = pgg$ if we assume that one "building block" is a pair of closely touching molecules mutually related by a center of inversion. Using translation $t_1$, a closely-packed row of such pairs can be formed (shown by solid lines in

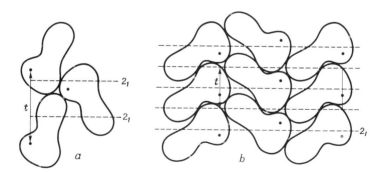

*Fig. 23.* Layer *pg* with different (a and b) orientations with respect to elements of symmetry.

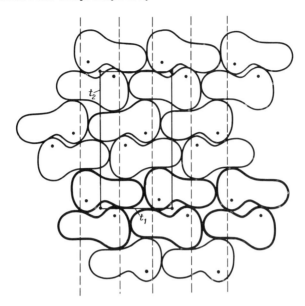

**Fig. 24.** Layer *pgg* formed by pairs of figures related by inversion centres.

Fig. 24). The origin in this group is fixed—it is the center of inversion. Varying the value of translation $t_2$ normal to $t_1$ and by the operation of the glide line perpendicular to $t_1$ and spaced $\frac{1}{4}t_1$ from the origin we may move the next row of molecules relative to the initial row so that each molecule in this row touches two molecules in the initial row, and a layer with coordination 6 is formed. The second glide line perpendicular to the first one and passing $\frac{1}{4}t_2$ from the center of inversion originates as a derivative of the combined action of the center and the glide line normal to the latter

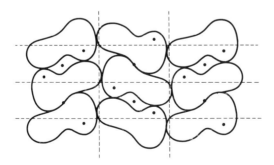

**Fig. 25.** Layer *pgg* with coordination 4.

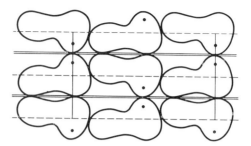

**Fig. 26.** Layer with symmetry *cm*.

The close-packed layer so obtained is, nevertheless, not a layer of maximum density for unsymmetrical molecules in the sense assigned to this term at the beginning of this section, because the minimum area of the mesh arises only if the requirement of the "optimum" packing of a pair of molecules related by a center of symmetry is satisfied. For example, if a figure has the orientation shown in Fig. 25, coordination 6 will not be obtained for this symmetry.

In plane groups $C_s^{III} = cm$ and $C_{2v}^{IV} = cmm$, formation of a layer with coordination 6 is impossible in the general case when the molecule has an arbitrary shape. Indeed, unlike groups *pg* and *pgg* where the glide line "promotes" close packing, in these groups mirror reflection lines pass parallel to the glide lines (Figs. 26 and 27), as a result of which the molecules are packed on a "projection-to-projection" principle and layers with coordination 4 arise. The packing pattern with coordination 6 is also unfeasible for plane group *pmg* (Fig. 28) in the general case of arbitrarily, shaped molecules.

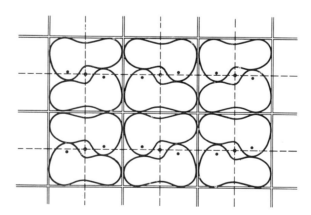

**Fig. 27.** Layer with symmetry *cmm*.

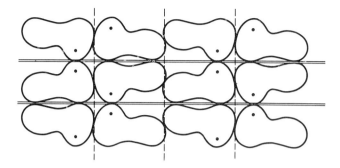

*Fig. 28.* Layer with symmetry *pmg*.

In the plane groups of higher crystal systems, tetragonal (*p4*, *p4m*, *p4g*) and hexagonal (*p3*, *p3m1*, *p31m*, *p6*, *p6m*), the requirement of the equality of the unit cell axes imposed by symmetry ($a = b$) renders impossible the packing of arbitrarily shaped figures without overlapping (see Fig. 29).

Thus, of the nine plane groups only four are suitable for forming closest packing in molecules of arbitrary shape: *p1*, *p1̄*, *pg*, and *pgg*; the latter is fit only for centrosymmetric pairs of molecules.

It now remains to consider the packing patterns that we called layers of maximum density.

If a figure retains a symmetry line, in principle it can be accommodated in one of the three plane groups of symmetry: *pm*, *pmg*, and *cm*. However, in the first group (Fig. 30) coordination 6 will not be feasible for a figure of arbitrary shape. In the two other groups coordination 6 is always achievable. It can be

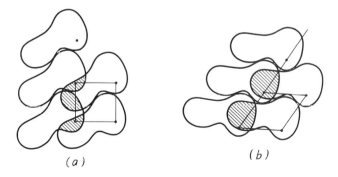

*(a)*                    *(b)*

*Fig. 29.* Inapplicability of tetragonal (a) and hexagonal (b) cells for stacking arbitrarily shaped figures.

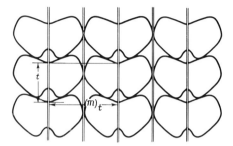

**Fig. 30.** Layer *pm* built up from figures with a symmetry line.

seen from Fig. 31 that with the aid of a translation perpendicular (in group *cm*) or parallel (in group *pmg*) to the glide line of the figure, one can always form a row of figures closely touching each other. Selecting a suitable value of the second translation normal to the initial one, the next row can be brought in contact with the first row. The glide line ensures a displacement in the stacking of the rows so that each figure touches two figures in the adjacent row. This gives coordination 6 in the layer. For the formation of such layers, it is requisite that the molecules are adjusted to each other in an appropriate manner regardless of their shape and depending only on the position of the line of symmetry. If this stringent requirement is met, then both groups *cm* and *pmg*

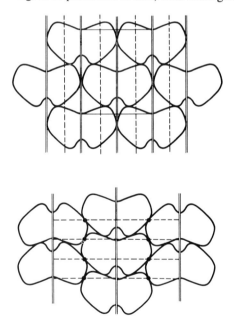

**Fig. 31.** Layers *cm* and *pmg* built up from figures with a symmetry line.

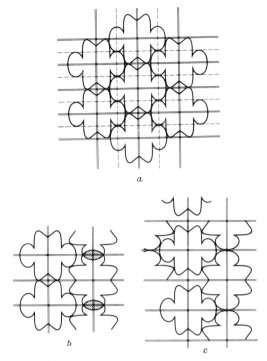

*Fig. 32.* Layers of figures with two lines of symmetry.

are suitable for forming a layer of maximum density. The figures included in the text show that, depending on the shape of the figures, either one or the other group can provide the conditions for packing of maximum density.

If a figure has two symmetry lines and takes a special position in the lattice, it loses all its degrees of freedom. From Fig. 32 it is obvious that the figures of such symmetry and arbitrary shape form a packing pattern with coordination 6 in plane group *cmm* with a centered cell. The figures can also be packed in symmetry layer *pmm* (Fig. 32b,c), however, in this case coordination 6 will not be provided for a figure of arbitrary shape; the figures will either overlap in the second row or fail to touch each other in the first row.

Let us consider all possible packing patterns for figures having a center of inversion. For an oblique cell (plane group *p*2) all the above statements pertaining to a layer of unsymmetrical figures are also applicable to centro-symmetric figures. In fact (Fig. 33) after forming a row of centrosymmetric figures by means of translation $t_1$ we see that the operation of the intermediate inversion center (the one located between the figures) is identical to the operation of translation $t_1$.

Consequently, using the second translation with no restrictions imposed on

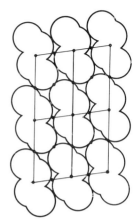

*Fig. 33.* Layer with symmetry p2 built up from centro-symmetric figures.

its magnitude and direction, it is always possible to provide the contact of a given figure with two figures of the adjacent row. Here coordination 6 is achieved without any reservations.

In plane groups *pmg* and *cmm*, centrosymmetric figures will be packed by means of lines of symmetry, which fails to provide coordination 6 in a layer; consequently, these two centrosymmetric groups should be rejected.

The only centrosymmetric group that remains to be considered is *pgg*. We shall proceed from a row of centrosymmetric figures (Fig. 34). The glide line cannot be positioned arbitrarily; it passes midway between two adjacent centers of inversion. Now, would it be possible to attain close packing in a layer with coordination 6 under these conditions? The answer is yes, and this

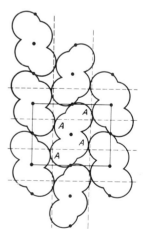

*Fig. 34.* Layer with symmetry *pgg* built up from centro-symmetric figures.

without any restrictions on the packing of the figures in the first row, despite the fixed position of the glide line. Indeed, let us form the second row of figures from the first one with the help of the glide line and move both rows toward each other by decreasing the amount of the glide. Each figure of the second row will simultaneously come into contact (a) with two figures in the first row, because the second row is shifted with respect to the first by half the unit translation along this row. As a result, coordination 6 in this layer is provided without any restrictions with respect to the packing in the initial row of centrosymmetric figures.

Summarizing the above, packing of maximum density is feasible in a layer under the following conditions: in plane groups *pmg* and *cm*—if the figure retains a line of symmetry; in plane group *cmm*—if the figure retains two lines of symmetry, and in plane groups *p2* and *pgg*—for any orientation of arbitrarily shaped centrosymmetric figures. The latter statement means that for figures with a center of inversion, packing of maximum density is equivalent to the closest packing.

## 6. SPACE GROUPS SUITABLE FOR CLOSE PACKING OF MOLECULES

Crystals with the lower syngonies typical of organic compounds are built up by stacking layers of three-dimensional figures, the postulated requirement being that the layers must be closest packed. This implies, first, that layers must not be stacked with the aid of the mirror plane of symmetry. The closest stacking of layers can be provided either through a monoclinic displacement–translation that forms an arbitrary angle with the layer plane, or by inversion centers, glide planes, or screw axes. The formation of closest packing by stacking of layers related through twofold axes is hardly probable, but not totally excluded, because there may be cases when molecules related by twofold axes will not be in contact and, consequently, will not prevent closest packing by other symmetry elements.

Analysis of possible arrays of molecules in each space group made on the basis of these considerations has demonstrated [1] that in molecules without symmetry elements, closest packing is attainable in the following space groups: P1, $P2_1$, $P2_1/c$, Pca, Pna, $P2_12_12_1$.† For centrosymmetric molecules, the number of such groups is still smaller: P$\bar{1}$, $P2_1/c$, C2/c, Pbca.

In these cases of closest packing, coordination 6 in a layer can be provided with any mutual orientation of the molecules.

All the above groups exhibit the following peculiar feature: The close-packed

†Group P1 stands apart; no special proofs are needed to show that the case of all molecules being parallel is in conflict with the requirements of close packing.

plane layers considered above are to be found in one, two, or three systems of planes.

Similar relationships are observed between plane layers that have maximum density for molecules of inherent symmetry and space groups of maximum density. For instance, if a molecule has twofold symmetry and retains this symmetry in the crystal, such molecules can be packed with maximum density in groups $C2/c$, $P2_12_12$, and $Pbcn$. Table 2 summarizes all possible packing patterns according to molecular symmetry in a crystal.

*Table 2*

CLOSEST-PACKED SPACE GROUPS AND SPACE GROUPS OF MAXIMUM DENSITY FOR ORGANIC CRYSTALS

| Molecular symmetry in crystal | 1 | 2 | $m$ | $\bar{1}$ | $mm$ | $2/m$ | $222$ | $mmm$ |
|---|---|---|---|---|---|---|---|---|
| Closest-packed space groups | $P\bar{1}$ $P2_1$ $P2_1/c$ $Pca$ $Pna$ $P2_12_12_1$ | none | none | $P\bar{1}$ $P2_1/c$ $C2/c$ $Pbca$ | none | none | none | none |
| Space groups of maximum density | none | $C2/c$ $P2_12_12$ $Pbcn$ | $Pmc$ $Cmc$ $Pnma$ | none | $Fmm$ $Pmma$ $Pmnn$ | $C2/m$ $Pbaa$ $Cmca$ | $C222$ $F222$ $I222$ $Ccca$ | $Cmmm$ $Fmmm$ $Immm$ |

The six space groups that have thus been found possible for molecules occupying general positions in a crystal are the groups that are most frequently encountered among the structures studied.

From the point of view of close-packing requirements, a remarkable predominance of the space group $P2_1/c$ in organic chemical crystallography becomes obvious. It is only in this group that closest-packed layers can be built on all three coordinate planes of the cell.

Groups $P2_1$ and $P2_12_12_1$ without symmetry center are also among the closest packed. Since their potential abilities for close packing are lower as compared with group $P2_1/c$, they are quite logically to be found in cases when molecules are in either their right-handed or left-handed configuration.†

---

†These three groups rank first in the list of space groups.

As is known, molecules of high-symmetry quite frequently form crystals in which these molecules occupy special positions with a decreased number of degrees of freedom or even without degrees of freedom (hexamethylentetramine, di-*p*-xylylene). It is clear that in this case close packing is somewhat hampered because the symmetry of a special position implies certain mutual orientations of the molecules. However, close packing is completely feasible in a special case—when a molecule occupies a centrosymmetric position in a crystal. In this case, similarly to a molecule in a general position, close-packed layers can be formed with coordination number 6 for an arbitrary tilt of the molecular axes to the axes of the layer mesh. Here, we have again an oblique layer mesh with symmetry centers, and a rectangular layer mesh displaying the symmetry of the layer *bc* of the space group $P2_1/c$.

The space group $P2_1/c$ retains its dominant position in this case too, but now there are two molecules in the cell.

Since the preservation of the center of symmetry in a crystal does not involve a lower packing density, this symmetry element is always present in a crystal. For example, molecules with *mmm* symmetry lose, as a rule, their symmetry planes in a crystal and retain the inversion center (a common naphthalene structure).

The preservation of other symmetries in a crystal is indubitably associated with a lower packing density. The analysis of concrete examples shows that the retention of twofold axial symmetry in a crystal, or occupation of the mirror symmetry plane by a molecule, may cause a 0.02–0.03 decrease in the packing coefficient. The geometrical model obviously cannot explain this preference given by Nature to a symmetrical arrangement. Using this model, however, it is possible to deduce space groups that provide the highest packing density under the supplementary condition that a molecule retain its symmetry in a crystal. We have derived these groups and found that, in full agreement with experiment, for molecules located on symmetry planes, the first place is held by the group P*nma*, whereas for molecules that exhibit the symmetry of the twofold rotation axis the best groups are $C2/c$ and P*bcn*.

It is not difficult to compile a list of space groups suitable for close packing of molecules that retain the symmetry of any of the point groups in a crystal.

It should be emphasized that in all known cases when a molecule preserves high symmetry in a crystal at the expense of a certain decrease in the packing density, this decrease is not large. If this sacrifice in the packing density must be large, the symmetry is partially or completely lost.

Therefore, the symmetry cases encircled by a solid line in the above table are only exceptional cases. It is not very difficult to preserve in a crystal one "inconvenient" symmetry element—plane *m* or axis 2. Conversely, the retention of higher symmetry does not compensate for a considerable decrease in the packing density.

The fact that molecular symmetry 2 or *m* is always retained in a crystal for molecules with *mm* symmetry naturally suggests that a symmetrical arrangement of molecules is advantageous thermodynamically.

The molecular array in a crystal is such as to give the minimum free energy. But

$$F = U + F^{\text{vib}}$$

where *U* is the lattice energy. The minimum of the potential energy of molecular interaction, i.e., the lattice energy minimum, corresponds to the closest packing (see Section II.10). The minimum of the vibrational energy component depends on the freedom of molecular motion, which evidently must be larger if molecular symmetry is retained in the crystal. This qualitative consideration substantiates the above rules for selecting space groups.

There are practically no exceptions to these rules. In other words, we do not know any organic crystal structures with packing coefficients below $\sim 0.65$; in most cases, the closest packings are realized; if a molecule preserves high symmetry in a crystal, crystallization occurs in one of the space groups providing the best conditions for packing. As an example, we have considered all structures of organic crystals described in the "Structure Reports" of 1952–1956. Of the 315 structures considered, 268 (86%) belong to the closest-packed structures, 14 retain high symmetry in the optimum space groups, and nine are disordered structures. There are ten structures in which the close-packing requirements are, so to speak, overfulfilled. These structure have more than one molecule in a general position. Seven or eight structures are doubtful cases which need verification, and the remaining structures are peculiar rarely encountered packings in which molecules either occupy two nonequivalent positions of different symmetry, or are in general positions in crystals showing tetragonal or hexagonal symmetry.

Whereas the latter examples enable close packing despite the restrictions imposed by high symmetry of the crystal, the instances when molecules occupy two or more symmetrically independent positions (e.g., biphenylene: four molecules in general positions, two at symmetry centers) entail an increase in the number of degrees of freedom per cell and, consequently, can ensure packing of the highest density.

It would be of interest to discover the failure of "simple" packing for these elegant structures, e.g., to show that all conceivable cells of biphenylene with symmetry $P2_1/c$, $Z = 2$ give packings of lower density as compared with that of the experimentally observed structures.

Thus, we can state that the geometrical model produces results which are in excellent agreement with experiment. The reasons for this excellent agreement of the geometrical model with experiment need some explanation. As will be shown subsequently, the geometrical model describes with a sufficient degree of accuracy the effects of the tendency for minimization displayed by the potential energy of molecular interaction. The trend to retain symmetry must be accounted for by the trend of the crystal free energy toward minimization, and must be principally governed by entropy requirements. It can be stated that at absolute zero crystal structures should be in much better agreement with the geometrical model than at 300°K at which experiments are usually made. It is true that in a number of instances we observe phase transitions at low temperatures with the result that the molecule loses its symmetry in a crystal. But it is equally obvious that very often no polymorphic changes occur at low temperatures for kinetic reasons, to say nothing of the fact that the behavior of substances at temperatures below the temperature of liquid nitrogen has been examined in very rare cases only.

The geometrical model is unable to explain why a molecule preserves symmetry in a crystal (except for symmetry $\bar{1}$), neither can it explain why it is only very rarely that more than one molecule occupies a general position.† The geometrical model fails to explain the scatter of intermolecular distances within one structure and from crystal to crystal. This model relates the structure to the crystal's properties only in terms of a packing coefficient, i.e., qualitatively. Nevertheless, the model is valuable because, first, it provides a graphic representation of the mutual orientations of the molecules in a crystal, and, secondly, it is an aid in calculating unknown structures.

A number of workers have successfully used the analysis of the molecular packing in a crystal in formulating a rough structural model. Preliminary calculations of a rough model can be carried out before the analysis of the X-ray diffraction intensities. At the present time it has become possible to employ electronic computers and molecular models for this geometrical analysis; in other words, for calculating the possible orientations of molecules compatible with their shape from the known size and symmetry of the unit cell. For simple structures, a unique solution can be obtained; in more complicated instances, several solutions are possible, but, at any rate, the number of possible orientations can always be quite definitely determined. The calculated structures possible from the geometrical point of view can be used to calculate intensities and, thereby, a correct solution may be obtained. There is also another alternative: minimization of a certain function which is the sum of the conventional $R$-factor (see p. 241) and an expression that reaches a minimum for geometrically correct solutions. This procedure facilitates and speeds up the solution of the problem.

---

†This does not refer to numerous structures in which molecules are dimeric due to hydrogen bonds.

## B. Typical Structures

### 7. LINEAR AROMATIC SYSTEMS

*a. Condensed Systems. Condensed systems are substances with molecules of the following shapes:*

(V)

The first representative of this class of compounds is naphthalene. The naphthalene molecule has the inherent symmetry *mmm*.

*Naphthalene* [15]. In the crystal the molecule retains only an inversion center: space group $C_{2h}^5 = P2_1/a$, $Z = 2$; the dimensions of the unit cell are $a = 8.235$, $b = 6.003$, $c = 8.658$ Å; $\beta = 122° 55'$. The volume of the molecule is $V_0 = 126.8$ Å, $k = 0.706$. The closest-packed layer coincides with the *ab* plane of the unit cell; the layers are stacked with a monoclinic displacement by translation along the *c* axis. Figure 35a,b shows the shortest intermolecular distances. The adjacent layers make contact through hydrogen stems lying at distances $H \cdots H = 2.40$ Å. This gives the intermolecular radius of a hydrogen atom $R_H = 1.20$ Å. In this case, the shortest distances $C \cdots H$ inside the *ab* layer which are equal to 2.82 Å, yield an appreciably underestimated value of the intermolecular radius of the carbon atom: $R_C = 2.82 - 1.20 = 1.62$ Å. However, the contradiction with the hypothesis of the additivity of intermolecular radii appears insignificant for the geometrical model. In fact, if instead of the Eulerian angles $\phi_1{}^0, \phi_2{}^0, \phi_3{}^0$ that describe the position of a molecule in the real structure, we use angles $\phi_1, \phi_2, \phi_3$ which are derived as a result of joint solution of the four equations determining four principal contacts in the structure, we find [16] that the only general solution can be obtained on condition that $R_H = 1.17$ Å and $R_C = 1.74$ Å, and the derived values of the Eulerian angles differ from the actual values by not more than 2–3°.

*Anthracene.* The crystalline structure of anthracene [17] is similar to the naphthalene structure (IV) (unit cell: $a = 8.561$, $b = 6.036$, $c = 11.163$, $\beta = 124.7°$; space group $C_{2h}^5 = P2_1/a$, $Z = 2$). The packing of molecules in

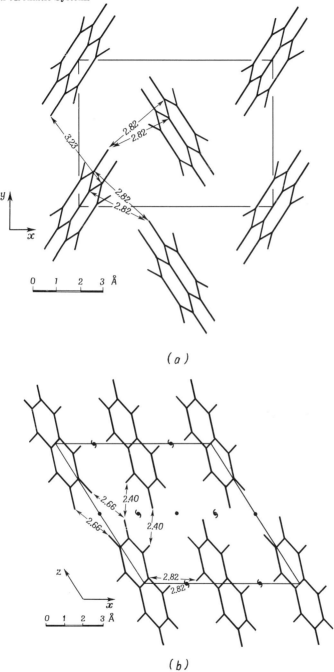

Fig. 35. Projections ab (a) and ac (b) of naphthalene structure.

(VI)

the *ab* layer is fully identical to that in the naphthalene structure (Fig. 36). The increased number of aromatic rings in the anthracene molecule, as compared with naphthalene, causes a corresponding increase of the *c* axis along which the long molecular axis is directed. Stacking of the layers by translation *c* is less close than the molecular packing inside the layer. This explains the perfect cleavage typical of naphthalene and anthracene crystals, which permits a ready splitting of the crystals along the *ab* planes.

Crystals whose molecules contain four or more aromatic rings belong to the triclinic syngony, the molecules being positioned on two families of symmetrically unrelated centers, thus forming a two-layer packing.

All the crystallographic data for this class of substances are given in Table 3 [18]. Noteworthy is the considerable growth of the packing coefficient with increasing number of aromatic rings in the molecule.

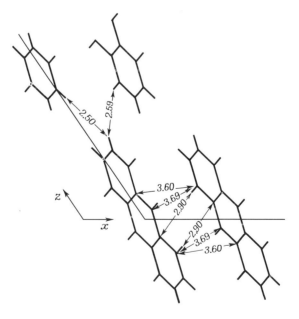

***Fig. 36.*** Projection *ac* of anthracene structure.

*Table 3*

| Substance | Naphthalene $C_{10}H_8$ | Anthracene $C_{14}H_{10}$ | Tetracene $C_{18}H_{12}$ | Pentacene $C_{22}H_{14}$ | Hexacene $C_{26}H_{16}$ |
|---|---|---|---|---|---|
| Syngony Parameters | Monoclinic | Monoclinic | Triclinic | Triclinic | Triclinic |
| $a$ (Å) | 8.24 | 8.56 | 7.90 | 7.90 | 7.9 |
| $b$ (Å) | 6.00 | 6.04 | 6.03 | 6.06 | 6.1 |
| $c$ (Å) | 8.66 | 11.16 | 13.53 | 16.01 | 18.4 |
| $\alpha$ (°) | 90.0 | 90.0 | 100.3 | 101.9 | 102.7 |
| $\beta$ (°) | 122.9 | 124.7 | 113.2 | 112.6 | 112.3 |
| $\gamma$ (°) | 90.0 | 90.0 | 86.3 | 85.8 | 83.6 |
| $V$ (Å³) | 360 | 474 | 583 | 692 | 800 |
| $Z$ | 2 | 2 | 2 | 2 | 2 |
| Space group | $P2_1/a$ | $P2_1/a$ | $P\bar{1}$ | $P\bar{1}$ | $P\bar{1}$ |
| $V_{mol}$ | 126.8 | 170.2 | 213.6 | 257.0 | 300.4 |
| $k$ | 0.704 | 0.718 | 0.733 | 0.743 | 0.751 |

*b. Polyphenyls. Polyphenyls are substances whose molecules have the form* (III):

(VII)

The first representative of these substances, diphenyl, has a monoclinic unit cell with parameters: $a = 8.12$, $b = 5.64$, $c = 9.47$ Å, $\beta = 95.4°$, space group $P2_1/a$, $Z = 2$ [19]. Figure 37 shows the projection of the diphenyl structure on the $ac$ face. Packing in the $ab$ layer is typical of aromatic molecules; the following atoms contact: $H2 \cdots C1 = 2.93$ Å, $H2 \cdots C'1 = 2.80$ Å, $C3 \cdots H6 = 2.91$ Å. The stacking of the layers by translation along $c$ is less close: $H4 \cdots C3 = 3.16$ Å; $H4 \cdots C5 = 3.26$ Å; $H4 \cdots H3 = 2.49$ Å. It should be pointed out that the authors, who have refined the structure down to $R = 8.9\%$ (for the magnitude of $R$ in structure studies, see Chapter IV), introduced the H atoms geometrically, assuming that the length of the C–H bond is 1.08 Å, and used the coordinates of all the H atoms only for reducing the $R$-factor. Their works give only intermolecular $C \cdots C$ distances, which all exceed the sum of the intermolecular radii (3.81 Å is the shortest distance). The packing

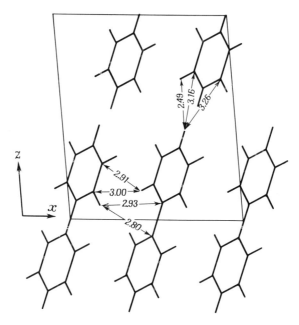

**Fig. 37.** Projection *ac* of the diphenyl structure.

of the molecules is undoubtedly defined by C ⋯ H contacts in the *ab* layer
and between these layers.

The crystalline structures of *p*-diphenylbenzene (terphenyl) and *p*,*p*′-bis-
(phenyl)-diphenyl (quaterphenyl) (IX) change only in accordance with the

(VIII)

(IX)

elongation of the molecule, even though the symmetry center of the terphenyl
molecule is situated in the center of the middle ring, unlike molecules with an
even number of aromatic rings. The crystallographic data for this class of
compounds are listed in Table 4.

Two series of structures given in this section may serve as examples of a
specific phenomenon that may be called "homologous isomorphism":
Geometrically similar molecules inside homologous series form identical
packings. Indeed, in each series one of the cross sections of the unit cell remains

*Table 4*

| Substance | Diphenyl $C_{12}H_{10}$ | Terphenyl $C_{18}H_{14}$ | Quaterphenyl $C_{24}H_{18}$ |
|---|---|---|---|
| Syngony \ Parameters | Monoclinic | Monoclinic | Monoclinic |
| $a$ (Å) | 8.12 | 8.12 | 8.05 |
| $b$ (Å) | 5.64 | 5.62 | 5.55 |
| $c$ (Å) | 9.47 | 13.62 | 17.81 |
| $\beta$ (°) | 95.4° | 92.4° | 95.8° |
| $V$ (Å³) | 431.8 | 621.0 | 791.7 |
| $Z$ | 2 | 2 | 2 |
| Space group | $P2_1/a$ | $P2_1/a$ | $\cdot P2_1/a$ |
| $V_{mol}$ | 155.4 | 233.1 | 310.8 |
| $k$ | 0.720 | 0.751 | 0.785 |

approximately the same; one of the parameters increases by a value equal to the "elongation" of the molecule. In the naphthalene–anthracene–naphthacene series the elongation of the molecule equals the width of the benzene ring, i.e., 2.4 Å; the $c$ axis of the anthracene cell is 2.50 Å longer than the same axis of the naphthalene cell, while the $c$ axis of the naphthacene cell is 2.37 Å longer than that of anthracene, etc. In the triclinic structures of this series, molecules form an *ab* layer almost as if these molecules were related by a twofold axis; in other words, the change of symmetry does not impair the homologous isomorphism. A still more perfect similarity is observed in the polyphenyl series (see Table 4).

8. NONLINEAR CONDENSED AROMATIC MOLECULES
   OF SYMMETRY *mm* AND *mmm*

*a. 1,12-Benzperylene* (X) [20]

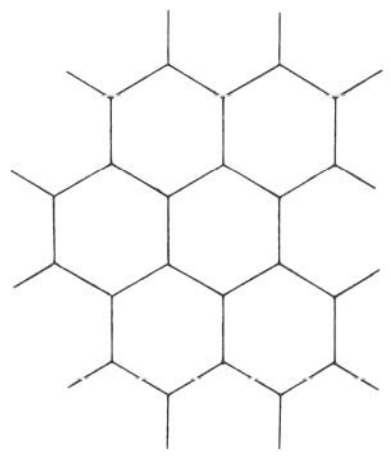

(X)

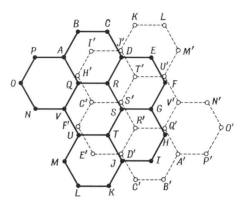

*Fig. 38.* 1,12-Benzperylene. Stacking of molecules related by $\bar{1}$.

Unit cell $a = 11.72$, $b = 11.88$, $c = 9.89$ Å, $\beta = 98.5°$ space group $C_{2h}^5 = P2_1/a, Z = 4$. The molecular symmetry is probably 2 (because of steric hindrance (see Chapter 7)), but in the crystal the molecule occupies a general position, and thereby loses its element of symmetry. The adjacent parallel molecules are stacked on top of each other with a displacement (Fig. 38). The distances between their planes are 3.38 Å. The intermolecular distances are as follows:

|  |  |  |
|---|---|---|
| C(S)···C(S') | | $\bar{1}'$ 3.38 Å |
| C(D)···C(J') | | $\bar{1}$ 3.39 |
| CH(F)···C(U') | | $\bar{1}$ 3.39 |
| C(Q)···CH(H') | | $\bar{1}$ 3.39 |
| C(D)···CH(O') | glide on $a$ | 3.91 |
| CH(N)···CH(F') | | $2_1$ 3.84 |
| CH(N)···CH(E') | | $2_1$ 3.52 |
| CH(N)···C(S') | | $2_1$ 3.74 |
| CH(O)···CH(E') | | $2_1$ 3.72 |
| CH(O)···CH(F') | | $2_1$ 3.49 |
| CH(O)···CH(I') | | $2_1$ 3.76 |
| CH(P)···C(J) | translation | 4.00 |
| | along $c$ | |

Some of the intermolecular C···C contacts are somewhat short.

## b. *1,14-Benzbisanthrene* (XI) [20]

Unit cell: $a = 36.13$, $b = 10.26$, $c = 4.68$, $\beta = 90° \pm 0.5°$, space group $C_s^2 = Pa$, $Z = 4$; i.e., the structure contains two independent molecules in general positions, and the molecules totally lose their symmetry. The perpendicular spacing between the planes of molecules related by translation along the $b$ axis is 3.4 Å, a common figure for aromatic hydrocarbons with a crystalline structure of this type.

(XI)

c. *Pyrene* (XII) [21]

(XII)

Unit cell: $a = 13.60$; $b = 9.24$; $c = 8.37$; $\beta = 100.2°$; space group $C_{2h}^5 = P2_1/a$, $Z = 4$. The molecular symmetry is *mmm*, but in the crystal the molecule occupies a general position. In the structure, the role of one "crystallographic" molecule is played by a pair of molecules related by a center of inversion (Fig. 39). The distance between the molecular planes is 3.53 Å. Molecules

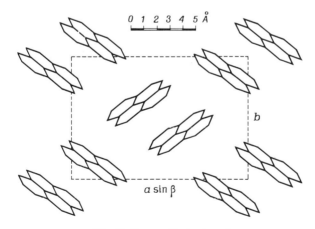

0 1 2 3 4 5 Å

b

a sin β

**Fig. 39.** Pyrene. Projection *ab*.

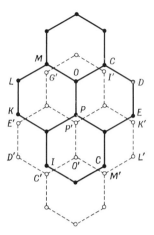

*Fig. 40.* Pyrene. Stacking of molecules related by Ī.

forming an "island" are stacked with a displacement (Fig. 40). The intermolecular distances are: $C(P) \cdots C(P') = C(M) \cdots C(G') = C(C) \cdots C(I') = C(K) \cdots C(E') = 3.54$ Å. The other distances are larger: $C(D) \cdots C(O') = 3.61$, $C(D) \cdots C(P') = 3.64$, $C(A) \cdots C(N') = 3.96$ Å.

This is a rare (if not the only) case of a packing in which a molecule of symmetry *mmm* in the free state takes up a general position in the crystal.

### d. Coronene (XIII) [22]

Unit cell: $a = 16.10$, $b = 4.695$, $c = 10.15$ Å, $\beta = 110.8°$, space group: $C_{2h}^5 = P2_1/a$, $Z = 2$. The distance between the planes of molecules related by translation along $b$ is 3.40 Å. However, these molecules are stacked with a displacement (Fig. 41), so that the $C \cdots C$ distances are larger than the above value: $J' \cdots J1 = G \cdots D1 = 3.43$ Å; $K \cdots B1 = J \cdots L1 = 3.44$ Å. Other $C \cdots C$ distances between these molecules are above 3.5 Å. For molecules

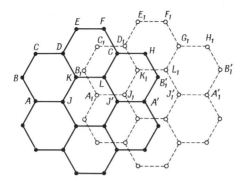

*Fig. 41.* Coronene. Stacking of molecules related through translation *b*.

related by a translation along the *c* axis, $E \cdots E' = 3.87$, $E \cdots G' = 3.93$ Å. For molecules related by the glide plane, $I \cdots B = 3.77$, $I \cdots I' = 3.97$ Å. All other intermolecular separations exceed 4 Å.

*e. Ovalene* (XIV)

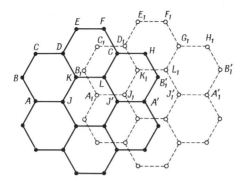

Unit cell: $a = 19.47$, $b = 4.70$, $c = 10.12$ Å, $\beta = 105°$, space group: $C_{2h}^5 = P2_1/a$, $Z = 2$.

The molecular symmetry of ovalene, similar to coronene, is *mmm*. The molecules retain only a center of inversion in the crystal. The position of the molecule in a unit cell is illustrated in Fig. 42.

In all the structures described in this section it would be of interest to analyze the types of stacking of plane molecular layers. It can be seen from the figures that due to the monoclinic displacement, no "atom-on-atom" stacking occurs in the crystal. But the figures also show the absence of a displacement that would place certain atoms in optimum conditions (the middle of the hexagons of the preceeding layer), while other atoms would stack accurately one upon another (as in graphite). The solution chosen by Nature is that all atoms are placed in approximately equal conditions.

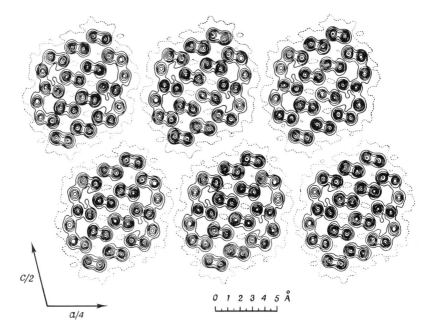

*Fig. 42.* Ovalene. Projection *ac* of the electron-density series.

## 9. STRUCTURE OF NORMAL PARAFFINS

Normal paraffins belong to a class of substances occupying an intermediate position between low polymers and high polymers. Paraffins can with full justification be called low-molecular weight polymers, or "oligomers."

In the structures of *n*-paraffins the axes of all molecules always run parallel to each other, regardless of the crystalline modification [23].

Studies of *n*-paraffins are of major importance for disclosing the principles and laws of organic chemical crystallography. It is obvious that these simple structures are highly suitable for the purpose.

An interest in *n*-paraffins is also motivated by the fact that they exibit a great variety of polymorphous forms. An analysis of the molecular packing in the various forms can provide a better insight into the causes and factors that render one crystalline modification more stable as compared with another. It thus becomes possible to study polymorphism as a function both of the temperature and the number of carbon atoms in the *n*-paraffin chain [24].

### *a. Configurations of Aliphatic Chains*

In all the compounds in which carbon atoms are valence-bonded with four atoms of one element, e.g., $CH_4$, $CCl_4$, etc., ideal tetrahedral angles between

$$
\cdots - \underset{\underset{H}{|}}{\overset{\overset{H}{|}}{C}} - \underset{\underset{H}{|}}{\overset{\overset{H}{|}}{C}} - \underset{\underset{H}{|}}{\overset{\overset{H}{|}}{C}} - \underset{\underset{H}{|}}{\overset{\overset{H}{|}}{C}} - \cdots
$$

(XV)

the bonds are observed. In the case of the simple aliphatic chains (XV) we are discussing, each carbon atom joins two other C atoms and two hydrogen atoms; but, despite the difference of these four atoms, all the angles are close to the tetrahedral angle, although there are certain systematic departures from this rule to be described below.

This circumstance, however, does not fully determine the shape of the chain being considered, since each of its links may rotate around single bonds. From spectroscopic and thermal data obtained in investigations of vapors, Pitzer [25] and Taylor [26] concluded that not all rotations involve the minimum of energy. Thus, Taylor in his thermodynamic calculations used a formula describing the dependence of potential energy $V$ on rotation angle $\phi$ around a single bond:

$$
V = V_0 [x(1 - \cos \phi) + (1 - x)(1 - \cos 3\phi)]
$$

where $x$ is a parameter. Taylor's equation defines the configuration corresponding to the minimum of potential energy as a flat zigzag of carbon atoms (for more details, see Chapter VII).

In Fig. 43, the axis of the chain passes through the midpoints of the C–C bonds. The hydrogen atoms (not shown) lie in planes normal to the chain

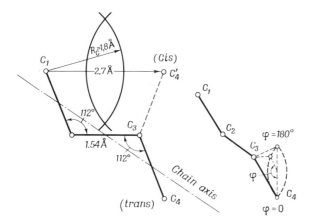

**Fig. 43.** Configuration of an aliphatic chain and its determining interactions of nonbonded atoms.

axis. The H–C–H angles are approximately tetrahedral. The minimum of potential energy is associated with the trans configuration ($\phi = 0$).

The structural justification of the advantages offered by such a configuration is given by an analysis of the interactions between atoms unrelated through valence bonds. The angles between the bonds in the aliphatic chain are fixed as a result of the mutual interaction between all the four atoms bonded to each given carbon atom. It is of interest to note that according to the X-ray diffraction data [27, 28] the C–C–C angle is always larger than the ideal tetrahedral angle ($\sim 112°$ instead of 109.5°). This certainly must be so from the viewpoint of the mechanical model of the molecule (see Chapter VII) offered by the author [29], in which all the deviations from ideal angles in aliphatic compounds are ascribed to the repulsion of atoms unrelated by valence bonds. The same specific features of intermolecular interaction define the plane zigzag shape of a chain of carbon atoms. As can be seen from Fig. 54, the cis configuration ($\phi = 180°$) results in a C1–C'4 distance of 2.7 Å, which is substantially smaller than $2R_C = 3.6$ Å. It is natural, therefore, that the minimum of energy is associated with the trans configuration ($\phi = 0°$). The same conclusion is suggested by calculations of conformation energy (Chapter VII).

The hydrogen atoms of methylene groups are positioned in planes passing through the carbon atoms perpendicular to the chain axis. They were clearly detected in the electron-diffraction study of Vainshtein and Pinsker [30]. The H–C–H angle is about 108–109°.

### b. Results of Experimental Structure Studies

The study of the structures and properties of the elementary chain compounds under discussion entails a number of special difficulties. These are, primarily, the difficulty of obtaining individual hydrocarbons in a pure form without admixture of adjacent homologues, and also the difficulty of growing perfect single crystals for suitable X-ray diffraction analysis.

*Classification by Crystal Modifications.* The outstanding investigator of the structure of $n$-paraffins is Müller [31]. His principal findings have been confirmed by later researchers. His first systematic studies used the Debye X-ray diffraction technique. Müller was able to grow a sufficiently good single crystal of $n$-$C_{29}H_{60}$ and to determine its structure completely. He found that the molecules in the crystal have the form of plane-zigzag chains of carbon atoms and that their axes are parallel. Planes normal to the axes can be drawn through the end groups which separate one layer of molecules from another. Every second layer is translationally identical to the previous ones; i.e., the structure is composed of two layers. The unit cell is orthorhombic, with parameters $a = 7.45$, $b = 4.97$, $c = 77.2$ Å. Müller found the space group to be

$V_h^{16} = Pnam$, however, much later Smith [32] performed diffraction studies on specimens with a higher degree of purity and found another space group: $V_h^{1i} = Pcam$.

Figure 44 shows the $ab$ face of the unit cell perpendicular to the chain axes; it was called by Müller a basic (main) plane. The transverse arrangement of the molecular chains shown in this figure is highly characteristic of most $n$-paraffins and many other chain compounds, including a number of high-polymers. In the latter case there are no difficulties due to the location of the end groups. The simplest polymer is polyethylene—an infinite plane zigzag of methylene groups. Its structure has been investigated by Bunn [33]. It is practically identical to the structure of the single layer of paraffin molecules considered above if we imagine that the number of carbon atoms in the chain $n \to \infty$, in other words, if the specific features of end group packing are ignored.

In order to generalize the similarity detected between distinct structures of crystals composed of chain molecules, Vand [34] has introduced the concept of a subcell determined by the positions of the methylene groups. The most widely encounted type of subcell is the orthorhombic polyethylene cell with the parameters $a_0 = 7.45$, $b_0 = 4.97$, $c_0 = 2.54$ Å (the length of a single plane aliphatic zigzag), whose $a_0 b_0$ plane is shown in Fig. 44. This is not, however, the only type of subcell. Already Müller had previously noticed a certain difference between even-numbered $n$-paraffins with long chains and those with shorter chains. To interpret this fact, Müller assumed that the crystal energy is composed of two components, one of which is associated with the side interaction of the molecular chains, and the other with the interaction of the end groups. The transition to shorter molecules changes the proportions of these two components and leads to alterations in the crystalline structure.

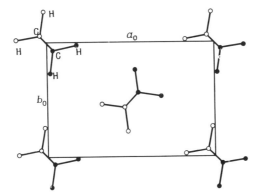

**Fig. 44.** Face $ab$ of orthorhombic paraffin cell. Mutual arrangement of molecules with axes normal to drawing plane is shown. Atoms designated by white and black circles are positioned at different levels along $c_0$; distance between levels is $c_0/2 = 1.27$ Å.

In 1948 Müller and Lonsdale [35] published the results of their studies on the unit cell of an $n$-$C_{18}H_{38}$ crystal which was found to be triclinic. Thus, the structure of $n$-$C_{18}H_{38}$ is based on another type packing. The presence in this subcell of only one molecule indicates that the planes passing through the carbon atoms of the zigzag chains are parallel. The triclinic $n$-$C_{18}H_{38}$ crystals have the same type of structure as other members of the homologous series with an even number of carbon atoms from $C_6$ to $C_{24}$.

Apart from the existence of two types of crystalline structure in the series of even $n$-paraffins, physicochemical differences have been established between even and odd $n$-paraffins. Müller quite rightly pointed out that these lie in the structure of the chain itself. He noted a difference in the symmetries of even- and odd-numbered molecules, although he does not use this term. Müller arrived at the conclusion that as we move along the $c$ axis, every second molecule in an odd $n$-paraffin crystal will be translationally identical, while in an even $n$-paraffin crystal all molecules are successively related by simple translation.

Along with the identification of the structures of the individual members of the homologous series of $n$-paraffins, certain authors studied the dependence of various parameters on the number $n$ of C atoms in the chain. These studies helped to classify $n$-paraffins by their crystalline modifications, without detailed investigation of all the members of the homologous series. Thus, Piper *et als.* [36] continued the work on the precise measurement of the large spacing started by Müller. A "large" or "long" spacing is the distance between the planes of the crystal lattice passing through the end groups of the chain molecules. This distance is of the order of the chain length of one molecule and is readily measured because the corresponding lines on the powder X-ray diagram can be easily identified, because they lie in the region of comparatively small reflection angles $\theta$.

The results of these studies appear in Fig. 45 which clearly illustrates the

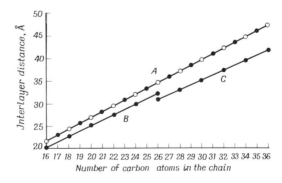

*Fig. 45.* Dependence of interlayer distance on number $n$ of carbon atoms in the chain.

various crystal modifications of the members of the homologous series of $n$-paraffins. The points for all the odd $n$-paraffins lie on the straight line A. These paraffins have a structure with an orthorhombic subcell and the rectangular layers interpreted by Müller [31] ($n$-$C_{29}H_{60}$) and Smith [32] ($n$-$C_{23}H_{48}$). The crystals of the triclinic paraffins (even members of the homologous series up to $C_{24}$) are built up from oblique layers and the line joining the points associated with these layers passes somewhat lower ($B$). It has already been noted that the unit cell of $n$-$C_{18}H_{38}$ as determined by Müller and Lonsdale [35] has the following parameters: $a = 4.28$, $b = 4.82$, $c = 23.07$ Å, $\alpha = 91° 6'$, $\beta = 92° 4'$, $\gamma = 107° 18'$, $d_{001} = 23.04$ Å, $Z = 1$. Even $n$-paraffins from $C_{28}$ and above have orthorhombic subcells and a packing with even more oblique layers, since their respective line in Fig. 45 lies still lower (C). In the case of $n$-$C_{36}H_{74}$ the true cell is monoclinic and contains two molecules: $a = 5.57$, $b = 7.42$, $c = 48.35$ Å, $\beta = 119° 6'$; space group $C_{2h}^5 = P2_1/a$. The orthorhombic subcell and the coordinates of the carbon atoms in the subcell are in close agreement with those found for polyethylene.

*Phase Transformations.* Apart from the crystalline forms that have already been discussed, there is one more stable form in which $n$-paraffins can exist. At a certain temperature close to the melting point Müller discovered a reversible phase transition at which the angle between the diagonals of the $ab$ face in the orthorhombic unit cell shown in Fig. 44 becomes 60° upon which the symmetry increases to hexagonal. This form corresponds to the close packing of cylinders, and is a partial case of the so-called gaseous crystalline or rotational crystalline state (see below) which is characterized by rigorous order in the arrangement of the centers (axes) of the molecules and disorder in their azimuthal rotations. This is a rather frequent occurrence.

The most complete description of the phase behavior of $n$-paraffins is given by the graph (Fig. 46) compiled by Schaerer, *et al.* [37]. The graph shows the melting points and the temperatures of the reversible phase transitions in a solid state.†

The domain of existence of the $n$-modification is designated by a vertical line on the graph: T, triclinic structure of the type $n$-$C_{18}$; O, orthorhombic structure of the types $n$-$C_{23}$ and $n$-$C_{29}$; M, monoclinic structure with O-subcell type $n$-$C_{36}$; H, hexagonal gaseous–crystalline structure.

As is obvious from the graph, in the members of the series below $C_{20}$ the melting points alternate, while in the members above $C_{20}$ there is no such alternation, but in the latter case we see alternation of the points of phase

---

† As compared with other $n$-paraffins, $C_{24}$ and $C_{26}$, which occupy an intermediate position according to their structure, reveal a more complex phase behavior that has not yet been fully studied.

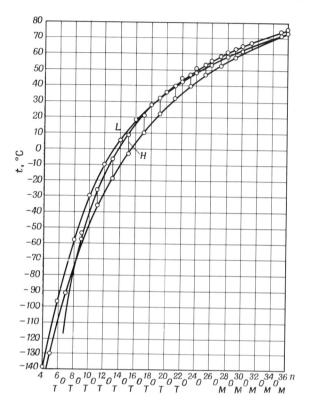

**Fig. 46.** Diagram compiled by Schaerer with collaborators summarizing all reliable data on phase changes in $n$-paraffins.

transition in the solid state. In fact, in the region from $n$-$C_6$ to $n$-$C_{20}$ the melting points of the even members of the homologous series lie on a curve which is above the curve representing the melting points of the odd members; only odd members of the series have a gaseous crystalline structure in the vicinity of the melting point. Now, passing on to $n$-paraffins with longer chains, we see that they all, regardless of whether they are odd or even, display a rotational crystalline modification in an appropriate temperature range, and that all their melting points fall on one curve. On the other hand, the phase transition points in the solid state give different curves depending on whether $n$-paraffins are odd or even numbered. Alternation of the physical and chemical properties between even and odd $n$-paraffins is caused by the alternation of the types of their crystalline structure: for shorter paraffins O, H of odd members alternate with T of even members, for longer ones O and M alternate respectively. As to the absence of alternation in the properties of the H modifications of *n*-paraffins above $C_{22}$ this seems quite natural, since the structures for all the members of the series, whether odd or even, are of the same type.

### c. Close Packing of Chain Molecules

In the case of chain molecules geometrical analysis can be employed for predicting a small number of possible crystalline structures. This set includes all the structures experimentally found so far. The analysis is greatly facilitated by the following basic statement: The structure always consists of layers of flat zigzag molecules whose axes run parallel. The molecular packing problem can be broken up into two parts: the packing of the molecules in a layer and the packing of the layers. Below follows a general discussion of these two types of packing.

If the chains are azimuthally chaotic (rotate around their axes), their average cross sections are circular, which gives rise to a circular (hexagonal) packing. For molecules of an arbitrary cross section, two types of close-packed layers are possible: one with an oblique and the other with a rectangular cell. If we represent the cross section of the chain by an arbitrary contour, all the three types of packing will be as shown in Fig. 47.

The geometrical analysis of a structure implies, primarily, that a three-dimensional shape must be attributed to an aliphatic molecule following the

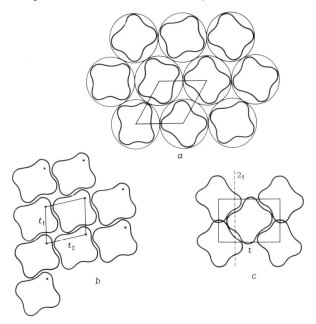

*Fig. 47.* Three possible types of close packing of chain molecule (a) hexagonal, (b) oblique, (c) rectangular cell.

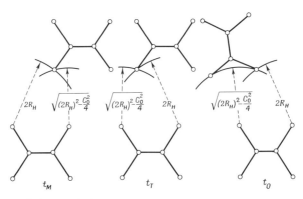

**Fig. 48.** Types of close-packed arrays of a pair of aliphatic chains (chain axes are perpendicular to drawing plane).

rules of organic chemical crystallography, i.e., using the standard intermolecular radii of carbon and hydrogen atoms. The analysis starts with the consideration of the closest-packed arrangements of two infinite molecular chains. These arrangements occur only if an H atom of one molecule fits into a "hollow" formed by three H atoms of the adjacent molecule. Two types of such close contacts can be obtained between parallel molecules shifted by simple translation. These are the $t_M$ and $t_T$ arrays given in Figs. 48 and 49. Configuration $t_O$ (Fig. 48) corresponds to case b in Fig. 47 when the zigzag planes are not parallel. Each of the three above arrangements of the chain pair results in one closest packing. Array $t_T$ gives a structure with a triclinic subcell† (T), $t_M$ with a monoclinic (M) and $t_O$ with an orthorhombic subcell (O).

The further analytical procedure will be briefly illustrated by the derivation of the structure obtained from the $t_T$ array. Two molecules related by translation $t_T$ define a way of packing an infinite series of molecules in one of the crystal directions. Close stacking of two such rows can readily be found by geometrical construction or with the aid of molecular models. It is effected by a translation $b_0$ perpendicular to the molecule axes. We thus find the only arrangement yielding coordination 6. This is an ideal structure, i.e., a structure in which all intermolecular distances H ··· H are equal.

If we put $R_H = 1.3$ Å, the resulting cell will contain two $CH_2$ groups and have the following parameters:

$$a_0 = 4.3 \text{ Å}, \qquad \alpha = 90°$$

$$b_0 = 4.45 \text{ Å}, \qquad \beta = 107.5°$$

$$c_0 = 2.54 \text{ Å}, \qquad \alpha = 102°$$

---

†The subcell is built up on the vectors which repeat the methylene groups.

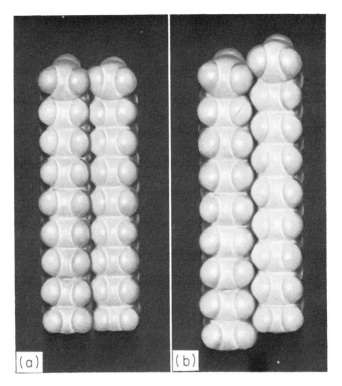

*Fig. 49.* Array of a pair of chains types $t_M$ (a) and $t_T$ (b) illustrated by means of models.

The ideal parameters of the O and M cells are found similarly. The results are given in the first column of Table 5. For molecular chains of infinite length, these unit cells are true cells. But a layer of finite-length molecules has only a two-dimensional periodicity and, obviously, is not a crystal. In this case the cells O, M, and T identified in a layer of molecules are subcells, i.e., they characterize the packing of methylene groups.

For layers composed of finite-length molecules, a new parameter must be taken into consideration—the angle of tilt between the plane passing through the end groups and the molecular axes. If this angle is 90°, the layers are rectangular, in all other cases they will be oblique. This angle cannot assume any arbitrary value. The reason is that parallel displacements along the axes of the molecules which give rise to different versions of a true cell without changing the packing patterns of methylene groups, must be equal to $mc_0$, where $m$ is an integer. The idea that all versions of the structure for chain molecules can be obtained from the initial structure by the above discrete shifts of the molecules was put forth 30 years ago by Schoon [38], but quite naturally it could not be verified by using the structural data available at that

*Table 5*

| Subcell | Layer | Layer cell (Å) | $\phi_a$ | $\phi_b$ | $\phi_c$ |
|---|---|---|---|---|---|
| H | H[00] | $a = 4.8$ | 90 | 90 | 0 |
| O | O[0,0] | $a = 7.42; b = 4.96; \gamma = 90°$ | 90 | 90 | 0 |
| | O[0, ±2] | $a = 9.0; b = 4.96; \gamma = 90°$ | 90 | 55.5 / 124.5 | 34.5 |
| $a_0 = 7.42$ | O[±1,0] | $a = 7.42; b = 5.57; \gamma = 90°$ | 63 / 117 | 90 | 27 |
| $b_0 = 4.96$ | O[0, ±1] | $a = 7.85; b = 4.96; \gamma = 90°$ | 90 | 71 / 109 | 19 |
| $c_0 = 2.54$ | O[±1, ±1] | $a = 7.85; b = 5.57; \gamma = 81.5°$ for O[1, 1] and O[1, 1] = 98.5° for O[$\bar{1}$, 1] and O[1, $\bar{1}$] | 63 / 117 | 71 / 109 | 31.5 |
| M | M[0,0] | $a = 4.2; b = 4.4; \gamma = 111°$ | 90 | 90 | 0 |
| $a_0 = 4.2$ | M[±1,0] | $a = 4.9; b = 4.4; \gamma = 107°$ | 59 / 121 | 90 | 32 |
| $b_0 = 4.4$ | M[0, ±1] | $a = 4.2; b = 5.1; \gamma = 107°$ | 90 | 60 / 120 | 32 |
| $c_0 = 42.54$ $\gamma_0 = 111°$ | | | | | |
| T | T[±½,0] | $a = 4.3; b = 4.5; \gamma = 103°$ | [73] / 107 | [90] | [18] |
| $a_0 = 4.3$ $b_0 = 4.45$ $c_0 = 2.54$ $\alpha_0 = 90°, \beta_0 = 107.5°$ $\gamma_0 = 102°$ | T[½, 1] T[−½, 1] | $a = 4.3; b = 5.2; \gamma = 109°$ | 73 / 107 | 120 / 60 | 36 |

time. At present this assertion needs no proof—it suffices to look at Fig. 49. If any molecule is shifted along its axis by a whole number of "stories" (one "story" equals the chain repeat length 2.54 Å), the convex parts of one molecule will fit again the hollows of another, and the pattern of arrangement of methylene groups will remain unchanged.

Let us designate the displacement of molecules related by translation $a_0$ by $m$ and a displacement of molecules related by translation $b_0$ by $n$. The two-dimensional cell of the layer is determined by the subcell and numbers $m, n$. In contrast to the O and M subcells, a T subcell cannot give rectangular layers by virtue of its structure, since the methylene groups related by the $a_0$ axis are shifted by $c_0/2$. Therefore, possible two-dimensional cells are described by the symbols

$$O[m,n], \qquad M[m,n], \qquad T[m + \tfrac{1}{2}, n]$$

It is also necessary to recognize that layers with different signs of $m$ and $n$ are not equivalent; each pair of numbers $m, n$ gives four layers: $m, n$; $m, \bar{n}$; $\bar{m}, n$;

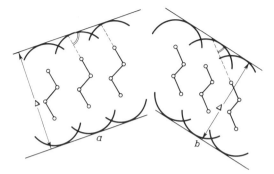

**Fig. 50.** Layers differing only in a displacement sign.

$\bar{m}, \bar{n}$. This difference is illustrated in Fig. 50. There is no need to consider all possible chain displacements. If the displacement of two adjacent chains exceeds one period $c_0$, the end groups will make contact with methylene groups, thus resulting in lower packing density. Therefore, one will be well advised to restrict consideration only to the layers given in the second column of Table 5.

All the layers with a T subcell are oblique layers. If the *ab* plane of the two-dimensional cell of the layer coincides with the $a_0 b_0$ plane of the subcell, the symbols of these layers will be $T[\frac{1}{2}.0]$ and $T[-\frac{1}{2}.0]$. In layers with a T subcell each molecule has six neighbors translated by vectors $\pm \mathbf{a_0}$, $\pm \mathbf{b_0}$, and $\pm \mathbf{c_0}$, assuming, as usual, that the angle $\gamma$ is obtuse. Since it is postulated that a shift of any of these molecules cannot be above unity, only two more layers $T[\frac{1}{2}.1]$ and $T[-\frac{1}{2}.1]$ can exist.

Table 5 gives all the layers thus derived with the parameters of their two-dimensional cells and the inclination angles $\phi_a$, $\phi_b$, and $\phi_c$ of the molecular axes to the axes $a, b$, and to the normal of the plane *ab*, respectively.

Furthermore, we have considered the possible orthorhombic, monoclinic, and triclinic structures arising from layer stacking. In particular, structures with triclinic symmetry can be obtained by stacking any types of layers O, M, and T with arbitrary displacements. At this point it should be recalled that layers are never stacked precisely one upon another, i.e., they cannot be related by a mirror symmetry plane, because then the "projections" of the top layer would have got onto the "projections" of the bottom layer, and the packing would not have been dense as shown in Fig. 51a. Since adjacent layers are always stacked with a displacement, the *c* axis of the true unit cell, which passes through identical points of the molecules of adjacent layers, will not lie in the same direction as the molecular axis (Fig. 51b). These axes will be aligned only in a two-dimensional orthorhombic structure in which the third layer is located over the first one (Fig. 51c).

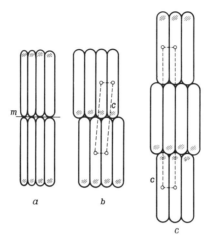

*Fig. 51.* Stacking of layers: (a) adjacent molecular layers are never related as shown through mirror symmetry plane as it does not produce close packing; (b) in single-layer structures axis $c$ of unit cell does not coincide with molecular axis $c_0$; (c) in orthorhombic (double-layer) structures axes $c$ and $c_0$ are aligned.

From the point of view of close packing an arbitrary displacement is more advantageous than that preserving the monoclinic and, still more so, the orthorhombic symmetry. Nevertheless, a triclinic displacement will not always take place, since the attainment of minimum free energy involves not only the tendency to a maximum packing density but also the tendency of the molecule to preserve the highest possible symmetry in the crystal. The same applies to the molecular layers whose stacking forms the crystal. It has been shown experimentally that crystals $n\text{-}C_{18}H_{38}$ are not built from layers $T[+\frac{1}{2}.0]$, but from layers $T[-\frac{1}{2}.0]$. The structures of these crystals formed by the only possible close stacking of these layers is shown on models in Fig. 52. It should be remembered that the other $n$-paraffins shown by the straight line $B$ (Fig. 45) are of the same type as $n\text{-}C_{18}H_{38}$.

A complete description of the derived structures is given in the author's book "Chemical Organic Crystallography" referred to above. Among these are structures investigated by Müller, Schaerer and V. V. Smith, and A. Smith. The derived types of structures can probably be discovered not only in $n$-paraffins but also in other long-chain compounds where the aliphatic zigzag plays a major role in packing. The geometrical considerations based on the close packing principle explain the sources of the structural diversity in chain molecules and shows that this diversity is mainly typical of the even members of the homologous series of $n$-paraffins, while for odd numbers only two versions of structures with orthorhombic cells can be expected.

***Fig. 52.*** Packing of molecules in triclinic *n*-paraffins.

*Density of O and T Packings.* The experimental data enable comparison of the packing density of molecules in O and T subcells. The accuracy with which the parameters are known is sufficient for the purpose; however, it is necessary to point out that this comparison cannot be considered rigorous, since different substances are compared, though of a single-type.

The parameter $c_0$ in both cases is 2.54 Å; therefore, the molecular packing density in the subcell depends on the area per molecule in the cross section perpendicular to the chains. The cross-section area to be found for an O subcell is

$$S_O = a_0 b_0 / 2 = 18.56 \text{ Å}^2$$

In order to find a corresponding value for a T subcell, it is necessary to project the face *ab* on the plane normal to the chains and calculate the projected area:

$$S_T = b_0 a_{\perp 0} \cos(\gamma_0 - 90°) = b_0 [a_0{}^2 - (c_0/2)^2]^{1/2} \cos(\gamma_0 - 90°)$$

Substituting values $a_0 = 4.28$ Å, $b_0 = 4.82$ Å, $\gamma_0 = 107° 18'$ in this formula, we have $S_T = 18.73$ Å$^2$.

The small difference in density (less than 1%) probably accounts for the natural occurrence of both types of subcell, O and T, since the energetic advantage of one or the other structure depends primarily on the molecular packing density.

It is now pertinent to state a problem which is of importance for understanding polymorphism in $n$-paraffins. How can one explain the stability of triclinic structures for even $n$-paraffins with a number of carbon atoms $n < 26$ and the transition to structures with an O subcell at $n > 26$? The reply can be obtained by proceeding from the assumption that the free energy of the T subcell itself is somewhat larger than that of the O subcell. In this case triclinic structures will have advantages in terms of energy, because the triclinic layer stacking is in general denser, as has already been mentioned. On the other hand, the longer the molecules, the relatively smaller becomes the effect of this gain in the packing density of the ends, and, finally, for a sufficiently large $n$ it becomes insufficient to render a structure with a T subcell energetically more advantageous than the structure containing an O subcell. We shall return to this problem when discussing the experimental data pertaining to the solid solutions of $n$-paraffins (p. 116).

At any rate it is quite clear that crystals with packing patterns O and T do not display a large difference in the amount of crystal free energy. This is suggested both by the comparison of their densities given above and by the fact that a several percent admixture of adjacent homologues converts a triclinic structure into an orthorhombic structure.

Kabalkina [39] has carried out an X-ray study of the paraffins $n$-$C_{30}H_{67}$, $n$-$C_{32}H_{66}$, $n$-$C_{34}H_{70}$ at pressures up to 15,000 atm. This work established that a certain amount of the triclinic phase appeared in the structure and that it persisted for several days even after the pressure had been removed.

## 10. ORGANO–IRON COMPOUNDS

### a. Sandwich-Type Molecules

An interesting representative of this class of compounds is ferrocene [40]. Ferrocene crystallizes in the monoclinic space group $C_{2h}^5 = P2_1/a$, $a = 5.91$, $b = 7.59$, $c = 10.51$ Å, $\beta = 121.1°$, $Z = 2$; i.e., the centers of the molecules (Fe atoms) occupy special positions and the molecule has symmetry $\bar{1}$ in the crystal. Consequently, the ferrocene molecule exhibits an antiprismatic arrangement of the five-membered rings. Figure 53 illustrates the configuration of a ferrocene molecule with the point group $D_{5d}$. The significant electron density in the space between the carbon atoms observed in the differential

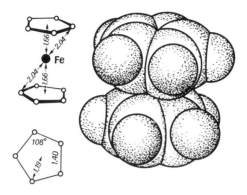

**Fig. 53.** Configuration of ferrocene molecule.

electron-density series is probably due to considerable torsional vibrations of the carbon cycles.

As can be seen from Table 6, unusual bonds of the "metal five-membered ring" type in the molecule do not inhibit the crystallization of cyclopentadiene metal compounds possesing both the antiprismatic (ferrocene type) and

## Table 6

CRYSTALLOGRAPHIC DATA FOR DICYCLOPENTADIENYL METAL COMPOUNDS

| Parameters<br>Compounds | $a(\text{Å})$ | $b(\text{Å})$ | $c(\text{Å})$ | $\beta°$ | $V_{cell}$ | $Z$ | Space group |
|---|---|---|---|---|---|---|---|
| $Fe(C_5H_5)_2$ | 5.91 | 7.59 | 10.51 | 121.1 | 403.6 | 2 | $P2_1/c$ |
| $Co(C_5H_5)_2$ | 5.90 | 7.71 | 10.60 | 121.1 | 412.8 | 2 | $P2_1/c$ |
| $Ni(C_5H_5)_2$ | 5.88 | 7.86 | 10.68 | 121.1 | 422.2 | 2 | $P2_1/c$ |
| $V(C_5H_5)_2$ | 5.88 | 8.02 | 10.82 | 121.2 | 436.0 | 2 | $P2_1/c$ |
| $Cr(C_5H_5)_2$ | 5.92 | 7.88 | 10.72 | 121.2 | 427.6 | 2 | $P2_1/c$ |
| $Mg(C_5H_5)_2$ | 5.98 | 8.04 | 10.98 | 121.9 | 448.0 | 2 | $P2_1/c$ |
| $Be(C_5H_5)_2$ | $5.9_1$ | $7.7_2$ | $10.8_6$ | 122 | 420 | 2 | $P2_1/c$ |
| $Ru(C_5H_5)_2$ | 7.13 | 8.59 | 12.81 | — | — | 4 | $Pnma$ |
| $Os(C_5H_5)_2$ | 7.159 | 8.988 | 12.800 | — | — | 4 | $Pnma$ |
| $Hg(C_5H_5)_2{}^a$ | $13.1_1$ | $13.4_1$ | $5.8_4$ | $\alpha = 99.8$<br>$\beta = 101.4$<br>$\gamma = 88.8$ | 993.0 | 4 | P1 or P$\bar{1}$ |
| $Mn(C_5H_5)_2$ | 15.20 | 11.74 | 9.96 | $\gamma = 114.46$ | — | 8 | $P2_1/b$ |

$^a$ At $-90°C$.

prismatic (ruthenocene type) molecular configuration following the general regularities of organic chemical crystallography.

As an example consider the crystalline structure of diferrocenyl [41]. The diferrocenyl molecule (XVI) occupies a centrosymmetric position in the

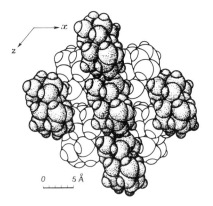

(XVI)

monoclinic unit cell with symmetry $P2_1/c$ ($a = 10.17$, $b = 7.86$, $c = 12.62$ Å, $\beta = 132°$, $Z = 2$); in other words, it has a trans configuration with respect to the $C_1$–$C'_1$ bond between the ferrocene residues. The bond lengths (Fe–C = 2.03; C–C = 1.42 Å) and valence angles (C–C–C = $108° \pm 2.5°$) in these residues coincide with those found in ferrocene itself. The length of bond $C_1$–$C'_1$ = 1.48 Å between the ferrocene residues is somewhat shortened. It is interesting to note that rotation angle $\phi = 16°$ of one cyclopentadiene ring in the sandwich with respect to the other appears to be intermediate between the prismatic ($\phi = 0°$) and antiprismatic ($\phi = 36°$) configurations inherent in these rings.

*Fig. 54.* Packing of molecules in diferrocenyl structure.

Figure 54 shows the packing of molecules in the structure. The structure is a close packing of layers with coordination 6 which run parallel to (10$\bar{1}$). It is evident that the trans conformation of the molecule ensures a convenient shape close to a sphere and provides possibilities for a rather dense packing. As a result, each molecule in the structure is surrounded by twelve neighbors and the packing coefficient is 0.71. The shortest intermolecular distance C $\cdots$ C equals 3.60 Å. Some of the C $\cdots$ H distances are reduced. The shortest distance (2.52 Å) would become still smaller ($\sim$2.3 Å) if the ferrocene sandwiches acquired an antiprismatic conformation by rotation of the external rings, without changing the cell parameters.

### b. Structures of Some Iron Carbonyls

In the iron pentacarbonyl molecule $Fe(CO)_5$ [42] the plane passing through Fe and three of the carbonyl groups is perpendicular to the plane drawn

(XVII)

through Fe and the two remaining carbonyl groups. In the unit cell of monoclinic symmetry ($a = 11.71, b = 6.80, c = 9.28$ Å, $\beta = 107.6°$; space group C2/c, $Z = 4$) the molecules are positioned on twofold axes. Since the entire molecule is bounded by oxygen atoms, O $\cdots$ O contacts alone are created by the molecular packing. Figure 55 represents the intermolecular distances in the structure, the determining ones being the contacts O $\cdots$ O = 3.15, 3.19, 3.21 Å. The mean intermolecular radius of an O atom is thus 1.59 Å; this is somewhat larger than that taken above (1.52 Å).

The same O $\cdots$ O distances (the minimum distance equals 3.1 Å) are to be found in the $Fe_2(CO)_9$ structure [43]. The crystals of this compound are hexagonal: $a = 6.45, c = 15.98$ Å; space group C6$_3$/m, $Z = 2$.

(XVIII)

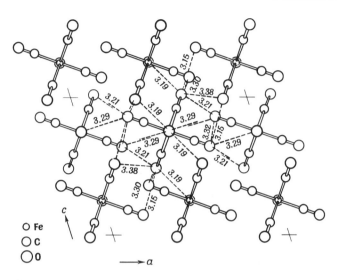

*Fig. 55.* Intermolecular distances in iron pentacarbonyl structure.

The structure of iron tetracarbonyl (acrylonitryl) $(CH_2CHCN)$ $Fe(CO)_4$ [44] is monoclinic $(a = 12.09,\ b = 11.45,\ c = 6.585\ \text{Å},\ \beta = 110.4°$; space group $P2_1/a,\ Z = 4)$· The molecules have the form

$$
\begin{array}{c}
\text{O}\!\!\equiv\!\!\text{C} \quad \overset{\overset{\text{O}}{\|}}{\text{C}} \quad \text{H}\diagdown_{\text{C}}\diagup^{\text{C}\!\equiv\!\!\text{N}} \\
\qquad\diagdown_{\text{Fe}}\text{------}\!\!\mid \\
\text{O}\!\!\equiv\!\!\text{C}\diagup \quad \underset{\underset{\text{O}}{\|}}{\text{C}} \quad \text{H}\diagup^{\text{C}}\diagdown_{\text{H}}
\end{array}
$$

(XIX)

The intermolecular spacings appear in Fig. 56. The adjacent molecules make contact via atoms $O \cdots O$ (the shortest distance is 3.04 Å), $N \cdots O$, $C \cdots N$. The authors ignore H atoms; therefore, the analysis of the contacts is far from complete.

The general conclusion following from consideration of the structures of crystals built up from molecules with "anomalous atom-group of atoms" bonds is beyond any doubt: The packing of identical molecules into a crystal follows the same rules as the packing of molecules with "ordinary" valence bonds.

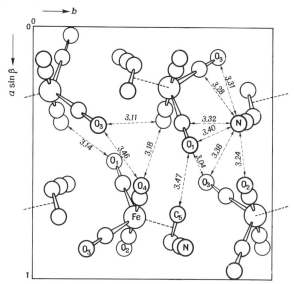

*Fig. 56.* Intermolecular distances in iron tetracarbonyl (acrylonitrile).

## 11. TETRAARYL COMPOUNDS

Tetraaryl compounds are composed of molecules $X(C_6H_5)_4$, where X is a tetravalent element. The interest in these compounds is dictated, primarily, by their high symmetry ($\bar{4}$) retained by the molecule in the crystal and, secondarily, by the possibilities for studying the steric hindrances between the ortho-atoms of two benzene rings lying on one plane.

It can be seen from Table 7 that the structures of the tetraaryl compounds so far known form an isomorphous series; therefore, it is sufficient to discuss the conformation problem for one member of this series, e.g., for tetraphenyl methane.

*Table 7*

| Compound | Syngony | Cell parameters, (Å) | Z | Space group | Molecular symmetry in crystal | $d_{x-c}$ (Å) |
|---|---|---|---|---|---|---|
| $C(C_6H_5)_4$ | tetr. | $a = 10.86; c = 7.26$ | 2 | $P\bar{4}2_1c$ | $\bar{4}$ | 1.50 |
| $Si(C_6H_5)_4$ | tetr. | $a = 11.30; c = 7.08$ | 2 | $P\bar{4}2_1c$ | $\bar{4}$ | 1.94 |
| $Ge(C_6H_5)_4$ | tetr. | $a = 11.60; c = 6.85$ | 2 | $P\bar{4}2_1c$ | $\bar{4}$ | 1.98 |
| $Pb(C_6H_5)_4$ | tetr. | $a = 12.03; c = 6.55$ | 2 | $P\bar{4}2_1c$ | $\bar{4}$ | 2.23 |
| $Sn(C_6H_5)_4$ | tetr. | $a = 11.85; c = 6.65$ | 2 | $P\bar{4}2_1c$ | $\bar{4}$ | 2.17 |

The structure of tetraphenylmethane [45] was determined from a projection on the basal plane and by geometrical analysis. The symmetry of the molecule is $\bar{4}2m$, 222, $m$ or 1 (depending on the rotation angles of the benzene rings); in the crystal, it is $\bar{4}$. Figure 57 gives the $ab$ projection of the structure (the H atoms are not shown). Prior to analyzing the packing of molecules in the crystal, consider the configuration of a free molecule. Let the molecule have symmetry $\bar{4}$. The short distance between the H ortho-atoms in the benzene rings does not allow arbitrary rotation of these rings around the C–X bond. When the rings lie on vertical planes (planes which pass through the $\bar{4}$ axis), the distance between the centers of the H ortho-atoms in the molecule benzene rings related by the twofold axis is very small, and with X = C (X–C = 1.50 Å) it does not exceed 0.58 Å (Fig. 58).

Let us take the configuration of the aryl rings when they are positioned in pairs on one plane for the initial position and describe the configuration by the angle $\psi$ between the ring plane and the vertical plane. As the ring rotates about the C–X bond, the centers of the hydrogen ortho-atoms (positions 2 and 6) draw circumferences with the radius:

$$R = (d_{C-C} + d_{C-H})\cos 30° = 2.15 \text{ Å}$$

where $d$ is a bond length. The same circumferences are drawn by hydrogen atoms in positions 3 and 5; the $H_4$ atom lies on the rotation axis.

The center of the circumference traced by the $H_2$ and $H_6$ atoms lies on the diameter of the benzene ring at distance $\rho$ from the central X atom:

$$\rho = [d_{X-C} + (d_{C-C} - d_{C-H})]\sin 30° = \tau_{X-C} + 0.16 \text{ Å}$$

Using the C–X bond lengths given in Table 7, we obtain the following different values of $\rho$:

$$\rho_{C-C} = 1.7, \qquad \rho_{Si-C} = 2.1, \qquad \rho_{Sn-C} = 2.33, \qquad \rho_{Pb-C} = 2.39 \text{ Å}$$

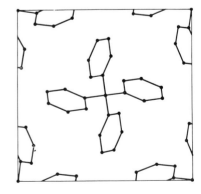

*Fig. 57.* Tetraphenylmethane. Projection *ab*.

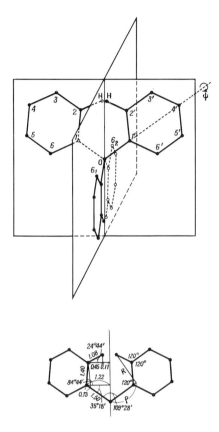

**Fig. 58.** Molecule of tetraaryl compound $XR_4$. Initial position of two benzene rings in one plane is shown.

If the tetrahedral valence angles of the central atom are not distorted, the squared distance from the hydrogen atom in an ortho-position to the $\bar{4}$ axis passing through the central atom is

$$D^2 = (\rho \sin 54° \, 44' - R \cos \psi \cos 54° \, 44')^2 + R^2 \sin^2 \psi$$

where $54° \, 44'$ is half of the terahedral angle. For the four X's given above we have

$$D^2 = \left[ \begin{Bmatrix} 1.388 \\ 1.715 \\ 1.903 \\ 1.952 \end{Bmatrix} - 1.24 \cos \psi \right]^2 + 2.15^2 \sin^2 \psi$$

Distance $D$ must not be less than the intermolecular radius of the hydrogen atom $R_H = 1.17$ Å, otherwise the hydrogen atoms of different benzene rings situated in ortho-positions (2 and 2'; 6 and 6') interfere with each other. For this reason, the values of $\psi$ from 0° to 23°–31.5° (when passing from C to Pb) appear to be forbidden.

However, another restriction is also imposed: The H ortho atoms of two benzene rings separated by the minimum distance at $\psi = 0°$ move apart as $\psi$ increases, but each of them starts to come closer to the H ortho-atom of another benzene ring, i.e., $H_2$ comes nearer to $H_6'$, and $H_6$ to $H_2'$. Since, due to the $\bar{4}$ symmetry, all the rings rotate similarly, with $\psi = 90°$ the distance between these H atoms will be $2\rho \sin 54° 44'$.

At any $\psi$ atoms 2 and 6' (2' and 6) are at an equal distance from the initial plane and on one side of the plane. The projection of the distance between these atoms onto a horizontal plane (normal to axis $\bar{4}$) is constant and equals $2\rho \sin 54° 44'$; only the projection of the distance onto the $\bar{4}$ axis, equal to $2\rho \cos \psi \sin 54° 44'$, varies. Calculations show that the values of $\psi$ are forbidden in the following ranges: for X = C, 44–90°; Si, 66–90°; Sn, 71–90°; Pb, 72–90°.

Thus, a tetraphenylmethane molecule can have configurations with $\psi = 32$–44° and, of course, $\psi = 136$–148°; for a tetraphenyllead molecule, $\psi = 23$–90° and 90–157° are possible.

The angle $\delta$ between the normal to the aryl ring and the $\bar{4}$ axis will be found from

$$\cos \delta = \cos \psi \cos 54° 44'$$

In all the tetragonal structures discussed here the translation along the $c$ axis is comparatively small. If molecules related by this translation touch the carbon atoms of the benzene rings, then $c \cos \delta = 3.6$ Å or $c \sin \psi = 2.9$ Å. The parameter $c$ ranges from 6.5 to 7.2 Å. Consequently, a contact is associated with $\psi$ equal to 24–28°; at larger values of $\psi$ there will be no contacts between the benzene rings of these molecules. However, as has been shown above, these small values of $\psi$ are forbidden.

Consequently, there will be no contacts between molecules related by translation along the $c$ axis in any of the tetragonal crystals of compounds of the type XAr$_4$.

Detailed investigations of the structures of tetra-$p$-tolyl tin and tetra-$p$-methoxyphenyl tin [46] indicate that the contacts here exist between molecules related by the $a$ translations and also by the $2_1$ screw axes.

Figure 59 shows the structure of $Sn(C_6H_5)_4$. In all the crystalline structures of XAR$_4$ type compounds studied, the molecular coordination number is 12. Experimental data show that in these crystals, the angle $\psi$ corresponds approximately to the middle of the allowed interval: for $C(C_6H_5)_4$, $\psi = 55°$; for

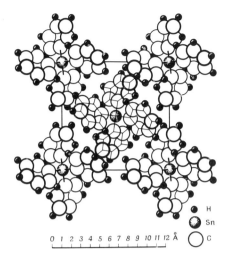

*Fig. 59.* Projection of Sn(C₆H₅)₄ on face *ab*.

Si(C$_6$H$_5$)$_4$, 37°; for Sn(C$_6$H$_5$)$_4$, 42°; and for Pb(C$_6$H$_5$)$_4$, 50°. The orientation of the aryl rings has been established experimentally to within several degrees. Nonetheless, the results of the structure determinations are somewhat uncertain in that the valence angles may deviate from the tetrahedral angle; this in no way contradicts the $\bar{4}$ symmetry.

The packing coefficient of the tetraphenylmethane molecules is 0.73. As the X–C bond length increases, the packing coefficient drops with the transition to silicon, tin, and lead compounds. The values of the packing coefficient are 0.71, 0.70, and 0.69.

## 12. POLYMORPHIC MODIFICATIONS

The majority of organic substances seem to be capable of crystallizing in several different structures. Most frequently encountered are monotropic modifications of the same substance. Suppose that two different crystals, A and B, can be obtained by crystallizing a substance from different solvents. If by bringing these crystals into contact, we can, sometimes varying only the temperature, convert modification A into B, but under no conditions convert B into A, then the transition A into B is monotropic. In other words, for any parameters of state, the free energy of crystal B is lower than the free energy of crystal A.

An unstable crystal can be quite stable in practice. For molecular crystals, this circumstance need not seem surprising, since transition into a stable state involves overcoming high barriers: it is easily seen that self-diffusion in a perfect defect-free molecular cr ystal built up from rather large molecules is next to impossible.

Besides, the differences between the free energies of monotropic modifications are in all probability also very small and, therefore, the thermodynamic pressure difference is insignificant. For example, for normal propylbenzene, the free energies of the stable and unstable modifications do not differ by more than 70 cal/mole. This is, certainly, a low value, since it amounts to a mere fraction of a percent of the total free energy of the crystal (including lattice energy).

Consider a few examples. Several polymorphous modifications are to be found for acridine (XX), a substance whose molecule (XX) has a "hollow" in

(XX)

the middle and is not, as a consequence, very convenient for packing. By X-ray diffraction studies we have measured the unit cells of at least five acridine modifications [47] obtained either from different solvents, by sublimation (acridine-IV), or from the melt (acridine-II). Table 8 summarizes the crystallographic data of four acridine modifications designated in a different manner by different authors; we give Phillips' notations, the notations of other researchers are bracketed. The most stable of these modifications is acridine-II, which is stable from room temperature to the melting point. Acridine-I has

### Table 8

POLYMORPHOUS MODIFICATIONS OF ACRIDINE AND PHENAZINE

| | $a$ (Å) | $b$ (Å) | $c$ (Å) | $\beta$ | $Z$ | Space group | X-ray density |
|---|---|---|---|---|---|---|---|
| Acridine-II ($\alpha$) | 16.34 | 18.90 | 6.08 | 95°5′ | 8 | $P2_1/a$ | 1.27 |
| Acridine-III | 11.41 | 5.99 | 13.69 | 98°48′ | 4 | $P2_1/n$ | 1.29 |
| Acridine-IV ($\delta$) | 15.75 | 29.43 | 6.20 | — | 12 | $P2_12_12_1$ | 1.24 |
| Acridine ($\beta$) | 16.37 | 5.95 | 30.01 | 141°20′ | 8 | $Aa$ | 1.29 |
| $\alpha$-Phenazine | 13.22 | 5.061 | 7.088 | 109°13′ | 2 | $P2_1/a$ | 1.33 |
| $\beta$-Phenazine | 11.64 | 11.58 | 6.88 | 99°19′ | 4 | $P2_1/n$ | 1.28 |

been found to be a monohydrate; the number of molecules of $C_{13}H_9N \cdot H_2O$ in the cell is $15.75 \pm 0.15$. When heated, acridine-I loses one $H_2O$ molecule and is converted into acridine-II. Acridine-III and acridine-IV are also converted into acridine-II at 45° and 75°C, respectively. It is significant that only acridine-III has a "normal" number of molecules in the unit cell ($Z = 4$). That the shape of the molecule is inconvenient for packing seems to explain why the unit cells of other modifications contain more than one independent molecule in general positions. It is interesting to point out that all the polymorphic modifications have completely different structures.

An example of a polymorphic substance with "similar" structures is phenazine [48]. The phenazine (XXI) molecule is more symmetric than the

(XXI)

acridine molecule and is probably more suitable for packing, since its two "hollows" are located symmetrically. Two polymorphic phenazine modifications have been reported; their crystallographic data are given in the same table.

In the case of an enantiotropic polymorphic change, a single crystal can be converted into a single crystal from both sides. The phase boundary in a single crystal can be immobilized and two polymorphous modifications of the same substance can indefinitely exist in contact under equilibrium conditions.

The heats of isothermal change have been measured for a large number of crystals. If a crystal changes into a rotational crystalline state, the heat of this transition is high and exceeds the heat of melting. When one crystal form is converted into another the heat of the transition is about 500 cal/mole, while the gain in entropy makes up tenths of units of cal/deg mole.

With only rare exceptions, all polymorphic transitions in organic crystals involve sudden changes in the volume. The structures of different modifications of the same substance may be "similar" or completely different.

For example, the triclinic modification of $p$-dichlorobenzene [49], ($a = 7.32$, $b = 5.95$, $c = 3.98$ Å, $\alpha = 93° \, 10'$, $\beta = 113° \, 35'$, $\gamma = 93° \, 30'$; space group P $\bar{1}$, $Z = 1$) "resembles" the monoclinic modification [50] ($a - 14.80$, $b - 5.78$, $c = 3.99$ Å, $\beta = 113° \, 0'$; space group $P2_1/a$, $Z = 2$). The tilts of the molecules in a layer are practically similar. The only difference is that in the triclinic

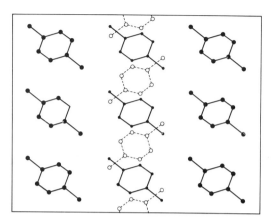

*Fig. 60.* Orientation of molecules in the triclinic and monoclinic modifications of *p*-dichlorobenzene.

crystal (Fig. 60) the layers are translationally identical, while in the monoclinic crystal the adjacent layers are related by a twofold screw axis.

At this point it should be emphasized that this "similarity" of structures is in no way equivalent to a certain cooperative course of a polymorphic change. Mnyukh [51] has convincingly demonstrated that the growth of one crystal from another crystal follows the same laws as does the growth of a crystal from the melt, regardless of whether the polymorphic modifications are similar.

No special significance should be assigned to the similarity of structures. It is evident that the free energies of two structures at a certain *p*, *T* may be equal both in "similar" and "unlike" structures.

## 13. HYDROGEN BONDS IN CRYSTALS

This field of organic chemical crystallography has not yet been systematically studied. Information about the nature of hydrogen bonds in certain substances has been obtained, so to speak, as a by-product of other research.

Although some spectral data seem to indicate hydrogen bonds between atoms of a number of elements, molecular crystals display such bonds only with oxygen and nitrogen atoms. These bonds are sufficiently strong: there is indirect evidence to the effect that the energy of each bond is of the order of 3 kcal/mole.

Now that significant progress has been made in the computation of lattice energies by means of atom–atom potentials it should be quite in order to carry out a systematic investigation of the heats of sublimation of hydrogen-bonded organic crystals. The experimental value of the lattice energy minus the interaction energy calculated for all atom pairs yields the total energy of the hydrogen bonds. (For the purpose of these calculations we "forget" the protons responsible for the hydrogen bonds.) An interesting program of systematic research into a class of compounds which contain hydroxyl, hydroxyl and carbonyl, hydroxyl and amino groups, etc. is still awaiting its enthusiast.

It seems well justified to extend the atom–atom approximation for lattice energies to the case of hydrogen bonds as well (see Chapter II).

The hydrogen bond parameters appear to vary insignificantly from one molecule to another. Therefore, it can reasonably be expected that accurate results on molecular packing will be obtained even if we assume these parameters to be practically constant or, at most, differentiate between hydrogen atom parameters in the hydroxyl and carbonyl groups (the literature data suggest a sufficiently pronounced difference in the equilibrium values of the hydrogen bonds in the two groups).

The formation of each hydrogen bond involves a significant loss of energy and it is, therefore, obvious that such structures will display a tendency to form all possible hydrogen bonds. If any such bond fails to form, this can be due only to geometric restrictions.

Experimental evidence indicates that, as a rule, nature succeeds in building up the unit cell of a crystal so that the hydrogen bonds would affect but slightly the molecular packing density. The simultaneous satisfaction of the requirement of maximum packing density (or the equivalent minimization condition for the sum of atom–atom potentials associated with conventional interactions) and the tendency for maximum formation of hydrogen bonds is facilitated by the fact that minor bending of a hydrogen bond entails low energy losses. This conclusion is confirmed by the shape of the atom–atom potential for hydrogen bonds (see Section II.6) and the absence of a correlation between proton deviations from the bond link and the bond length.

In order to develop organic chemical crystallography for organic substances whose molecules are combined into crystals by means of hydrogen bonds, it would be worthwhile to consider all types of hydrogen bonds, and also all types of "islands" formed by H bonds inside each type of hydrogen bonding.

We believe that classification according to types of bonds must be based on the groups of atoms forming the hydrogen bonds.

A. Hydrogen bond $O–H \cdots O$ occurs between groups:

1. COOH   and   COOH

2. C=O   and   COOH

3. OH   and   COOH

    4. C=O   and   OH

    5. OH    and   OH

B. Hydrogen bond N–H $\cdots$ O

    a. Nitrogen atom in $NH_2$ group.

Bonding exists between groups:

    1. $NH_2$   and   COOH

    2. $NH_2$   and   OH

    3. $NH_2$   and   C=O

    b. Nitrogen atom in NH group.

Bonding is effected between groups:

    1. NH   and   COOH

    2. etc.

The geometries of saturated hydrogen bonds may be quite different depending on the arrangement of the donor and acceptor groups in the molecule.

While a pair of molecules usually cannot produce more than two mutual hydrogen bonds, hydrogen-bonded molecules can form either rings of two, three, or more molecules, or infinite chains. Formally speaking, the latter case corresponds to the formation of a large crystalline "molecule." If there are three such bonds possible, three-dimensional frame structures may be built up from any number of molecules.

When hydrogen bonds are responsible for the formation of rings or frame structures and the interaction between these structures is due to dispersion forces, hydrogen bonding may lose some of its characteristic properties (e.g., the melting temperature may not rise).

If molecules can have three hydrogen bonds, the resultant structure may fall into clusters of chains or two-dimensional layers. In this case, too, a number of crystal properties will be determined by van der Waals forces rather than by hydrogen bonds as the former bind clusters of chain and layers of molecules to form a crystal.

Finally, with four hydrogen bonds, double layers and three-dimensional arrays can be built up. In the latter case the hydrogen bonds become critical for the crystal properties since any disintegration of a crystal requires their breaking.

The program of regular studies on intermolecular H bonding is far from being completed. Strictly speaking, it would be required to know the structure and properties of at least two or three representatives of each geometrical type for all types of bonding. At the present time majority of the boxes of the suggested classification are not filled up. Therefore, pending generalizations, which will be made possible only by at least partial fulfillment of this program, we shall give only a few typical examples.

Hydrogen bonds form infinite bands of molecules in the structures of dicarboxylic acids with a common formula $HOOC(CH_2)_nCOOH$ (XXII)

$$(XXII)$$

($n = 0$ corresponds to oxalic acid, $n = 1$ to malonic acid, $n = 2$ to succinic acid, $n = 3$ to glutaric acid, and $n = 4$ to adipic acid). These acids can exist in two polymorphic modifications with small differences in the internal energy. By way of example we shall describe the structures of $\alpha$- and $\beta$-forms of oxalic acid.

$\alpha$-form of oxalic acid HOOC-COOH [7, p. S.671; 52]. Unit cell: $a = 6.546$, $b = 7.847$, $c = 6.086$ Å; space group $V_h^{15} = Pcab$, $Z = 4$. Molecular symmetry: point group $2/m$, in crystal $\bar{1}$. The molecule is planar and has the configuration shown in Fig. 61.

Intermolecular distances $O \cdots O$ are (Å):

$$O_1(I) \cdots O_2(II) = 2.71 \qquad O_1(I) \cdots O_2(III) = 3.54$$

$$O_1(I) \cdots O_2(II') = 3.21 \qquad O_1(I) \cdots O_2(III') = 3.46$$

$$O_1(I) \cdots O_1(IV) = 3.29 \qquad O_1(I) \cdots O_2(IV) = 3.29$$

$$O_1(I) \cdots O_1(IV') = 3.28 \qquad O_1(I) \cdots O_2(IV') = 3.24$$

$$O_1(I) \cdots O_1(IV'a) = 3.28 \qquad O_2(I) \cdots O_2(III) = 3.22$$

$$O_1(I) \cdots O_2(II') = 3.11 \qquad O_2(I) \cdots O_2(IIIa) = 3.32$$

$$O_2(I) \cdots O_2(IIa') = 3.11$$

The first of these distances undoubtedly indicates a hydrogen bond. The other values are larger than $2R_0$. Each of the O atoms is involved in only one hydrogen bond $-O-H \cdots O=$. These bonds bind molecules into infinite crimped layers parallel to $bc$ (Fig. 62).

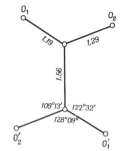

**Fig. 61.** α-Form of oxalic acid. Molecular configuration.

β-form of axalic acid [7, p. S.672]. Unit cell: $a = 5.30$, $b = 6.09$, $c = 5.51$ Å, $\beta = 115° 30'$, space group $C_{2h}^5 = P2_1/c$, $Z = 2$.

Among the intermolecular distances $O_1 \cdots O_2' = 2.71$: 3.11; 3.48; $O_1 \cdots O_1' = 3.45$ Å, there is one reduced distance (2.71 Å), which is the evidence of hydrogen bonding. Similarly to the structures of the β-form of succinic acid and the higher dicarboxylic fatty acids, but unlike the structure of the α-form of oxalic acid, hydrogen bonds bind the molecules into the infinite chains shown above. The chains are separated by normal intermolecular distances.

The structures of the α- and β-forms of resorcinol (XXIII) employ hydrogen bonds of the type O–H $\cdots$ O.

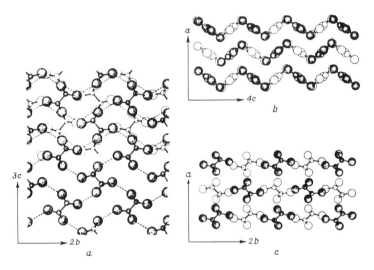

**Fig. 62.** α-Form of oxalic acid. (a) projection $bc$, dashed lines hydrogen bonds; top, two superposing layers, bottom, one layer; (b) projection $ac$, layers of molecules formed by hydrogen bonds can be seen well; (c) projection $ab$.

(XXIII)

α-form of resorcinol [53]. Unit cell: $a = 10.53$, $b = 9.53$, $c = 5.66$ Å; space group $C_{2v}^9 = Pna$, $Z = 4$. Molecular symmetry: point group $mm$, $m$, 2, or 1 (depending on the positions of the OH groups); in the crystal it is 1.

The molecules are planar, their packing is determined by stable hydrogen bonds. Figure 63 shows the projection of the structure onto $ab$. The intermolecular distances $O \cdots O = 2.66$ and 2.75 Å; the shortest $C \cdots O$ distance is 3.49 Å and the shortest $C \cdots C$ distance is 3.59 Å.

The β-form of resorcinol [54] is a high-temperature form, the temperature of the $\alpha \rightarrow \beta$ transformation is 74°. Unit cell: $a = 7.91$, $b = 12.57$, $c = 5.50$ Å; space group $C_{2v}^9 = Pna$, $Z = 4$.

The difference from the structure of the α-form lies in a somewhat different pattern of the hydrogen bond system (Fig. 64). Due to the presence of two hydrogen bonds (2.70 and 2.75 Å) the molecules form a three-dimensional array of hydrogen bonds O–H $\cdots$ O. The shortest intermolecular distances C---O = 3.41 Å and C---C = 3.54 Å.

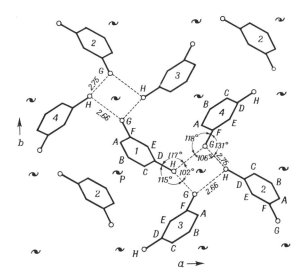

*Fig. 63.* α-Form of resorcinol. Projection *ab*.

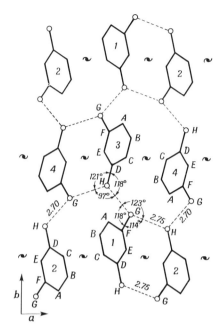

*Fig. 64.* β-Form of resorcinol. Projection *ab*.

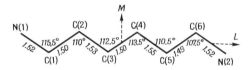

*Fig. 65.* Configuration of the hexamethylenediamine molecule.

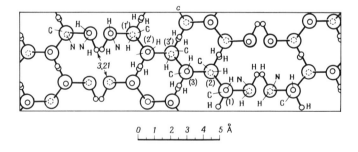

*Fig. 66.* Hexamethylenediamine. Projection *bc*.

The above examples show that the distance between the oxygen atoms of different molecules involved in the formation of hydrogen bonds $-O-H\cdots O-$ varies insignificantly near 2.70 Å (2.66–2.75 Å).

An example of a structure with the hydrogen bond type $\rangle N-H\cdots N\langle$ is the structure of hexamethylenediamine [55] $H_2N(CH_2)_6NH_2$ (unit cell: $a = 6.94$, $b = 5.77$, $c = 19.22$ Å; space group $V_h^{15} = Pbca$, $Z = 4$; molecular symmetry: point group $2/m$, $2$, $m$, $\bar{1}$ (depending on the rotation around the bonds); in the crystal it is $\bar{1}$.

The configuration of the molecule (without hydrogen atoms) is shown in Fig. 65. The molecule is planar. One hydrogen bond (Fig. 66), 3.21 Å long, is formed between the $NH_2$ groups of adjacent molecules. Other intermolecular distances have conventional values: $NH_2 \cdots NH_2 = 3.68$ and $C_1 \cdots NH_2 = 3.92$ Å. All intermolecular distances in the direction of the $a$ axis exceed 4 Å.

The structure of 2-amino-4,6-dichloropyrimidine (XXIV) ($a = 16.457$, $b = 3.845$, $c = 10.283$ Å, $\beta = 107°\ 58'$; $P2_1/a$, $Z = 4$) has hydrogen bonds

$$
\begin{array}{c}
NH_2 \\
| \\
N{\diagup}^{C}{\diagdown}N \\
\| \qquad | \\
Cl{\diagup}^{C}{\diagdown}C{\diagup}^{=}C{\diagdown}Cl \\
| \\
H
\end{array}
$$

(XXIV)

$\rangle N-H\cdots N\langle$ with lengths 3.21 and 3.37 Å (Fig. 67) which are almost coplanar with the rings. Among the other intermolecular distances one finds slightly shortened distances $Cl \cdots Cl$: 3.44 and 3.56 Å instead of $2R_{Cl} = 3.60$ Å. The remaining intermolecular distances exceed 4 Å.

There exists a large number of structures with hydrogen bonds $\rangle N-H\cdots O-$. Glycine and other amino acids have such structures.

**Table 9**

CRYSTALLOGRAPHIC DATA FOR THREE FORMS OF GLYCINE[a]

| Form of glycine | $a$ (Å) | $b$ (Å) | $c$ (Å) | $\beta$ | $V$ | Space group | $Z$ |
|---|---|---|---|---|---|---|---|
| $\alpha$ | 5.102 | 11.971 | 5.458 | 111°42′ | 309.7 | $P2_1/n$ | 4 |
| $\beta$ | 5.077 | 6.268 | 5.380 | 113°12′ | 157.4 | $P2_1$ | 2 |
| $\gamma$ | 7.037 | — | 5.483 | — | 235.1 | $P3_1$ or $P3_2$ | 3 |

[a] Y. Iitaka. *Acta Crystrallogr.* **13**, 35, 1960.

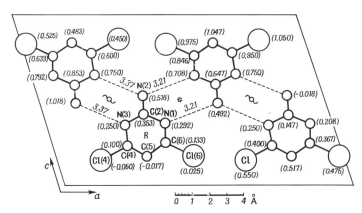

*Fig. 67.* 2-Amino-4,6-dichloropyrimidine. Projection *ac*. In brackets, *y* coordinates in Å. Dashed lines, hydrogen bonds.

Glycine is available in three crystalline modifications: α-, β-, and γ-glycine [56]. Crystallographic data for these modifications are given in Table 9. Unlike other amino acids, the molecules of glycine $H_2NCH_2COOH$ do not contain a nonsymmetric carbon atom and therefore display no optical activity in solutions. However, the molecules in glycine crystals are not planar and can exist as two mirror-related forms.

Figure 68 shows the configuration of the molecules in the structure of α-glycine. The nitrogen atom goes out of the plane of the carboxyl group and $C_2$ atom by 0.436 Å. The molecular configurations in the crystals of the β- and γ forms differ only by a somewhat different value of the deviation of the N atom from the above plane: 0.583 and 0.309 Å, respectively. Each molecule has four hydrogen bonds with the molecules of four adjacent unit cells, which

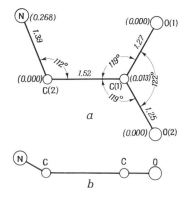

*Fig. 68.* Configuration of molecules in the α-glycine crystal.

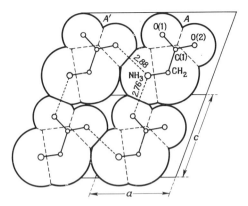

*Fig. 69.* Glycine. Close-packed layer *ac* formed by hydrogen bonds N–H···O (dashed lines).

results in the formation of infinite close-packed *ac* layers (Fig. 69). The packing pattern corresponds to cleavage along (010). Hydrogen bonds N–H···$O_1$ = 2.76 Å and N–H···$O_2$ = 2.88 Å in the layer are very strong and form an angle of 108.5° with one another. The angles with the C–N bonds are, respectively, 118 and 120°, i.e. close to the tetrahedral angle.

The stacking of the *ac* layers related by inversion centers at $Z = 0$ and $\frac{1}{2}$ is also found to be very dense due to the hydrogen bonds between these layers. Conversely, the distances between the *ac* layers related by the *n* glide planes with $z = \frac{1}{4}$ and $\frac{3}{4}$ equal the sums of intermolecular radii, which indicates that there are no hydrogen bonds between these layers. Thus, the structure is built up from double *ac* layers formed by hydrogen bonds (Fig. 70).

The binding of single layers into a double layer (Fig. 71) is effected by the bifurcated hydrogen bonds $>$N—H$:\begin{smallmatrix} \cdot\cdot O \\ \cdot\cdot O \end{smallmatrix}$ formed by the H(3) atom of the $NH_3^+$ group which is actually involved in two hydrogen bonds N–H(3)···$O_1$ = 2.93 Å and N–H(3)···$O_2$ = 3.05 Å (the primes denote the atoms of the adjacent layer).

In the structure of β-glycine [56] hydrogen bonds N–H···O with lengths 2.758 and 2.833 Å bind the molecules into single layers which run parallel to the (010) plane. Single layers are bound by bifurcated hydrogen bonds (3.002 and 3.022 Å), but not into pairs as in α-glycine, but into a three-dimensional array.

In the structure of γ-glycine [56] hydrogen bonds, 2.801 and 2.817 Å long, bind the molecule into spiral chains around triad screw axes parallel to the *c* axis. The individual spirals are mutually related by hydrogen bonds, 2.970 Å long, as well as by the electrostatic attractive forces between adjacent $NH_3^+$ and $COO^-$ groups.

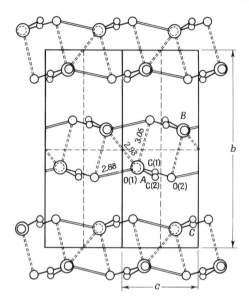

*Fig. 70.* Glycine. Projection *ab* showing doubling of layers and large distances between double layers.

In the structure of d,1-alanine $H_2NCH(CH_3)COOH$ each $NH_3^+$ group forms three hydrogen bonds with the O atoms of adjacent molecules; here the distances are equal to 2.88, 2.84, and 2.78 Å. These bonds bind the molecules into a three-dimensional array (Fig. 72).

As has already been shown above, organic chemical crystallography has not, so far, produced enough data for making general conclusions about inter-molecular hydrogen bonds. So far only one significant conclusion suggests

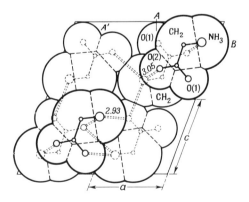

*Fig. 71.* Glycine. Stacking of molecules of two single layers to form one double layer.

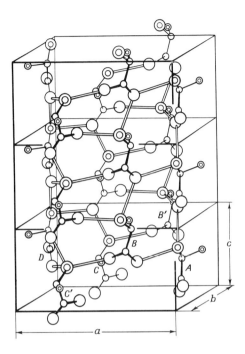

**Fig. 72.** d,1-Alanine. Association of molecular chains parallel to axis *c* into a "tube."
Double lines, hydrogen bonds.

itself: The formation of hydrogen bonds does not handicap the layout of
molecules in conformity with the general rules of the packing of molecules in
the crystals discussed above.

## C. Crystals with Elements of Disorder†

### 14. RIGID DISORDER

The arrangement of molecules in the cell established by X-ray diffraction
analysis is the arrangement averaged over all the crystal cells. If there is a
certain disorder in the molecular arrangement, inherent in the crystal, the
X-ray technique detects "the average molecule" by the superposition of all the

---

†It is recommended that Sections C, D, and E be read after the subsequent chapters
of the book.

molecules located in the same crystallographic position of the average unit cell.

One of the most elementary and common examples of this kind of disorder is the formation of centrosymmetric crystals by molecules without a symmetry center. Such crystals are, for instance, *p*-chlorobromobenzene, *p*-nitrochlorobenzene, and azulene (a hydrocarbon consisting of a five-membered ring conjugated with a seven-membered ring).

In such crystals, the average molecule is actually obtained by centrosymmetric superposition of two molecules with half-weight atoms.

If we construct a hypothetical space lattice of average cells (for all the three crystals this is a $P2_1/c$ lattice with two molecules in a unit cell), the space lattice will have an equal number of points with molecules facing the opposite directions. It is quite evident that such a disordered crystal can be formed only if both arrangements have similar energies.

As an illustration, we have made simple calculations for the azulene crystal by means of the atom–atom model. The crystalline structure of azulene [57] (XXV) is characterized by a monocinlic unit cell with parameters: $a = 7.884$,

(XXV)

$b = 5.988$, $c = 7.840$ Å, $\beta = 101° 33'$; space group $C_{2h}^5 = P2_1/a$. The unit cell contains two molecules; consequently, the noncentrosymmetric azulene molecules must be positioned on inversion centers. Thus, the azulene structure is disordered in the sense that the center of inversion is occupied statistically.

Figure 73 represents the average molecule of the average cell of an azulene crystal in the form given in the original publication. We have added the H atoms to this drawing. The crystal structure is very similar to that of naphthalene. It can be easily seen that the inversion of the molecule (a half of the average molecule) by the symmetry center practically does not change the contacts between the molecules of the *ab* layers. We have considered all possible versions of the mutual arrangement of azulene molecules in the nearest-neighbor approximation on the *ab* plane. The calculations have demonstrated that 23 arrangement patterns are not favorable because they are associated with a higher energy than the interaction energy of molecules located so as to form a regular noncentrosymmetric lattice ($P2_1$). Seven arrangement versions, however, have been found to be more favorable, and two of these cases involved energy which was 0.4 kcal/mole lower than that of an ordered crystal.

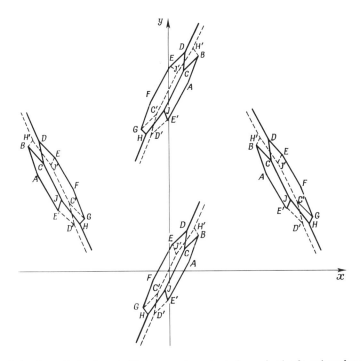

***Fig. 73.*** Azulene. Projection *ab*. Hydrogen atoms (not shown in the figure) replace each other when half of the molecules are inverted.

The following point is also worthy of note: All the molecular arrangement patterns are associated with interaction energies lying within 1 kcal/mole of one another. Furthermore, the configurational entropy contribution for distribution over two positions $RT \ln 2$ at room temperature is 0.42 kcal/mole.

The calculations show that, first, no second-order phase change should be expected, because preservation of order in the same lattice of molecular centers is much less favorable. Secondly, the loss in lattice binding energy due to unavoidable unfavorable arrangements in an azulene crystal may apparently be compensated for by an increase of entropy.

Of course, the above example in no way proves that azulene must necessarily crystallize into disordered crystals. To prove this, it would be required to apply the usual procedure i.e., to show that all other conceivable azulene structures are "worse" than the experimentally observed disordered structure. Nevertheless, it has been demonstrated that our physical model offers a correct interpretation of the phenomenon of disordered structure formation.

Another interesting example of a disordered structure is *p*-chloronitrobenzene [58]. The monoclinic unit cell with parameters: $a = 3.84$, $b = 6.80$, $c = 13.35$ Å, $\beta = 97°\ 31'$ (space group: P2$_1$/c) contains two molecules. The

(XXVI)

molecules are noncentrosymmetric and, consequently, being located on the centers of inversion, they are in a disordered arrangement in a crystal: roughly speaking, the nitro groups of one-half of the crystal molecules face, for instance, rightward, and those of the other half leftward. The volume increments of the $-Cl$ and $-NO_2$ groups are 29 Å and 23 Å$^3$, respectively. In this way, the volume of a certain average substituent equals 26 Å$^3$. The packing coefficient of such "average" molecules in the real structure is found equal to 0.62. It can be easily understood that disorder in the arrangement of groups of different volumes results in a rather loose packing which must probably be compensated for by a gain in entropy.

A very interesting class of disordered structures are those of the hexasubstituted benzene derivatives. Using for substitutents halogen atoms, methyl groups, and other small radicals, one can obtain a large number of isomorphous crystals of the hexachlorobenzene type (an ordinary centrosymmetric monoclinic structure with two molecules in the unit cell).

According to the results of the X-ray diffraction and nuclear quadrupole analyses conducted in the author's laboratory [59], the type of orientational distributions of molecules may differ considerably from one crystal to another. Since the average molecule is centrosymmetric, there are three independent positions in a crystal over which the benzene ring substitutents are distributed.

In some cases, the stituation is quite simple. For instance, a monobromopentachlorobenzene molecule has one-sixth of the Br atoms and five-sixths of the Cl atoms in each position. The picture is equally clear for a hexachlorobenzene molecule: If two ortho- or meta-Cl atoms in a dibromotetrachlorobenzene molecule are substituted by Br atoms, there will be one-third of the Br atoms and two-thirds of the Cl atoms in each position. As in many other similar cases, the average molecule displays $6/mmm$ symmetry here.

The behavior of tetrachloro-$p$-dibromobenzene molecules is specific. The average molecule has $mmm$ symmetry. One of the orientations is strongly preferred: Two-thirds of the molecules are oriented similarly, while the remaining molecules are equally distributed between two other positions.

These examples show that molecules in disordered crystals of this type may exhibit both uniform and nonuniform positional distributions.

Another subject of interest is the investigation of short-range order in the arrangement of differently oriented molecules. However, no work has been done to study this problem.

Molecules of disordered organic crystals perform the same motions as do those in quite ordered crystals. Some molecules only vibrate about an equilibrium position (*p*-nitrochlorobenzene, azulene), others rotate about one of their axes. In the case of small substituents, all hexasubstitution benzene derivatives are probably capable of reorientation. The phenomenon of reorientation in such molecules containing protons has been demonstrated by French workers [60] using the NMR technique.

## 15. ROTATIONAL CRYSTALLINE STATE

Crystals with the disorder elements which have been discussed in the foregoing section give good-quality X-ray pictures containing hundreds of reflections. This means that deviations of the size and shape of the unit cells from the average value do not exceed too much those for conventional ordered systems. This has been confirmed by our calculations for the azulene crystal: There is but a slight difference in the energy of various mutual arrangements of molecules in unit cells of the same size. A characteristic structural feature of disordered crystals of this type is that the inversion of a molecule causes complete or partial transposition of the atoms; at any rate, certain important geometrical elements of the molecule—planes or axes—are aligned.

We have considered structures in which disorder is brought about by a relatively small number of different orientations, while order is retained with respect to one of the molecule axes, namely, a normal to its plane. It is quite probable that other structures of the same type exist, e.g., low-symmetry structures in which disorder is caused by rotations about the long molecular axis or several different spatial orientations.

It should be pointed out that there may be both static and dynamic versions of disordered crystals of this kind. In this case, molecular reorientations do not generate any structural peculiarities. More than that, no effect is produced on the thermodynamic properties of the crystals. If the NMR method were not available, no proofs of molecular reorientation could have been furnished.

Any classification is always arbitrary to a certain extent. Nevertheless, one can safely distinguish between the above disordered crystals and a relatively wide class of substances commonly called plastic crystals by British authors. What we would like to emphasize is that we are speaking about a very peculiar state of the substance which always has its own domain of existence in the phase diagram. Therefore, we prefer the term *rotational crystalline state*. The specific features of these substances discovered by Timmermans have been described in detail in various reviews [61], therefore, we only point out the following two peculiarities: high-symmetry cells typical of sphere and cylinder packings, and the existence of not more that ten reflections on X-ray photographs, i.e., a scatter of the unit cell sizes within 1 or even 2 Å.

The rotational crystalline state is characteristic of molecules of an almost spherical shape, for instance, methane and ethane derivatives with small substituents, or molecules of a shape close to that of a cylinder, e.g., paraffin-type molecules.

The data available on the studied substances of this type are summarized in Table 10.

*Table 10*

CRYSTALLINE STRUCTURES OF CERTAIN PLASTIC CRYSTALS

| Substance | Type of lattice or space group[a] | Number of molecules in unit cell | Lattice parameter (Å) | $\dfrac{D_{eff}}{D_{max}}$ |
|---|---|---|---|---|
| Tetrahedral molecules | | | | |
| $C(SCH_3)_4$ | bcc | 2 | 8.15 | |
| $(CH_3)_3CCl$ | fcc | 4 | 8.40 | |
| $(CH_3)_3CBr$ | fcc | 4 | 8.70 | |
| $(CH_3)_3CSH$ | fcc | 4 | 8.82 | |
| $C(NO_2)_4$ | $T_d^3(I\bar{4}3m)$ | 2 | 7.08 | |
| $SiF_4$ | $T_d^3(I\bar{4}3m)$ | 2 | 5.41 | |
| $SiI_4$ | $T_h^6(Pa3)$ | 8 | 11.99 | |
| Ditetrahedral molecules | | | | |
| $Cl_3C–CCl_3$ | $O_h^9(Im3m)$ | 2 | 7.43 | |
| $(CH_3)_3C–C(CH_3)_3$ | bcc | 2 | 7.69 | |
| $(CH_3)_3C–C(CH_3)Cl_2$ | bcc | 2 | 7.38 | |
| $(CH_3)_3C–C(CH_3)_2Cl$ | bcc | 2 | 7.62 | |
| $(CH_3)_2ClC–C(CH_3)_2Cl$ | bcc | 2 | 7.58 | |
| $(CH_3)_2ClC–CCl_3$ | bcc | 2 | 7.4 | |
| $Br_3C–CBr_3$ | bcc | 2 | — | |
| $(CH_3)_3Si–Si(CH_3)_3$ | bcc | 2 | 8.47 | |
| $(CH_3)_3C–COOH$ | $O_h^5(Fm3m)$ | 4 | 8.82 | |
| Cyclic molecules | | | | |
| Cyclobutane | bcc | 2 | 6.06 | 5.25/6.22 |
| Cyclopentane | hexagonal | 2 | 5.86 | 5.83/6.81 |
| Cyclohexane | fcc | 4 | 8.76 | 6.19/6.57 |
| Thiacyclohexane | fcc | 4 | 8.59 | |
| Cyclohexanol | $O_h^5(Fm3m)$ | 4 | 8.83 | 6.24/7.01 |
| Cyclohexanone | fcc | 4 | 8.61 | |
| Chlorocyclohexane | bcc | – | 9.05 | |
| Cycloheptatriene | cubic | 8 | 10.6 | |
| Terpene molecules | | | | |
| Quinuclidine | fcc | 4 | 8.977 | 6.35/7.22 |
| Symmetric-Tricyclodecane | $T_d^2(F\bar{4}3m)$ | 4 | 9.426 | |
| dl-Camphor | fcc | 4 | 10.1 | 7.14/8.45 |
| dl-Camphene | bcc | 2 | 8.00 | |
| Borneol | fcc | 4 | 10.25 | |
| Bornyl chloride | fcc | 4 | 10.39 | |

[a] Types of lattices: fcc, face-centered cubic lattice; bcc, body-centered cubic lattice.

There may be two cases when a low-symmetry molecule occupies a high-symmetry position in a crystal: first, when all orientations with respect to the crystal axes are equiprobable (the average molecule displays spherical or cylindrical symmetry), and secondly, when a molecule has several fixed orientations derived by the crystal symmetry elements from one arbitrary orientation. In the latter case, for instance, a nonsymmetric molecule might be located at the points of a face-centered cubic lattice in 48 distinct orientations, a centrosymmetric molecule in 24 orientations, a molecule showing $D_{3h}$ symmetry in eight positions, etc.

No matter which of the above hypotheses may be correct, plastic crystals obviously have pronounced short-range order in the arrangement of the molecules or even a domain structure in which individual ordered domains merge at the expense of "erroneously" located molecules. It is also evident that the short-range order is quite labile, because reorientation of molecules takes place in absolutely all substances in the rotational crystallin state.

The existence of short-range order can be deduced from many facts. First, the maximum diameter of a molecule is always much larger than the average size of an effective sphere, the difference reaching 1 Å. Secondly, according to thermodynamic theory, the crystalline part of the heat capacity of a freely rotating molecule should be close to 9 cal/deg mole, while actually it is about 12, as in ordered crystals. Finally, NMR analysis shows that the barrier to reorientation amounts to several kcal/mole, and that most of the time each molecule exists in a state of vibration around an equilibrium position and only sometimes jumps from one equilibrium state to another.

We now come back to the molecular packing problems. Although much work has been done in this field, the following two problems of prime importance have not yet been solved: The type of short-range order remains unknown, and no data are available for proving the presence or absence of orientations fixed with respect to the crystal axes.

Satisfactory compatibility of the entropy involved in transition from the ordered state into the rotational crystalline state with a configurational gain in the entropy is often taken as evidence of the finiteness of the number of molecular orientations in a crystal. Thus, from symmetry considerations, a bicycloheptane molecule must occupy 24 different positions in order to simulate the symmetry of a cubic lattice point. The transition entropy has been estimated at $R \ln 24 = 6.3$ cal/deg mole which agrees with the experimental figure. In our opinion, we can hardly rely on such comparisons. It is clear that the vibration amplitudes of molecules in the rotational crystalline state are considerably larger than those in an ordinary crystal. Therefore, a substantial portion of the transition entropy should be attributed to the freedom of motion.

Nonetheless, the hypothesis about certain preferred orientations with respect to the crystal axes seems to be sufficiently logical, despite the fact that the more spherical a molecule is, the lower is the probability of orientations fixed with respect to the axes of the crystal.

For solving the short-range order problem within the framework of any of the two models described in the text it would be profitable to state the problem of mutual molecular arrangement in a group of several molecules. It would be worthwhile to consider this problem for two molecules, three molecules, and for one coordination sphere, i.e., a group of 13 molecules (one central molecule and 12 neighbors). Appropriate calculations can be carried out in terms of the geometrical model or by evaluating the minimum energy with the help of the atom–atom potential method.

The problem for two (as well as for three) molecules can be formulated as follows: To find the optimum arrangement of a pair of molecules whose centers are located at the average distance established by an X-ray experiment.

Calculations for a group of molecules making up a coordination sphere will be illustrated on linear paraffins. As is known, at temperatures approaching the melting points, paraffin crystals undergo a polymorphic change and pass into the rotational crystalline state which is a close packing of cylinders with an effective diameter of 4.8 Å.

We know what arrangement of a group of seven molecules (one central molecule and six neighbors) is optimum from the point of view of potential energy. This arrangement is packing into a rectangular cell with a screw symmetry axis typical of organic chemical crystallography. In other words, it is the packing that existed with long-range order before the polymorphic change had occurred. Let us superimpose this packing on a hexagonal packing. From Fig. 74 it follows that there may be six different ways of such a superposition. Thus, the proposed model is constructed of identical domains oriented in six different ways with respect to the crystal axes. The average hexagonal cell

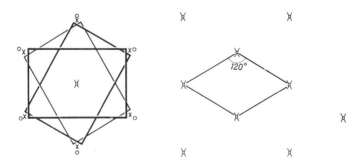

*Fig. 74.* Rotational disorder in paraffins.

formed by superposition of six rectangular cells will naturally be somewhat smaller than the experimentally observed cell. It can be supposed that a material increase of the cell volume which accompanies the transition of the molecule to the rotational crystalline state is associated with the transition layers which must necessarily exist in such a model of a crystal. Knowing the potential energy and free energy of the domains, we could try to predict the most probable molecular arrangement by constructing possible transition layers.

The same procedure could be applied to determine the most probable arrangement of molecules in the closest spherical packing under the assumption that on the packing plane (the (111) plane of the face-centered cubic lattice) there exist small ordered regions where molecular arrangement is governed by the same law of symmetry as on the $ab$ plane of the space group $P2_1/a$ or $P2_1$.

A more rigorous solution of the short-range order problem is given in our recent work [62, 62a].

Since rotational transitions from one position into another are rather rare, we may treat the crystal as a system of molecules whose centers occupy lattice points while their orientations are distributed in a certain manner between $v$ different orientational states. In the model considered in this book the interactions of the molecules are pairwise. Therefore, the configurational part of the crystal Hamiltonian can be written as

$$H = \tfrac{1}{2} \sum_{\substack{\mathbf{R},\mathbf{R}' \\ m,l}} V_{ml}(\mathbf{R}-\mathbf{R}')\, C_m(\mathbf{R})\, C_l(\mathbf{R}')$$

where the subscripts $m$, $l$ denote the $v$ different orientational states of each molecule, $\mathbf{R}$ is the radius vector of the crystal lattice point at which the molecule's center of gravity is positioned, and

$$C_m(\mathbf{R}) = \begin{cases} 1 & \text{if the molecule at point } \mathbf{R} \text{ is in state } m \\ 0 & \text{otherwise} \end{cases}$$

The random variables $C_m(\mathbf{R})$ obey the normalization condition

$$\sum_{m=1}^{v} C_m(\mathbf{R}) = 1 \quad \text{(for all } \mathbf{R}\text{)}$$

$V_{ml}(\mathbf{R}-\mathbf{R}')$ is the energy of pairwise interaction of the molecules at points $\mathbf{R}$, $\mathbf{R}'$ and in states $m$, $l$, respectively ($m, l = 1, 2, ..., v$). The following relationships are valid

$$V_{ml}(\mathbf{R}-\mathbf{R}') = V_{lm}(\mathbf{R}'-\mathbf{R}), \qquad V_{mm}(\mathbf{R}-\mathbf{R}') = V_{ll}(\mathbf{R}-\mathbf{R}')$$

i.e., a matrix built up from the values $V_{ml}(\mathbf{R}-\mathbf{R}')$ is a Hermitian matrix.

For determining thermodynamic quantities, it is required to calculate the statistical sum

$$Z = \sum_{\{C_m(\mathbf{R})\}=0}^{1} \exp\left[-\tfrac{1}{2}\beta \sum_{\substack{\mathbf{R},\mathbf{R}' \\ m,l}} V_{ml}(\mathbf{R}-\mathbf{R}')\,C_m(\mathbf{R})\,C_l(\mathbf{R}')\right]$$
$$\times \prod_{\mathbf{R}} \delta\left(\sum_m C_m(\mathbf{R}) - 1\right)$$

taken over all possible configurations, where $\beta = (kT)^{-1}$, $k$ is the Boltzmann constant, and $T$ is the absolute temperature. The normalization condition is introduced in the statistical sum by means of the function $\delta(x)$ determined by the equation

$$\delta(x) = \begin{cases} 1 & \text{for } x = 0 \\ 0 & \text{otherwise} \end{cases}$$

The calculation of the statistical sum given above is a more complicated process than its calculation for the case $m, l = (1,2)$, i.e., for the Ising model. It is necessary to use approximate methods to evaluate it. The statistical calculations in the publication referred to above are based on the self-consistent-field method. This method yields results that seem to be asymptotically accurate both at high and low temperatures.

Calculations show that only at the highest temperatures (when $V \ll kT$) are all orientations equiprobable. The crystals discussed in this section melt at not above 300°K as a rule ($RT = 0.6$ kcal/mole). The interaction energy of two organic molecules, say of the ethane type, is of the same order. Thus, complete full disorder in the rotational crystalline state is probably not to be found.

Theory demonstrates that for lower temperatures one can obtain a number of solutions corresponding to partially ordered systems. This means that certain orientations exist with probability above $1/v$, and others with probability below $1/v$. It has been found that with decreasing temperatures, the transition from complete disorder to full order may involve several partially ordered phases.

Theoretically, one can in principle determine the parameters of the polymorphic changes. It is also found possible to estimate the short-range order in the arrangement of the molecules in different orientations, namely, the probability that any two orientations exist in pairs. These formulas are fully identical to the formulas describing the probabilities of neighborhoods for solid solutions and will be given below.

### D. Binary Systems

16. CONDITIONS FOR FORMATION OF SOLID SOLUTIONS

The representation of an organic crystal as a packing of solids possessing certain shapes and sizes has enabled research into the mechanisms of mutual solubility of organic substances in the solid state. In 1956, the author formulated geometrical conditions for the formation of solid solutions [63], which in their general form proved inapplicable to other classes of compounds where interatomic electron bonds can fully suppress the effects of the symmetry and close-packing factors.

It is clear that by mixing two organic substances we may expect the formation of solid solutions only by substitution. Indeed, the arrangement of molecules with packing coefficient 0.6–0.8 shows that the voids in the structures are so small as compared with the sizes of the molecules themselves, that filling up these voids by the solute molecules, i.e., by forming interstitial solid solutions will probably be feasible only in the rare cases (so far unknown) when the solute consists of extremely small molecules. In the following text we shall speak only about substitutional solid solutions.

A necessary and sufficient condition for formation of solid solution crystals by two or more organic substances is similarity of the shapes and sizes of the component molecules. Only if this condition is satisfied will the substitution of certain molecules in the lattice of the matrix (solvent) by alien molecules of the solute not cause appreciable changes in the number of contacts of a solute molecule with adjacent molecules and also in the intermolecular distances, i.e., will not lead to a substantial rise in the free energy of the solution crystal as compared with a pure crystal. It has been shown experimentally, for example, that a substance rapidly stops being soluble if two or three of its intermolecular distances are reduced by about 30% (anthracene in acridine). At the same time an increase of some intermolecular distances, i.e., replacement of larger molecules by smaller molecules occurs much more easily (acridine in anthracene, naphthalene in β-chloronaphthalene), as is suggested by the asymmetric curve of the potential energy of interaction between nonbonded atoms.

It is obvious that apart from packing factors, other requirements arising from symmetry must be met for the formation of a continuous series of solid solutions by two substances: The structures of the substances being mixed must be isomorphous, in other words, they must not only have an identical space group and the same number of molecules in the unit cell, but also exhibit a similar packing of molecules. If, however, these conditions are not satisfied, there will necessarily be a discontinuity in the solubility curve. By way of an example, consider the system: diphenyl-α, α-dipyridyl (XXVII, XXVIII), with

(XXVII)             (XXVIII)

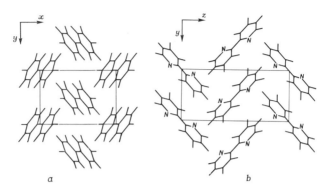

**Fig. 75.** Projections onto the rectangular face of (a) the diphenyl structure; (b) the α,α′-dipyridyl structure.

very similar molecules. Both substances crystallize with two molecules in a unit cell with symmetry $C_{2h}^5$. However, as is evidenced by the projections of the structures onto the rectangular face (Fig. 75a, b), the packing of the molecules is extremely different. It has been established experimentally that this system reveals only limited solubility (the phase diagram belongs to Roseboom's type V, Fig. 76) [64].

Continuous solubility in the anthrone–anthraquinone system is not an exception from the above rule. Though the symmetries of the molecules of anthrone (XXIX) and anthraquinone (XXX) are different, the crystalline

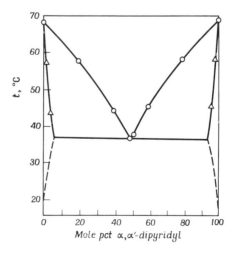

**Fig. 76.** Phase diagram of the diphenyl-α,α′-dipyridyl system.

(XXIX)                    (XXX)

structures of these substances are isomorphous (common space group $P2_1/a$, $Z = 2$), the anthrone structure being disordered so that the average of two non-centrosymmetric molecules become centrosymmetric. Since the isostructural nature of these substances has been confirmed by detailed X-ray diffraction studies [66] there can be no doubts as to the capability of these substances to mix in all proportions without changing the molecular packing.

The exceptions from this rule are substances with nonsymmetric molecules. The addition of even the first molecule in its right- or left-handed configuration to the racemate must change the symmetry of the racemate crystal so that further solubility can be continuous at all concentrations.

When analyzing the possibilities of formation of solid solution crystals by two organic substances within a certain range of concentrations, it is, of course, necessary to take into account more than merely the packing co-efficient, i.e., not only the losses in the molecular interaction energy. In the general case, changes in the free energy of crystal A when some of its molecules are substituted by molecules of solute B are caused by the following factors [67]: (1) the difference in the sizes and shapes of the molecules being mixed causes "strains" in the lattice which change the lattice energy by $\Delta U$; (2) the free energy of one mole drops by

$$\Delta F_{\text{mix}} = T|\Delta S_k| \quad \text{where} \quad \Delta S_k = R[(1-x)\ln(1-x) + x\ln x]$$

is the entropy of mixing; (3) as a molecule is substituted, this substitution will cause a change in the free energy of the lattice vibrational spectrum by $\Delta F^{\text{vib}}$; (4) it is necessary to take into account possible variations in the conformation of the solute molecule as it intrudes into the solvent lattice ($\Delta E$).

Let $n_A$ and $n_B$ be the variable numbers of moles of the components in the first phase with structure A so that concentrations

$$x = \frac{n_B}{n_A + n_B} \quad \text{and} \quad (1-x) = \frac{n_A}{n_A + n_B}$$

The free energy of a solid solution composed of $n_A$ and $n_B$ moles of components A and B can be written as

$$F_1 = (n_A + n_B)F_A + \Phi(n_B) + (n_A + n_B)\Delta F^A_{\text{mix}}$$

Here $F_A$ and $\Delta F^A_{mix}$ refer to one mole of the substance. The same line can be given as

$$F_1 = F_A + \phi(x) + RT[(1-x)\ln(1-x) + x\ln x]$$

Here $F_1$ and $\phi(x)$ also refer to one mole.

The change of the free energy of the matrix due to the solute $\phi(x)$ (strictly speaking, the change minus mixing energy given here in its simplest form which disregards the possibility of short-range order in the arrangement of the solute molecules) is expandable into a series

$$\phi(x) = \phi'(0)x + \tfrac{1}{2}\phi''(0)x^2 + \cdots$$

If $x \ll 1$, i.e., for dilute solutions when the molecules of the solute are surrounded in the majority of cases by matrix molecules only, the approximation

$$F_1 = F_A + x\phi' + RTx\ln(x/e)$$

is fulfilled for $F_1$ sufficiently rigorously.

Measuring $F_1$ and $F_A$ calorimetrically, we can also find $\phi'$. Such measurements have not yet been carried out. Below it will be shown that conclusions about $\phi'$ can be made by analyzing the type of the phase diagram.

We can also try to calculate $\phi'$ by representing this value as a sum of the components contributed by the above factors which cause variations of the free energy during substitution. It would be logical to put

$$x\phi' = \Delta V + \Delta E + \Delta F^{vib}$$

Using the above simple representation of the free energy of a solid solution, the solubility conditions and equilibrium conditions can be written down.

The solubility conditions will be obtained by taking into consideration that at a given temperature a mechanical mixture with free energy $xF_B + (1-x)F_A$ can also be formed; consequently, for a solid solution to form it is necessary that

$$F_1 < xF_B + (1-x)F_A$$

After simple transformations and excluding from both sides of the inequality the intramolecular free energy which is not influenced by molecular interaction, the thermodynamic condition for the existence of a solid solution can be written as

$$T|\Delta S_k| > x(F^{Cr}_A - F^{Cr}_B) + x\phi'$$

A similar inequality can be given for a solid solution of A in B.

It is obvious that the reasons for limited solubility, i.e., the presence of boundaries in the phase diagram, are determined by the solubilities from both ends of the diagram.

Let us introduce appropriate designations for a solid solution of A in B: $n_A'$ and $n_B'$ are the variable numbers of moles of the components so that $x' = n_B'/n_A' + n_B'$ is the concentration of molecules B in crystal B, i.e., is $x < x'$.

The free energy of a solid solution crystal with structure B may be written as

$$F_2 = (n_A' + n_B') F_B + \Psi(n_A') + (n_A' + n_B') \Delta F_{mix}^B$$

In a melt of the eutectic composition at temperature $t_E$ two phases are in equilibrium—solid solutions with structures A and B. Sometimes under eutectic conditions the crystals of these two phases can be grown simultaneously on one inoculating needle (see the following section).

The equilibrium conditions for these two phases are

$$\left(\frac{\partial F_1}{\partial n_A}\right)_{n_B = \text{const}} = \left(\frac{\partial F_2}{\partial n_A'}\right)_{n_B' = \text{const}}$$

$$\left(\frac{\partial F_1}{\partial n_B}\right)_{n_A = \text{const}} = \left(\frac{\partial F_2}{\partial n_B'}\right)_{n_A' = \text{const}}$$

After a number of transformations the equilibrium conditions become

$$\frac{1-x}{1-x'} = e^\varepsilon e^\alpha; \qquad \frac{x}{x'} = e^\varepsilon e^{-\beta}$$

where

$$\varepsilon = \frac{F_B - F_A}{RT}, \qquad \alpha = \frac{1}{RT}\frac{\partial \psi}{\partial(1-x')}, \qquad \beta = \frac{1}{RT}\frac{\partial \phi}{\partial x}$$

From these equations we find

$$x = \frac{e^{(\alpha + \varepsilon)} - 1}{e^{(\alpha + \beta)} - 1}, \qquad x' = \frac{e^{-(\alpha - \varepsilon)} - 1}{e^{-(\alpha + \beta)} - 1}$$

Thus, the limits of solubility $x$ and $x'$ depend on the solubilities $\alpha$ and $\beta$ simultaneously from both ends of the phase diagram, and also on the difference in the free energies of the components. It is for this reason that we cannot "explain" why A dissolves in B irrespective of the extent of solubility of B in A; the solubility depends on which is energetically more advantageous: the solution of A in B or B in A.

It is worthwhile to analyze in detail the conditions of equilibrium of the two solid solutions written above. In fact, if the values of $\alpha$, $\beta$, $\varepsilon$ are known for any two substances with similar molecules, our equations would allow the prediction of the phase boundaries of their composition–temperature diagram without resorting to experiment.

It would probably be easier to determine the phase diagram experimentally, i.e., find $x$ and $x'$. In this case the two equations describing the equilibrium conditions for two solid solutions can be used for estimating the three unknown values $\varepsilon$, $\alpha$, $\beta$.

Parameter $\varepsilon$ is directly proportional to the difference in the free energies of the pure components A and B: $\Delta F = F_B^{Cr} - F_A^{Cr}$. This parameter can be calculated if the heat capacities and heats of sublimation are measured for the components, and if the intramolecular vibration frequences are known. Thus, the two equations describing solubility, in the absence of direct calorimetric data on the solid solutions, can be used to determine $\phi$ and $\psi$: the changes in the free energies of the limiting solid solutions (from both sides of the phase diagram) compared with the pure components. Now what is peculiar about these values? The three terms composing $\phi$ and $\psi$ include $\Delta U$, $\Delta E$, and $\Delta F^{vib}$. The increment of lattice energy $\Delta E$ due to a possible change in the conformation of the solute molecule as it is incorporated into the matrix lattice may be neglected if the molecules are rigid (naphthalene, anthracene); for more flexible molecules (dibenzyl, stilbene), a value $\Delta E$ can be calculated. $\Delta F^{vib}$ can be roughly assessed if the thermal factors for the structures of the pure component $B_p$ and of the limiting solid solution with the structure of this component $B_s$ are known from the results of X-ray intensity analysis. After calculating from the formula $\overline{u^2} = B/8\pi^2$, the corresponding values of mean-square deviations (see Section IV. 2), we may put the ratio between the characteristic temperatures of the crystals equal to

$$\Theta_p/\Theta_s = (\overline{u_s^2}/\overline{u_p^2})^{1/2}$$

Then, using the Debye approximation we obtain

$$\Delta F^{vib} = 6RT \ln(\Theta_p/\Theta_s).$$

This procedure gives an undoubtedly correct estimation of the order of magnitude of this quantity.

After substituting the estimated values into the initial formulas we obtain values for $\Delta U_A$ and $\Delta U_B$; these are the variations in the lattice energy. Under the assumption that the solutions are dilute this implies the changes caused by one solute molecule in the solvent lattice.

The values for $\Delta U_A$ and $\Delta U_B$ obtained by means of this empirical procedure can be used to analyze the intermolecular interactions, for instance, for verifying the correctness of the atom–atom potential method.

Using atom–atom potentials, $\Delta U$ can be calculated as follows. Let us accommodate in the matrix structure instead of one of its molecules a solute molecule in an optimum manner, i.e., with approximately the same number of

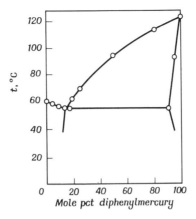

*Fig. 77.* Phase diagram of tolane–diphenylmercury system.

contacts, and then calculate the energy of interaction of this "foreign" molecule with all the matrix molecules, e.g., within a radius of 15 Å. The energy $U_A$ thus obtained will differ from the lattice energy of the pure component (matrix) $U_M$ by the value $U_A - U_M$. If the matrix lattice contains $x$ percent of the solute, the total lattice energy is

$$xU_A + (1-x)U_M = \{x(U_A - U_M) + U_M\}$$

i.e.,

$$\Delta U = x(U_A - U_M)$$

This, however, is a limiting value, because the calculations have not taken any account of the lattice distortions, which unavoidably arise during the formation of a solid solution, nor of the interactions between the solute molecules. Consequently, the comparison of the calculated value of $\Delta U$ with the experimentally derived $\Delta U_A$ (or $\Delta U_B$) allows one to judge, first the correctness of the selected atom–atom potentials and, secondly, the nature of the redistribution of the stresses in the structure of the solid solution.

By way of example let us make a detailed analysis of the solubility conditions in the tolane–diphenylmercury system.

The phase diagram of this system has been established by growing single crystals of solid solutions [68] (Fig. 77). At the eutectic temperature $t_\varepsilon = 56°C$ the solid solutions of (a) the tolane phase, $x = 0.14$ and $1 - x = 0.86$ and (b) the diphenylmercury phase, $x = 0.92$, $1 - x = 0.08$, are in equilibrium.

The tolane (XXXI) and diphenylmercury (XXXII) molecules are planar and have practically the same geometrical dimensions; it can be assumed that no changes occur in the solute molecule's conformation, i.e., $\Delta E = 0$.

(XXXI)                                          (XXXII)

The crystalline portions of the free energies of the components included in $\Delta F$ can be represented as the sum of the heats of sublimation and the vibrational parts of the free energies. The heats of sublimation for tolane and diphenylmercury have been estimated [69]: $\Delta H_A = 21.2$ and $\Delta H = 27.0$ kcal/mole. The values of the free energies of the translational and vibrational motions of the tolane and diphenylmercury molecules are given by Koreshkov [70]: $F_A^{\text{Cr vib}} = -6.3$ and $F_B^{\text{Cr vib}} = -6.6$ kcal/mole. Consequently, $F_A^{\text{Cr}} = -\Delta H_A + F_A^{\text{Cr vib}} = -21.2 - 6.3 = -27.5$ kcal/mole, and $F_B^{\text{Cr}} = -\Delta H_B + F_B^{\text{Cr vib}} = -27.0 - 6.6 = -33.6$ kcal/mole, whence $\Delta F = F_B^{\text{Cr}} - F_A^{\text{Cr}} = -6.1 \simeq -6$ kcal/mole.

Substituting these values into the equilibrium conditions, we obtain $\phi' = -4.9$ kcal/mole and $\psi' = +7.4$ kcal/mole.

Thus, the free energy of the limiting solid solution of the tolane phase will differ from the free energy of pure tolane by $\phi(x) = x\,(-4.9\text{ kcal/mole}) = -0.69$ kcal/mole, while for the phase with the diphenylmercury structure the difference from the free energy of a pure diphenylmercury crystal is

$$\psi(x') = (1-x')\,(+7.4\text{ kcal/mole}) = 0.59\text{ kcal/mole}$$

In order to estimate the vibrational component of the free energy of a solid solution crystal we shall take data on the variations of the thermal factor of the structure with increase of solute concentration. For the tolane structure phase $B_p = 4.7$ Å$^{-2}$ (pure tolane), and $B_s = 3.7$ Å$^{-2}$ (solid solution containing 14% of diphenylmercury molecules). Since in this example the $B(x)$ curve passes through a minimum (Fig. 78), it is possible to separate static shifts $\overline{u_{\text{stat}}^2}$ from dynamic lattice vibrations $\overline{u_{\text{dyn}}^2}$. To a first approximation, however, $\overline{u_{\text{stat}}^2}$ can be considered a negligible value because the geometrical dimensions of the molecules being mixed are practically equal. Then, $\overline{u_{\text{dyn}}^2} = B/8\pi^2$ and approximately

$$\frac{\Theta_p}{\Theta_s} = \left[\frac{(\overline{u_{\text{dyn}}^2})_s}{(\overline{u_{\text{dyn}}^2})_p}\right]^{1/2} = \left(\frac{B_s}{B_p}\right)^{1/2} = \left(\frac{3.7}{4.7}\right)^{1/2} = 0.89$$

It follows that $\Delta F_B^{\text{vib}} = 6RT \ln \Theta_p/\Theta_s = -0.043$ kcal/mole. Thus, assuming $\Delta E = 0$, we obtain $\phi(x) = -0.69$ kcal/mole $= \Delta U_B - 0.43$ kcal/mole, whence $\Delta U_B = -0.69 + 0.43 = -0.26$ kcal/mole.

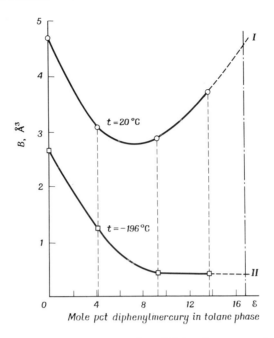

**Fig. 78.** Thermal factor $B$ versus concentration of diphenyl–mercury admixture in tolane phase: I—at room temperature; II—at liquid nitrogen temperature.

No experimental data are available for making similar estimates for the phase with the diphenylmercury structure. The maximum values of $\Delta U$ have been calculated by the atom–atom potential method. Calculations that have been carried out on electronic computers used curves of potential interaction between C–C, C–H, and H–H atoms which are already sufficiently reliably established [71]; the same also applies to the interactions between Hg–Hg, Hg–C, and Hg–H atoms determined from the crystalline structure of diphenyl-mercury [72]. In the geometrical substitution of molecules in matrix structures it was assumed that the coordinates of the benzene ring atoms do not change so that the replacement amounts to that of a $-C{\equiv}C-$ bridge for $-Hg-$, or vice versa. These calculations have been carried out for both phases of the system under consideration.

*The Solid Solution with the Tolane Structure.* Direct X-ray diffraction analysis has demonstrated [73] that diphenylmercury solute molecules replace tolane molecules only in layer 1, one of the two molecular layers in the tolane structure that are not related symmetrically. Figure 79 shows the corresponding electron-density projections for whose computation, signs calculated over only the 12 C atoms of two benzene rings have been assigned to the experimental

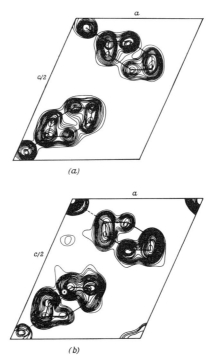

*Fig. 79.* Projections ac of electron density: (a) pure tolane structure; (b) structure of solid solution of diphenylmercury in tolane phase.

values of $F_{h0l}$, The high maximum on the $b$ projection at the point 00 corresponds to the Hg atom; there is no maximum at the point $0\frac{1}{2}$. Energy calculations have confirmed the results of the X-ray experiment: the replacement of one tolane molecule by a diphenylmercury molecule in layer I changes the lattice energy by $-4.88$ kcal/mole, whereas the same substitution in layer II only by $-2.52$ kcal/mole, the difference is $\Delta = 2.36$ kcal/mole. Using the Boltzmann distribution formula, we can estimate the ratio of the probabilities of the solute molecules' distribution over the nonequivalent layers

$$\omega_I/\omega_{II} = e^{-\Delta/RT} \simeq 50$$

Since these estimations show that the probability of replacement in layer I is 50 times as high as in layer II, it shows that practically all the solute goes into layer I. As has been pointed out above, the formation of solid solution changes the lattice energy by $x(U_A - U_M)$. The calculated value of $U_A - U_M = -4.88$ kcal/mole (solute molecules appear only in layer I), whence $\Delta U = 0.14\,(-4.88)$ $= -0.68$ kcal/mole. This maximum value, disregarding the fact that stresses arising around the solute molecule are distributed over the entire crystal volume, is approximately 2.5 times larger than the value found experimentally from the solubility conditions.

*The Solid Solution with Diphenylmercury Structure.* Diphenylmercury molecules form a single-layer packing. According to the results of energy calculations, the replacement of some molecules in the structure by tolane solute molecules alters the lattice energy of the most concentrated solid solution ($x' = 0.08$) by $\Delta U = +0.30$ kcal/mole, as compared with the pure component. Compare this figure with value $+0.59$ obtained from the analysis of the solubility conditions.

Thus, the extimates of $\Delta U$ in both solutions seem to be quite reasonable.

## 17. DETERMINATION OF PHASE DIAGRAMS

Physicochemical analysis employs various methods for constructing phase diagrams: recording of time-versus-temperature curves (thermography) on a Kurnakov pyrometer (often with a differential thermocouple), studies of the micro- and macrostructures of the alloys under investigation, X-ray phase analysis, measurement of the changes in the physical characteristics (specific gravity, thermal expansion, electrical resistance, etc.) and mechanical properties as the composition and temperature vary. These methods allow determination of the number and structure of the phases in a system, as well as their domains of existence, with a high degree of accuracy.

It should be pointed out that organic substances feature, first, a pronounced tendency toward overcooling and, secondly, the thermal effects during phase transitions in these substances are small. It is for this reason that the sensitivity of the conventional thermographic technique proves to be insufficient for studying organic systems. Several dozens of binary systems can be listed whose phase diagrams have been shown to be Roseboom type I or III by thermographic analysis (Fig. 80 depicts five types of Roseboom phase diagrams of solid solution systems). They include, in particular, the following systems: acridine–anthracene, anthracene–phenanthrene, naphthalene–$\beta$–chloronaphthalene, naphthalene–$\beta$–naphthylamine, dibenzyl–stilbene; the components of these systems have different symmetries, which makes a continuous series of solid solutions in them impossible. Later, after differential recording had been introduced in the thermographic method, and after the microscope with a heating glass had been developed, some of these phase diagrams were corrected.

Thermal analysis with the aid of a microscope with a heating glass is much more senstive than thermographic analysis. Microscopic examinations are carried out on specimens placed between the microscope slide and cover glass. Two operating procedures may be applied. One consists in studying first the

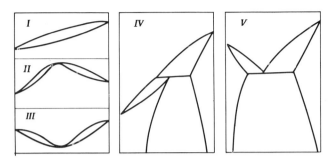

*Fig. 80.* Five Rooseboom types of phase diagrams of systems with solid solutions: I, II, III—unlimited solubility in solid state; IV—limited solubility with peritectic point; V—limited solubility with eutectic point.

melting process on a powder specimen, i.e., on a finely ground mixture, after which eutectic temperature determination is made. The other procedure involves examination of a crystalline film of the melt which has solidified between the slide and the cover glass and determination of the primary crystallization points. A detailed description of this method is given in the book of L. Kofler and A. Kofler [74]. The authors developed the contact specimen method for constructing phase diagrams which is described as follows. Small amounts of finely ground pure components are placed side by side on the microscope slide, covered by the cover glass and heated. When the component with the lower melting point has melted, a contact zone with an unfused component arises, a certain amount of this component can mix or react with the melt, and finally the entire mass melts. Rapid cooling and then reheating of the mixture make it possible to study all the phenomena: phase changes in the solid state, number of phases in the system, presence of solid solutions or eutectic mixture, etc. on one specimen. Thus, using the contact specimen technique, with simultaneous observation of the melting of mixtures with different compositions, we may plot the phase diagram of a binary system.

   This method gives a quite reliable qualitative estimate of the phase changes in a system, though it cannot claim high accuracy. Thus, the rather complex phase diagram of the *p*-dibromobenzene–*p*-chloronitrobenzene system (Fig. 81) obtained by growing single crystals [75] is qualitatively confirmed by examinations on the heating glass of a microscope, however, quantitative discrepancies reach 10% in composition and 5–10° in temperature. This probably can be ascribed to the impossibility of obtaining volumetric equilibrium between the liquid and solid phases under the cover glass.

   The method of plotting phase diagrams by growing single crystals used to construct most phase diagrams shown in this chapter was developed in the

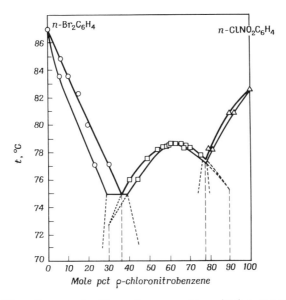

*Fig. 81.* Phase diagram of *p*-dibromobenzene-*p*-chloronitrobenzene system.

author's laboratory in 1956–1957 specially for experimental verification of the conditions for formation of solid solution crystals by two organic substances.

The method depends on maximum approximation of the crystallization conditions in the melt to equilibrium conditions. A crystallizer containing the mixture of two substances of a given concentration is immersed in a thermostat with transparent walls to enable observation of crystal growth. The melting point of crystal nuclei forming on an inoculating needle plunged into the melt of this mixture determines the liquidus line point corresponding to this concentration. Any slight overcooling of the melt (0.1–0.2°) (±0.05°) causes growth of single crystals of the inoculating needle, the maximum size of the crystals reaching 1–2 mm within 5–10 hr of thermostat operation. The number of phases in the system is defined both from the external shape of the growing crystals and by X-ray diffraction.

Three to four grams of the mixture produce not more that 10–15 mg of crystals, so that, to a first approximation, they may be considered compositionally homogeneous. These single crystals are used for various purposes. Since binary crystals melt in a certain temperature range, the determination of this range, and of the melting end-point in particular, makes it possible to draw a solidus line after the liquidus line of the phase diagram has already been determined. The solidus line, i.e., the composition of the grown crystals, can also be found by microanalysis for some element present in the molecule of one component only. Observation of the disintegration of these crystals with decreasing temperature helps establish the solubility limits in the solid state. Finally, solid solution crystals are also employed for the detailed X-ray study to be discussed in the next section.

The above method of determining phase diagrams is rather laborious, but its accuracy is high: $\pm 0.05°$ in temperature and not worse than $\pm 0.5\%$ in composition even in the eutectic or peritectic point regions, where it is often required to extrapolate the liquidus lines. Correctness of such extrapolation is frequently confirmed when crystals of both phases can be grown simultaneously on the inoculating needle in a melt of a given composition. The composition of the solid solution crystals is estimated with an accuracy not worse than 1–2%.

## 18. X-Ray Diffraction of Solid Solution Crystals

Solid solution crystals of organic substances usually display low symmetry (as a rule, monoclinic), and therefore X-ray diffraction studies cannot be carried out on them using the conventional methods of Debye powder patterns so widely used in metallography. When low-symmetry substances are dealt with, single crystals are necessary even for measuring the parameters of the unit cells. It is for this reason that, though laborious, the method of determining phase diagrams by growing single crystals proves indispensable for studying mixed organic crystals.

The measurement of the unit cell parameters of solid solution crystals by the high-precision X-ray method allows finding for each phase the function $V_{mol}(x)$, where $x$ is the concentration of solute molecules, and $V_{mol} = V_{cell}/Z$ is the volume per molecule in the unit cell.

In most cases the form of the function $V_{mol}(x)$ can be interpreted from geometrical considerations. For example, the molecules of acridine (XXXIII)

(XXXIII)

in the structure of acridine-II [76]—a modification whose crystals grow from the melt—are so packed that the "hollow" near $\diagdown$N is "filled up" by an –H atom of the adjacent molecule. On account of the "inconvenient" shape of the molecules, their packing is unusual: The unit cell of symmetry $C_{2h}^5$ contains eight molecules, i.e., two independent molecules in a general position; the packing coefficient is rather larger $k = 0.713$.

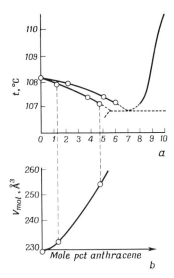

**Fig. 82.** (a) Acridine side of phase diagram; (b) variation of volume per molecule with growing concentration of anthracene admixture.

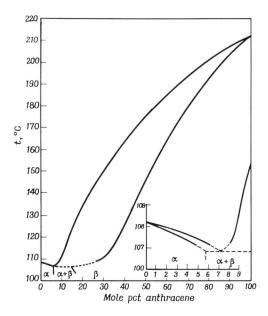

**Fig. 83.** Phase diagram of acridine–anthracene system. (Dashed line, diagram determined thermographically previously).

When the acridine molecule is geometrically replaced by an anthracene molecule (XXXIV) the $H \cdots H$ and $C \cdots H$ distances are considerably reduced. Experiment has demonstrated a substantial "swelling" of the unit cell

(XXXIV)

of a solid solution crystal with the acridine structure (Fig. 82), the packing coefficient drops to 0.644 and a discontinuity in solubility appears when the melt contains 7% of anthracene molecules [77] (Fig. 83).

Thus, in the acridine–anthracene system, the high solubility of anthracene in acridine is handicapped by a sharp rise of $\Delta U$. On the other hand, the crystals of the solid solution with the anthracene structure may contain up to 72% of acridine molecules; in this case the dimensions of the unit cell change accordingly; the packing coefficient drops from 0.722 to 0.707. Consequently, "hollows" increase the free energy of a crystal much less than do densely overpacked places.

The phase diagram of the dibenzyl–stilbene system is shown in Fig. 84 [78]. Unlike the stilbene molecule (XXXV) the dibenzyl molecule (XXXVI) is not

(XXXV)                                        (XXXVI)

planar and has two additional H atoms. Construction of electron density series demonstrates that due to "swelling" of the unit cell (Fig. 84b) probably no appreciable changes occur in the conformation of the solute molecules in these solid-solution structures [79].

Replacement of molecules in the $\beta$-chloronaphthalene structure by smaller-volume naphthalene molecules results in decreased dimensions of the "mean" unit cell (Fig. 85), and in a drop of $V_{mol}(x)$ in the dibenzyl phase (Fig. 84b). A simple geometrical explanation can be furnished to account for changes of unit cells in the $\alpha$- and $\beta$-phases of the p-dibromobenzene–p-diiodobenzene system [80]. Figure 86a illustrates the phase diagram of this system, in which

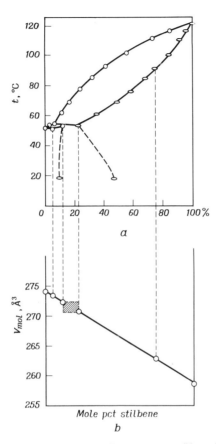

**Fig. 84.** (a) Phase diagram of dibenzyl–stilbene system; (b) variation of volume per molecule with growing concentration of stilbene admixture.

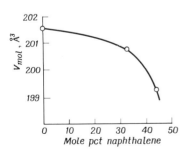

**Fig. 85.** Variation of volume per molecule in phase with $\beta$-chloronaphthalene structure depending on naphthalene admixture concentration.

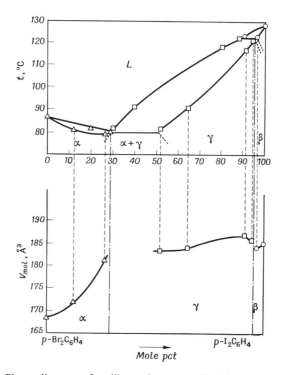

**Fig. 86.** (a) Phase diagram of *p*-dibromobenzene-*p*-diiodobenzene; (b) $V_{mol}$ versus p-diiodobenzene concentration in all three phases.

three phases of different symmetries are separated by one eutectic and one peritectic point (it should be mentioned that the latter was not detected thermographically). The phase diagram is followed by the graph of the $V_{mol}(x)$ relationship for all the three phases (Fig. 86b).

A maximum on the $V_{mol}(x)$ curve for the γ-phase is noteworthy. A similar maximum is to be found in the tolane phase of the tolane–diphenylmercury system, where the geometrical factor is practically absent (Fig. 87). It is rather diffcult to account for such fine effects. At any rate, phenomena like these do not have a simple geometrical interpretation.

It would seem reasonable to analyze the dependence of the thermal vibrations on concentration. No detailed investigations of this kind have been carried out so far. In several works of the author and co-workers attempts have been made to analyze the mean thermal factor of X-ray interference. The same investigations were also conducted for solutions of *p*-dibromobenzene–*p*-diiodobenzene which exhibit a very specific shape of the $V(x)$ curve. An analysis of the intensities on X-ray diagrams obtained from crystals of these solid solutions shows that the replacement of the small *p*-dibromobenzene molecules

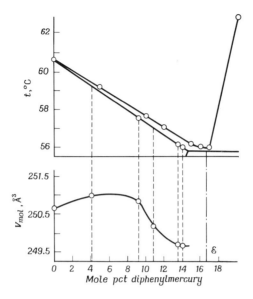

**Fig. 87.** (a) Tolane side of phase diagram of tolane–diphenylmercury system; (b) $V_{mol}(x)$ relationship.

by larger molecules causes a growth of the thermal factor ($\alpha$-phase). In the $\beta$-phase it is interesting that a diminished thermal vibration energy is observed (Fig. 88); this is already obvious from the better quality of the X-ray picture.

In the $\gamma$-phase the $B(x)$ curve passes through a minimum. For the tolane phase, the $B(x)$ curves have been obtained at room temperature and at liquid

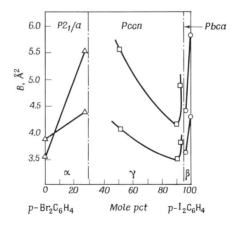

**Fig. 88.** Values of thermal factor $B$ for $hk0$ and $h0l$ zones as functions of concentration in $p$-dibromobenzene-$p$-diiodobenzene system at room temperature.

nitrogen temperature (Fig. 78). The minimum on the $B(x)$ curve at room temperature is much more conspicuous [81].

This result is not trivial. Indeed, assuming that the mean square of atomic displacements

$$\overline{u^2} = \frac{B}{8\pi^2} = \overline{u_{stat}^2} + \overline{u_{dyn}^2}$$

where $\overline{u_{stat}^2}$ depends only on the concentration of solute molecules and $\overline{u_{dyn}^2}$ depends only on the temperature, one could expect a rise of $B$ with increasing $x$ as well as an identical shape of the $B(x)$ curves at two different temperatures. $u_{dyn}^2$ appears to be dependent not only on the temperature but also on the concentration of the solute molecules. Apparently, dynamic effects are the key factors that determine the conditions for forming solid solutions both in the tolane phase of the tolane–diphenylmercury system and in the $\gamma$-phase of the $p$-$Br_2C_6H_4$-$p$-$I_2C_6H_4$ system.

Apart from a large number of cases of formation of true solid solutions when the replacement of matrix molecules by solute molecules brings about a noticeable change in the dimensions of the unit cell, examples testifying to the opposite are also available. Substitution of naphthalene molecules by molecules of $\beta$-chloronaphthalene (XXXVII) and also molecules of 1,5-dinitro-

(XXXVII)                    (XXXVIII)                    (XXXIX)

naphthalene (XXXVIII) by 1,8-dinitronaphthalene molecules (XXXIX) does occur, as is confirmed both microanalytically and by observation of the melting of crystals of these solid solutions. However, $V_{mol}(x)$ remains practically unchanged (Fig. 89). It can be imagined that when one molecule differs from another by some additional volume this substitution does not occur uniformly over the entire volume of a single mosaic block but only in the boundary cells so that these "additions" are left outside the mosaic blocks and do not influence the dimensions of the unit cell.

In contrast to true solid solutions, this type of solubility has been termed interblock solubility. It occurs where there is partial geometrical conformity between the matrix and solute molecules. The role of this partial conformity in the production of new luminiscent crystals of naphthalene and anthracene with various additions has been convincingly demonstrated by Belikova and Belyaev [82].

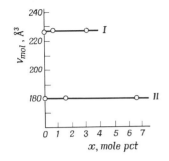

**Fig. 89.** $V_{mol}(x)$ relationships: I—solution of 1,8-dinitronaphthalene in 1,5-dinitronaphthalene; II—solution of $\beta$-chloronaphthalene in naphthalene.

Indirect data obtained when studying the luminiscence and absorption spectra of the solute show that a molecule unable geometrically to be incorporated into the matrix crystal lattice is nevertheless somewhat adapted to the lattice by using (or forming) defects in the matrix lattice. Particularly interesting in this respect are results obtained in the studies of aromatic molecules added in small amounts to melted paraffins. When such a system is frozen, the molecules of the solute are always surrounded in the same manner by matrix molecules, as is shown by the appearance of sharp quasi-line spectra (Shpolskij effect) [83]. The role of partial geometrical conformity is quite apparent in this case. Thus, for instance, the sharpest spectra are produced by anthracene dissolved in heptane (the linear dimensions of the molecules are the same) [84].

## 19. GEOMETRICAL ANALYSIS AND ENERGY CALCULATIONS

X-ray measurements are known to display, for a given crystal, the "mean" pattern of the lattice distortions which have appeared in the course of solid solution formation. The stress set up by the differences in the dimensions and shapes of the molecules being mixed is distributed over the entire volume of the single crystal block. Indeed, investigations of solid solutions by the nuclear quadrupole resonance method allow the estimation of the distortions in the matrix lattice caused by the admixture molecules. Due to the high sensitivity of the nuclear quadrupole reasonance technique to deviations from the perfect shape of the crystal lattice of the specimen, the amplitude of the nuclear quadrupole resonance signal diminishes almost exponentially with increasing solute concentrations. It has been shown that, depending on the dimensions and shape of the solute molecule, the latter upsets the resonance conditions in a varying number of adjacent molecules [85]. According to our estimations, for example, one $p$-diiodobenzene molecule in the $p$-dichlorobenzene structure causes static distortions in a sphere containing approximately 450 matrix molecules. The smaller-size solute molecules of $p$-bromochlorobenzene, $p$-bromoiodobenzene and $p$-chloraniline in the same matrix produce distortions in about 105, 330, and 160 surrounding molecules. Thus, the nuclear quadrupole resonance technique also confirms that smaller solute molecules give rise to smaller distortions in the matrix lattice, in other words, they are "better dissolved" in this lattice.

Geometrical analysis of the process of finding the optimum geometric location of a solute molecule on the site of one of the matrix molecules (without distorting its lattice), followed by complete calculation of all the intermolecular distances, yields a picture of the maximum distortions. For example, in the case of the geometrical replacement of an acridine molecule by an anthracene molecule which has no "hollow" in the middle ring, the structure shows shortened $H \cdots H$ distances: 1.63 and 2.07 Å (instead of 2.34 Å) and also $C \cdots H$ distances: 2.57 and 2.81 Å (instead of 2.97 Å). The cumulative action of the stress caused by one anthracene molecule changes the interaction energy by $\Delta U = 1.02$ kcal/mole.

Consider another example: a 1,8-dinitronaphthalene molecule in the 1,5-dinitronaphthalene structure is strongly compressed by adjacent molecules: two $O \cdots O$ distances equal 1.89 and 1.99 instead of 2.72 Å; two $O \cdots H$ distances are 1.69 instead of 2.53 Å; two $C \cdots O$ distances equal 2.91 instead of 3.16 Å; and one $N \cdots O$ distance is 1.87 instead of 2.93 Å. The total energy energy loss for all these shortened distances would reach about 16 kcal/mole if the system had true solubility. Similarly, the substitution of one naphthalene molecule by a $\beta$-chloronaphthalene molecule would require an energy of about 9–10 kcal/mole.

Thus, geometrical analysis allows a classification of true and interblock solid solutions in terms of energy: If the substitution of a solvent molecule by a solute molecule needs about 1–2 kcal/mole, there will be true solubility in the system; conversely, if about 10–20 kcal/mole is required by such a replacement, only interblock solubility, if any, can be expected.

The gentle slope of the atom–atom potential curve toward attraction is still further proof that but little energy is required to retain hollows in the packing, so that the solubility of small molecules among molecules of a somewhat larger volume will always be true solubility.

A convincing example that testifies to this same effect are the solid solutions of normal (linear) paraffins [24, 86]. The crystalline structure of $n$-paraffins exhibits a dense parallel array of hydrocarbon chains of the type shown in (XL). Figure 52 shows an example of paraffin molecule packing. Free rotation

(XL)

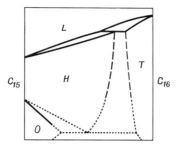

*Fig. 90.* Phase diagram of the *n*-paraffin system $C_{15}H_{32}$–$C_{16}H_{34}$.

of the end $-CH_3$ groups enables close stacking of layers formed by parallel molecules.

Simple conditions can be formulated for the formation of solid solutions of *n*-paraffins. Here, the requirement that the shapes and sizes of the molecules being mixed must be approximately the same merely implies a small difference in their length. Since even- and odd-numbered paraffins have different symmetries, continuous solubility in the "even + odd" system is not possible; an example is the phase diagram of $C_{15}H_{32}$–$C_{16}H_{34}$ (Fig. 90). In the "odd + odd" paraffin systems the type of solubility will depend on the relative length of the molecules, because all odd *n*-paraffins are mutually isomorphous. Even paraffins of different symmetry fail to form continuous solid solutions. Examples are given in Fig. 91.

An analysis of a large number of the phase diagrams of *n*-paraffins suggests the conclusion illustrated by the diagram in Fig. 92: Other conditions being equal, dissolution of *n*-paraffins with a shorter chain in longer-chain paraffins involves preservation of a larger amount of the solvent structure than in the

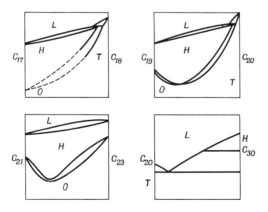

*Fig. 91.* Examples of phase diagrams of *n*-paraffins.

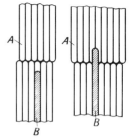

**Fig. 92.** Schematic representation of packing of linear molecule ends showing that "projections" are less advantageous in terms of energy than "hollows."

opposite case. The diagram shows that with $B < A$ undistorted packing is feasible; the solubility limit is achieved when the number of "hollows" becomes large and the packing density appreciably drops. Conversely, if $B > A$, the replacement of even a small number of $A$ molecules is bound to cause distortions which quickly stops solubility.

It should be pointed out that when applied to the replacement of a matrix molecule by a solute molecule, the fundamental statement about the resulting noticeable loss in lattice energy holds true only when the replacement results in a sharp decrease in some intermolecular distances. If, however, there are but minor changes in the distances, then only calculation of the total interaction energy can give a correct idea of whether the replacement is advantageous in terms of energy. Such a calculation has been carried out, for example, for the system diphenyl-$\alpha$, $\alpha'$-dipyridyl. It has been found that differences in energy depend but slightly on the molecular composition of the first coordination sphere: In the case of substitution of a diphenyl molecule by an $\alpha,\alpha'$-dipyridyl molecule, $U_A - U_M = 1.23$ kcal/mole with one solute molecule in the first coordination sphere, and 1.20 kcal/mole with two, three, and four molecules. In the structure of pure $\alpha,\alpha'$-dipyridyl the molecules are so packed that hollows near the $\diagdown$N group remain unfilled; therefore, the substitution may be said to result in a closer packing, so that the lattice energy of the solid solution crystal is even somewhat reduced: $U_A - U_M = -0.25$ kcal/mole. Thus, losses in the potential interaction energy cannot hinder solubility. Nonetheless, solubility in the system is low from both sides in the diagram. There can be only one reason for this phenomenon: a significant effect of the substitution on the vibration spectrum of the crystal.

Now let us turn to the problem of the distribution of solute molecules in the matrix lattice, i.e., the problem of short-range order in a solid–solution crystal.

Experimental investigations of this kind are totally lacking for molecular crystals. As to the possibility of theoretical calculations, the theory developed recently permits computation of the short-range order of molecules on the basis of the atom–atom potential technique [62, 87].

As is known, the distribution of particles in a binary solid solution can be described by the simple Ising statistical model in which atoms of both components occupy points of a certain crystal lattice and where interatomic interaction is pairwise. The Ising problem, however, has a precise solution only for one- and two-dimensional lattices in the nearest-neighbor approximation. For studying the more interesting three-dimensional case, various approximate methods must be used.

At the present time there are several well-known approximate statistical techniques used for the purpose: the Bragg–Williams, Gorsky method, which disregards correlation in the solid solution; the Bethe–Peierls theory; the Guggenheim and Fowler quasichemical approximation; and also the Kirkwood technique associated with the expansion of the statistical sum in $W/T$ ($W$, mixing energy for the first coordination sphere; $T$, absolute temperature in units of energy). The limitations of these theories are that they take into account atomic interaction only in the nearest-neighbor approximation, make preliminary assumptions about the structure of the ordered phase (the crystal points are a priori broken up into "own" and "strange" nodes), they are applicable for a rather limited type of crystal lattices (actually, for an alloy AB in which, in the ordered state, atoms of one species are surrounded by atoms of the other species).

Some of these limitations have been recently overcome in several of the publications which have been mentioned above.

The object of these publications is the development of particle arrangement theory (1) which would take account of the short-range order with particles interacting over an arbitrary number of coordination spheres and with an arbitrary form of the interaction potential, (2) which would not make a priori assumptions about the structure of the ordered state of the crystal, and (3) which would be applicable for describing the arrangement of particles for any symmetry of the Ising lattice and any type of structure of the ordered phase (if any). This theory is a version of the thermodynamical pertrubation theory.

The configurational Hamiltonian of a binary system with pairwise interaction is given by

$$H = \tfrac{1}{2} \sum_{r,r'} V_{rr'}\, C_r\, C_{r'}$$

where the $C_r$ are function that can assume the two values 1 and 0, depending on whether point **r** contains the A or B species molecule, $V_{rr'} = V_{AA}(r,r') + V_{BB}(r,r') - 2V_{AB}(r,r')$ are the mixing energies connected with the pairwise interaction potentials between the molecules of species A, of species B and species A and B, $V(0)=0$. The summation is carried out over all the points of the crystal lattice. The tsatistical sum of such a system has the form

$$Z = \sum_{C_r=0}^{1} \exp(-\tfrac{1}{2}\beta \sum_{r,r'} V_{rr'}\, C_r\, C_{r'})$$

where $Z$ is calculated under the assumption that the total number of molecules in the system is retained, and $\beta = 1/T$.

It can be shown that for a disordered solid solution of molecules A in structure B the probability that the molecules A are located simultaneously at points $\mathbf{r}$ and $\mathbf{r}'$ is

$$
\langle C_r C_{r'} \rangle = c^2 - \frac{V(\mathbf{r}-\mathbf{r}')}{T} f_2{}^1 + \frac{1}{2T^2}\left[ V^2(\mathbf{r}-\mathbf{r}') f_1{}^2 f_2{}^2 \right.
$$

$$
+ 2f_1{}^3 \sum_{r''} V(\mathbf{r}''-\mathbf{r}')V(\mathbf{r}'-\mathbf{r}'') \Bigg] - \frac{1}{6T^3}\Bigg[ V^3(\mathbf{r}-\mathbf{r}') f_1{}^2 f_3{}^2
$$

$$
+ 6f_1{}^4 \sum_{r''r'''} V(\mathbf{r}''-\mathbf{r}''')V(\mathbf{r}'''-\mathbf{r})V(\mathbf{r}'-\mathbf{r}'')
$$

$$
- 12f_1{}^4 V(\mathbf{r}-\mathbf{r}')\sum_{r''} V^2(\mathbf{r}'-\mathbf{r}'') + 6f_1{}^3 f_2{}^2 \sum V(\mathbf{r}''-\mathbf{r})
$$

$$
\times V^2(\mathbf{r}'-\mathbf{r}'') + 6f_1{}^3 f_2{}^2 V(\mathbf{r}-\mathbf{r}')\sum_{r''} V(\mathbf{r}''-\mathbf{r})V(\mathbf{r}'-\mathbf{r}'') \Bigg]
$$

Here $C$ is the concentration of the molecules of species A,

$$
f_1 = C(1-C), \qquad f_2 = 1 - 2C, \qquad f_3 = 1 - 6C + 6C^2
$$

and $\mathbf{r}''$ and $\mathbf{r}'''$ denote all the points except $\mathbf{r}$ and $\mathbf{r}'$.

If the binary solution has the concentration $\frac{1}{2}$, then

$$
\langle C_r C_{r'} \rangle = \frac{1}{4} - \frac{V(\mathbf{r}-\mathbf{r}')}{16T} + \frac{\sum_{r''} V(\mathbf{r}''-\mathbf{r})\,V(\mathbf{r}'-\mathbf{r}'')}{64T^2}
$$

$$
- \frac{1}{768T^3}\Bigg[ 2V^3(\mathbf{r}-\mathbf{r}') + 3\sum_{r'',r'''} V(\mathbf{r}''-\mathbf{r}''')\cdot V(\mathbf{r}'''-\mathbf{r})\cdot V(\mathbf{r}'-\mathbf{r}'')
$$

$$
- 6\,V(\mathbf{r}-\mathbf{r}')\sum_{r''} V^2(\mathbf{r}'-\mathbf{r}''') \Bigg]
$$

The theory for multicomponent organic crystals can be stated in the same approximation. At sufficiently high temperature ($V \ll T$) the simultaneous probability that the particles of the species $\alpha$ and $\beta$ are located at points $\mathbf{r}$ and $\mathbf{r}'$, respectively, is given by the approximate formula

$$
\langle C_\alpha(\mathbf{r}) C_\beta(\mathbf{r}') \rangle = C_\alpha C_\beta - \frac{C_\alpha C_\beta}{T}\Bigg[ V_{\alpha\beta}(\mathbf{r}-\mathbf{r}')
$$

$$
- 2\sum_{\gamma=1}^{\upsilon} V_{\alpha,\beta}(\mathbf{r}-\mathbf{r}') C_\gamma + \sum_{\gamma,\delta=1}^{\upsilon} V_{\gamma\delta}(\mathbf{r}-\mathbf{r}') C_\gamma C_\delta \Bigg]
$$

where $\upsilon$ is the number of particles of the different species; $c_i$ is the concentration of the particles of species $i$; and $V_{ij}(\mathbf{r}-\mathbf{r'})$ is the pairwise interaction energy of two particles (atoms, molecules) of species $i$ and $j$ at points $\mathbf{r}$ and $\mathbf{r'}$, respectively.

In conclusion it should be emphasized that the problem of the structure and properties of solid solution crystals is far from being solved. Crystals with hydrogen bonds have been omitted from consideration. It is clear that solute molecules must be selected for such substances so as to adapt these molecules into the hydrogen bond net of the crystal matrix. Thus, an attempt to dissolve $p$-dichlorobenzene in hydroquinone has so far failed. The short-range order in solid solution crystals has not yet been studied, i.e., no data on the distribution pattern of the solute molecules are available. The process of solid solution unmixing does not readily lend itself to study because of the sluggish diffusion of molecules in the solid state—a supersaturated solid solution can persist for years as a single-phase system. If, on the other hand, it becomes possible to induce separation into two phases by cooling, or by mechanical effects, the unmixing process can hardly be stopped at a certain stage in order to obtain an X-ray diffraction photograph, and, even worse, the crystal cannot be brought back to its initial state as is often done in metallic alloy investigations.

## 20. MOLECULAR COMPOUNDS

In a sufficiently large number of cases it is possible to grow crystals composed of molecules of two distinct species that are arranged, not in statistical disorder, but in mutually identical positions in all the crystal's cells. Such crystals are called molecular-compound crystals. They occur in the compositions 1:1 and 2:1; more complex compositions are fairly rare.

It should be first stressed that the formation of such a crystal does not necessarily point to the action of some kind of specific forces between the "compound" molecules. The very term "compound" cannot be considered quite adequate to describe such crystals.

Quite obviously, the free energy of a crystal composed of two molecular species is lower than the average energy of a monomolecular crystal. It may occur that the packing of a bimolecular crystal comes out to have a better density (the interaction potential energy is lower) than the packing of its components. The molecular compound of triphenylmethane and benzene can be taken as an example of this effect.

The structure of this crystal is unknown, but from purely geometric considerations it appears logical to assume that the benzene molecules help inconveniently shaped triphenylmethane molecules to pack more closely.

Most frequently, the formation of bimolecular crystals may be expected where the molecules are capable of producing hydrogen bonds. Here, as was shown above, the optimum structure must obey both the close-packing requirements and the trend to the maximum saturation of the hydrogen bonds.

It may be thought natural that the solution of this difficult problem in many cases in nature is obtained more successfully by combining two different molecules in one crystal. The examples of such crystals (in which again no new component is added to the interaction energy) are many, both among aliphatic compounds (fatty acids) and among aromatics, e.g., the system meta- and $p$-cresol (1:1), dichloroethylene-tetrachloroacetylene (1:1), chloroform-ethyl ether ($-100°C$, 1:1), aldehyde-ethyl alcohol ($-120°C$, 1:1), aniline-phenol (1:1), urea-nitrophenol (1:1) [88], $p$-dibromobenzene–$p$-chloronitrobenzene (3:2) [75].

A body of physical data has been obtained by optical spectroscopy and radiospectroscopy on the specific interactions of molecules in bimolecular crystals which contain as one of the components a molecule with nitro groups [89]. In this case a molecule possessing a strong dipole moment is assumed to polarize the molecule of the other component. This molecular pair will have a lower electrostatic interaction energy. According to other data, the transfer of an electric charge can be effected between the molecules of such a pair.

Many bimolecular crystals of this type have been subjected to comprehensive structural analysis. The main result of these investigations is as follows: The close packing principle is never violated in the geometry of the mutual arrangement of the molecules. Often the molecules of the components alternate to form columns. This is the case of the complex formed by the molecules of anthracene and symmetric trinitrobenzene in the ratio of 1:1. The structure has been established at two temperatures, room temperature and $-100°C$ [90]. The crystals are monoclinic, contain four molecular complexes in a unit cell with the parameters: at room temperature: $a = 11.70$, $b = 16.20$, $c = 13.22$ Å, $\beta = 132.8°$; at $-100°C$: $a = 11.35$, $b = 16.27$, $c = 13.02$ Å, $\beta = 133.2°$; space group $C2/c$. Figure 93 illustrates molecular packing along the $c$ axis: The molecules are arranged one after another, each in two distinct orientations, forming infinite columns along $c$ so that the intermolecular contacts are closest parallel to this directions (Fig. 94).

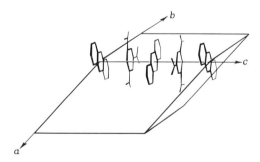

*Fig. 93.* Molecular compound anthracene-symmetric-trinitrobenzene. Packing of molecules along axis $c$.

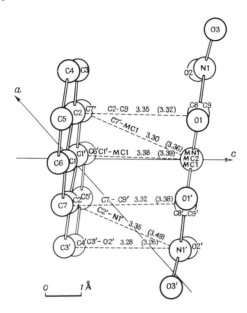

**Fig. 94.** Shortest intermolecular distances between anthracene and symmetric-trinitro-benzene molecules. (In brackets, values at room temperature.)

Similar columns of molecules are found in the crystal containing $p$-iodoaniline instead of anthracene. The crystals of this molecular compound are also monoclinic (space group $P2_1/c$, $Z = 4$), unit cell parameters: $a = 7.43$, $b = 7.39$, $c = 28.3$ Å, $\beta = 103°\ 44'$. The structure is built up from rows of alternating planar molecules of both components; the normals to the planes of the molecules are somewhat tilted with respect to the general direction of the row. The cell contains four rows parallel to the $a$ axis (Fig. 95). There are also other examples of molecular compounds whose molecules are arranged in columns of alternating components, while the columns form close packings of a nearly hexagonal shape.

The recently published work [91] presenting many interesting data and useful references is devoted to detailed investigations of the pyrene–pyromellite dianhydride system. The C ··· C distances between the column molecules are slightly shortened: two distances equal 3.29 Å, i.e., they are only several hundredths of an Angstrom smaller than in graphite. This is a common peculiarity of binary systems of this kind which evidences that the additional interactions are rather weak.

The authors of the publication referred to above mentioned the following interesting phenomenon. There exist such column-type molecular compounds whose repeat period along the column axis is close to 7 Å (twice the thickness

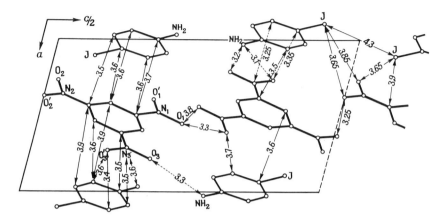

*Fig. 95.* Symmetric-trinitrobenzene + $p$-iodoaniline. Projection $ac$.

of an aromatic molecule); there is also a group of binary systems with repeat period 14 Å. It is shown that the system in question has two modifications, one with repeat period 7 Å, and the other with period 14 Å.

The low-temperature modification (14 Å) changes into the other modification in such a way that the crystal remains intact and the direction of the column axis is preserved unchanged. The authors suppose that a phase transition of the order–disorder type occurs which may be brought about by rotations in their plane of the molecules of one or both components of the given compound. The inference of the authors that similar transitions are sufficiently typical of this class of substances seems fully justified. Most common will probably be the case in which molecules of one species assume a fixed position with respect to the crystal axes, while the molecules of the other component can be in two orientations—ordered at a low temperature and disordered at a high temperature.

In bimolecular crystals in which the polarizing component is 4,4'-dinitrodiphenyl, the component molecules are located at an angle of 90°. Let us give, for example, a more detailed consideration to the very astonishing structure of the molecular compound of 4,4'-dinitrodiphenyl with 4-hydroxy-diphenyl. Unit cell: $a = 20.06$, $b = 9.46$, $c = 11.13$ Å, $\beta = 99°$, $39'$; space group $C_s^3 = Cm$, $Z = 2$ ($6 \cdot O_2NC_6H_4C_6H_4NO_2$ and $2 \cdot C_6H_5C_6H_4OH$).

The 4-hydroxy-diphenyl molecule is planar and positioned in the symmetry plane at $y = \frac{1}{2}$. The 4,4'-dinitrodiphenyl molecules also have the symmetry $m$ in the crystal, but in this case the $m$ is normal to the molecular plane (the molecular plane is here the plane of the coplanar benzene rings of the molecule); the planes of the three molecules run approximately parallel to each other and coincide with plane $20\bar{6}$. (X-ray photographs show diffuse reflections that

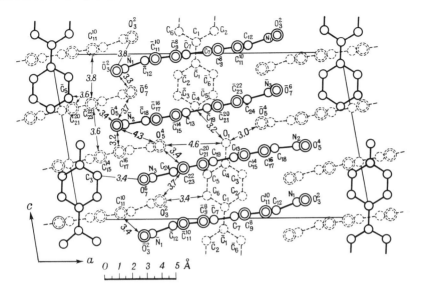

*Fig. 96.* 4,4′-Dinitrodiphenyl + 4-hydroxy-diphenyl. Projection *ac*. Molecules shown by dashed line are positioned above or below solid-line molecules by *b*/2. Double circles, superposing atoms related by symmetry plane.

indicate strong vibrations of the long planar molecules positioned on these planes.)

Figure 96 shows the *ac* projection and Fig. 97, the *ab* projection of the structure of this molecular compound. The structure can be regarded as a face-centered packing of 4,4′-dinitrodiphenyl molecules. This packing has long hollows that also form a face centered pattern and are filled with the hydroxy-diphenyl molecules, with their long axes approximately perpendicular to the planes in which the molecules of 4,4′-dinitrodiphenyl lie. Each hydroxydiphenyl molecule is surrounded by 2 dinitrodiphenyl molecules at about the same distances from the molecule. Thus, it is impossible to single out a "molecular complex" in the structure. All the intermolecular spacings have conventional values and the contacts between molecules of the same element involve the same distances as the contacts between the molecules of both components.

There is no doubt that molecular compounds with some other geometry of mutual arrangement of the component molecules will also be found. Evidently, there are numerous ways to diminish the electrostatic interaction energy.

An important factor must be mentioned: In all these cases the ordinary distances between the atoms of adjacent molecules not connected through valence bonds are not appreciably reduced (a decrease in the C ⋯ C distance by 0.1 Å seems to be the limit).

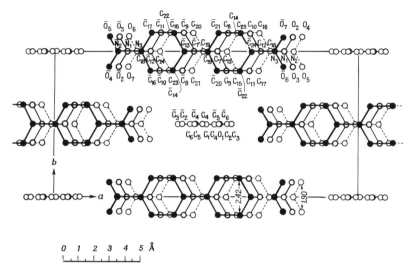

**Fig. 97.** 4,4′-Dinitrodiphenyl+4-hydroxy-diphenyl. Projection *ab*. Designations of molecules: ● $z = \frac{2}{3}$; ○ $z = \frac{1}{3}$; ○ $z = 0$.

The fact that the molecular packing principle, even in such bimolecular crystals which strongly point to additional molecular interaction energy, is no different from the packing principle in monomolecular crystals, was long ago demonstrated by the author together with Frolova [92] by calculating the packing coefficients for bimolecular crystals and components. Table 11 taken from this little-known publication is given below.

So, the main energy component, the sum of the universal atom–atom potentials, appears to remain practically unchanged. As to the formation of a bimolecular crystal, its cause is probably the fact that mixing of dipole molecules with molecules with an induced dipole moment results in a decrease of the electrostatic part of the interaction energy. It seems doubtful that a transfer of the charge would significantly affect the lattice energy.

It would be of interest to verify the above conclusions experimentally, i.e., to make thermodynamical measurements and calculations of lattice energy.

A large group of crystals composed of molecules of two or more elements are the so-called clathrate compounds. Here, hydrogen bonds play a decisive role. Hydrogen-bonded molecules form extremely different framework structures with hollows of different sizes. These hollows may be filled by molecules of other species. The interaction between the molecules filling the hollows with the molecules forming a framework structure can be analyzed in terms of both the geometrical and the physical models.

Thus, the packing coefficient of a small molecule in a framework structure

## Table 11

PACKING COEFFICIENT OF BINARY CRYSTALS

| Compound | | Packing coefficient of bimolecular crystal | Packing coefficient | |
|---|---|---|---|---|
| A | B | | A | B |
| Naphthalene | +trinitrobenzene | 0.72 | 0.70 | 0.64 |
| Naphthalene | +trinitrochlorobenzene | 0.68 | 0.70 | 0.65 |
| Naphthalene | +trinitrophenol | 0.68 | 0.70 | 0.65 |
| Naphthalene | +α-trinitrotoluene | 0.68 | 0.70 | 0.63 |
| Naphthalene | +γ-trinitrotoluene | 0.78 | 0.70 | 0.62 |
| Acenaphthene+trinitrophenol | | 0.68 | 0.70 | 0.65 |
| Acenaphthene+α-trinitrotuluene | | 0.64 | 0.70 | 0.63 |
| Acenaphthene+β-trinitrotoluene | | 0.63 | 0.70 | 0.70 |
| Acenaphthene+γ-trinitrotoluene | | 0.63 | 0.70 | 0.62 |
| Phenanthrene | +trinitrobenzene | 0.71 | 0.68 | 0.64 |
| Phenanthrene | +trinitrochlorobenzene | 0.7 | 0.68 | 0.65 |
| Phenanthrene | +trinitrophenol | 0.56 | 0.68 | 0.65 |
| Phenanthrene | +α-trinitrotoluene | 0.70 | 0.68 | 0.63 |
| Phenanthrene | +β-trinitrotoluene | 0.65 | 0.68 | 0.70 |
| Fluorene | +α-tronitritoluene | 0.67 | 0.69 | 0.63 |

can be calculated. Calculations show that the incorporated molecules do not move apart the "walls of the jail" in which they are confined. Sometimes incorporated molecules are conveniently packed in the cell hollow and make only minor motions about the equilibrium position. In other cases, as is revealed by both X-ray diffraction analysis and thermodynamic studies, an incorporated molecule is in a free rotational state.

By way of example consider the molecular compound of hydroquinone with sulfur dioxide $SO_2$ (unit cell: $a = 16.29$, $c = 5.81$ Å; space group $C_{3v}^2 = R\bar{3}$). It contains three formula units with composition $3 \cdot C_6H_4(OH)_2M$ where M can be the small molecule, in our case $SO_2$, but also $H_2S$, HCOOH, HCN, HCl, HBr, $CH_3CN$, or $CH_3OH$.

In the structure the hydroquinone molecules are connected by hydrogen bonds, 2.75 Å long, and form an infinite three-dimensional skeleton. The six O atoms of six hydroquinone molecules form a plane hexagon parallel to $ab$; the central $O \cdots O$ lines of these six hydroquinone molecules are directed alternately upward and downward from the plane of the hexagon. The angle $\omega$ between the above $O \cdots O$ lines and $ab$ equals 44.5°; the angle between the benzene ring plane and $ac$ has the same magnitude. Figure 98 gives a representation of such a system of hydrogen bonds.

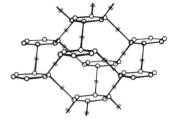

*Fig. 98.* Perspective representation of the hydrogen bond system.

The second O atom of each of the six hydroquinone molecules is included in a similar hexagon and forms a second skeleton parallel to that shown in Fig. 98 but shifted with respect to the former along the *c* axis by half the distance between the two hexagons lying one above the other. Both skeletons are interlaced and penetrate into each other without interfering with intermolecular distances. In this interlaced structure there are three hollows per cell; the hollows being framework structures ("cages") of an approximately spherical shape and diameter 7.5 Å as measured from the centers of the atoms surrounding the hollow. Taking into account intermolecular radii it can be easily seen that a molecule filling a hollow is allowed free space with a diameter about 4 Å, i.e., a volume quite sufficient for accommodating a small molecule.

Figure 99 shows the projection on (110) of the structure of the clathrate compound of hydroquinone with $SO_2$. The size of the hollow prevents free rotation of the $SO_2$ molecule but large thermal motions are possible.

It is interesting to point out that the cages of this skeleton structure are completely closed. A molecule, however small, can neither enter the cage of a formed crystal nor leave it. Therefore, although $SO_2$ is volatile and there are no additional bonds between the hydroquinone and $SO_2$ molecules, they form a stable molecular compound.

Very interesting investigations of clathrate compounds in which water layers surround large molecules, and the crystal hydrate problem in general,

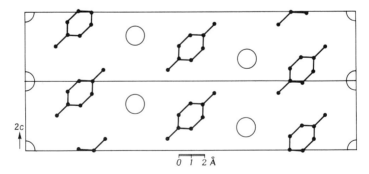

*Fig. 99.* Clathrate compound 3 $C_6H_4(OH)_2 \cdot SO_2$. Projection on (110).

are intimately linked with the clathrate problem and deserve special attention. Interesting and important regularities in this sphere have been established
by Jeffrey *et al.* [93].

We believe that the consideration of this problem will be still more in-
triguing if simple examples are used for studying the applicability of the atom–
atom potential technique for describing hydrogen bonding.

Consider in more detail the structure of a clathrate compound of the
"gaseous hydrate" type. The crystalline hydrate of composition $6,4C_2H_4O \cdot 46H_2O$ [94] displays high symmetry: space group $Pm3n$, unit cell parameter
$a = 12.03$ Å at $-25°C$. The 46 water molecules are hydrogen-bonded into a
skeleton of polyhedrons shown in Fig. 100. The O$\cdots$O distances in the hydrogen
bonds range from 2.766 to 2.844 Å. The ethylene oxide molecules occupy six
equivalent hollows in this skeleton and, judging from the electron-density
distribution, there may be two statistically equiprobable orientations (Fig. 101)
of this molecule on a $\bar{4}$ axis. The structure is a normal crystalline hydrate in
the sense that the atoms of the main skeleton and of the molecule incorporated
into this skeleton are spaced at normal intermolecular distances, the shortest
ones being O$\cdots$O $= 3.18$ Å, O$\cdots$O $= 3.70$ Å, so that we cannot speak about
any specific bond between them.

Both in "gaseous hydrate" structures and in structures of the type of per-
alkylated cation hydrates, e.g., $5(n\text{-}C_4H_9) N^+ F^- \cdot 16H_2O$ [95] the matrix lattices formed by the hydrogen bonds include pentagon-dodecahedrons $H_{40}O_{20}$,
without hydrogen bonds between these lattices and the incorporated molecules
or cations. Crystalline hydrates of another type can also exist [93]; e.g., in the
$12(C_2H_5)_2NH \cdot 104H_2O$ structure the amine is hydrogen-bonded in the
hollows of the water lattice containing polyhedrons with 12 pentagonal and

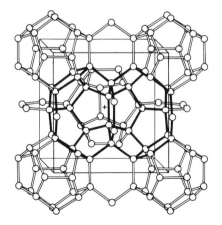

**Fig. 100.** Skeleton of polyhedrons formed by water molecules.

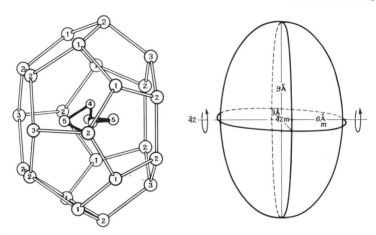

*Fig. 101.* Arrangement of an ethylene oxide molecule in one "hollow" of the skeleton. Symmetry and dimensions of this polyhedron are shown in the right-hand figure.

six hexagonal faces but including no dodecahedrons; in the $(CH_3)_4NOH \cdot 5H_2O$ structure the cations are accommodated in the hollows of a water–anion lattice composed of distorted truncated octahedrons.

The studies of the binary crystals of organic substances are a key for studying intermolecular ineractions. There can be no doubt that investigation in this sphere will grow and intensify.

REFERENCES

1. A. I. Kitaigorodsky, "Organic Chemical Crystallography." Consultants Bureau, New York, 1961.
2. J. Trotter, *Acta Crystallogr.* **13**, 95 (1960).
3. I. N. Streltsova and Yu. T. Struchkov, *Zh. Strukt. Khim.* **2**, 312 (1961).
4. A. I. Kitaigorodsky, *Izv. Akad. Nauk SSSR Ser.* Khim. No. 6, 587 (1946).
5. T. L. Khotzianova, A. I. Kitaigorodsky, and Yu. T. Struchkov, *Zh. Fiz. Khim.* **27**, 647 (1953).
6. W. Nowacki, *Helv. Chim. Acta* **28**, 1233 (1945).
7. *Strukurbericht* 3, S.688 (1900).
8. Z. A. Akopyan, Yu. T. Struchkov, and V. G. Dashevsky, *Zh. Strukt. Khim.* **7**, 408 (1966).
9. W. Bolton, *Acta Crystallogr.* **17**, 147 (1964).
10. A. Bondi, *J. Phys. Chem.* **68**, 441 (1964).
11. A. I. Kitaigorodsky, K. V. Mirskaya, and V. V. Nauchitel, *Kristallografiya* **14**, 900 (1969).
12. V. M. Koshin, *Zh. Strukt. Khim.* **9**, 671 (1968).
13. "Modern Problems of Physical Chemistry." Vol. I. State Univ. Publ., House Moscow, 1968.

14. P. Vaughan, and J. Donohue, *Acta Crystallogr.* **5**, 530 (1952).
15. S. C. Abrahams, J. M. Robertson, and I. G. White, *Acta Crystallogr.* **2**, 233 (1949); D. W. I. Cruickshank, *Acta Crystallogr.* **10**, 504 (1957).
16. A. I. Kitaigorodsky, Arrangement of molecules in organic crystals. Doctoral Thesis, Moscow, 1946.
17. V. C. Sinclair, J. M. Robertson, and A. M.L. Mathieson, *Acta Crystallogr.* **3**, 245, 251 (1950).
18. J. M. Robertson and J. Trotter, *Acta Crystallogr.* **15**, 289 (1962).
19. J. Trotter, *Acta Crystallogr.* **14**, 1135 (1961); **15**, 365 (1962).
20. J. Trotter, *Acta Crystallogr.* **11**, 423 (1958).
21. J. M. Robertson and J. G. White, *J. Chem. Soc.* p. 358 (1947).
22. J. M. Robertson and J. G. White, *J. Chem. Soc.* p. 607 (1945).
23. A. Müller, *Nature (London)* **164**, 1002 (1949); *Proc. Roy. Soc. Ser. A* **207**, 101 (1951).
24. Yu. V. Mnyukh, *Zh. Strukt. Khim.* **1**, 370 (1960).
25. K. S. Pitzer, *Discuss. Faraday Soc.* **10**, 00 (1951).
26. W. J. Taylor, *J. Chem. Phys.* **16**, 257 (1948).
27. H. M. M. Shearer and V. Vand, *Acta Crystallogr.* **9**, 379 (1956).
28. J. R. Brathovde and E. C. Lingafelter, *Acta Crystallogr.* **11**, 729 (1958).
29. A. I. Kitaigorodsky, *Tetrahedron* **9**, 183 (1960)
30. B. K. Vainshtein, and Z. G. Pinsker, *Dokl. Akad. Nauk SSSR* **72**, 53 (1950).
31. A. Müller, *J. Chem. Soc.* **127**, 599 (1925); *Proc. Roy. Soc. Ser. A* **120**, 437 (1928); **127**, 805 (1930); **138**, 514 (1932).
32. A. E. Smith, *J. Chem. Phys.* **21**, 2229 (1953).
33. C. W. Bunn, *Trans. Faraday Soc.* **35**, 482 (1939).
34. V Vand, *Acta Crystallogr.* **4**, 104 (1951).
35. A. Müller and K. Lonsdale, *Acta Crystallogr.* **1**, 129 (1948).
36. S. H. Piper *et al.*, *J. Chem. Soc.* **127**, 2194 (1925); *Biochem. J.* **25**, 2072 (1931).
37. A. A. Schaerer, G. G. Bayle, and W. H. Mazel, *Rec. Trav. Chim. Pays-Bas* **75**, 513 (1956).
38. T. Schoon, *Z. Phys. Chem. Abt. B* **39**, 385 (1938).
39. S. S. Kabalkina, *Dokl. Akad. Nauk SSSR* **125**, 114 (1959).
40. J. D Dunitz, L. E. Orgel, and A. Rich, *Acta Crystallogr.* **3**, 373 (1956).
41. Z. L. Kaluski, Yu. T. Struchkov, and R. L. Avoyan, *Zh. Strukt. Khim.* **5**, 743 (1964).
42. J. Donohue and A. Caron, *Acta Crystallogr.* **17**, 663 (1964).
43. H. M. Powell and R. N. G. Ewens, *J. Chem. Soc.* p. 286 (1939).
44. A. R. Luxmoore and M. R. Truter, *Acta Crystallogr.* **15**, 1117 (1962).
45. H. T. Sumsion and D. McLachlan, *Acta Crystallogr.* **3**, 217 (1950).
46. G. S. Zhdanov and I. G. Ismail-Zade, *Dokl. Akad. Nauk SSSR* **68**, 95 (1949).
47. D. C. Phillips, *Acta Crystallogr.* **7**, 649 (1954); F. H. Herbstein, *Ibid.* **8**, 401 (1955).
48. F. H. Herbstein and G. M. J. Schmidt, *Acta Crystallogr.* **8**, 399 (1955).
49. J. Housty and J. Clastre, *Acta Crystallogr.* **10**, 695 (1957).
50. E. Frasson, C. Garbuglio, and S. Bezzi, *Acta Crystallogr.* **12**, 126 (1959).
51. Yu. V. Mnyukh, *J. Phys. Chem. Solids* **24**, 631 (1963).
52. E. G. Cox, M. W. Dougill, and G. A. Jeffrey, *J. Chem. Soc.* p. 4854 (1952).
53. J. M. Robertson, *Proc. Roy. Soc. Ser. A* **157**, 79 (1936).
54. J. M. Robertson, *Proc. Roy. Soc. Ser. A* **167**, 122 (1938).
55. W. P. Binne and J. M. Robertson, *Acta Crystallogr.* **3**, 424 (1950).

56. G. V. Gurskaya, "Amino Acid Structures." Nauka Publ. House, Moscow, 1966.
57. J. M. Robertson, H. M. M. Shearer, G. A. Sim, and D. G. Watson, *Acta Crystallogr.* **15**, 1 (1962).
58. T. C. W. Mak and J. Trotter, *Acta Crystallogr.* **15**, 1078 (1962).
59. T. L. Khotzianova, T. A. Babushkina, and G. K. Semin, *Zh. Strukt. Khim.* **7**, 634 (1966); T. L. Khotzianova, V. I. Robas, and G. K. Semin, "Solid-State Radiospectroscopy," p. 233. Atomizdat, Moscow, 1967.
60. M. Eveno and I. Meinnel, *J. Chim. Phys. Physiochim. Biol.* **63**. 108 (1966); Y. Balcon, and I. Meinnel, *Ibid.* **63**, 114 (1966); B. Lassier and C. Brot, *Ibid.* **65**, 1723 (1968).
61. D. Fox, M. M. Labes, and A. Weissberger, eds., "Physics and Chemistry of the Organic Solid State," Vol. I. Wiley (Interscience), New York, 1963.
62. D. A. Badalyan, A. G. Khachaturyan, and A. I. Kitaigorodsky, *Kristallografiya* **14**, 404 (1969). 62a. D. A. Badalyan, *Kristollografiya* **14**, 48 (1969).
63. A. I. Kitaigorodsky, *Dokl. Akad. Nauk SSSR* **113**, 604 (1957).
64. S. A. Remyga, R. M. Myasnikova, and A. I. Kitaigorodsky, *Kristallografiya* **12**, 900 (1967).
65. I. W. Harris, *Nature (London)* **206**, 1038 (1965).
66. S. N. Stivastava, *Acta Crystallogr.* **17**, 851 (1964).
67. A. I. Kitaigorodsky, *Zh. Strukt. Khim.* **1**, 324 (1960).
68. A. I. Kitaigorodsky, R. M. Myasnikova, and V. D. Samarskaya, *Kristallografiya* **8**, 393 (1963).
69. A. S. Carson, D. R. Stranks, and B. R. Wilmshurst, *Proc. Roy. Soc. Ser. A* **244**, 72 (1958); K. L. Wolf and H. Weghofer, *Z. Phys. Chem. Abt. B* **39**, 194 (1938).
70. B. D. Koreshkov, Thesis, Moscow, 1968.
71. A. I. Kitaigorodsky and K. V. Mirskaya, *Kristallografiya* **9**, 174 (1964).
72. G. W. Frank, R. M. Myasnikova, and A. I. Kitaigorodsky, *Kristallografiya* To be published.
73. V. D. Samarskaya, R. M. Myasnikova, and A. I. Kitaigorodsky, *Kristallografiya* **13**, 616 (1968).
74. L. Kofler and A. Kofler, "Thermo-Micro-Methoden." 1954.
75. S. A. Remyga, R. M. Myasnikova, and A. I. Kitaigorodsky, *Kristallografiya* **10**, 875 (1965).
76. D. C. Phillips, F. R. Ahmed, and W. H. Barnes, *Acta Crystallogr.* **13**, 365 (1960).
77. R. M. Myasnikova and A. I. Kitaigorodsky, *Kristallografiya* **3**, 160 (1958).
78. N. Ya. Kolosov, *Kristallografiya* **3**, 700 (1958).
79. N. Ya. Kolosov, *Zh. Strukt. Khim.* **6**, 926 (1965).
80. A. I. Kitaigorodsky and L. Dun-chai, *Kristallografiya* **5**, 238 (1960).
81. V. D. Samaraskaya, R. M. Myasnikova, and A. I. Kitaigorodsky, *Kristallografiya* **11**, 855 (1966).
82. G. S. Belikova and L. M. Belyaev, "Crystal Growth," Vol. III. Akademizdat, Moscow, 1961.
83. E. V. Shpolskij, *Usp. Fiz. Nauk* **68**, 51 (1959); **71**, 215 (1960); **77**, 321 (1962); **80**, 255 (1963).
84. T. N. Bolotnikova, L. A. Klimova, G. N. Nercesova, and L. F. Utkina, *Opt. Spektrosk.* **21**, 420 (1966).
85. E. I. Fedin and A. I. Kitaigorodsky, *Kristallografiya* **6**, 406 (1961).
86. A. I. Kitaigorodsky and Yu. V. Mnyukh, *Vysokomol. Soedin.* **1**, 128 (1959).
87. D. A. Badalyan and A. G. Khachaturyan, *Fiz. Tverd. Tela.* **12**, 439 (1970).
88. J. Timmermans, "Les Solutions Concentrées." Paris, 1936.
89. S. P. McGlynn and J. D. Boggus, *J. Amer. Chem. Soc.* **80**, 5096 (1958).

90. D. S. Brown, S. C. Wallwork, and A. Willson, *Acta Crystallogr.* **17**, 168 (1964).
91. F. H. Herbstein and I. A. Snyman, *Phil. Trans. Roy. Soc. London Ser. A* **264**, 635 (1969).
92. A. I. Kitaigorodsky and A. A. Frolova, *Izv. Sekt. Fiz. Khim. Anal. Inst. Obshch. Neorg. Khim. Acad. Nauk SSSR* **19**, 306 (1949).
93. G. A. Jeffrey, T. Jordar, and R. K. McMullan, Abstracts of papers. *Int. Crystallogr. Congr., 7th, Moscow, 1966*, **8**, p. 54.
94. R. K. McMullan and G. A. Jeffrey, *J. Chem. Phys.* **42**, 2725 (1965).
95. R. K. McMullan, M. Bonamico, and G. A. Jeffrey, *J. Chem. Phys.* **30**, 3295 (1963).

# Chapter II
# Lattice Energy

## A. Interactions of Molecules

1. VAN DER WAALS FORCES

The interactions between uncharged atoms and molecules, in the absence of an appreciable exchange of electrons, consist of weak forces which are generally known as van der Waals forces. In most organic crystals the molecules are held together by these forces.

The structures of organic crystals in which the van der Waals forces operate is characterized by close packing. When in the solid state, the inert elements form crystals of a closely packed cubic or hexagonal structure.

The heats of sublimation of such crystals are very small as compared with those of ionic and valency crystals. The heats of sublimation of the inert gases Ne, Ar, Kr, and Xe are 0.6, 2.0, 2.8, and 3.9 kcal/mole, respectively [1]. For organic substances, the heats of sublimation are also very low, of the order of 10–20 kcal/mole. Even in the case of crystals with very large molecules, the heat of sublimation is only about 40–50 kcal/mole (e.g., 37.5 and 53.5 kcal/mole for 9,10-diphenyl anthracene $C_{26}H_{18}$ and violanthrene $C_{34}H_{18}$, respectively). A large number of data on the heats of sublimation of organic substances may be found in a review by Westrum and McCullough [2] (see also p. 335).

*a. London's Theory of van der Waals Forces*

The origin of the van der Waals forces was not completely understood for a long time although information about this type of interaction was not difficult to obtain by studying such simple physical phenomena as, for example, the capillarity in nonpolar liquids and the nonideal behavior of real gases.

Some simple ideas about the nature of the van der Waals forces were developed long before the appearance of a rigorous theory (concerning this subject see Margenau's review [3]). The interaction between polar molecules was partly explained in the theories of the Keesom orientation effect, also called the alignment effect (the electrostatic attraction of permanent multipoles) and the Debye induction effect (caused by the polarizability of molecules). The principal effect, which consists of the mutual attraction between neutral particles, independently of their polarity, still needed explanation. A strict and consistent interpretation of the van der Waals forces first appeared in the works of London [4] who reduced them to electrical interactions.

The physical meaning of the attraction of neutral particles separated from one another by large distances may be described as follows. Even in such atoms or molecules, whose electrical multipole moments are on the average equal to zero, there exist certain fluctuating multipole moments which depend on the instantaneous positions of the moving electrons in the atoms. The instantaneous electric field associated with these moments leads to the appearance of induced multipole moments in neighboring atoms. The averaged interaction between the electrical moments of the initial atom or molecule and the induced moments of neighboring atoms or molecules gives rise to attractive forces between the particles. London has shown that the dominating part of this effect is due to the outer, most loosely bound electrons—the same electrons that are responsible for optical dispersion. This explains why the effect of interplay between electrically neutral particles was called the dispersion effect. This term is now generally accepted.

*The Interaction of a Pair of Particles.* London has derived a rigorous formula for the energy of the dispersion interaction of two molecules on the basis of the quantum-mechanical theory of second-order perturbations. The two interacting molecules are regarded as quasielastic oscillators having an energy different from zero even in their lowest energy states (the quantum-mechanical zero-point energy). If the electrical potential energy of the interaction of the two molecules is expanded in a multipole series with respect to the coordinates of the charge elements and only the dipole–dipole term used, the interaction energy averaged over various orientations of the molecules with respect to the vector $\mathbf{R}$ joining them varies with the inverse sixth power of the distance $R$ between the centers of the two molecules:

$$U(R) = -A/R^6$$

where $A$ is a constant given by

$$A = \frac{2}{3} {\sum_{i'}}' {\sum_{j'}}' \frac{|\langle i|\mu_1|i'\rangle|^2 |\langle j|\mu_2|j'\rangle|^2}{E_{i'}+E_{j'}-E_i-E_j}$$

Here we designate the ground states of the unperturbed molecules by $i, j$ and the excited states by $i', j'$ respectively; $E_i, E_j, E_{i'}, E_{j'}$ are eigenvalues of the molecules, and $\langle i|\mu_1|i'\rangle$, $\langle j|\mu_2|j'\rangle$ are the matrix elements of the dipoles between the states $i, i'$ in molecule 1 and between the states $j, j'$ in molecule 2. The primes on the summation signs exclude $i' = i$ and $j' = j$. We see that the energy $U(R)$ is negative. Hence the corresponding van der Waals force is attractive.

In addition to the rigorous formula, which is too complicated for actual calculations, London proposed an approximate formula correlating the van der Waals force with the polarizabilities of the molecules ($\alpha_1, \alpha_2$) and the ionization potentials ($I_1$ and $I_2$):

$$E_{\text{dip-dip}} = -\frac{3}{2}\alpha_1\alpha_2 \frac{I_1 I_2}{I_1+I_2} \frac{1}{R^6}$$

Slater and Kirkwood [5] obtained a similar relationship for long-range interactions by using the variational method. This method does not possess the same generality as the London perturbation method, but it is applicable to many nonpolar molecules in the normal state. The Slater–Kirkwood formula has the form

$$E_{\text{dip-dip}} = -\frac{1}{R^6} \frac{3eh}{8\pi} \left(\frac{N\alpha^3}{m}\right)^{1/2}$$

where $N$ is the number of electrons in the outer shell, and $\alpha$ the polarizability of the atom (or molecule). The interacting systems are assumed to be spherically symmetrical.

The agreement between these two formulas is not always good; it is, however, difficult to give preference to either one of them since both contain errors caused by approximations; these errors cannot be estimated.

Similar considerations of the dipole–quadrupole and quadrupole–quadrupole interactions lead to additional van der Waals potentials $-A'/R^8$, $-A''/R^{10}$ varying, respectively, with the inverse eighth and tenth powers of the distance. From estimates of the constants $A'$ and $A''$ by Margenau it follows that even for nearest neighbors the dipole–quadrupole potential is only a relatively small fraction of the dipole–dipole potential. For molecules farther apart, all terms in the interaction potential other than the dipole–dipole term rapidly become insignificant. For instance, let us regard the $CH_4$ molecule as a center of attraction: If the distance between the centers of the molecules were 1.66 Å, the dipole–dipole and dipole–quadrupole energies would be equal; but it is known that the actual distances between nearest neighboring molecules are never less than about 4 Å.

The expression for the van der Waals force appearing in the second-order approximation of the perturbation theory is quite satisfactory for the case of atoms (molecules) farther apart. For distances comparable with the size of particles, the multipole expansion proves to be less rigorous, although in this case too the potential of the dipole–dipole interaction probably gives the correct order of magnitude of the full interaction. When the particles approach each other to a distance smaller than the equilibrium distance, the electron shells begin to overlap one another, which gives rise to a strong repulsion (overlap force). These additional repulsive forces due to the overlap are not considered in London's theory. There is as yet no simple theoretical expression for calculating the repulsive energy, and therefore for the van der Waals interactions, as well as for the short-range interactions of ions, one has to resort to simple representations of the energy of repulsion in the form of inverse power $(B/R^n)$ or exponential $(ae^{-R/\rho})$ dependences. The constants entering into these expressions must be determined empirically.

*The Interaction of a Group of Particles and the Additive Rule.* When London applied his method to a group of molecules, he obtained the sum of the energies of interaction between the molecules taken in pairs. In other words, in the second-order perturbation the dispersion effect possesses the property of additivity.

It was shown later that by carrying on the perturbation procedure for nonoverlapping molecules to an approximation of order higher than the second, the strict additivity of the van der Waals forces vanishes. The dispersion forces in that third-order perturbation for three particles were first deduced by Axilrod and Teller [6]. These authors made an attempt to determine the role of the nonadditivity effect in calculations of the conditions for stability of the crystal lattices of rigid inert gases and of the third virial coefficient of gases. This effect appeared to be negligible. Similar results for rigid inert gases were obtained by Jansen [7], who, by introducing a number of simplifying assumptions, estimated the nonadditivity effect in the region of overlap of electron clouds.

A somewhat more rigorous calculation of the nonadditivity for small distances has been carried out by Wojtala [8]. From the results of his estimate for hydrogen atoms it follows that the nonadditive contribution to the repulsion potential would be great if the particles were separated by distances of the order of 0.5–1 Å (for a group of particles separated by a distance of 1 Å, this contribution would constitute up to 20%). In real liquids and solids the distances between the hydrogen atoms of neighboring molecules are nerve less than 2 Å. For such distances, the rapidly decreasing nonadditive contribution reduces to zero.

A greater effect may be expected for three bodies when one of these is of a very large size. The presence of a surface in the immediate vicinity of the two interacting molecules can change the energy of their interaction so that either a stronger attraction or a stronger repulsion may occur.

*b. The Potential Formulas Including Repulsion between Particles*

The first field of application of the London theory of intermolecular forces was an investigation into the equation of state for an ideal gas. As is known, the second virial coefficient appearing in this equation depends on the potential of interaction of a pair of gas particles. Calculation of the second virial coefficient and comparison of the results obtained with the experimental data have made it possible to establish the shape of the potential.

As far back as 1903 Mie [9] suggested writing the interaction energy shared between two atoms as the sum of two terms, repulsion $\Phi_1$ and attraction $\Phi_2$, varying inversely with distance. Later on the following formula was proposed:

$$\Phi = -\frac{A}{r^m} + \frac{B}{r^n} \qquad (1.1)$$

where $n > m > 0$. This form was called the $(m:n)$ interaction for short.

In subsequent years other analytic expressions were also proposed. They are often written in the form

$$\Phi = \varepsilon u(z), \qquad z = r/r_m$$

where $\varepsilon$ and $r_m$ are convenient "scale" factors, and the function $u(z)$ describes, in fact, the shape of the curve and can itself contain one or several parameters.

After the works of London the exponent $m$ in Eq. (1.1) was taken equal to 6. Lennard-Jones [10] has shown that if the potential curve is taken in the form of $6:n$ and the second virial coefficients of gases are calculated with the help of that curve, good agreement with experiment can then be secured for a wide range of the values of $n$ (from 8 to 15). Buckingham [11] tried to use the potential in the form of 6-exp (the exponential law for the repulsive potential), but he also found that the sensitivity of the second virial coefficient to the shape of the potential curves is low.[†] This was also confirmed by the results of other later works.

---

[†] The potential curve in the form of $6:n$ is generally known as the Lennard-Jones potential, and that in the form of 6-exp, the Buckingham potential.

Attempts have been made to apply the theory of the van der Waals forces to the solid state for the purpose of obtaining additional data on the form of the potentials. This makes it possible, first, to investigate the nature of the interaction between particles at so-called intermediate distances and, secondly, to employ new experimental quantities, mainly the lattice parameters and the heat of sublimation of crystals, for comparing theory with experiment.

*c.  Early Applications of the Theory to Molecular Crystals*

The fact that the theory of van der Waals forces was first applied to the simplest crystals possible, namely crystals consisting of atoms of the rare gases, is quite natural. These crystals are highly symmetric, and the atoms situated at their lattice points satisfy, with great accuracy, the requirements of London's theory. By the way, such crystals continue to draw attention up to the present time.

Approximate estimates of the cohesive energies of rare gas crystals had already been carried out by London, who had in mind obtaining only a correct order of magnitude.

The first consistent work devoted to the description of the behavior of crystals over a wide temperature range with the aid of atomic interaction curves was performed by Rice [12]. He wrote down the full energy of an argon crystal in the form of a sum of the attraction energy

$$U_{\text{attr}} = -c_6 r^{-6} - c_8 r^{-8}$$

calculated by him with the help of the lattice sums of Lennard-Jones and Ingham [13], the repulsion energy

$$U_{\text{rep}} = U_0 + \sum_{i=2,3,4} b_i (r - r_0)^i$$

taken into account only for the 12 nearest neighbors, a term depending on the temperature change of the lattice parameters, and a term taking account of the energy of zero-point vibrations. To describe his potential Rice had to choose five constants $(c_6, c_8, b_2, b_3, b_4)$. As a result, he obtained a curve that described satisfactorily the then known properties of the solid phase of argon within the temperature range from $0°K$ to the melting point, and also the second virial coefficient for the gaseous phase.

As new experimental data accumulate, potential curves taking account of the properties of the solid phase of the rare gases are made more exact by various authors using various modifications of the procedures commonly employed and lead to quite reasonable results over rather wide limits.

Crystals composed of di-, tri-, and polyatomic molecules bound together by van der Waals forces have been studied but little.

As we have seen, the London theory and all the succeeding supplements to it are based on the assumption of central forces. London himself has applied his theory not only to crystals built up of atoms of the rare gases, but also to the simplest molecular crystals ($N_2$, $O_2$, $CO_2$, $CH_4$, NO), assuming that each molecule is a point center of attraction. The central force model has been applied to the interaction between the molecules $N_2$, $H_2$, $O_2$, and $CH_4$ by Slater and Kirkwood [5].

It was, however, clear that this approximation is applicable only in the simplest cases. The interaction energy even for $H_2$ molecules depends on mutual orientation; the orientation effect may amount to 25% of the mean energy. A very much greater effect may be expected in the case of short distances where the interaction is extremely sensitive to the degree of overlapping of electron clouds.

All these limitations are evidently present, to a very great extent, in crystals in which the molecules are in fixed positions and in close contact with one another.

As far as a molecular organic crystal is concerned, the consideration of a molecule as a point center is obviously completely meaningless.

2. CALCULATION OF THE LATTICE ENERGY OF MOLECULAR CRYSTALS

A rigorous approach to the problem should consist in investigating the Schrödinger wave equation which is obtained by substituting the Hamiltonian of the system of interacting particles into the general wave equation

$$i\hbar\,\partial\Psi/\partial t = \hat{H}\Psi, \qquad \hat{H} = \hat{H}_0 + \hat{U}(r_1, r_2, \ldots) \qquad (2.1)$$

The first term is the sum of the Hamiltonians of the individual molecules, and the second term is the potential energy of their interaction in the crystal lattice.

Molecular crystals possess certain specific features which facilitate, to some extent, their theoretical consideration. In view of the weak binding forces between the particles, one may regard the potential energy in the Hamiltonian (2.1) as a small correction to the operator $H_0$ and use the quantum-mechanical theory of perturbations in order to solve the Schrödinger equation.

In the general case of nonsymmetrical molecules the desired correction to the energy of an unperturbed system appears already in the first approximation of the perturbation theory and is determined by the mean values of the multipole moments of interacting particles. If the dipole or quadrupole moments of the molecules are different from zero, the interaction energy then decreases, respectively, with the third $(1/R^3)$ or the fifth $(1/R^5)$ power of the distance and depends on the relative orientation of the molecules.

If the interacting particles do not exhibit permanent electrical moments, their interaction energy in the first approximation of the perturbation theory vanishes. In this case a nonzero result, which is quadratic with respect to the matrix elements of the potential energy operator, appears in the second order of the perturbation theory. Here the term most slowly decreasing with distance in the expression for the energy of interaction between the particles is proportional to $1/R^6$.

At shorter distances, when the electron clouds of the atoms comprising the molecules overlap, the first approximation of the perturbation theory leads, by the exclusion principle, to a strong repulsive interaction which depends on the symmetry of the interacting particles. For spherically symmetrical particles at short distances, the quantum theory leads to a decrease of the energy with distance according to an exponential law.

What are the principal terms? Computation of the energy of the electrostatic interactions in a crystal lattice, taking account of the multipole moments of the molecules, does not yield even approximately, as we shall see below, the values of the cohesive energy that are observed in experiment; this energy appears to be extremely small. Thus the major role in the cohesive energy of molecular crystals must be played by the effects of the second-order approximation of the perturbation theory for large distances and by the repulsion due to the overlap of electron clouds for short distances; we call these interactions the van der Waals interactions.[†] We shall see below to what extent this supposition is justifiable.

The methods of perturbation theory appear to be of little use for any detailed calculation aiming at correlating the lattice energy with crystal structure. Significant progress becomes impossible without information about the potential energy of interaction.

Interpretation of the potential energy of interaction between molecules is realistic only on the basis of the adiabatic approximation introduced into quantum mechanics by Born and Oppenheimer.

These authors have shown, first, that the Schrödinger equation for a system of $N$ atomic nuclei and $n$ electrons can be transformed to the form

$$\left(H_{\text{nuc}}^k + E_\alpha^{\text{elec}} + \sum_k \frac{P_{k,\alpha\alpha}^2}{2M_k}\right)\psi_\alpha + \sum_{k,\alpha'(\neq\alpha)}\left(\frac{P_{k,\alpha\alpha'}^2}{2M_k} + \frac{P_{k,\alpha\alpha'}P_k}{M_k}\right)\psi_{\alpha'} = i\hbar\dot{\psi}_\alpha$$

where $H_{\text{nuc}}^k$ is the kinetic energy of the nuclei; $E_\alpha^{\text{elec}}$ is the eigenvalue of the electronic state for a system of nuclei with the coordinates $X_k$ ($k=1,...,N$); $\psi_\alpha$ is the wave function of the nuclear coordinates; $P_{k,\alpha\alpha'}^2$ and $P_{k,\alpha\alpha'}$, are the matrix elements of the relatively intrinsic electronic functions.

---

† It should be noted that some authors use the term van der Waals forces to denote jointly the electrostatic interactions, dispersion forces, and repulsive forces.

The adiabatic approximation consists in neglecting the second term of the left-hand side of the equality. Such simplification is permissible if we ignore the connection between the states of the nuclei belonging to different excited electronic states.

Further, if, because of the large mass of the nuclei, we also ignore the sum in the first parentheses (as compared with $E_\alpha$), the Schrödinger equation then assumes the form

$$(H^k_{\text{nuc}} + E^{\text{elec}}_\alpha)\psi_\alpha = i\hbar\dot{\psi}_\alpha$$

Thus the electron–nuclear system is transformed into an atomic one. The role of the potential energy of interaction $\Phi(X_k)$ is played by $E_\alpha$. In the case of the ground state of the electronic configuration

$$\Phi(X_k) = E^{\text{elec}}_0(X_k)$$

i.e., the potential interaction energy of the atoms is equal to the eigenvalue of the lowest electronic state.

There exist theoretical calculations showing that for dielectrics in which the energy difference between the zeroth and first states is of the order of 1 eV, the approximation holds quite well.

The adiabatic approximation allows us to "forget" about electrons and to reduce the problem to atomic interactions. The configuration of the potential $\Phi_k$ however does not follow from theory. Further steps along this line can be made only with the aid of new simplifications and additional hypotheses.

The fact that the molecules in an organic crystal form close packings and that in this case the atoms of each molecule tend to arrange themselves between the atoms of the neighboring molecules and as close to them as possible, suggests the idea of additivity of the interaction energy not only with respect to the interaction between the members of each pair of molecules, but also with respect to the interactions of atoms with one another.

The possibility of representing the energy of interaction of molecules as the sum of the interactions of their constituent atoms will of course be of heuristic value if it appears that an energy increment is characteristic of two species of atoms independently of what molecules these atoms are part of. Indeed, in this case the main peculiarity of organic compounds—that enormous classes of compounds, such as, for example, the hydrocarbons, are made up of only two species of atoms—makes it possible to predict the properties and structures of thousands of substances with the help of empirical constants determined on the basis of the analysis of the structure and the properties of a few reference substances.

The increment of interaction must of course be found from a curve of the atom–atom potential rather than by stating a number. The principal question to be answered is the degree of universality of this curve. This approach would be of little value if these curves were essentially different, for example, for each hydrocarbon. The proof of the universality of such curves would already be of sufficient interest for aliphatic hydrocarbons, aromatic hydrocarbons, etc. And, finally, of greatest interest would be a result showing that the structure and properties of all hydrocarbons could be described, with reasonable accuracy, with the aid of identical curves.

May we hope that in calculations of the interaction energy, the peculiarity in the electron distribution and the absence of the spherical symmetry of the electron distribution in the atom which occurs strictly speaking, in an organic molecule, may prove unimportant in computing the interaction energy? It is difficult to give an answer to this question beforehand, and the final solution can be provided only by experiment. However, two circumstances are encouraging as regards the solution of this question. Precision X-ray structure studies show that excellent agreement with experiment can be secured by calculations using isotropic atomic factors of X-ray scattering, and that the patterns of electron density can be represented, with experimental accuracy, as the superposition of spherically symmetrical atoms. And, further, the residual electrical charges at the atoms which result in a molecule's possessing a certain electrostatic multipole moment, lead to electrostatic energies of interaction negligible as compared with the heats of sublimation of such substances as the hydrocarbons, and to relatively small values even in the cases of molecules with large dipole moments. The latter circumstance, however, is not obvious and must be proved by straightforward calculations.

On the basis of the foregoing qualitative considerations we have proposed the following model of the potential energy of interaction between molecules:

1. The energy of interaction between molecules is equal to the sum of the interactions of the constituent atoms (additivity).

2. Central forces operate between atoms. (A somewhat more complicated model taking account of the ellipticity of the atoms may be introduced, if necessary.)

3. The interaction potential of two atoms $\varphi_{ik}$ can be broken up into two terms

$$\varphi_{ik} = u_{ik} + \varphi_{ik}^{\text{electrost}}$$

The term $u_{ik}$ takes account of the forces of repulsion and dispersion attraction.

4. The potentials $u_{ik}$ are universal, i.e., depend only on the species of the atoms, no matter what molecule they are part of and what their valence states may be.

5. For the $u_{ik}$, we may adopt various analytical expressions of the type of

6-exp or $6:n$ potentials with arbitrary parameters which must be determined by experiment. The most direct verification of the method is likely to be provided by the calculation of the lattice energy of a crystal as a function of the lattice parameters. The coordinates of the minimum of such an energy surface give the values of the heat of sublimation and of the structure's parameters at the absolute zero of temperature. This method can be rather convincingly verified by establishing the fact that the real structure at the absolute zero of temperature corresponds to a minimum on the energy surface. It is also natural that it is this method that can be employed for selecting the best atom–atom potential curves, by the use of which we can proceed, if we are successful, to the calculation of the properties of crystals.

## B. Electrostatic Energy

### 3. The Dipole–Dipole Interaction in a Molecular Crystal[†]

In expanding the intermolecular potential energy for electrically neutral molecules into a series with respect to the coordinates of the charge elements, the terms with the lowest nonvanishing multipole moment must be most important. In a lattice consisting of dipole molecules, the contribution of electrostatics to the lattice energy is determined mainly by the dipole–dipole interactions.

The energy of interaction of two identical point dipoles with moments $\mu$ joined by the vector $\mathbf{r}_{12}$ and having directions $\mathbf{S}_1$ and $\mathbf{S}_2$ is equal to

$$E_\mu = \mu^2 \left( \frac{\mathbf{S}_1 \cdot \mathbf{S}_2}{|r_{12}|^3} - \frac{3(\mathbf{S}_1 \cdot \mathbf{r}_{12})(\mathbf{S}_2 \cdot \mathbf{r}_{12})}{(r_{12})^5} \right) \tag{3.1}$$

Taking the dipole $\mu \mathbf{S}_1$ as the origin of coordinates, we find its energy in the field of the dipole lattice as the sum of the interactions with all the other dipoles of the lattice

$$E_\mu = \frac{\mu^2}{2} \sum_{n=2}^{\infty} \left[ \frac{\mathbf{S}_1 \cdot \mathbf{S}_n}{|r_{1n}|^3} - \frac{3(\mathbf{S}_1 \cdot \mathbf{r}_{1n})(\mathbf{S}_n \cdot \mathbf{r}_{1n})}{(r_{1n})^5} \right] \tag{3.2}$$

The sum (3.2) is a conditionally convergent alternating infinite series. The magnitude of the sum and the rapidity of convergence of the series depend on the choice of the order of summation.

The summation of the series (3.2) can be performed most simply in the general case for lattices of cubic symmetry. The procedures for calculating

---

[†] The problems of this section and Section 4 are treated in Kitaigorodsky and Mirskaya [14].

these sums for dipole and quadrupole cubic lattices have been discussed earlier by Kornfeld who accepted the order of summation proposed by Ewald [15] for ionic lattices. However, as has been shown by Campbell [16], Kornfeld's reasoning in carrying over the summation method from ionic lattices to multipole ones contains an error in logic leading to wrong results. In particular, Kornfeld came to the conclusion that the energy of a simple cubic dipole lattice is different from zero, whereas it is, in fact, strictly equal to zero (see below).

Calculation of lattice sums for molecular crystals in a general form is very difficult because of the large diversity in the modes of symmetrical arrangement of the molecules in the lattice, and in the ratios of the parameters describing the unit cell. As a rule, computations of the multipole energy are made for concrete crystals and are performed according to general formulas (3.1) and (3.2) with the aid of electronic computers [17, 18]. An attempt at a more general consideration of this problem has recently been made by De Wette and Schacher [19]. Here, in each particular case, one has to find the whole set of distances and the orientations of the dipoles at all the lattice points.

However, nothing can be said beforehand about the set of distances in an arbitrary lattice. As regards the relative orientations of the dipoles, we know that they obey quite definite laws; namely, they depend on the space group of the symmetry of a crystal. The question of the dependence of the dipole energy of a crystal lattice on the orientation of the dipoles can therefore be stated in a general form.

In an attempt to solve this problem we shall compare different dipole lattices, in which the centers of the dipoles are arranged identically, namely, according to the mode of spherical close packing. Keeping the mutual arrangement of the centers unchanged, we assign different space symmetries to the arrangements of dipole vectors. Such considerations make it possible to obtain for each space group an analytic dependence of the electrostatic lattice energy on the orientation of the dipole vector taken at the origin of the coordinates. The resulting analytic dependences can be used to construct polar maps that will give a graphic representation of the orientation dependence of the dipole energy, including the regions of the most favorable orientations of the molecules in the lattice from the viewpoint of its electrostatic energy. A comparison of the results obtained with real structures will show the contribution of the electrostatic interactions to the establishment of the equilibrium structure of the crystal. Moreover, by comparing the analytical expressions with the polar maps for different space groups, it is possible to elucidate the effect of the symmetry of a crystal lattice on its electrostatic energy in "pure form."

We shall limit ourselves (if necessary, this limitation can easily be eliminated and the suggested procedure of calculation can be worked out in a completely general form) to a symmetry of arrangement of dipole vectors not higher than the orthorhombic one and consider cells in which there is only one independent vector. In this case all the possible symmetric arrangements of dipole vectors can be obtained on the basis of the two simplest closest packings—cubic and hexagonal.

For lattices built up according to the principle of close packing of spheres we may suggest a summation procedure which, making successful use of the symmetry inherent in such lattices, turns out to be rather simple and provides good convergence of the series (3.2.)

Let us consider a rectangular lattice with periods $a(3)^{1/2}/6$ along the $x$ axis, $a/2$ along the $y$ axis and $a(2/3)^{1/2}$ along the $z$ axis ($a$ is the shortest distance between the centers of the dipoles). We denote the lattice points by the subscripts $m$, $n$, and $p$; the radius vector of any lattice point will be equal to $\mathbf{r} = a[(3^{1/2}/6)m\mathbf{i} + (1/2)n\mathbf{j} + (2/3)^{1/2}p\mathbf{k}]$. The length of the vector is $r = [a/(12)^{1/2}](m^2 + 3n^2 + 8p^2)^{1/2}$. The dimensionless ratio $r/a$ is designated by $R$.

By filling some of the points of this lattice we can obtain closest sphere packings with any recurrence along the $z$ axis (dense layers are formed in the $xy$ plane). The closest hexagonal (two-layer) packing (Fig. 1a) is formed by filling even layers ($p = 0, 2, 4, ...$) according to the pattern $m$ as a multiple of 6 for $n$ even and of 3 for $n$ odd, and odd layers ($p = 1, 3, 5$) according to the pattern $m = -2, 4, 10, ...$ for $n$ even and $m = 1, 7, 13, ...$ for $n$ odd. The closest cubic (three-layer) packing (Fig. 1b) arises from the filling of layers with $p = 0, 3, 6, ...$ according to the same mode as the even layers in the hexagonal close-packed structure, of layers with $p = 1, 4, 7, ...$ according to the same mode as the odd layers in the hexagonal packing, and of layers with

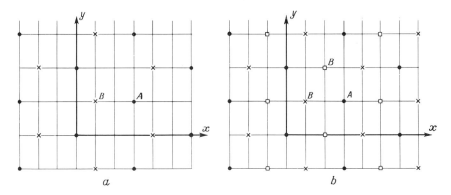

**Fig. 1.** Selection of coordinate axes in two- and three-layer spherical packing schemes. Identical symbols designate points of the same layer.

$p = 2, 5, 8, \ldots$ according to the pattern $m = -4, 2, 8, \ldots$ for $n$ even and $m = -1, 5, \ldots$ for $n$ odd. In this way one can describe any more complicated closest packing. We place dipoles in the lattice points, and on summation the end of the vector $\mathbf{r}$, the origin of which is at the zero point, runs through all the filled lattice points.

Let us break up all the dipoles in the lattice into groups connected with the zero dipole by one family of elements of symmetry. The dipoles joined by parallel elements of symmetry—the axes 2 and $2_1$ or reflection and glide planes—are included in one group. The reason for this division will immediately be clear.

If the basic vector is $\mathbf{S}_1 = x\mathbf{i} + y\mathbf{j} + z\mathbf{k}$, where $x, y, z$ are direction cosines, then the vectors connected with it by various elements of symmetry are written down in the form: for the center of inversion $\bar{1}$: $\mathbf{S}_2 = -x\mathbf{i} - y\mathbf{j} - z\mathbf{k}$; for the axis $2\|Y$: $\mathbf{S}_2 = -x\mathbf{i} + y\mathbf{j} - z\mathbf{k}$; for the plane $m \perp Y$: $\mathbf{S}_2 = x\mathbf{i} - y\mathbf{j} + z\mathbf{k}$; for the axis $2\|Z$: $\mathbf{S}_2 = -x\mathbf{i} - y\mathbf{j} + z\mathbf{k}$, and so on. The distinctions between the axes 2 and $2_1$ and the reflection and glide planes are unimportant.

According to the general formula (3.1) the energy of interaction between two dipoles—the zero dipole and the dipole located in the point $mnp$—when they are, for example, translationally identical, is equal to (we write the energy referred to one dipole):

$$E_{\mu,t} = \frac{\mu^2}{a^3} \left\{ \frac{1}{2R^3} - \frac{3}{2R^5} \left[ \frac{3^{1/2}}{6} mx + \frac{1}{2} ny + \left(\frac{2}{3}\right)^{1/2} pz \right]^2 \right\} \qquad (3.3)$$

In case the dipole situated in the point $mnp$ is connected with the zero dipole by a second-order axis parallel to $Y$, the energy of their interaction is equal to

$$E_{\mu,2\|Y} = \frac{\mu^2}{a^3} \left\{ \frac{2y^2 - 1}{2R^3} - \frac{3}{2R^5} \left[ \frac{1}{4} n^2 y^2 - \left(\frac{\sqrt{3}}{6} mx + \sqrt{\frac{2}{3}} pz\right)^2 \right] \right\} \qquad (3.4)$$

Similar formulas can be obtained for other cases as well.

Now we remark that in the lattices under consideration, we can, owing to their intrinsic symmetry, single out lattice points of two types (see Fig. 1): lattice points of type A having in the general position a multiplicity of eight (this means that there are eight points, the indices of which differ only in signs—$mnp, \bar{m}np, m\bar{n}p, mn\bar{p}, \bar{m}\bar{n}p, \bar{m}n\bar{p}, m\bar{n}\bar{p}, \bar{m}\bar{n}\bar{p}$); in special positions these lattice points have a multiplicity of four or two; lattice points of type B exhibit a multiplicity of four in the general position ($mnp, m\bar{n}p, \bar{m}\bar{n}p, \bar{m}np$ for cubic packing and $mnp, m\bar{n}p, mn\bar{p}, m\bar{n}\bar{p}$ for hexagonal packing), and in special positions they have a multiplicity of two.

If we designate the totality of points whose indices differ only in sign as "a point of the form $\{mnp\}$," then the contribution of such a "point" to the dipole energy of the lattice when, for example, the dipoles of the "point" are translationally identical with the zero dipole, is found, on the basis of (3.), to be equal to

$$E_{\mu,t}^{\{mnp\}} = \frac{\mu^2}{a^3}\left(\frac{4\gamma}{R^3} - \frac{\gamma m^2}{R^5}x^2 - 3\frac{\gamma n^2}{R^5}y^2 - 8\frac{\gamma p^2}{R^5}z^2 - 4(2)^{\frac{1}{2}}\frac{\gamma mp}{R^5}xz\right)$$ (3.5)

The coefficient $\gamma$ takes into account the multiplicity of the point $\{mnp\}$ and is equal to 1, $\frac{1}{2}$, or $\frac{1}{4}$ if the multiplicity equals eight, four, or two, respectively. We notice that for lattice points of type A and also for those of type B in the hexagonal framework, the last term in the sum (3.5) is equal to zero. The contribution of the entire group of dipoles, translationally identical to the zero dipole, to the energy is equal to

$$E_{\mu,t} = (\mu^2/a^3)(4A_0 - A_m x^2 - 3A_n y^2 - 8A_p z^2 - 4(2)^{\frac{1}{2}}A_{mp}xz)$$

where

$$A_0 = \sum_i \gamma_i/R_i^3; \qquad A_m = \sum_i \gamma_i m_i^2/R_i^5; \qquad A_n = \sum_i \gamma_i n_i^2/R_i^5$$

$$A_p = \sum_i \gamma_i p_i^2/R_i^5; \qquad A_{mp} = \sum_i \gamma_i m_i p_i/R_i^5$$ (3.6)

The summation is carried out over the points constituting the group. Here it suffices to write only one representative of each point of the form $\{mnp\}$. This means that for lattice points of type A, we may limit ourselves to the positive octant; and for lattice points of type B, to the positive quadrant ($n > 0, p > 0$). The sum $A_{mp}$ is calculated only for lattice points of type B in the cubic framework.

Proceeding in a similar way, that is, introducing $\gamma$ and summing over the points of the form $\{mnp\}$ in case the group of dipoles is connected with the zero dipole by the $2\|Y$ axis, we obtain, on the basis of (3.4),

$$E_{\mu,2\|Y} = (\mu^2/a^3)[4A_0(2y^2 - 1) + A_m x^2 - 3A_n y^2 + 8A_p z^2 + 4\sqrt{2}A_{mp}xz]$$

As we see, the contribution of this group of dipoles to the energy is expressed with the aid of similar sums (3.6), with the only difference that the summation is performed over other groups of indices. It is evident that the contributions of other groups of dipoles is expressed with the help of the same sums.

Now we write the contributions of all possible groups of dipoles to the energy:

$$E_{\mu,t} = (\mu^2/a^3)(4A_0 - A_m x^2 - 3A_n y^2 - 8A_p z^2 - 4(2)^{\frac{1}{2}}A_{mp}xz)$$

$$E_{\mu,\bar{1}} = -E_{\mu,t}$$

$$E_{\mu,2\|X} = (\mu^2/a^3)[4A_0(2x^2 - 1) - A_m x^2 + 3A_n y^2 + 8A_p z^2]$$

$$E_{\mu,m\perp X} = -E_{\mu,2\|X} \tag{3.7}$$

$$E_{\mu,2\|Y} = (\mu^2/a^3)[4A_0(2y^2-1) + A_m x^2 - 3A_n y^2 + 8A_p z^2 + 4(2)^{\frac{1}{2}}A_{mp}xz]$$

$$E_{\mu,m\perp Y} = -E_{\mu,2\|Y}$$

$$E_{\mu,2\|Z} = (\mu^2/a^3)[4A_0(2z^2-1) + A_m x^2 + 3A_n y^2 - 8A_p z^2]$$

$$E_{\mu,m\perp Z} = -E_{\mu,2\|Z}$$

Let us now try to split the points of the cubic and hexagonal lattices into the following four groups:

$$
\begin{array}{lll}
\text{Group} \quad \text{I} & p \text{ even,} & n \text{ even} \\
\text{Group} \quad \text{II} & p \text{ even,} & n \text{ odd} \\
\text{Group III} & p \text{ odd,} & n \text{ even} \\
\text{Group IV} & p \text{ odd,} & n \text{ odd}
\end{array}
\tag{3.8}
$$

If the points of these groups are filled with differently oriented dipoles, we can obtain a rather large number of examples of symmetrical arrangement of dipoles characteristic of molecular crystals.

It is clear that in the case of such a division the set of values of $R$ and indices $mnp$ within the same group for three- and two-layer packings will be different. Therefore the sums $A_0$, $A_m$, $A_n$, $A_p$ and $A_{mp}$ should be computed for each of the four groups for both the two- and three-layer variants. The limits of summation are determined by the rapidity of convergence of the series (3.2). In the work cited use has been made of the distances exceeding fivefold the shortest distance between the dipoles.

Having calculated the sums, we immediately obtain, from formulas (3.7), the expressions for the energy of dipole interaction for all the space groups of symmetry, not higher than the orthorhombic ones, with a multiplicity of the general position of not higher than four. Some symmetrical arrangements can be obtained on the basis of both the two- and the three-layer packings. Others can be effected only in one of them (with the limitations mentioned above).

The results for space groups satisfying the above requirements are collected in Table 1. It has been found that for each case of symmetrical arrangement of dipoles the lattice sums $E_\mu' = E_\mu/(\mu^2/a^3)$ can be represented as $E_\mu' = Kf(x,y,z)$ where the quantities $x,y,z$ characterize the orientation of dipoles in the lattice. The form of the functions $f(x,y,z)$ does not depend on the radius of summation, and the values of $K$ reduce to certain limiting values given in the corresponding column of Table 1. The rapidity of convergence is illustrated in Fig. 2a and b. The extremal values of the lattice sums for each space group, corresponding to the extremal values of the functions $f(x,y,z)$, are presented in the last two columns of Table 1.

## Table 1

DIPOLE ENERGY OF CRYSTAL LATTICE VERSUS
ORIENTATION OF DIPOLES AND SPACE SYMMETRY

| Lattice symmetry | Pack-ing | Symmetry operations by groups | | | | $f(x,y,z)$ | $K$ | $E'_{\mu,\,min}$ | $E'_{\mu,\,max}$ |
|---|---|---|---|---|---|---|---|---|---|
| | | I | II | III | IV | | | | |
| P1 | $C^a$ | t | t | t | t | 0 | 0 | 0 | 0 |
| | $H^b$ | t | t | t | t | $3z^2-1$ | 0 | 0 | 0 |
| P$\bar{1}$ | C | t | t | $\bar{1}$ | $\bar{1}$ | $3z^2-1$ | 2.57 | $-2.57$ | 5.14 |
| | H | t | t | $\bar{1}$ | $\bar{1}$ | $3z^2-1$ | 2.57 | $-2.57$ | 5.14 |
| | C | t | $\bar{1}$ | t | $\bar{1}$ | $(\sqrt{2}x-z)^2-1$ | 1.50 | $-1.50$ | 3.00 |
| | H | t | $\bar{1}$ | t | $\bar{1}$ | $9x^2-6y^2-1$ | 0.21 | $-1.50$ | 1.71 |
| | C | t | $\bar{1}$ | $\bar{1}$ | t | $(\sqrt{8}x+z)^2-3$ | 0.86 | $-2.57$ | 5.14 |
| | H | t | $\bar{1}$ | $\bar{1}$ | t | $11x^2+z^2-4$ | 0.64 | $-2.57$ | 4.50 |
| P2$_1$ | C | t | 2$\|Y$ | t | 2$\|Y$ | $x^2-2\sqrt{2}xz$ | 1.50 | $-1.50$ | 3.00 |
| Pa | C | t | $m\perp Y$ | t | $m\perp Y$ | $-y^2$ | 1.50 | $-1.50$ | 0 |
| A2 | C | t | 2$\|Y$ | 2$\|Y$ | t | $5x^2-2z^2+4\sqrt{2}xz$ | 0.86 | $-2.57$ | 5.14 |
| Aa | C | t | $m\perp Y$ | $m\perp Y$ | t | $-y^2$ | 2.57 | $-2.57$ | 0 |
| P2$_1/c$ | C | t | 2$\|Y$ | $\bar{1}$ | $m\perp Y$ | $(\sqrt{8}x+z)^2-3$ | 0.86 | $-2.57$ | 5.14 |
| P2$_1$2$_1$2$_1$ | H | t | 2$\|X$ | 2$\|Z$ | 2$\|Y$ | $11z^2-12$ | 0.21 | $-2.57$ | $-0.21$ |
| P2$_1$2$_1$2$_1$ | H | t | 2$\|X$ | 2$\|Y$ | 2$\|Z$ | $5y^2+3z^2-12$ | 0.21 | $-2.57$ | $-1.50$ |
| Pca2$_1$ | H | t | $m\perp Y$ | $m\perp X$ | 2$\|Z$ | $-(12x^2+7y^2)$ | 0.21 | $-2.57$ | 0 |
| Pna2$_1$ | H | t | $m\perp Y$ | 2$\|Z$ | $m\perp X$ | $z^2-1$ | 2.57 | $-2.57$ | 0 |

$^a$ C, cubic packing.
$^b$ H, hexagonal packing.

It is seen from Table 1 that the symmetry P1 and P$\bar{1}$ can be achieved both in the two- and in the three-layer packing. Calculations have shown that for the three-layer packing each coordination sphere consisting of translationally identical dipoles gives a zero contribution to the energy, no matter whether the dipoles of this sphere are parallel or antiparallel to the zero dipole.Therefore, in the case of P1 symmetry the dipole energy of the lattice built up on the basis of the three-layer packing is strictly equal to zero. In the case of other symmetrical arrangements of dipoles (as, for instance, in the example given in Fig. 2b) some coordination spheres also have no contribution to $E_\mu$ since their points belong only to one of the four groups (3.8) and hence contain mutually parallel dipoles. In the case of the two-layer packing the dipole energy of a lattice of symmetry P1 is not strictly equal to zero, but tends to zero as the number of coordination spheres taken into account increases (Fig. 2a). Passing over in formulas $E_\mu' = Kf(x,y,z)$ from cartesian

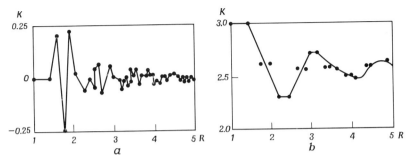

*Fig. 2.* Convergence of lattice sums for dipole lattices: (a) symmetry P1, hexagonal packing; (b) symmetry P$\bar{1}$, cubic packing.

co-ordinates to spherical ones ($x = \sin\theta\cos\varphi$, $y = \sin\theta\sin\varphi$, $z = \cos\theta$), we can construct polar maps showing the dependence of the dipole energy on the orientation of dipoles relative to the symmetry elements. Such maps are presented in Fig. 3. Here the polar angle $\theta$ is measured from the $Z$ axis perpendicular to the plane of the drawing, and the angle $\varphi$ from the $X$ axis which is directed leftward. The "equal-potential" lines are clearly seen. The directions in which the dipole energy has a maximum or minimum value are designated by dots. Comparing the polar maps we notice that in lattices of different space symmetry the orientations of dipoles leading to the extremal values of the energy are not equal, but the extremal values of the lattice sums $E_\mu'$ themselves often coincide even for lattices with different recurrence of dense layers. The dipole energies of lattices of orthorhombic symmetry for almost all orientations of the dipoles are negative, which, as a rule, is not observed in lattices of lower symmetry.

The formulas given in Table 1 make it possible to estimate approximately the values of the dipole contribution to the lattice energy of various molecular crystals without carrying out a complete calculation on the basis of formulas (3.1) and (3.2). This method of estimation is convenient to use in those cases when no exact result is required. Of course, the quantity $a$ entering into the energy formula cannot be calculated directly from structure data, because the ratios of the parameters in real unit cells do not coincide with those ratios which are obtained on the basis of spherical packings. We may, however, proceed as follows. We first calculate the volume of a unit cell per one molecule ($V_{mol}$) and then assume that we deal with a close packing of spherical molecules, each of which has the same volume $V_{mol}$. Taking into account the packing factor for spheres, we find that the quantity $a^3 = 2^{\frac{1}{2}} V_{mol}$. In this way we arrive at a certain "average" lattice which crudely corresponds to the original one. We do not know the exact direction of the dipoles in this lattice, but it is clear that the value of the dipole energy should be within the range between the maximum and minimum values corresponding

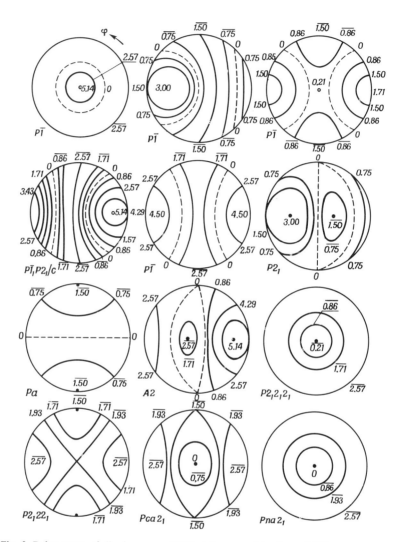

**Fig. 3.** Polar maps of dipole energy $E_\mu'(\theta, \varphi)$ for crystal lattices of different space symmetry (compare Table 1).

to the given space group. Some examples of such estimates for crystals composed of molecules with large dipole moments are given in Table 2. An exact calculation based on formulas (3.2) has been made for nitrobenzene: $E_\mu = -1.4$ kcal/mole (it has been assumed that the center of the dipole coincides with the position of the carbon atom to which the $NO_2$ grouping is linked).

*Table 2*

APPROXIMATE VALUES OF DIPOLE INTERACTION ENERGY
IN SOME CRYSTAL LATTICES

| Substance | Symmetry | $\mu$, D | $V_{mol}$ ($\text{Å}^3$) | $E_\mu$ (kcal/mole) min | max |
|---|---|---|---|---|---|
| Nitrobenzene $C_6H_5NO_2$ | $P2_1/c$ | 3.96 | 148.2 | $-2.8$ | 5.6 |
| Nitroanthracene $C_{14}H_9NO_2$ | $P2_1/a$ | 3.96 | 267.0 | $-1.5$ | 3.0 |
| m-Dinitrobenzene $C_6H_4(NO_2)_2$ | $Pbn2_1$ | 3.96 | 175.2 | $-2.3$ | 0 |
| Acenaphthenequinone $C_{10}H_2(CO_2)_2$ | $P2_12_12_1$ | 4.90 | 203.0 | $-3.6$ | 0.3 |

The heats of sublimation of the substances listed in Table 2 are of the order of 20 kcal/mole. It is seen that the values of the dipole energies constitute a relatively small fraction of this figure even if the dipole moment of the molecule is considerable. For crystals whose molecules have moments $\mu$ lower than 1 D, the value of the dipole contribution to the lattice energy must be quite negligible since this contribution is proportional to the square of the dipole moment.

From what has been said above it does not follow, however, that the dipole interactions do not affect the properties of a molecular crystal, in particular, the establishment of its equilibrium structure. Since, as is seen from Table 2, the differences between the extremal values of the dipole energy are not negligible, though not very large, if may be supposed that the equilibrium structure will be set up near the minimum $E_\mu$. At the same time, according to the principle of close packing, the real structure should satisfy the minimum of energy $U$ due to the van der Waals interactions (see below). The dipole energy $E_\mu$, however, varies slowly with the structure parameters, and it therefore cannot bring the structure out of the position of the minimum energy of van der Waals interactions. As an example, let us make extreme estimates of the angle by which dipole interactions could change the position of minimum energy $U$ for a structure with strongly dipolar molecules. To do this, we make use of the method employed by Craig [20] for similar estimates in the lattice of quadrupoles. If it is assumed that the variations of $U$ in the vicinity of the minimum are described by the quadratic law $U = \frac{1}{2}k(\theta - \theta_0)^2$, and the electrostatic term varies linearly $E_\mu = \lambda\theta$, the change in the equilibrium angle $\theta$ due to electrostatics will then be equal to $-\lambda/k$. The values of the force constants $k$ for "nondipole" structures of the naphthalene type are evaluated approximately as $20\ \mathrm{cm}^{-1}\ \mathrm{deg}^{-2}$. From the formula given in Table 1 we find that the greatest value of $(a^3/\mu^2)(\partial E_\mu/\partial\theta)$ for the group $P2_1/c$ is equal to $0.13\ \mathrm{deg.}^{-1}$. Let us suppose that we deal with a molecule of the nitro-naphthalene type ($\mu = 3.96$ D). The largest value of $\lambda$ will then be about $30\ \mathrm{cm}^{-1}\ \mathrm{deg}^{-1}$ and the shift of the minimum about $1.5°$, which is, of course, an unimportant effect on the packing.

Evidently, the molecules in real crystals are arranged so that the structure is in one of the lowest possible minima of the energy of the van der Waals interactions, with $E_\mu$ also simultaneously as close as possible to the minimum value. Calculations verifying this assertion have not yet been made.

In calculating the dipole energy it is quite legitimate to consider a lattice consisting of point dipoles, since in crystal lattices even the shortest distances between the dipoles exceed their size by three or four times at least. As calculations based on exact formulas show, with these ratios the error committed is insignificant. The difficulty lies not in this, but in the localization of the dipole.

### 4. THE QUADRUPOLE ENERGY

The quadrupole interactions become important in those cases when the sum of charges in a molecule and its dipole moment are equal to zero. The quadrupole interactions for some simple molecules ($N_2$, $N_2O$, $CO_2$, $CO$), crystallizing at low temperatures in the face-centered cubic lattice, have been considered by Nagai and Nakamura [21]. The regions of lowest quadrupole energy have been found and have turned out to be close to the actual orientations of the molecules in these crystals.

Let us consider a more general case of the quadrupole interactions in a crystal using the method described above for dipole lattices. We shall employ the results so obtained to assess approximately the role of these interactions in a molecular crystal.

As is known, the energy of interaction between two quadrupoles for an axial or cylindrically symmetrical distribution of charges in the molecules is given by

$$E_Q = \frac{3}{16}\frac{Q^2}{r^5}\{1 + 2(\mathbf{S}_1 \cdot \mathbf{S}_2)^2 - 5(S_{1r}^2 + S_{2r}^2) - 20 S_{1r} S_{2r}(\mathbf{S}_1 \cdot \mathbf{S}_2) + 35 S_{1r}^2 S_{2r}^2\}$$

$$(4.1)$$

where $\mathbf{r}$ is the vector joining the centers of the quadrupoles; $\mathbf{S}_1$ and $\mathbf{S}_2$ are those unit vectors with components $x, y, z$, characterizing the directions of the axes of quadrupoles relative to the lattice axes; $S_1$, and $S_2$, are the projections of $S_1$ and $S_2$ onto $\mathbf{r}$; and $Q$ is a scalar quantity completely determining the tensor of the quadrupole moment of the axial quadrupole (in the generally accepted notations $Q = Q_{xz}$).

We shall number, as previously, the points of a rectangular lattice with periods $a(3)^{1/2}/6$ along the $X$ axis, $a/2$ along the $Y$ axis and $a(2/3)^{1/2}$ along the $Z$ axis ($a$ is the shortest distance between the lattice points) as indices $m, n, p$, and write the radius-vector of any point in the form

$$\mathbf{R} = \frac{\mathbf{r}}{a}\left[\frac{3^{1/2}}{6}m\mathbf{i} + \frac{1}{2}n\mathbf{j} + \left(\frac{2}{3}\right)^{1/2}p\mathbf{k}\right]$$

For groups of quadrupoles joined with the zero quadrupole by different symmetry elements, we shall then obtain formulas for contributions to the energy, analogous to formulas (3.7). In distinction to the dipole variant, the interaction energies of the quadrupoles joined by the $2\|X$ axes and $m\perp X$ planes (analogously for $Y$ and $Z$) coincide, whereas for dipoles they differ in sign. This is natural because a quadrupole is an axial vector, and a dipole is a polar vector. The differences between the axes 2 and $2_1$ and the reflection and glide planes are unimportant, as before.

The formulas for the contributions to the energy of all possible groups of quadrupoles in the case of triclinic, monoclinic, and orthorhombic symmetrics have the form:

$$E_{Q,t} = E_{Q,\bar{1}} = \frac{Q^2}{a^5}\left\{\frac{9}{2}B_0 - \frac{15}{4}B_m x^2 - \frac{45}{4}B_n y^2 - 30B_p z^2\right.$$

$$- 15(2)^{1/2}B_{mp}xz + \frac{35}{96}B_{mm}x^4 + \frac{105}{32}B_{nn}y^4 + \frac{70}{3}B_{pp}z^4$$

$$+ \frac{105}{16}B_{mn}x^2y^2 + \frac{35}{2}B_{mp}x^2z^2 + \frac{105}{2}B_{np}y^2z^2$$

$$+ \frac{35(2)^{1/2}}{12}B_{m(mp)}x^3z + \frac{70(2)^{1/2}}{3}B_{(mp)p}xz^3$$

$$\left. + \frac{105(2)^{1/2}}{4}B_{n(mp)}xy^2z\right\}$$

$$E_{Q,2\|X} = E_{Q,m\perp X} = \frac{Q^2}{a^5}\left\{\frac{3}{2}B_0[1+2(2x^2-1)^2]\right.$$

$$\left. - \frac{5}{4}B_m x^2[1+2(2x^2-1)] - \frac{15}{4}B_n y^2[1-2(2x^2-1)]\right.$$

$$- 10B_p z^2 [1-2(2x^2-1)] + \frac{35}{96} B_{mm} x^4 + \frac{105}{32} B_{nn} y^4$$

$$+ \frac{70}{3} B_{pp} z^4 - \frac{35}{16} B_{mn} x^2 y^2 - \frac{35}{6} B_{mp} x^2 z^2 + \frac{105}{2} B_{np} y^2 z^2 \bigg\}$$

(4.2)

$$E_{Q,2 \parallel Y} = E_{Q,m \perp Y} = \frac{Q^2}{a^5} \bigg\{ \frac{3}{2} B_0 [1 + 2(2y^2-1)^2]$$

$$- \frac{5}{4} B_m x^2 [1-2(2y^2-1)] - \frac{15}{4} B_n y^2 [1+2(2y^2-1)]$$

$$- 10B_p z^2 [1-2(2y^2-1)] - 5(2)^{\frac{1}{2}} B_{(mp)} xz[1-2(2y^2-1)]$$

$$+ \frac{35}{96} B_{mm} x^4 + \frac{105}{32} B_{nn} y^4 + \frac{70}{3} B_{pp} z^4 - \frac{35}{16} B_{mn} x^2 y^2$$

$$+ \frac{35}{2} B_{mp} x^2 z^2 - \frac{35}{2} B_{np} y^2 z^2 + \frac{35(2)^{\frac{1}{2}}}{12} B_{m(mp)} x^3 z$$

$$+ \frac{70(2)^{\frac{1}{2}}}{3} B_{(mp)p} xz^3 - \frac{35(2)^{\frac{1}{2}}}{4} B_{n(mp)} xy^2 z \bigg\}.$$

$$E_{Q,2 \parallel Z} = E_{Q,m \perp Z} = \frac{Q^2}{a^5} \bigg\{ \frac{3}{2} B_0 [1 + 2(2z^2-1)^2]$$

$$- \frac{5}{4} B_m x^2 [1-2(2z^2-1)] - \frac{15}{4} B_n y^2 [1-2(2z^2-1)]$$

$$- 10B_p z^2 [1+2(2z^2-1)] + \frac{35}{96} B_{mm} x^4 + \frac{105}{32} B_{nn} y^4$$

$$+ \frac{70}{3} B_{pp} z^4 + \frac{105}{16} B_{mn} x^2 y^2 - \frac{35}{6} B_{mp} x^2 z^2 - \frac{35}{2} B_{np} y^2 z^2 \bigg\}$$

In these formulas

$$B_0 = \sum_i \gamma_i / R_i^5, \qquad B_m = \sum_i \gamma_i m_i^2 / R_i^7, \qquad B_n = \sum_i \gamma_i n_i^2 / R_i^7$$

$$B_p = \sum_i \gamma_i p_i^2 / R_i^7, \qquad B_{(mp)} = \sum_i \gamma_i m_i p_i / R_i^7, \qquad B_{mm} = \sum_i \gamma_i m_i^4 / R_i^9$$

$$B_{nn} = \sum_i \gamma_i n_i^4 / R_i^9, \qquad B_{pp} = \sum_i \gamma_i p_i^4 / R_i^9, \qquad B_{mn} = \sum_i \gamma_i m_i^2 n_i^2 / R^9$$

$$B_{mn} = \sum_i \gamma_i m_i^2 p_i^2 / R_i^9, \qquad B_{np} = \sum_i \gamma_i n_i^2 p_i^2 / R_i^9, \qquad B_{m(mp)} = \sum_i \gamma_i m_i^3 p_i / R_i^9$$

$$B_{(mp)p} = \sum_i \gamma_i m_i p_i^3 / R_i^9, \qquad B_{n(mp)} = \sum_i \gamma_i m_i n_i^2 p_i / R_i^9$$

(4.3)

The summation is carried out over the points constituting a group. The coefficient $\gamma$ takes account of the multiplicity of a point of the form $\{mnp\}$. The sums $B_{(mp)}$, $B_{(mp)p}$, $B_{n(mp)}$, $B_{m(mp)}$ are evaluated only for fourfold points in the cubic packing.

In order to obtain, with the help of formulas (4.2), expressions for the quadrupole energy in the case of lattices having different space symmetry, it is convenient, as before, to break up the points of cubic and hexagonal close packings into four groups (I—$p$ even, $n$ even; II—$p$ even, $n$ odd; III—$p$ odd, $n$ even; IV—$p$ odd, $n$ odd) and work out the sums (4.3) for each of these groups, for the two- and three-layer variants separately. In working out the sums we can restrict ourselves, as in the case of dipoles, to distances not exceeding five times the shortest distance between the lattice points.

With the help of the sums (4.3), the calculation of which is a rather time-consuming task, we can easily obtain expressions for the energy of quadrupole lattices of different space symmetry as a function of the angles $\theta$ and $\varphi$ formed by the quadrupoles, situated in the point [000], with the axes of the rectangular lattice. These expressions have the form

$$E_Q = \tfrac{1}{2}(Q^2/a^5)f(\theta,\varphi)$$

where the function $f(\theta,\varphi)$ depends on the symmetry of the mutual arrangement of the quadrupoles. In Table 3 are given formulas for the functions $f(\theta,\varphi)$ and also the extremal values of these functions for various space groups.

*Table 3*

QUADRUPOLE ENERGY VERSUS ORIENTATION OF AXIAL QUADRUPOLE

| Lattice symmetry | Packing | $f(\theta,\varphi)$ | $f_{max}(\theta,\varphi)$ | $f_{min}(\theta,\varphi)$ |
|---|---|---|---|---|
| P1, P$\bar{1}$ | cubic | $1.33 - 6.67\sin^2\theta + 5.84\sin^4\theta$ $+ 4.70\sin^3\theta\cos\theta\cos\varphi(1 - 4\sin^2\varphi)$ | 1.33 | $-2.00$ |
|  | hexag. | $1.35 - 6.75\sin^2\theta + 5.90\sin^4\theta$ | 1.35 | $-0.58$ |
| P2$_1$, Pa, P2$_1$/c | cubic | $1.33 - \sin^2\theta(6.67 + 11.47\sin^2\varphi)$ $+ \sin^4\theta(5.84 - 4.22\sin^2\varphi + 15.67\sin^4\varphi)$ $+ \sin^3\theta\cos\theta\cos\varphi(4.70 - 6.91\sin^2\varphi)$ | 1.33 | $-3.98$ |
| A2, Aa | cubic | $1.33 - \sin^2\theta(6.67 - 7.04\sin^2\varphi)$ $+ \sin^4\theta(5.84 - 21.46\sin^2\varphi + 14.42\sin^4\varphi)$ $+ \sin^3\theta\cos\psi\cos\varphi(4.70 - 11.89\sin^2\varphi)$ | 1.41 | $-3.15$ |
| P2$_1$2$_1$2$_1$ | hexag. | $1.35 - \sin^2\theta(4.90 + 11.72\sin^2\varphi)$ $+ \sin^4\theta(4.05 - 3.97\sin^2\varphi + 15.69\sin^4\varphi)$ | 1.35 | $-3.52$ |
| P2$_1$22$_1$ | hexag. | $1.35 - \sin^2\theta(-13.49 + 30.11\sin^2\varphi)$ $+ \sin^4\theta(-14.33 + 15.70\sin^2\varphi + 14.41\sin^4\varphi)$ | 4.51 | $-3.10$ |
| Pca2$_1$ | hexag. | $1.35 - \sin^2\theta(16.62 + 1.59\sin^2\varphi)$ $+ \sin^4\theta(15.77 - 12.82\sin^2\varphi + 14.41\sin^4\varphi)$ | 1.35 | $-4.50$ |
| Pna2$_1$ | hexag. | $1.35 - \sin^2\theta(16.62 - 16.92\sin^2\varphi)$ $+ \sin^4\theta(15.77 - 32.62\sin^2\varphi + 15.69\sin^4\varphi)$ | 1.35 | $-3.42$ |

As seen in Table 3, some coefficients in the formulas for $f(\theta, \varphi)$ coincide for different lattices. The equality of the coefficients means degeneration of the symmetry in some definite directions (e.g., in the direction perpendicular to dense layers, for which $\theta = 0$). Rapid convergence of the values of $f(\theta, \varphi)$ is observed when the radius of summation is increased. It is clear from Fig. 4 that in practice it is sufficient to carry out the summation up to distances three or four times the shortest distance.

Polar maps (Fig. 5) have been constructed according to the formulas for $E_Q' = f(\theta, \varphi)$, which help us to find immediately the regions of the lowest quadrupole energy for each group of symmetry; comparison of these maps reveals the effect of the symmetry of a lattice on its quadrupole energy.

The evaluation of the order of magnitude of $E_Q$ for a particular molecular crystal can, of course, be made only approximately, as in the dipole case. For this purpose, the true cell is replaced by a face-centered cubic or hexagonal cell of equivalent volume. Besides, to calculate $E_Q$ it is necessary to know the quadrupole moment of the molecule. The quadrupole moments have, however, been determined experimentally only for some simple molecules (on the basis of data on the widening of the microwave spectra due to pressure [22–24]). If the spatial distribution of charges in a molecule were known, the quadrupole moment could be calculated from the formula

$$Q_{ij} = 3 \sum_k e_k(x_i x_j - \tfrac{1}{3} r^2 \delta_{ij})$$

where $e_k$ is the magnitude of charge $k$; $\mathbf{r}$ are the radius vectors of the individual charges; $x_i$ are the components of the vector $\mathbf{r}$; and $\delta_{ij}$ is the Kronecker symbol.

Unfortunately, only very crude estimates are possible in this case too. For the benzene molecule, for instance, if we assume that the dipole moment of the C–H bond is equal to $0.4 \times 10^{-18}$ esu [25] and suppose that the positive and negative charges are concentrated on the carbon atoms and on the

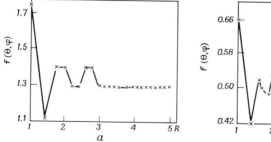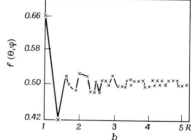

*Fig. 4.* Quadrupole lattice energy versus summation radius: (a) $f(\theta, \varphi)$ for space group $P2_1/c$ ($\theta = 0$); (b) $f(\theta, \varphi)$ for group $P2_12_12_1$ ($\phi = 0$, $\theta = 90°$).

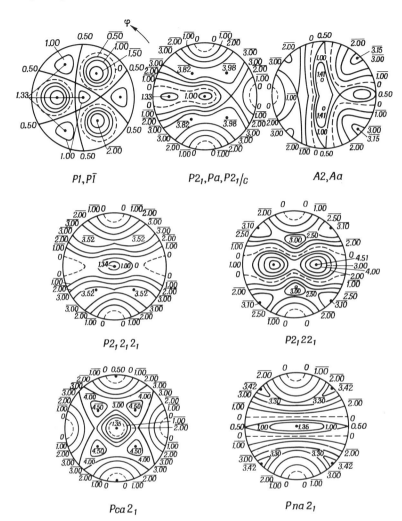

**Fig. 5.** Polar maps of quadrupole energy $E_Q'(\theta, \varphi)$ for crystal lattices of different space symmetry.

hydrogen atoms, respectively, we shall have $Q \approx 2\,\text{Å}^2$. According to the experimental data obtained by Hill and Smith [22], this value is equal to $1.3\,\text{Å}^2$. If we adopt an analogous model for the naphthalene molecule, i.e., if we suppose that the nondiagonal components of the tensor $Q_{ij}$ are equal to zero and also that $Q_{xx} = Q_{yy} = -\frac{1}{2}Q_{zz}$, we shall obtain approximately $Q = 3\,\text{Å}^2$. These estimates provide, in any case, the correct order of magnitude.

Using a naphthalene crystal (space group $P2_1/a$) as an example, we can gain an idea of the order of magnitude of the quadrupole part of the lattice energy for an organic crystal. Knowing the size of the unit cell of a naphthalene crystal, we find the dimensions of an equivalent face-centered cubic cell ($a \approx 6.3 \,\text{Å}$). In Table 3 we find the extremal values of the function $f(\theta, \varphi)$ for the group $P2_1/a$. From these data it follows that the value of $E_Q$ for naphthalene must be within the range of $-0.7$ to $0.2$ kcal/mole. This result agrees well with the value recently obtained by Craig et al. [20] ($-0.4$ kcal/mole) which was calculated for a naphthalene crystal with the aid of exact lattice sums and approximate values of the components of the tensor $Q_{ij}$. It is clear that the quadrupole contribution to the lattice energy, which is equal to $-16.7$ kcal/mole according to the experimental data, is negligible.

The quadrupole energy varies much more slowly with the structure parameters than does the dipole energy. For naphthalene, for example, the largest value is $\partial E_Q / \partial \theta \approx 7 \,\text{cm}^{-1} \text{deg}^{-1}$, which can shift the minimum of the energy of the van der Waals interactions quite insignificantly, by not more than $0.3°$, according to approximate estimates.

5. CONCLUDING REMARKS

It is obvious that the role of the electrostatic interactions, caused by the presence in molecules of moments higher than the quadrupole moment, is even less significant. Thus, we see that the electrostatic multipole interactions play a very insignificant part in a crystal lattice, though, for instance, in the case of dipole lattices, they decrease with distance much more slowly than the nondirected interactions of the dispersion type. This can be explained by the strong dependence of such interactions on direction. In a crystal lattice each multipole is surrounded by a great number of neighbors having very different orientations; but, as is known, the static interactions when averaged over various orientations reduce to zero.

That the electrostatic interactions appear to be insignificant for many problems and many substances is a quite fortunate circumstance. It is not the computational difficulties that present a problem; they are easily overcome by the use of electronic computers. The main obstacle is the absence of information about the site of localization of the dipole, in the case of dipole molecules, which is associated with the fact that we know nothing about the nature of the charge distribution in a molecule.

This uncertainty is still greater when we are dealing with molecules containing several bonds, to which dipole moments may be ascribed, since experiment yields only the general moment of the molecule. It should also be recalled that the distribution of dipole moments over bonds is conventional, and rather arbitrary.

The schemes for calculation of the electrostatic energy may be different. If it is possible to ascribe a "residual" charge $e$ (i.e., an excess or deficiency of electrons compared with the neutral atom) to each atom, then

$$U^{\text{el}} = \sum_{ik} -e_i e_k / r_{ik}$$

Errors that occur in such calculations due to small deviations of the electronic distributions of the atoms from the spherical are negligible. The difficulty lies solely in the absence of information on $e_i$. Attempts to calculate the residual charges by the methods of quantum chemistry seem not to be convincing.

It is, of course, possible to evaluate the dipole–dipole interactions for dipole molecules and the quadrupole–quadrupole interactions for other molecules. In the first case the main difficulty consists in localizing the center of the dipole. It is also unclear how significant will be the other terms of the series expansion of the electrostatic energy (the dipole–quadrupole and quadrupole–quadrupole interactions).

The most rigorous calculation should be based on the integral representation of the energy,

$$-\int_{(6)} \frac{\rho_1 \rho_2 \, dV_1 \, dV_2}{r}$$

Here again the difficulties arise not in the calculation, but are due to our lack of knowledge of the charge distribution.

## C. The Device of Atom–Atom Potentials

### 6. The van der Waals Interactions in a Molecular Crystal†

Since the electrostatic contribution to the lattice energy of a molecular crystal is not a determining factor, the problem of calculating this energy must be reduced mainly to the investigation of the van der Waals interactions—the dispersion attractions of the London type and repulsion caused by the overlap of electron clouds.

The $-A/R^6$ law for attraction and the exponential law for repulsion are obeyed only when the interactions between particles are averaged over all possible orientations, i.e., when it is assumed that the symmetry of the particles is spherical. If the complete molecules in a crystal are regarded as interacting

---

† Here and elsewhere, the review article "General View on Molecular Packing" by the author [26] is widely used.

particles, as was done by London in his first calculations, it becomes clear that the assumption of spherical symmetry will be too crude for the overwhelming majority of molecules. If we try, as Muller [27], to break up a molecule into force centers, the most rigorously averaged London formula must then probably be satisfied by individual atoms of which the molecules are composed, rather than by groups of atoms or bonds. (The latter approach has been adopted in a few works devoted to the discussion of this problem.) The arguments in favor of the former approach are as follows: First, as is known, the average magnitude of the dipole moment for the electronic system of an atom in a stationary state is always equal to zero—a statement that cannot be made with respect to a group of atoms [28]. Secondly, as the results of crystallochemical studies show, the electron density of the atoms in molecular crystals can be represented, with high accuracy (0.1–0.2 electron) by spherically symmetrical functions. As regards their polarizabilities, atoms may also be regarded as being almost isotropic, as has been indicated in Pitzer's review [29]. Moreover, the application of perturbation theory to individual atoms does not require special assumptions as to the form of the function for the electron density of a molecule, or of its constituent parts. And, finally, the ratio of particle size to particle separation is always more favorable for atoms, even for the shortest distances encountered in a crystal, than for entire molecules or any groups of atoms.

The interaction potential for a pair of atoms $i$ and $j$ separated by the distance $r_{ij}$ can be written in the form

$$\varphi_{ij} = -Ar_{ij}^{-6} + Be^{-\alpha r_{ij}}$$

The way this formula is written means that in the second order of the perturbation theory we limit ourselves to the dipole–dipole approximation and adopt the exponential law for the repulsion potential. Then, following the model described in Section 2, we must find the constants $A$, $B$, and $\alpha$ for each species of atoms, in order to calculate the interaction energy of the molecules as the sum of the interaction energies of their constituent atoms.

A theoretical formula for the coefficient $A$ is too complicated for actual calculations in all but the simplest cases. It requires a knowledge of the eigenvalues of the unperturbed atoms which could be obtained by solving the Schrödinger equation for each atom separately. It is known, however, that in the case of multielectron systems even the best functions, resulting from the solution of the Schrödinger equation by any approximate method usually yield values of the energy differing noticeably from the experimental data. Therefore, in practice, in order to calculate the constants in the attraction potential one has to resort to more or less crude approximations, the relative accuracy of which is difficult to gauge.

We did not employ theoretical methods at all in calculating the constants

*B* and α in the repulsion potential; instead, these constants are chosen empirically. The theoretical calculation of the interaction energy for a crystal lattice is further complicated by the presence of the region of so-called intermediate distances, for which no rigorous quantum-mechanical calculations exist that would permit expressing this energy with sufficient accuracy.

From what has been said above it seems expedient to apply a purely empirical approach to the choice of the constants of the potential curves for both the long and the short interatomic distances. Molecular crystals, especially organic ones, can furnish an abundance of data for this purpose.

## 7. POTENTIAL CURVES

The principal advantage of the atom–atom potential method for studying organic substances is that it permits the selection of atom–atom potentials from the experimental data for only a few representatives of a class of compounds; the curves thus obtained can then be used for predicting the properties of all the other compounds of this class. Thus, for example, if C–C, C–H, and H–H interaction potentials are known, it is possible to calculate the properties of the vast class of hydrocarbons.

A tremendous body of experimental data can be employed to verify the method and the potential curves obtained. Such physical parameters as the elasticity tensor, expansion tensor, and normal crystal vibration spectrum can be used for this purpose. However, the most straightforward verification and, consequently, the initial checking stage, is calculation of the heat of sublimation and the crystal structure. This technique may be applied to select the optimum atom–atom potential curves to be used in calculating the other properties of crystals.

The constants $A, B, \alpha$ of the 6-exp potential of neutral atom interaction can readily be expressed through the following three parameters: the equilibrium distance $r_0$, the potential well depth $U_0$, and the second derivative at the point of minimum $D$.

$r_0$ and $U_0$ can easily be included in the potential formula, which then assumes the form

$$U(r) = \frac{U_0}{6-\alpha r_0}\left(-\frac{\alpha r_0{}^7}{r^6} + 6e^{-\alpha(r-r_0)}\right)$$

As to α, it is related to all three physical parameters through the equation

$$\frac{Dr_0{}^2}{U_0} = \frac{6(\alpha r_0)^2 - 42\alpha r_0}{6-\alpha r_0}$$

It is worthwhile to compare the analytical and experimental data with respect to these physical parameters, because, first, their magnitudes lie within a rather narrow range, and, secondly, each parameter is particularly sensitive to certain properties. Thus, the equilibrium distances are mainly responsible for the lattice constants; the potential well depth for the heat of sublimation; and the parameter $D$ for the molecular vibration frequencies and the crystal elasticity tensor.

Preliminary calculations have shown that, as could well be expected, the values of $r_0$ exceed the mean distances between the nearest atoms of adjacent molecules which we find in crystals by about 15%. In the publications by the author and Mirskaya [30], the values used for $C \cdots C$, $C \cdots H$, and $H \cdots H$ interactions are, respectively, 3.8, 3.3, and 2.8 Å.

The parameters $U_0$ have not yet been carefully studied. Probably, their values do not differ much for different pairs of atoms. At any rate, for hydrocarbons the values of $U_0$ lie in the vicinity of 0.06 kcal/mole. The values of the product of the parameters $\alpha r_0$ also vary insignificantly, namely, within 12–16.

Figure 6a shows the atom–atom potentials used in several different experiments. Figure 6b gives a comparison between our curves and those determined

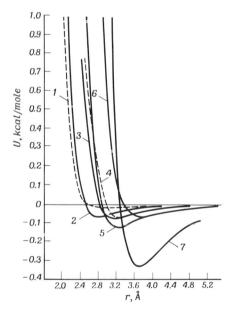

**Fig. 6a.** Pairwise interaction energy of atoms of different elements as a function of interatomic distance. (1) He–He [C. Vittorio, G. Colin, and S. Leopoldo, *Rev. Mexic. Fis.* **12**, 69–83 (1963)]; (2) H–H [K. V. Mirskaya. Thesis., Moscow, 1966]; (3) F–F [M. Lwasaki, *J. Polym. Sci. Part A.* **1**, 1099 (1963)]; (4) C–H [K. V. Mirskaya, Thesis, Moscow, 1966]; (5) O–O [A. I. Kitaigorodsky, K. V. Mirskaya, and V. V. Nauchitel, *Kristallografiya* **14**, 900 (1969)]; (6) C–C [K. V. Mirskaya. Thesis., Moscow, 1966]; (7) Cl–Cl [T. L. Hill, *J. Chem. Phys.* **16**, 399 (1948)].

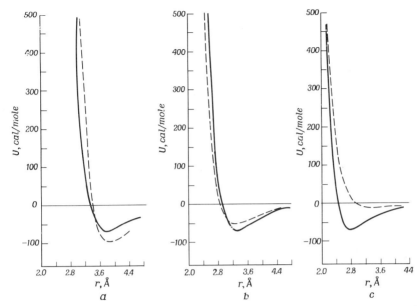

*Fig. 6b.* Potential curves of Kitaigorodsky (solid line) and Williams (dashed line): (a) C–C interactions; (b) C–H interactions; (c) H–H interactions.

by Williams [31], which are so far the only atom–atom potentials specially selected for calculating lattice energy.

At present there appears to be no need to complicate molecular interaction energy calculations by allowing for the noncentrality of the atom–atom potentials. Furthermore, attempts at using the atom–atom potentials for calculating the barriers to rotation of one part of a molecule with respect to the other about single bonds show that the atom interaction energy depends on the angle between the interatomic distance and the valence bonds.

This dependence compelled some authors to introduce into the calculation scheme an additional energy component which is interpreted as bond interaction energy. However, another procedure can also be applied; namely, it is possible to make a reasonable assumption (which is independently confirmed by organic chemical crystallography) about the dependence of the equilibrium radius of a monovalent atom on the angle with a valence bond.

Assuming $r_0$ equal to $a$ and $a/\varepsilon$ along and across the bond respectively, we have calculated the energy of interaction between the H atoms of an ethane molecule from our 6-exp potential expression. The only supplementary condition is: The equilibrium radius $r_0$ is assumed to be given by

$$r_0 = \tfrac{1}{2}a[(\varepsilon \sin^2 \psi_1 + \cos^2 \psi_1)^{-\frac{1}{2}} + (\varepsilon \sin^2 \psi_2 + \cos^2 \psi_2)^{-\frac{1}{2}}]$$

where $\psi_1$ and $\psi_2$ are angles between the interatomic vector and the bonds. A new parameter appears in the equation. For $\varepsilon = 0.82$, the experimental value of the barrier to rotation is estimated at 3 kcal/mole.

It is quite sufficient to introduce such ellipticity for monovalent atoms. Interaction between C atoms can be thought to remain central; there is no conflict with experiment.

As to the hydrogen bond, it is reasonable to include it in our atom–atom potential model in the following way. All interactions are determined according to the above formulas, except for the interaction of an H atom associated with an O or N atom through a hydrogen bond. For this atomic pair, we introduce a short-range potential.

A potential well for the hydrogen bond can be described by a parabola cut off at the point of intersection with the abscissa or a curve of the Morse potential type, etc.

It appears probable that the arbitrary configuration of the potential will affect but slightly the results of the calculation of optimum molecular conformations or packings. The coordinates of the potential-well bottom are the only important parameters.

In our atom–atom potential model the dependence of the hydrogen bond energy on the O–H $\cdots$ O angle is effected automatically. It can be easily seen that a deflection of this angle from 180° at constant O–H and H $\cdots$ O distances results in a sharp rise of energy due to O atom repulsion. In this way, using our "universal" potential and assuming the equilibrium radius for the oxygen atom to be 1.46 Å, we obtain an energy loss of 0.05 kcal/mole for a 160° angle, 0.15 kcal/mole for a 140° angle, and above 1 kcal/mole for 100–120° angles.

This calculation takes into account the steric interaction of the O atoms alone. Actual energy loss may be higher than the above figures. This calculation scheme has not yet been used for determining molecular interaction energy.

Another aspect of the hydrogen bond potential curves has been studied by Lippincott, Shroeder, and, particularly thoroughly, by Reid [32].

The O–H $\cdots$ O potential was treated as a sum of the three components: O–H interaction, H $\cdots$ O interaction, and O $\cdots$ O interaction. Thus, the authors were interested in a relation between the O–H and H $\cdots$ O distances.

Reid's potential has the form

$$V(r,R) = D_0 \left[ 1 - \exp\left( \frac{-n(r-r_e)^2}{2r} \right) \right] + CD_0 \left[ 1 - \exp \frac{-n(R_e - r - r_e)^2}{2c(R_e - r)} \right]$$

$$+ \frac{259.5}{R_e^6} - 4.55 \times 10^6 e^{-4.8R_e}$$

where $R_e$ is the equilibrium O $\cdots$ O distance; $r$ is the O–H distance; and $r_e$ is the H $\cdots$ O distance. The constants are selected empirically.

## 8. ENERGY AS A FUNCTION OF LATTICE PARAMETERS

The calculation of the lattice energy of an ordered crystal with a known structure by the atom–atom potential method amounts to the compilation of tables listing all possible distances between atomic pairs in different molecules and in the calculation of lattice sums $\sum r^{-6}$ and $\sum e^{-\alpha r}$. The lattice sums for simple cubic structures whose points contain atoms interacting according to the $\alpha r^{-n}$ law have been calculated by Lennard-Jones and Ingham [13] and generalized in publication [33] for the case when the initial point is not a lattice point. For molecular crystals, lattice sums can, in the general case, be computed only by straightforward summation, which is facilitated by the use of electronic computers. Since the dispersion energy and, in particular, the repulsive energy of atoms quickly decrease with distance, it is sufficient to use distances up to about 15 Å in such a summation. Calculations for a number of organic crystals show that the lattice energy error is below 1%. Figure 7 shows an example of the dependence of the lattice energy on the summation radius for crystalline benzene.

For calculating the lattice energies of molecular crystals, we use information on structures obtained by X-ray diffraction analysis. The coordinates of hydrogen atoms that cannot be found by this method can in many cases be fairly accurately computed from the arrangement of other atoms, the direction of the valences, and also the length of the C–H bonds determined by electron diffraction and neutron diffraction techniques.

The interaction energy of a pair of nonbonded atoms is a function of the distance $r$ between their centers. Distances in a crystal lattice of a given symmetry seem to depend on the lattice parameters and mutual disposition of the molecules in the crystal. Therefore, for the theoretical study of the crystal properties, it is convenient to take into account all possible changes in the

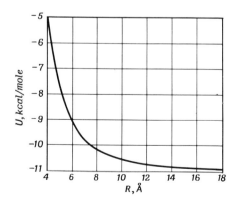

*Fig. 7.* Lattice energy of crystalline benzene versus summation radius.

lattice geometry while treating lattice energy as a function of the spacings $a, b, c$, the unit cell angles $\alpha, \beta, \gamma$, the Eulerian angles $\theta, \varphi, \psi$ describing the position of the main molecule with respect to the crystal axes, and the co-ordinates of the center of gravity of this molecule $x, y, z$. If there are only small variations in the above parameters, the arrangement of the atoms inside the molecule can in very many cases be taken as constant.

In this way, instead of the lattice energy at one single point, we obtain a multidimensional surface whose abscissae are the geometrical parameters of the lattice, and whose ordinate is the amount of its energy:

$$U = U(a, b, c, \alpha, \beta, \gamma, \theta, \varphi, \psi, x, y, z) \qquad (8.1)$$

In principle, for given dimensions and shapes of the molecules, such a surface has a multitude of minima, one of which must correspond to the equilibrium structure of the crystal. It cannot be asserted that this will necessarily be the deepest minimum, as far as its absolute value is concerned, because the energy surface may have several minima of similar values and the establishment of the final structure may be influenced both by electrostatic interactions and finer effects which we ignore in our proposed scheme of lattice energy calculations.

Investigation of the surfaces (8.1) requires computer programs which allow computation of the energy at any surface point, the amount of energy as a function of one, two, etc. geometrical parameters with other characteristics being unchanged—the so-called one-dimensional, two-dimensional, etc. cross sections of the energy surface, the positions of the surface minima in any of these cross-sections, and the total surface minimum over all variables.

The ordinate of the potential-well bottom furnishes information on the heat of sublimation. Under given ambient conditions the sublimation heat of a crystal equals the difference between the internal energies of the crystal and gas plus the expansion work. Since sublimation pressures are not large, the heat of sublimation is practically equal to the difference between the internal energies of the gas and the crystal. Of particular interest is the heat of sublimation extrapolated to the absolute zero of temperature. Since, according to quantum mechanics, at this temperature a lattice is not in a static equilibrium state, due to the zero-point vibrations, the heat of sublimation at absolute zero ($\Delta H_0$) is a sum of two terms: the potential energy of particle interaction in the lattice and the zero-point vibrational energy $K_0$,

$$-\Delta H_0 = U_0 + K_0 \qquad (8.2)$$

The contribution of the second term to $\Delta H_0$ depends on the particle sizes and on the nature of the cohesive forces in the crystal lattice. Zero-point energy is the largest for crystals built up of light-weight molecules with small moments of inertia; it can also be significant if strong directional interactions (hydrogen bonds, etc.) exist in the crystal. However, with increasing mass of the particles and growing moments of inertia, and also with decreasing intermolecular forces, particularly directional, the ratio of the zero-point energy to the heat of sublimation drops sharply. According to different estimates for such atomic crystals as He and Ne, and also in a crystal of molecular $H_2$, the zero-point energy contributes significantly to the heat of sublimation $\Delta H_0$ (for He and $H_2$ this contribution is found to be predominant). According to Walley [34], for water (ice), the intermolecular zero-point energy makes up 31% due to the presence of hydrogen bonds. But already for such crystals as $N_2$, $O_2$, and CO, the zero-point energy is only 10%, while in solid $CO_2$ it does not exceed 2% of the heat of sublimation $\Delta H_0$. It is obvious that for large molecules composed, for example, of ten or more atoms the contribution of the zero-point energy to the total crystal energy must be negligible.

For molecular crystals without hydrogen bonds, the problem of calculating the sublimation heat at the absolute zero of temperature must obviously amount to the calculation of the first term in (8.2), the potential energy of the crystal lattice. This energy must be minimum for the lattice in the static equilibrium state. Therefore, the heat of sublimation $\Delta H_0$ is calculated with an accuracy equal to the zero-point energy as the ordinate of the total surface minimum (8.1) over all the variables. Since no experiments can separate the zero-point energy from the total energy of the lattice, it is usually included in the definition of the lattice energy, and is added to the theoretical cohesive energy prior to comparing the theoretical and experimental results.

It must be noted that the heats of sublimation are not actually measured at $0°K$ but at higher temperatures. Nevertheless, as the temperature dependences of latent heats of transition are usually small, the differences between $\Delta H_T$ and $\Delta H_0$ are not large, and it is rather easy to allow for these differences in a sufficiently reasonable way (see, e.g., [35]).

The equilibrium crystal structure is determined by the abscissae of the potential-well bottom. The structure of a crystal at temperature $T$ is defined by the minimum of the free energy which can be written as $F = U_{lat} + E_{vib} - TS$. At lower temperatures the role of the vibrations and the entropy factor for lattice equilibrium is diminished, and at absolute zero the equilibrium structure of a crystal is determined by the minimum potential lattice energy accurate to the zero-point vibration effects.

The effect of the zero-point vibrations is that they somewhat expand the lattice as compared with the completely static equilibrium state. This expansion is usually not large, even for crystals composed of small light-weight particles. According to Herzfeld and Goeppert-Mayer [36] it equals about $0.05\,Å$ for argon and about $0.15\,\text{Å}$ for neon, of which the unit cells are 3.83 and $3.20\,\text{Å}$, respectively. The approximate estimates of the zero-point vibration effects can be given, provided one knows the characteristic temperature-versus-volume relationship and the crystal compressibility at low temperatures. The pressure required to compensate for the zero-point vibrations in naphthalene (see Chapter VI) must be of the order of $250\ \text{kg/cm}^2$. Such a pressure is able to change the unit cell volume by less than $1\,\text{Å}^3$ (the compressibility of naphthalene at low temperatures is approximately $10^{-5}\ \text{cm}^2\ \text{kg}^{-1}$) which practically does not affect the lattice parameters at all.

It is obvious that in other crystals composed of large molecules the effect of zero-point vibrations on the establishment of the equilibrium structure at $0°K$ must also be very small. Therefore, the crystal structure corresponding to the minimum of its potential energy must coincide very accurately with its structure at the absolute zero temperature.

9. CALCULATION OF THE STRUCTURE AND OF THE ENERGY SURFACES FOR
BENZENE, NAPHTHALENE, AND ANTHRACENE CRYSTALS

Calculations directed toward finding the position and depth of the lattice energy surface minimum as functions of the parameters of the unit cell and the Eulerian angles, and for investigating the shape of this surface in the vicinity of the minimum, have been carried out for benzene [38], naphthalene, and anthracene [38] crystals. All these calculations were based on the atom–atom potential method without taking into account electrostatic interactions, because such interactions, as has been shown in Sections 3 and 4, make a negligible contribution both to the lattice energy of hydrocarbon crystals and to the establishment of the equilibrium orientations of the molecules in the crystal. The C–C, C–H, and H–H potential curves we use are written in the form

$$\varphi = -Ar^{-6} + B\exp(-\alpha r)$$

The parameters $A$, $B$, and $\alpha$ have the following values:

|       | $A$ (kcal/mole/Å) | $B$ (kcal/mole) | $\alpha\ (\text{Å}^{-1})$ |
|-------|-------------------|-----------------|---------------------------|
| C–C   | 358               | $4.2 \cdot 10^4$ | 3.58                      |
| C–H   | 154               | $4.2 \cdot 10^4$ | 4.12                      |
| H–H   | 57                | $4.2 \cdot 10^4$ | 4.86                      |

The summation radius in the lattice energy calculation is assumed to be $15\,\text{Å}$. The procedure for selecting the energy surface minimum is described in publication [39].

BENZENE $C_6H_6$

The calculations used the structural data obtained by Cox *et al.* [40] at
$-3°$ and data on the thermal expansion tensor of benzene obtained by Kozhin
and Kitaigorodsky [41] and Bacon *et al.* [41].

*a. The Crystal Structure at Temperatures $T > 0°K$*

Let us find the position of the benzene molecules in the lattice of symmetry
P*bca* with four molecules in the unit cell and lattice parameters $a = 7.46$,
$b = 9.666$, $c = 7.033$ Å corresponding to the structure of this crystal at $-3°C$.
The C–H bond lengths in the molecule are assumed equal to 1.08 Å.

Let us suppose that we do not know anything about the magnitude of
angles $\theta$, $\varphi$, and $\psi$ which describe the rotation of the coordinate system
$x'$, $y'$, $z'$ associated with the molecule, with respect to the system of coordinates
of the crystal (Fig. 8). As a starting point we put these values equal to 0 (the
zero molecule will in this case lie totally on the *ab* plane). We change the
mutual arrangement of the molecules so that the angle $\theta$ assumes all values
from 0 to 180° and calculate the lattice energy variation curve for this process
(Fig. 9a). As can well be expected, this curve is symmetric with respect to
$\theta = 90°$ and has two equivalent minima with a depth of $-8.4$ kcal/mole, of
which we take, for example, the left-hand minimum with abscissa $\theta = 24°$.
Then we hold $\theta$ fixed at 24° and find the optimum arrangement of molecules

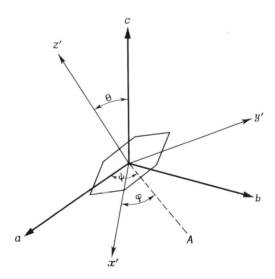

**Fig. 8.** Orientation of coordinate system associated with benzene molecule with respect
to crystal coordinate system. $\theta, \varphi, \psi$, Eulerian angles.

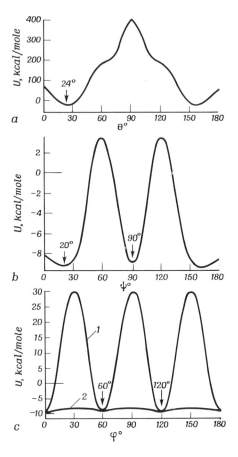

**Fig. 9.** Determination of equilibrium orientation of molecules in lattice with given parameters of unit cell: (a) $U(\theta)$ with $\varphi = 0$, $\psi = 0$; (b) $U(\psi)$ with $\theta = 24°$, $\varphi = 0$; (c) $U(\varphi)$ with $\theta = 24°$, $\psi = 20°(1)$ and $\psi = 90°(2)$.

as angle $\psi$ varies from 0 to 180° (Fig. 9b). For each of the two symmetrically independent minima of the curve $U(\psi)$ ($\psi = 20$ and 90°) which have similar depth ($-9.4$ and $-8.8$ kcal/mole, respectively), find curve $U(\varphi)$ (Fig. 9c, curves 1 and 2). In both cases the optimum molecular arrangements are associated with angles $\varphi$ close to 0, 60, 120°, etc. (within 0–360° six equivalent positions are derived due to the sixfold symmetry axis in the benzene molecule; select any value of this angle, e.g., $\varphi = 180°$).

In this way the steepest descent has yielded sets of Eulerian angles ($\theta = 24°$, $\varphi = 180°$, $\psi = 20°$ and $\theta = 24°$, $\varphi = 180°$, $\psi = 90°$), which we treat as approximations for a more accurate determination of the equilibrium structure. A further descent to the minimum will be carried out by varying all Eulerian

angles simultaneously. In the first case the minimum abscissae are 22, 180, 20°
and the ordinate is 9.6 kcal/mole, in the second case the equilibrium angles
$\theta$, $\varphi$, $\psi$ amount to 47.6, 178, and 104.7°, respectively, while the value of the
energy at the minimum is 11 kcal/mole. Thus, the second of the possible
structures has been found to be more advantageous in terms of energy. The
physically observed structure agrees very accurately with this more advan-
tageous second structure we have derived ($\theta_{exp} = 47.3°$, $\varphi_{exp} = 178°$, $\psi_{exp} =$
107.5°). This calculation can be regarded as a physical justification of the
determination of the molecular packing using the method of so-called geometric analysis carried out with the aid of models on a structure seeker
apparatus [43].

It is interesting to investigate which of the three types of interaction, CC,
CH, or HH, makes the basic contribution to the establishment of the equilibrium orientation of the molecules in the crystal. Analysis of the contributions made by the different atomic species shows that equilibrium in a
benzene crystal depends primarily on the CH and HH interactions (Fig. 10).
Even in cases when neither $U_{CH}$ nor $U_{HH}$ separately passes through a minimum

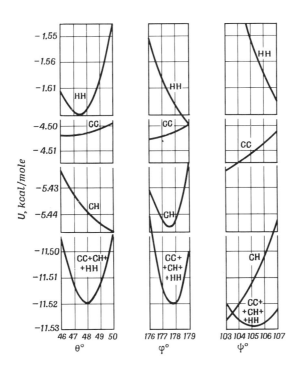

**Fig. 10.** Contributions of C–C, C–H, and H–H interactions to establishment of equilibrium molecular orientation in crystal.

in the vicinity of the equilibrium point, the equilibrium structure is determined by the sum of these interactions. As to the interaction energy of the carbon atoms near the equilibrium point, it varies so slowly that practically no effect is produced on the position of the total minimum of the energy curve. This result is in agreement with the well-known assertion that a benzene crystal has a larger number of intermolecular CH and HH contacts than carbon atom contacts. In this connection the conclusion of Craig *et al.* [20] that the primary role in the establishment of the intermolecular orientations in benzene is played by hydrogen atom repulsions cannot be considered indisputable, since the interaction of C and H atoms was not considered in this work.

### b.  The Crystal Structure at $0°K$

To determine a crystal structure at the absolute zero temperature it is necessary to find the surface minimum $U(a, b, c, \theta, \varphi, \psi)$ over all the six variables (the angles $\alpha, \beta, \gamma$ are held equal to $90°$ for a crystal of orthorhombic symmetry). It would be logical, from the physical point of view, to divide these variables into two groups: the Eulerian angles $\theta, \varphi, \psi$ and the lattice parameters $a, b, c$, and to carry out a minimization for each group separtealy.

Since the accuracy of the determination of the position of the energy surface minimum depends on the accuracy of calculating the energy at each surface point and on the derivatives of the parameters when finding the direction of the steepest descent, descents to the minimum from different points of the surface do not give absolutely identical values of the minimum coordinates. Therefore, for locating the minimum, the authors used a descent from different surface points close to the minimum with subsequent averaging of the results obtained. First, the mutual equilibrium orientation of the molecules was found for a certain, generally speaking, arbitrary point on the energy surface (in our case it coincided with one of the experimentally derived values of the structural parameters) for selected $a, b, c$ by minimizing the lattice energy as a function of the Eulerian angles $\theta, \varphi, \psi$. After that, keeping the obtained values of the Eulerian angles unchanged, the authors performed a descent to the surface minimum $(a, b, c)$. A similar procedure was carried out for an additional one or two starting surface points, and then the lattice parameters obtained from the various descents were averaged. By holding the average magnitudes of the parameters fixed, the authors determined corresponding values of equilibrium Eulerian angles and, finally, the position of the minimum was accurately specified by descent from the mean point first along $a, b, c$ and then along $\theta, \varphi, \psi$. The above procedure can be schematically shown as follows:

$$a_1\, b_1\, c_1\, \beta_1 \qquad\qquad\qquad a_2\, b_2\, c_2\, \beta_2$$
$$\downarrow_{\min(\theta,\varphi,\psi)} \qquad\qquad\qquad\quad \downarrow_{\min(\theta,\varphi,\psi)}$$

$$a_1\, b_1\, c_1\, \beta_1\, \theta_1\, \varphi_1\, \psi_1 \qquad\qquad a_2\, b_2\, c_2\, \beta_2\, \theta_2\, \varphi_2\, \psi_2$$

$$\downarrow_{\min(a,b,c,\beta)} \qquad\qquad\qquad \downarrow_{\min(a,b,c,\beta)}$$

$$a_1^0\, b_1^0\, c_1^0\, \beta_1^0\, \theta_1\, \varphi_1\, \psi_1 \qquad\qquad a_2^0\, b_2^0\, c_2^0\, \beta_2^0\, \theta_2\, \varphi_2\, \psi_2$$

$$\searrow \quad \text{averaging} \quad \swarrow$$

$$a_{12}^0\, b_{12}^0\, c_{12}^0\, \beta_{12}^0\, \theta_{12}\, \varphi_{12}\, \psi_{12}$$

$$\downarrow_{\min(\theta,\varphi,\psi)}$$

$$a_{12}^0\, b_{12}^0\, c_{12}^0\, \beta_{12}^0\, \theta_{12}^0\, \varphi_{12}^0\, \psi_{12}^0$$

$$\downarrow_{\min(a,b,c,\beta)(\theta,\varphi,\psi)}$$

$$a^0\, b^0\, c^0\, \beta^0\, \theta^0\, \varphi^0\, \psi^0$$

Table 4 lists the calculated equilibrium Eulerian angles at three surface points corresponding to the crystal structure at three temperatures. The lattice parameters for this calculation have been taken from the data of Cox *et al.* at $-3°C$ and Bacon *et al.* at $-55$ and $135°$. The bracketed figures are the experimentally obtained angles of the normal to the molecular plane with the axis $c$ (angle $\theta$). Calculations in full agreement with experiment show that the equilibrium Eulerian angles are but weakly dependent on the temperature variations of the lattice parameters. Values of $a, b, c$ averaged over the three subsequent descents with respect to the lattice parameters were taken as initial figures for the final location of the surface minimum over all the variables. The coordinates of this minimum are as follows: $a$, $b$, and $c$ are equal, respectively, to 7.26, 9.41, 6.75 Å, the equilibrium Eulerian angles are 47.3, 178.3, 104.3°.

In order to compare the calculated coordinates of the energy surface minimum with the experimentally derived data, it is necessary to extrapolate

*Table 4*

EQUILIBRIUM ORIENTATION OF MOLECULES IN
BENZENE CRYSTAL AT DIFFERENT TEMPERATURES

| $t(°C)$ | Cell parameters (Å) | | | Eulerian angles (°) | | |
|---|---|---|---|---|---|---|
| | $a$ | $b$ | $c$ | $\theta$ | $\varphi$ | $\psi$ |
| $-3$ | 7.46 | 9.67 | 7.03 | 47.6 (48.1) | 178.0 | 104.7 |
| $-55$ | 7.44 | 9.55 | 6.92 | 47.1 (47.1) | 178.2 | 104.0 |
| $-135$ | 7.39 | 9.42 | 6.81 | 46.7 (46.6) | 178.2 | 104.2 |

## Table 5

PARAMETERS OF UNIT CELL OF BENZENE CRYSTAL
AT ABSOLUTE ZERO TEMPERATURE (Å)

| Cell parameters | Theory | Experiment | |
|---|---|---|---|
| | | [41] | [42] |
| $a$ | 7.26 | 7.27 | 7.36 |
| $b$ | 9.41 | 9.43 | 9.32 |
| $c$ | 6.75 | 6.71 | 6.75 |

the structural data at low temperatures to the absolute zero of temperature. The results of such an extrapolation and the calculated lattice parameters at the potential surface minimum are given in Table 5.

It can be seen that the experimental and analytical results are consistent to within 1–1.5% accuracy.

The good agreement of the calculated and experimentally obtained cell parameters and Eulerian angles indicates that the equilibrium distances $r_0$ between the atomic pairs C–C, C–H, and H–H in the interatomic potential curves, which primarily determine the results of the equilibrium structure calculation, have evidently been selected sufficiently close to the actual values. As to the lattice energy of benzene, it has been estimated at $-11.73$ kcal/mole at the point of the energy surface minimum, but the experimental heat of sublimation is 10–10.5 kcal/mole [44, 45]. This result cannot be thought of as quite satisfactory, all the more so in that the latent energies at the potential surface minima for other hydrocarbons as well are found to be below the experimental values of the heats of sublimation.

It is sufficiently clear that the heats of sublimation depend mainly on the depths of the energy wells in the interatomic potential curves. In the above C–C, C–H, and H–H curves the depths of the potential wells were assumed equal to 0.067 kcal/mole. If we decrease the depth of the potential well of the C–H curve to 0.057 kcal/mole and of the H–H curve to 0.048 kcal/mole without changing the positions of the minima, and the value of $\alpha$ in the potentials, then for benzene, for example, we obtain 10.5 kcal/mole lattice energy at the potential surface minimum in good agreement with experiment.

### c. The Configuration of the Energy Surface in the Vicinity of the Minimum

Figure 11 shows the two-dimensional cross sections of the three-dimensional energy surface of benzene $U(\theta, \varphi, \psi)$ through the energy minimum point in the three main projections. The cross section is calculated with an accuracy of $3°$ for the angles $\theta, \varphi, \psi$ varying within $\pm 9°$ from the equilibrium values.

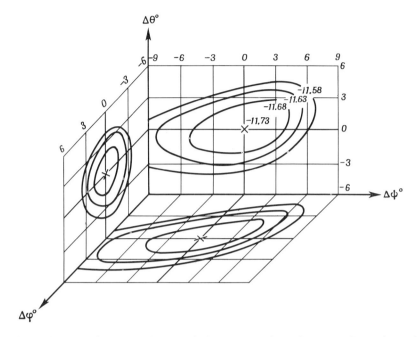

**Fig. 11.** Cross Section $U(\theta, \varphi, \psi)$ through the energy surface minimum point. Values of energy on equipotential lines are given in kilocalories per mole.

Besides, 1°-accuracy calculation has been made in the immediate vicinity of the minimum, with a view to studying in more detail the surface near the minimum. The three-dimensional potential well in this range is quite smooth and is gently sloping with respect to variations of the angle $\psi$ (near the minimum $\partial U/\partial\psi = 0.011$ kcal/mole deg), but is steeper in the directions of $\theta$ and $\varphi$ ($\partial U/\partial\theta = 0.018$, $\partial U/\partial\varphi = 0.023$ kcal/mole deg). The cross section has no special features (local minima, "rough" walls) in the entire range of Eulerian angle variations we have considered. This range, by the way, fully covers all possible amplitudes of the angular vibrations of the molecules in the lattice which, for example, vary at different temperatures from 2.5 to 7.9° [41] for benzene molecule vibrations in the ring plane.

As long as lattice distortions do not cause changes in the space group and in the mutual orientations of the molecules, the lattice energy is a three-dimensional function of the unit cell spacings $U = U(a, b, c)$. The three main cross sections of this surface $U(a, b)$, $U(a, c)$, $U(b, c)$ through the energy minimum point over all the variables calculated with 0.5 Å accuracy are shown in Fig. 12. The surface $U(a, b, c)$ differs from the surface $U(\theta, \varphi, \psi)$ in that, in principle, it can have only one minimum whose position depends on the size and orientation of the molecules in the lattice and on the selected interatomic

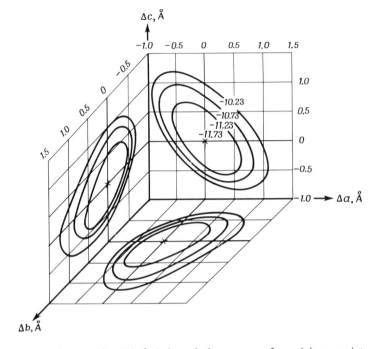

**Fig. 12.** Cross section $U(a,b,c)$ through the energy surface minimum point.

potential curves. The configuration of the surface is similar to that of the interatomic potential—steep ascent from the equilibrium point towards decreased lattice spacings, i.e., distances between the molecules, and more gentle ascent in the opposite direction. A cross section obtained with a higher accuracy (0.1 Å) indicates that near the minimum the surface $U(a,b,c)$ is rather flat; variations in the lattice parameters of about 0.1 Å are accompanied by changes in the lattice energy of the order of several hundredths of one kilocalorie per mole. Then the potential well becomes steeper. With a simultaneous increase of all the benzene lattice parameters corresponding to the transition from the surface minimum point to a point close to the melting temperature (270°K), the lattice energy, according to these calculations, varies by approximately 0.6 kcal/mole, i.e., within reasonable limits.

NAPHTHALENE $C_{10}H_8$ [46]

The crystal is monoclinic, space group $P2_1/a$, two molecules in a unit cell. Theoretical prediction of the equilibrium structure is based on the initial structure derived by Abrahams et al. [47] at room temperature. The co-ordinates of the H atoms were estimated by putting the length of the C–H bond equal to 1.08 Å. Data on the parameters of the lattice at low temperatures obtained by Ryzhenkov and Kozhin [48] were also used. The calculated equilibrium orientation of the molecules at three surface points corresponding to the structure of these crystals at three different temperatures is described by the Eulerian angles shown in Table 6. The results obtained at room temperature can be compared with experiment: $\theta_{exp} = 29°$, $\varphi_{exp} = 40°$, $\psi_{exp} = 296°$.

*Table 6*

EQUILIBRIUM ORIENTATION OF MOLECULES
IN NAPHTHALENE CRYSTAL AT DIFFERENT TEMPERATURES

| | Cell parameters (Å) | | | | Eulerian angles (°) | | |
|---|---|---|---|---|---|---|---|
| $T$ (°K) | $a$ | $b$ | $c$ | $\beta$ | $\theta$ | $\varphi$ | $\psi$ |
| 78 | 8.081 | 5.955 | 8.630 | 124°30′ | 30.3 | 47.0 | 292.6 |
| 173 | 8.138 | 5.971 | 8.649 | 124°11′ | 30.3 | 46.6 | 293.0 |
| 293 | 8.240 | 5.997 | 8.666 | 122°55′ | 28.8 | 44.2 | 294.7 |

A further descent along the lattice parameters and then determination of the Eulerian angles in accordance with the scheme in Section 9,b has yielded the following result: at the minimum point $a = 7.94$, $b = 5.95$, $c = 8.57$ Å; $\beta = 123°32′$, $\theta = 29.2$, $\varphi = 46.0$, $\psi = 293.8°$ (monoclinic angle $\beta$ has also been taken here as a minimization parameter). Experiment (extrapolation of the data obtained by Ryzhenkov and Kozhin to 0°K) gives the parameters of the unit cell, $a, b, c, \beta$ equal to 8.06, 5.95, 8.62 Å, 124°36′, respectively, at the absolute zero of temperature. Thus, for naphthalene, as well as for benzene, the equilibrium lattice parameters calculated by means of atom–atom potentials are consistent with the experimental data to an accuracy of about 1–2%.

The function $U(\theta, \varphi, \psi)$ in the vicinity of the minimum for naphthalene has been studied in the range of ±12° deviation of Eulerian angles from the equilibrium position. With minor deviations from the equilibrium (±4°), the potential well has an approximately equal slope (0.2–0.3 kcal/mole deg) in the direction of $\theta, \varphi, \psi$. If the angles $\varphi$ and $\psi$ change simultaneously, it is much steeper (0.06 kcal/mole deg). The cross section $U(\varphi, \psi)$ at an equilibrium value of $\theta$ is shown in Fig. 13. The same cross section with $\theta = \theta_0 + 8°$ (Fig. 13) shows that in the vicinity of the main minimum, −17.9 kcal/mole deep, there are two additional, much more shallow, minima (about −13 kcal/mole).

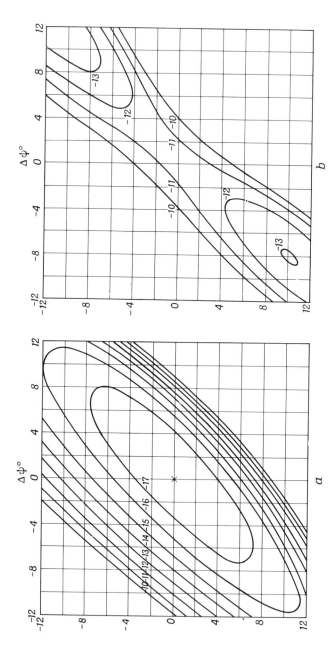

*Fig. 13.* Contour maps of naphthalene lattice energy as a function of Eulerian angles: (a) cross section $U(\varphi, \psi)$ through the energy surface's minimum point (the position of the minimum is designated by ×); (b) cross section $U(\varphi, \psi)$ at $\theta = \theta_0 + 8°$. Contours with energy above $-10$ kcal/mole are not shown.

Similar to benzene, the cross section $U(a, b, c^*)$ through the energy surface minimum point ($c^*$ in this case implies the distance between dense molecular layers) can be regarded as a three-dimensional analog of the interatomic potential function $\varphi(R)$. Variation of $c \cos \beta$, a displacement of the layers, adds nothing to the total picture; it only shifts the position of the total minimum of the crystal potential energy surface.

ANTHRACENE $C_{14}H_{10}$

Space group $P2_1/a$, two molecules in a unit cell. The structure at room temperature has been determined by Mathieson et al. [49] and specified by Cruickshank [50]. The coordinates of hydrogen atoms taken from Cruickshank's work are such that they give a 1.09 Å C–H bond length in the molecule. The Eulerian angles relating the molecular coordinate system to the crystallographic system are 66.3, 110.2, 61.7°, respectively. Variations of the lattice parameters at low temperatures for anthracene have been made in publications [51, 52] and by Myasnikova [53]. In his work Mason [52] gives the angles of inclination of the molecular coordinate axes to the crystallographic axes at two temperatures, 290 and 95°K. From these data it can be inferred that temperature variations cause insignificant changes in the Eulerian angles. Angle $\theta$ exhibits the largest change (1°). The calculations show the variations of the Eulerian angles with temperature (Table 7).

### Table 7

EQUILIBRIUM ORIENTATION OF MOLECULES
IN ANTHRACENE CRYSTALS AT DIFFERENT TEMPERATURES

| $T$ (°K) | Cell parameters (Å) | | | | Eulerian angles (°) | | |
|---|---|---|---|---|---|---|---|
| | $a$ | $b$ | $c$ | $\beta$ | $\theta$ | $\varphi$ | $\psi$ |
| 78 | 8.439 | 5.992 | 11.112 | 125°05′ | 65.2 | 109.4 | 64.0 |
| 300 | 8.550 | 6.028 | 11.173 | 124°35′ | 66.0 | 109.2 | 64.3 |
| 425 | 8.685 | 6.059 | 11.240 | 123°32′ | 67.7 | 109.4 | 63.8 |

The agreement of the theoretical and experimental values is satisfactory; the temperature dependence of the tilt of the molecules to the coordinate axes is not strong and is mainly manifested in a change of the angle $\theta$.

Applying the above procedure for locating the minimum, the authors have obtained the following values of the equilibrium lattice parameters and the Eulerian angles for anthracene: $a = 8.39$, $b = 5.95$, $c = 11.02$ Å; $\beta = 124°30′$,

$\theta = 66.3$, $\varphi = 109.6$, $\psi = 64.9°$. The extrapolation of the low-temperature experimental data to the absolute zero of temperature has given the following lattice parameters: 8.44, 5.99, 11.11 Å, 125°05'. The good agreement between the calculated equilibrium structure and the experimental one is another proof that the interatomic potential curves selected by the authors are a sufficiently good approximation for the prediction of the packing of hydrocarbon molecules in a crystal.

The cross section $U(\theta, \varphi, \psi)$ for anthracene calculated within $\pm 12°$ variations of the Eulerian angles from the equilibrium values has a peculiar shape. Figure 14a,b,c represents three main cross sections of this surface, and Fig. 14d gives the cross section $U(\varphi, \psi)$ at $\Delta\theta = -4°$. These contour maps show the main minimum of $-23.2$ kcal/mole, energy "hills" and "valleys" and also a local minimum (Fig. 14d) with a depth about 20 kcal/mole.

Italian and American researchers have carried out a number of experiments based on the above concept of representing lattice energy as a sum of atom–atom potentials and prediction of the packing of molecules by varying their positions in the cell.

In the light of these concepts, Giglio and Liquori [54] considered a structure of hexamethylbenzene convenient for calculations. The structure easily lends itself to calculations because the compound crystallizes in group PĪ with one molecule in the unit cell. They calculated the crystal's potential energy which is here a function only of the orientational rotational degrees of freedom. It was found that in the three-dimensional energy distribution the deepest minimum is quite consistent with the actual molecular arrangement.

Corradini et al. [55] have made structural analyses of a number of dicarboxylic acids and have calculated the lattice energies of these crystals. The objective of the authors was to verify the heuristic possibility of applying the densest-packing principle for predicting the type of molecular packing. As an example, they describe the case of meso-$\beta,\beta'$-dimethyladipic acid. The analysis gave the only version of the structure suitable from the viewpoint of the densest-packing principle. The same results have been obtained in calculating the lattice energy by means of the atom–atom potential technique. This result, confirmed by subsequent determination of the structure, shows that it is possible to predict the structure of an organic crystal.

In Williams [31, 56] the packing of the molecules in a crystal was calculated by minimizing the energy with respect to the positions of the molecules by using the method of steepest descent. The minimization involves the variation of the three translational parameters that determine the coordinates of the center of the molecule, three orientational parameters, six or fewer lattice constants and, if necessary, the parameters describing rotations around single bonds inside the molecule. The crystal lattice energy was calculated by means of the technique developed by the author and Mirskaya. Potential functions

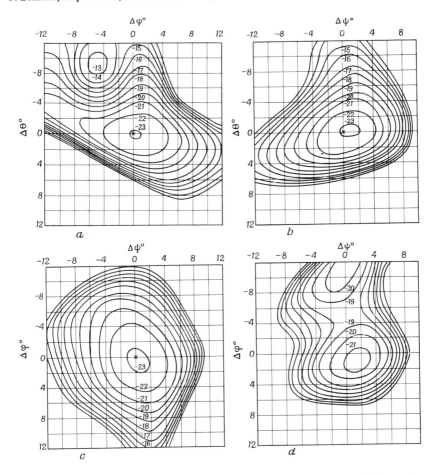

*Fig. 14.* Contour maps of anthracene lattice energy as a function of Eulerian angles: (a), (b), (c) cross sections $U(\theta, \varphi)$, $U(\theta, \psi)$, $U(\varphi, \psi)$ through the energy surface minimum point (the position of the minimum is designated by $\times$); (d) cross section $U(\varphi, \psi)$ at $\theta = \theta_0 - 4°$. Contours with energy above $-13$ kcal/mole are not shown.

in the form of 6-exp, with parameters given by the author and Bartell [57], were tried. In order to avoid cumbersome calculations, summing is terminated when interatomic distances exceed the position of the potential function minimum by more than 2 Å. It has been found that such a discontinuity in the series strongly affects the magnitude of the lattice energy, but only slightly the molecular packing. Good results have been derived for naphthalene crystals (the positions of the carbon atoms agree with the experimental figures within $\pm 0.04$ Å), anthracene, and phenanthrene crystals. Somewhat poorer results are available for 1,3,5-triphenylbenzene.

10. The Condition for the Structural Stability of an Organic
Crystal and the Principle of Close Packing

The actual structure of a crystal is characterized by minimum free energy. To explain why a crystal has a specific molecular packing pattern means to show that all packings other than the actual one display a higher free energy.

In this discussion I shall disregard the effects of the crystalline field on the shape of the molecule, and the intermolecular vibrations. Generally speaking, these effects are insignificant and will be considered in the next section.

If the mutual molecular orientation is given, the free energy can be calculated on the basis of the law governing intermolecular interaction. The crystal free energy is a function of the symmetry and size of the unit cell, the coordinates of the center of gravity of the molecule and the Eulerian angles which define the molecular orientation [58].

The condition of stability of a crystal structure which implies that the free energy variations must be zero can be written

$$\delta F = \delta U + \delta F^{\text{vib}} = 0$$

At absolute zero the intermolecular vibrational energy is about 0.2 kcal/mole. Variations of this value from structure to structure do not exceed hundredths of a kilocalorie per mole. Since the potential-well depth is much larger than that, a crystal structure at absolute zero must be consistent with the minimum energy of intermolecular interaction; in other words, a stable structure must satisfy the condition

$$\delta F = \delta U = \frac{\partial U}{\partial R_i}\delta R_i + \frac{\partial U}{\partial \varphi_i}\delta\varphi_i + \frac{\partial U}{\partial a_i}\delta a_i = 0$$

where the $R_i$ are the radius vectors of the centers of the molecules; $\varphi_i$ are the Eulerian angles; and $a_i$ are the unit cell parameters.

If our considerations are restricted to centrosymmetric structures of the naphthalene type, the first term of this equation is omitted. Crystals of monoclinic symmetry exhibit a seven-dimensional energy surface. Modern computers make the construction of such a surface quite feasible. This problem has been partially solved by us for orthorhombic benzene (see above), and also by Giglio and Liquori for triclinic hexamethylbenzene [54]. In both instances, the energy surface (meaning, of course, the part of the surface which corresponds to the symmetry of only one of the possible space groups) displays one deep minimum which is associated with the actual structure.

Comprehensive investigation of the energy surface, including consideration of different symmetries, has intrinsic interest, because it helps understand the phase diagram of a substance in the solid state. The energy surface must show a multitude of minima. Shallow minima are of no interest because, due to thermal fluctuations, they cannot generate a structure. First, we choose the lowest-lying minimum among all the deep minima. This minimum should correspond to the structure that is stable at absolute zero. Other deep minima lying above the lowest minimum represent actual high-temperature modifications only if they correspond to minimum free energy.

Let us consider two minima: the deepest minimum with energy $U_0$ and the minimum that follows the deepest one and has higher energy $U_1$. At absolute zero the stable structure $U_0$ and the metastable structure $U_1$ are characterized by the abscissae of their potential-well bottoms. As the temperature rises, the representative point shifts to the saddle point between the wells. At a certain temperature $T$, the lattice binding energies become $U_0 + \Delta U_0$ and $U_1 + \Delta U_1$, and the vibrational parts of the free energy $F_0^{vib}$ and $F_1^{vib}$. A polymorphic change occurs if the total free energy of structure 1 becomes equal to the total free energy of the structure stable at $0°$, i.e., if

$$U_0 + \Delta U_0 + F_0^{vib} = U_1 + \Delta U_1 + F_1^{vib}$$

Only very few potential wells, if any, can satisfy this condition. It is clear from the above discussion that a polymorphic change is possible only if the entropy increases with temperature faster in the metastable structure. This condition is, however, not sufficient because the crystal may melt before the loss in the potential energy has been balanced in this way.

It is known that the heat required to convert a crystal into another crystal (but not for the transition into a rotational crystalline state) rarely exceeds 1 kcal/mole. It may be supposed that the free vibrational energy also changes during this transition by approximately the same value. This means that if a structure stable at absolute zero is represented by a potential well lying deeper than the others by about the same value, no polymorphic changes can occur in this substance.

Because the potential energy of molecular interaction changes but slightly under the effects of crystal heating, the minima of the energy surface are representative of all structures possible in the substance, several of these being the actual structures. In other words, each actual structure is given by its own minimum on the energy surface of our model.

As was shown in Sections 3 and 4, the coordinates of a minimum are only slightly displaced by electrostatic interactions. This means that the mutual molecular arrangement is characterized by a minimum of the energy surface that can be calculated by means of the atom–atom potential model. Of course, this minimum may not necessarily be the deepest, but only one of the deepest. The fact that every crystal structure has its own minimum on the energy surface which represents the sum total of the atom–atom interactions is suggestive of the geometrical concept of the close-packing principle. Only one additional simplifying assumption is required to effect transition to the

geometrical model: It is necessary to consider the molecules as being absolutely rigid, i.e., to substitute the repulsive portion of the curve by a vertical
line. In this case, the trend toward minimization of the potential energy is
evidently analogous to the requirement of locating a maximum number of
atoms at the shortest distances from each other.

11. THE EFFECT OF THE CRYSTALLINE FIELD ON THE SHAPE OF A MOLECULE

In order to estimate correctly the objective nature of the X-ray diffraction
technique for studying organic crystals, and to understand the differences in
the experimental determinations of interatomic distances and valence angles,
one must have a clear idea of the effect of the crystalline field on the structure
of a molecule.

If we consider a molecule to be sufficiently labile, the condition of crystal
structure stability must be written not in the form given in the previous section,
but as

$$\delta F = \delta U + \delta F^{\text{vib}} + \delta E = 0$$

where $\delta E$ denotes the molecular energy variation. This means that when
evaluating the comparative advantages of several different packing patterns,
it is necessary, strictly speaking, to estimate the variations in the energy of
the molecules themselves.

It is self-evident that these variations cannot be very large. It may perhaps
seem at first glance that where high accuracies are implied, the effect of the
crystalline field must always be taken into account. If this were true, it would
immensely complicate the problem of calculating an appropriate structure by
energy minimizations; the number of parameters would greatly increase if
the parameters of the molecule itself had to be added to the lattice parameters.

The solution of our problem is considerably facilitated by the fact that in
most cases it is not necessary to allow for the effect of the crystalline field at all
while in certain specific instances it is sufficient to consider only the parameters
associated with the rotation of the parts of the molecule about single bonds.

There are two principal consequences of the influence of the crystalline field
on the shape of a molecule: its deviation from the optimum conformation,
and the selection of one of the possible conformations.

We shall first consider molecules having only one conformation and try to
answer the following question: What is the difference between the geometry
of these molecules when in a crystal from when they are free molecules?
A great body of experimental information now available can help evaluate
the extent of the molecular deformations under the effects of the crystalline
field. There are four approaches to the solution of this problem. First, we may
compare the structures of gaseous and crystalline molecules; secondly, we

may make comparative studies of the geometries of crystallographically independent molecules in the same crystal; thirdly, we may analyze the structure of a molecule whose symmetry in a crystal is lower than that of the free molecule; and, finally, we may compare molecules of different polymorphic modifications.

Consider several examples. Studies of gaseous benzene molecules have established that the structure of a benzene molecule is a regular plane hexagon formed by C atoms with a distance of $1.397 \pm 0.001$ Å between the carbon atoms, and the length of the C–H bond equal to 1.084 Å. X-ray diffraction analysis of crystals yields the following values for this molecule [40]: bond lengths: 1.379, 1.374, 1.378 Å; valence angles: $119°50'$, $120°42'$, $119°28'$.

A phenanthrene molecule occupies a general position in the crystal. It is, however, obvious that this molecule exhibits at least twofold symmetry. The direct experimental results [59] are shown in Fig. 15a. The equivalent angles differ by several degrees, the difference between equivalent distances ranges from 0.01 to 0.03 Å.

Figure 15b shows the experimentally determined sizes of the two halves of

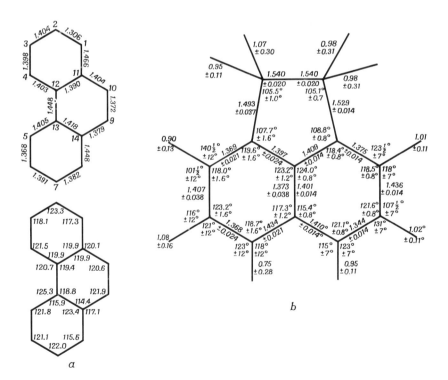

*Fig. 15.* The difference in bond length due to experimental errors.

an acenaphthene molecule [60]. The molecules in a crystal occupy two independent positions with mirror symmetry. The order of difference is here about the same as above.

The above data are typical enough. A question which now arises is as follows: Should these differences be ascribed to experimental errors or do they reflect actual differences in the molecular structure caused by the effect of the crystalline field?

It will be now shown by simple calculations that in all these cases the difference is due to experimental errors or to artifacts.

For each concrete case it would be necessary to point out the following. If the atoms are displaced so as to eliminate distortions, the molecular energy is bound to decrease by $\Delta E$, and the lattice binding energy to increase by $\Delta U$. The distortion of a molecule may be energetically justified if $|\Delta E| \ll |\Delta U|$. On the contrary, distortion is impossible if $|\Delta E| \gg |\Delta U|$. Strictly speaking, changes in free energy rather than in potential energy should be calculated. Nevertheless, the contribution of minor distortions to variations in molecular vibrations seems to be quite insignificant.

$\Delta E$ can be calculated for each particular case, for example, using a mechanical model; $\Delta U$ can be obtained by the atom–atom potential method. It is, however, by no means necessary to discuss concrete cases. It is quite sufficient to consider the curves given in Fig. 16. The figure shows the changes in the molecular energy which are caused by changes in the C–C bond length (a), displacement of an H atom in an aromatic molecule in the plane and per-

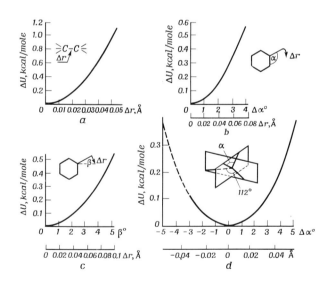

*Fig. 16.* The changes in molecular energy caused by different distortions.

pendicular to the plane (b, c), and changes in an HCH angle in an aliphatic molecule calculated from [61, 62].

For comparison, we have taken a typical example of a benzene srystal and calculated the variations in the lattice binding energy caused by the displacements of different C and H atoms in various directions from their true positions. The results have indicated that a 0.03 Å displacement of an atom from its true position always changes the lattice binding energy by less than 0.01 kcal/mole. In other words, in the scale of the figure which illustrates the energy of molecular distortion at the expense of variations in bond lengths, the curve describing lattice binding energy variations practically coincides with the X axis.

Thus, we arrive at the following conclusion: The crystalline field does not affect bond lengths. It should be pointed out, by the way, that from my point of view, the best measures of the accuracy of the X-ray diffraction analysis of organic crystals are the discrepancies in bond lengths that should be equal from the chemical viewpoint. With this criterion applied, there seems to be no experiment that can claim an accuracy above 0.01 Å.

It now remains to discuss the possible effects of the crystalline field on the valence angle energy and the nonbonded atom interaction energy.

From Fig. 16 it follows that the changes in the valence angles are not balanced by a gain in the lattice binding energy. The results are equally obvious a priori in cases where changes in a molecule are associated only with variations of the distances between nonbonded atoms, i.e., in the case of rotation about a single bond. Here, the major role is played by the steepness of the walls of the potential well representing the conformation of the molecule. Cases may be encountered when the nonbonded atoms of a molecule in an optimum conformation are located at distances close to equilibrium distances. The steepness of the potential well near an equilibrium will then be small, and the amount of interaction between the atoms of the two parts of the molecule will be of the same order as the interactions between them and the nonbonded atoms of adjacent molecules. In these conditions, the crystalline field is obviously quite able to cause considerable rotation of the parts of the molecule about the single bond. Therefore, calculation of an optimum structure should involve not only consideration of variations in the packing pattern, but also changes in the molecular energy which must be expressed in terms of angular rotation about single bonds. If the distance between nonbonded atoms decreases as the molecule rotates, the relevant energy quickly rises. After the parts of the molecule have rotated through a certain angle, intramolecular repulsion is balanced by the crystalline field forces.

Thus, the crystalline field can bring a molecule out of the optimum conformation by means of rotation about a single bond, but only if the free molecule has not been strained. On the other hand, if the optimum conformation of the free molecule exhibits an appreciable decrease in the distances between nonbonded atoms, this molecule is rigid with respect to the crystalline field, because the conformational well is very steep.

A classical example of the influence of the crystalline field on a molecular conformation is biphenyl which is nonplanar in the gaseous state and planar in the crystal. The flattening of the molecule involves loss of the ortho-atom repulsive energy, and an increase of the packing density (planar systems are more convenient for packing).

Molecules displaying relatively easy rotation of the benzene rings are generally good objects for studies of this problem.

An example appears in publication [63] which reports the results of the calculation of the structure of a 2-bromo-1,1-di-$p$-tolyl-ethylene crystal. The observation has been made that a phenyl ring located close to the bromine atom is turned 67° out of the ethylene plane, while the second phenyl ring is 24° out of plane. Calculations have shown that an optimum conformation is obtained at angles of 45° and 35°, respectively. Using the formula for the atom–atom potential model given above in Section 7, the authors have calculated the energy of interaction of a given molecule with adjacent molecules as a function of these rotation angles. It has been found that, with due allowance for molecular packing, the optimum ring rotation angles are 67 and 25°, which is fully in agreement with the experiment.

To summarize, it must be emphasized once again that the crystalline field does not change the bond lengths and valence angles of organic molecules. The major effect of the crystalline field on the shape of a molecule consists in provoking such rotations about single bonds as are most favorable for packing. However, these rotations can take place only if the optimum conformation of the molecule is unstrained.

Since the crystalline field compresses the molecule, then, other conditions being equal, a conformation is selected in which the molecule occupies the least volume.

If the energies of several conformations of one molecule differ to a more or less considerable degree, the crystal is composed of optimum molecules. The reason is that in this case large possibilities are provided for formation of different packing patterns, and, therefore, a decrease of the molecular energy by 2–3 kcal/mole can hardly be balanced by an increase of the lattice binding energy.

REFERENCES

1. O. K. Rice, "Electronic Structure and Chemical Bonding." McGraw-Hill, New York, 1940.
2. E. F. Westrum and J. P McCullough, *Phys. Chem. Org. Solid State* **1**, 1 (1963).

3. H. Margenau, *Rev. Mod. Phys.* **11**, 1 (1939).
4. F. London, *Z. Phys. Chem. Abt. B* **11**, 222, 236 (1930); *Z. Phys.* **63**, 245 (1930).
5. J. Slater and J. Kirkwood, *Phys. Rev.* **37**, 682 (1931).
6. B. M. Axilrod and E. Teller, *J. Chem. Phys.* **11**, 299 (1943).
7. L. Jansen, *Phys. Rev.* **125**, 1798 (1962).
8. V. J. Wojtala, *Acta Physica. Pol.* **25**, 27 (1964).
9. G. Mie, *Ann. Phys. (Leipzig)* **11**, 657 (1903).
10. J. E. Lennard-Jones, *Proc. Phys. Soc. London* **43**, 461 (1931).
11. R. A. Buckingham, *Proc. Roy. Soc. Ser. A* **168**, 264 (1938); R. A. Buckingham and J. Corner, *Ibid.* **189**, 118 (1947).
12. O. K. Rice, *J. Amer. Chem. Soc.* **63**, 3 (1941).
13. J. E. Lennard-Jones and L. Ingham, *Proc. Roy. Soc. Ser. A* **107**, 636 (1925).
14. A. I. Kitaigorodsky and K. V. Mirskaya, *Kristallografiya* **9**, 634 (1964); **10**, 162 (1965).
15. P. P. Ewald, *Ann. Phys. (Leipzig)* **64**, 253 (1921).
16. E. S. Campbell, *J. Phys. Chem. Solids* **24**, 197 (1963).
17. D. P. Craig and S. H. Walmsley, *Mol. Phys.* **4**, 113 (1961).
18. R. G. S. Arridge and C. G. Cannon, *Proc. Roy. Soc. Ser. A* **278**, 91 (1964).
19. F. W. De Wette and G. E. Schacher, *Phys. Rev. A* **137**, 78 (1965).
20. D. P. Craig, R. Mason, P. Pauling, and D. P. Santry, *Proc. Roy. Soc. Ser. A* **286**, 98 (1965).
21. O. Nagai and T. Nakamura, *Progr. Theor. Phys.* **24**, 432 (1960).
22. R. M. Hill and W. V. Smith, *Phys. Rev.* **82**, 451 (1951).
23. M. Mizushima, *Phys. Rev.* **83**, 94 (1951).
24. J. O. Hirschfelder, F. Curtiss, and R. Bird, "Molecular Theory of Gases and Liquids." Wiley, New York, 1954.
25. L. Pauling, "The Nature of the Chemical Bond." Cornell Univ. Press, Ithaca, New York, 1939.
26. A. I. Kitaigorodsky, *Advan. Struct. Res.* **3**, 173 (1970).
27. A. Müller, *Proc. Roy. Soc. Ser. A* **154**, 624 (1936).
28. L. D. Landau and E. M. Lifshitz, "Quantum Mechanics," Pergamon, Oxford, 1965.
29. K. S. Pitzer, *Advan. Chem. Phys.* **2**, 59 (1959).
30. A. I. Kitaigorodsky and K. V. Mirskaya, *Kristallografiya* **6**, 507 (1961); **9**, 174, (1964).
31. D. E· Williams *J. Chem. Phys.* **47**, 4680 (1967).
32. C. Reid, *J. Chem. Phys.* **30**, 182 (1959).
33. J. Hove and J. A. Krumhansl, *Phys. Rev.* **92**, 569 (1953).
34. E. Walley, *Trans. Faraday Soc.* **54**, 1613 (1958).
35. K. Suzuki, S. Onishi, T. Koide, and S. Seki, *Bull. Chem. Soc. Jap.* **29**, 127 (1956).
36. K. F. Herzfeld and M. Goeppert-Mayer, *Phys. Rev.* **46**, 995 (1934).
37. A. I. Kitaigorodsky, K. V. Mirskaya, and A. B. Tovbis, *Kristallografiya* **13**, 225 (1968).
38. K. V. Mirskaya and I. E. Kozlova, *Kristallografiya* **14**, 412 (1969).
39. N. P. Zhidkov, A. I. Kitaigorodsky, B. M. Shchedrin, and A. B. Tovbis, *Coll. Works of Univ. Comp. Centre* (1968).
40. E. G. Cox, D. W. J. Cruickshank, and J. A. S. Smith, *Proc. Roy. Soc. Ser. A* **247**, 1 (1958).
41. V. M. Kozhin and A. I. Kitaigorodsky, *Zh. Fiz. Khim.* **29**, 2075 (1955).
42. G. E. Bacon, N. A. Curry, and S. A. Wilson, *Proc. Roy. Soc. Ser. A* **279**, 98 (1964).
43. A. I. Kitaigorodsky, "Organic Chemical Crystallography," Consultants Bureau, New York, 1961.
44. K. Wolf and H. Weghofer, *Z. Phys. Chem. Abt. B* **39**, 194 (1938).
45. G. Milazzo, *Ann. Chim. (Rome)* **46**, 1105 (1956).

46. K. V. Mirskaya and I. A. Kozlova, *Kristallografiya* **14**, 412 (1969).
47. S. C. Abrahams, J. M. Robertson, and J. G. White, *Acta Crystallogr.* **2**, 233 (1949).
48. A. P. Ryzhenkov and V. M. Kozhin, *Kristallografiya* **12**, 1079 (1967).
49. A. M. L. Mathieson, J. M. Robertson, and V. C. Sinclair, *Acta Cryst. allogr* **3**, 245 (1950).
50. D. W. Cruickshank, *Acta Crystallogr.* **9**, 915 (1956).
51. V. M. Kozhin and A. I. Kitaigorodsky, *Zh. Fiz. Khim.* **27**, 1676 (1953).
52. R. Mason, *Acta Crystallogr.* **17**, 547 (1964).
53. A. P. Ryzhenkov, V. M. Kozhin, and R. M. Myasnikova, *Kristallografiya* **13**, 1028 (1968).
54. E. Giglio and A. M. Liquori. Abstracts of papers. 7th *Int. Crystallogr. Congr. Moscow, 1966,* 8.37, *Acta Crystallogr* **21**, A112 (1966).
55. P. Corradini, G. Avitabile, P. Ganis, and E. Martuscalli, Abstracts of papers. 7th *Int. Crystallogr. Congr. Moscow, 1966,* 8.26, *Acta Crystallogr* **21**, A108 (1966).
56. D. E. Williams, *Science* **147**, 605 (1965).
57. L. S. Bartell, *J. Chem. Phys.* **32**, 827 (1960).
58. A. I. Kitaigorodsky, *Acta Crystallogr.* **18**, 585 (1965).
59. J Trotter, *Acta Crystallogr.* **16**, 605 (1963).
60. H. W. W. Ehrlich, *Acta Crystallogr.* **10**, 699 (1957).
61. A. I. Kitaigorodsky and V. G. Dashevskij, *Theor. Exp. Chem.* (*USSR*) **3**, 35 (1967).
62. V. G. Dashevsky and A. I. Kitaigorodsky, *Theor. Exp. Chem.* (*USSR*) **3**, 43 (1967).
63. G. I. Casalone, C. Mariani, A. Mugnoli, and M. Simonetta, *Acta Crystallogr.* **22**, 228 (1967).

# Chapter III
## Lattice Dynamics

1. THE EQUATIONS OF MOTION

In most cases the molecules of an organic crystal execute slight libration about their equilibrium positions. The vibration amplitudes of the centers of gravity of molecules are of the order of a few tenths of an angstrom, while the amplitudes of angular librations are 2–3°.

Intramolecular displacements have much smaller amplitudes. One may therefore assume that the crystals of a fairly representative class of organic compounds have much in common with systems of rigid molecules. It follows then that the lattice dynamics of such a crystal are determined by small vibrations described in terms of six coordinates, e.g., three coordinates of the molecular center of gravity and three Eulerian angles that specify the spatial orientation of the molecule. Molecules shaped like a disk, cone, sphere, or cylinder may execute more complicated types of motion, such as reorientations. These phase states of organic compounds should be considered separately.

The problem to be discussed now is a special case of a well-known problem of classical mechanics. Let us recall that the acceleration calculated for each coordinate of a system is defined by the sum of the generalized forces acting on this coordinate; these forces result from coordinate displacements, provided that such displacements of the coordinates of the system from their equilibrium values are small [1]. Under these conditions, the contribution to the total force along each coordinate is proportional to the first power of its displacement $t$.

Analytically this theorem can be formulated as

$$\ddot{t}_I = -\sum_{I'} \lambda_{II'} t_{I'} \tag{1.1}$$

where the values of the subscript $I$ label all the coordinates of all molecules. As is well known, particular solutions of this set of equations are given by functions describing harmonic oscillations with the same frequency for all coordinates:

$$t_I = e_I e^{-i\omega t}$$

Substitution in Eq. (1.1) gives

$$\omega^2 e_I = \sum_{I'} \lambda_{II'} e_{I'} \tag{1.2}$$

If

$$|\lambda_{II'} - \omega^2 \delta_{II'}| = 0 \tag{1.3}$$

then, in accordance with a general theorem of algebra, $\sum \lambda_{II}$ is equal to the sum of all the roots of Eq. (1.3). It follows then that

$$\overline{\omega^2} = (1/n) \sum \lambda_{II} \tag{1.4}$$

where $n$ is the number of the system's coordinates. Thus the mean square of the system's frequencies depends only on the diagonal matrix elements, i.e., it is uniquely determined by the forces "acting on the coordinate" and causing deviations of the same coordinate from its equilibrium value.

Looking into the construction of the general solution of Eq. (1.2), we may note that its form (1.3) is of no interest; we should make use of the fact that we are concerned here with a crystal structure. If the periodic character of the structure is taken into account, it will be easily seen that the vibration amplitudes of equivalent coordinates for translationally coupled molecules are equal. In other words, it is the planewave type solutions of (1.1) that are to be sought.

Let the index $\mu$ refer to the molecules within one cell (all molecules having the same $\mu$ are translationally identical). We shall use the index $s$ to specify various displacements of molecules from their equilibrium positions. The values taken by this index are integers between one and six, and these values specify three linear displacements $x, y, z$ and three angular displacements $\varphi, \psi, \eta$. Since, in the case under discussion the displacements are small, the increments of the Eulerian angles are for all practical purposes equal to the angles of rotation about the corresponding axes. Now, the $s$th displacement of the $\mu$th molecule belonging to the $m$th cell [a particular solution of (1.1)] may be written as

$$t_{\mu s}^m = e_{\mu s} \exp[i(\mathbf{K} \cdot \mathbf{R}_{m\mu} - \omega t)] \tag{1.5}$$

where $e_{\mu s}$ is the vibration amplitude and $\mathbf{R}_{m\mu}$ is the vector extending from the center of inertia of molecule zero to the center of molecule $m\mu$. Substituting (1.5) into (1.1) leads to

$$\omega^2 e_{\mu s} = \sum_{\mu' s'} T_{\mu s \mu' s'} e_{\mu' s'} \tag{1.6}$$

where

$$T_{\mu s \mu' s'} = \sum_h \lambda_{\mu s \mu' s'}^h \exp(i\mathbf{K} \cdot \mathbf{R}_{h\mu\mu'})$$

Here the coupling coefficient refers to two molecules: molecule $\mu$ with its center at the origin and molecule $\mu'$ displaced from the position of equilibrium by a six-dimensional vector with the components $t_{\mu s}$ and $t_{\mu' s'}$. The coefficient $\lambda_{\mu s \mu s'}^0$ refers to one and the same molecule of the type $\mu$ that has two displacements $s$ and $s'$. The coefficient $\lambda_{\mu s \mu' s'}^0$ refers to two molecules in the same cell. The coefficient $\lambda_{\mu s \mu s'}^h$ couples two molecules of the same translationally identical set, the molecules zero and $h$.

The summation over $h$ is carried out in the following way: Taking the $\mu$th molecule as molecule zero we derive the terms corresponding to all pairs formed by the molecule with molecules $\mu'$ (of all sets) belonging both to the same and to all other cells of the crystal. Since the meaning of the vector $\mathbf{R}_{h\mu\mu'}$ is quite clear, we shall subsequently drop the indices $\mu$ and $\mu'$ wherever it is possible to do so without risk of ambiguity, so that the notation becomes simply $\mathbf{R}_h$. The secular equation is

$$|T_{\mu s \mu' s'} - \delta_{\mu s \mu' s'} \omega^2| = 0$$

The left-hand side is a polynomial of the order $6Z$ in $\omega^2$, $Z$ being the number of molecules in a cell. In general the equation should be solved for all $6Z$ roots corresponding to every wave vector $\mathbf{K}$. Thus we see that the normal vibration modes of a system are described by $6Z$ dispersion "hypersurfaces," which characterize the dependence of the vibration frequency on the wave vector.

It follows from general considerations that there will be three acoustic branches (see below) among the $6Z$ vibrational branches, which implies that if the wave vector is zero three frequencies out of all $6Z$ vanish.

There is no need to determine the type of hypersurfaces for comparison with the optical data on the frequencies of intermolecular vibrations. It is sufficient to solve the equations for the case of the zero wave vector and find the limiting values of the optical frequencies.

Besides the frequencies, the relative vibration amplitudes of the coordinates, i.e., the vibrational modes, may be determined for each wave vector. To do this one has to find the value of $e_{\mu s}$ from the set of homogeneous linear equations for all the $6Z$ values that $\omega$ can take. Then the displacement $t_{\mu s}$ (the general solution of the equation) is the sum of harmonic vibrations having frequencies $\omega_K$ $(K = 1, ..., 6Z)$. The harmonic (normal) vibration at the frequency $\omega_K$ is obviously oriented along the vectors with components $e_{\mu s}(\omega_K)$. Thus the vibration of molecules in the lattice is characterized in general by $6Z$ vectors in a $6Z$-dimensional vector space. A certain frequency $\omega_K$ corresponds to each vector. It is customary to normalize the components $e_{\mu s}(\omega_K)$ of the vectors in such a way that the length of each multidimensional vector is equal to unity.

## 2. SELECTION OF THE COORDINATE SYSTEM

In order that the equations of the coordinate vibrations may be written in the form of Eq. (1.1), each molecule should be described in a coordinate system having the inertial axes of the molecule in the equilibrium position: $u_i$ are displacements along and $\Omega_i$ are angular rotations about these axes. The axes of translationally identical molecules will be parallel. The axes of molecules which are not translationally coupled, but have common symmetry elements will transform in accordance with the symmetry elements of the crystal.

The equations of motion of a molecule now take the form

$$M\ddot{u}_i = -\partial U/\partial u_i, \qquad I_i\ddot{\Omega}_i = -\partial U/\partial \Omega_i, \qquad i = 1, 2, 3$$

or an equivalent form

$$\ddot{t}_{\mu s} = -\partial U/\partial t_{\mu s} \tag{2.1}$$

where reduced coordinate displacements $(s = 1, ..., 6)$ are introduced in such a way that

$$t_1 = M^{\frac{1}{2}}u_1, \qquad t_2 = M^{\frac{1}{2}}u_2, \qquad t_3 = M^{\frac{1}{2}}u_3$$

$$t_4 = I_1^{\frac{1}{2}}\Omega_1, \qquad t_5 = I_2^{\frac{1}{2}}\Omega_2, \qquad t_6 = I_3^{\frac{1}{2}}\Omega_3$$

In the special case of a molecule in a crystal the potential energy $U$ is a function of the displacements of all the molecules. We now expand $U$ in a power series of the displacements. In the harmonic approximation

$$\frac{\partial U}{\partial t_{\mu s}} = \sum_{h, \mu', s'} \frac{\partial^2 U}{\partial t_{\mu s}^0 \partial t_{\mu' s'}^h} t_{\mu' s'}$$

From (1.2), (1.6), and (2.1) we get

$$\lambda_{\mu s \mu' s'}^h = \partial^2 U/\partial t_{\mu s}^0 \partial t_{\mu' s'}^h$$

Here the second derivative is calculated at the equilibrium point. The following formulas are obvious:

$$\lambda^h_{\mu s \mu' s'} = \frac{1}{M} \frac{\partial^2 U}{\partial u^0_{\mu s} \, \partial u^h_{\mu' s'}} \qquad \text{for linear displacements}$$

$$\lambda^h_{\mu s \mu' s'} = \frac{1}{(I_{\mu s} I_{\mu' s'})^{\frac{1}{2}}} \frac{\partial^2 U}{\partial \Omega^0_{\mu s} \, \partial \Omega^h_{\mu' s'}} \qquad \text{for angular displacements}$$

$$\lambda^h_{\mu s \mu' s'} = \frac{1}{(M I_{\mu' s'})^{\frac{1}{2}}} \frac{\partial^2 U}{\partial u^0_{\mu s} \, \partial \Omega^h_{\mu' s'}} \qquad \text{for the mixed case}$$

We shall use these relations to rewrite Eq. (1.4) in an explicit form. Recalling that the summation in Eq. (1.4) may be restricted to the molecules of one cell and that the values of $\lambda_{II}$ are equal for molecules which are crystallographically equivalent, the following expression is derived for a crystal with any number of crystallographically equivalent molecules in a cell

$$\overline{\omega^2} = \frac{1}{6} \left\{ \frac{1}{M} \left( \frac{\partial^2 U}{\partial u_1^2} + \frac{\partial^2 U}{\partial u_2^2} + \frac{\partial^2 U}{\partial u_3^2} \right) + \frac{1}{I_1} \frac{\partial^2 U}{\partial \Omega_1^2} + \frac{1}{I_2} \frac{\partial^2 U}{\partial \Omega_2^2} + \frac{1}{I_3} \frac{\partial^2 U}{\partial \Omega_3^2} \right\} \quad (2.2)$$

## 3. THE COUPLING COEFFICIENTS

The calculation of the intermolecular coupling coefficients, which are by definition the second derivatives of the potential energy of interaction with respect to the displacements $t^0_{\mu s}$ and $t^h_{\mu' s'}$ calculated for the zero values of the other displacements, and for the equilibrium position taken as

$$\partial^2 U / \partial t^0_{\mu s} \, \partial t^h_{\mu' s'}$$

is rather straightforward if the atom–atom potentials are known. The calculation is facilitated by the existence of a number of relationships between the coupling coefficients. To start with, let us consider a crystal in which two molecules displaced by vectors with components $t_{\mu s}$ and $t_{\mu' s'}$ are at a distance $\mathbf{R}_h$ apart. If we apply to the crystal its symmetry operations, it is displaced or rotated as a whole, and the relative positions of the molecules will not be changed; the potential energy of their interaction will therefore remain constant. In consequence, various relationships arise among the coupling coefficients. The simplest of these is obtained as a corollary of the invariance of the potential energy under displacements by the translational vector defined as

$$\frac{\partial^2 U}{\partial t^0_{\mu s} \, \partial t^h_{\mu' s'}} = \frac{\partial^2 U}{\partial t^{\bar{h}}_{\mu s} \, \partial t^0_{\mu' s'}} = \frac{\partial^2 U}{\partial t^0_{\mu' s'} \, \partial t^{\bar{h}}_{\mu s}} \quad (3.1)$$

Symmetric relationships between the coupling coefficients are conveniently derived for the following two cases. In the first case the relationships result from the existence of symmetry elements coupling the molecules that are not translationally identical. All open-type symmetry operations will yield relationships of this kind; this is also true in the case of closed-type symmetries, provided the molecule occurs in a general position. In the second case the relationships arise from the existence of symmetry elements that couple translationally identical molecules. For simple Bravais groups, such symmetry transformations are effected only by closed-type symmetry operations, with the symmetry element belonging to the molecule. First, we shall specify two molecules $\mu$ and $\mu'$ in a crystal which are not translationally identical and are displaced from the position of equilibrium by vectors with the components $t_{\mu s}$ and $t_{\mu' s'}$ (the first case). The origin is at molecule $\mu$. The transformation produced by a symmetry element will shift the displacements $t_{\mu s}$ and $t_{\mu' s'}$ to two other molecules, the new vector $\mathbf{R}_{h\sigma}$ replacing the vector $\mathbf{R}_h$ which connected the centers of the displaced molecules. It is self-evident that $|\mathbf{R}_h| = |\mathbf{R}_{h\sigma}|$.

Now let us shift the origin, so that it coincides with the molecule obtained by the application of the symmetry element to former molecule zero. Thus the origin of the new system of coordinates of the crystal will be at molecule $\mu'$. Under a symmetric transformation, molecule $\mu'$ (which is now molecule zero) will obviously be displaced in exactly the same way as was its predecessor, i.e., by a vector with the components $t_s$. On the other hand, molecule $\mu$, which is at the end-point of the vector $\mathbf{R}_{h\sigma}$, will be displaced by a vector with components $t_{s'}$. In other words, a crystal in which molecules $\mu$ and $\mu'$ at distance $\mathbf{R}_h$ apart are displaced by vectors with components $t_s$ and $t_{s'}$ is identical to a crystal in which molecules $\mu'$ and $\mu$ at distance $\mathbf{R}_{h\sigma}$ apart are displaced by vectors with components $t_s$ and $t_{s'}$. Hence

$$\partial^2 U/\partial t_{\mu s}^0 \partial t_{\mu' s'}^h = \partial^2 U/\partial t_{\mu' s}^0 \partial t_{\mu s'}^{h\sigma} \tag{3.2}$$

The same discussion is also applicable to a crystal in which two translationally identical molecules, both having index $\mu$, are displaced by vectors with components $t_s$ and $t_{s'}$. The symmetric transformation converts them into molecules having the index $\mu'$. Similarly, the vector $\mathbf{R}_h$ becomes the vector $\mathbf{R}_{h\sigma}$ and the origin may be shifted in such a way as to coincide with the molecule that was molecule zero before the transformation. Thus

$$\partial^2 U/\partial t_{\mu s}^0 \partial t_{\mu s'}^h = \partial^2 U/\partial t_{\mu' s}^0 \partial t_{\mu' s'}^{h\sigma} \tag{3.3}$$

Symmetric transformations may be simplified for some special positions of the molecules.

This is the case, for instance, when the vector $\mathbf{R}_h$ is parallel to axes 2, or $2_1$, or to either an $m$ plane or a glide plane, where we have $\mathbf{R}_{h\sigma} = \mathbf{R}_h$, or when the vector $\mathbf{R}_h$ is perpendicular to the axes or to the planes, when $\mathbf{R}_{h\sigma} = \mathbf{R}_{\bar{h}}$ will hold. Thus, we have for the parallel case

$$\partial^2 U/\partial t^0_{\mu s}\, \partial t^h_{\mu's'} = \partial^2 U/\partial t^0_{\mu's}\, \partial t^h_{\mu s'}$$
$$\partial^2 U/\partial t^0_{\mu s}\, \partial t^h_{\mu s'} = \partial^2 U/\partial t^0_{\mu's}\, \partial t^h_{\mu's'} \qquad (\mu \ne \mu')$$

and for the perpendicular case [here we use Eq. (3.1)]

$$\partial^2 U/\partial t^0_{\mu s}\, \partial t^h_{\mu's'} = \partial^2 U/\partial t^0_{\mu s'}\, \partial t^h_{\mu's}$$
$$\partial^2 U/\partial t^0_{\mu s}\, \partial t^h_{\mu s'} = \partial^2 U/\partial t^0_{\mu's}\, \partial t^h_{\mu's'}$$

The simplification is quite substantial if there is an inversion center which may act as a transformation symmetry element. An arbitrary vector $\mathbf{R}_h$ is transformed by the inversion center into the vector $\mathbf{R}_{h_{\bar{1}}} = \mathbf{R}_{\bar{h}}$. Consequently Eqs. (3.2) and (3.3) in this case take the form

$$\partial^2 U/\partial t^0_{\mu s}\, \partial t^h_{\mu's'} = \partial^2 U/\partial t^0_{\mu's}\, \partial t^{\bar{h}}_{\mu s'} = \partial^2 U/\partial t^0_{\mu's'}\, \partial t^{\bar{h}}_{\mu's} = \partial^2 U/\partial t^0_{\mu s'}\, \partial t^h_{\mu's}$$

$$\partial^2 U/\partial t^0_{\mu s}\, \partial t^h_{\mu s'} = \partial^2 U/\partial t^0_{\mu's}\, \partial t^{\bar{h}}_{\mu's'} = \partial^2 U/\partial t^0_{\mu's'}\, \partial t^{\bar{h}}_{\mu s} = \partial^2 U/\partial t^0_{\mu s'}\, \partial t^h_{\mu's}$$

The second equation is also valid in the special case $h = 0$ ($h$ is always different from zero for the first equation).

We shall now take up a different type of symmetric relationship among the coupling coefficients, the relationships that arise owing to a special position of the molecule. According to the general principles of symmetry of the space lattice, if the symmetry elements $\bar{1}, m, 2$ are located at the lattice points, they are also located at midpoints in between. Here are some of the conclusions that may be drawn from this fact.

Let us consider a crystal in which two molecules designated by the indices $\mu$ and $\mu'$ are at distance $\mathbf{R}_h$ apart and have the displacements $t^0_{\mu s}$ and $t^h_{\mu's'}$. Apply a symmetry transformation, assuming for reasons of simplicity that the transformation is effected by a symmetry element passing through molecule zero. The molecules 0 and $h$ become molecules 0 and $h\sigma$, their indices remaining unchanged since the molecule $h$ is translationally identical to the molecule $h\sigma$. But the axes of the systems of coordinates connected with translationally identical molecules are parallel. For this reason, in contrast to the first case, when Eqs. (3.2) and (3.3) were obtained, the displacement vectors are affected by the symmetry transformation in such a way that some components $t_s$ will change their sign, while the signs of other components remain unchanged. Thus we see that in the general case the following equation holds true (the indices $\mu$ may coincide or differ)

$$\frac{\partial^2 U}{\partial t^0_{\mu s}\, \partial t^h_{\mu's'}} = S \frac{\partial^2 U}{\partial t^0_{\mu s}\, \partial t^{h\sigma}_{\mu's'}}$$

where $S$ is equal to $+1$ or $-1$ depending on whether the signs of appropriate displacements before and after the transformation are identical or different.

We have to consider in this case the behavior of all six components of the vector $t_s$, for which we adopt the notation $x, y, z, \varphi, \psi, \eta$; here $\varphi, \psi, \eta$ are the angles of rotation about the axes $x, y, z$. The transformation rules for three important cases of symmetry are as follows (Fig. 1)

$$\bar{1} \qquad x\ y\ z\ \varphi\ \psi\ \eta \to \bar{x}\ \bar{y}\ \bar{z}\ \varphi\ \psi\ \eta$$

$$2 \qquad x\ y\ z\ \varphi\ \psi\ \eta \to \bar{x}\ y\ \bar{z}\ \varphi\ \bar{\psi}\ \eta$$

$$m \qquad x\ y\ z\ \varphi\ \psi\ \eta \to x\ \bar{y}\ z\ \bar{\varphi}\ \psi\ \bar{\eta}$$

The last two transformations appear due to the fact that one of the molecular axes of inertia coincides with the $y$ axis of the crystal.

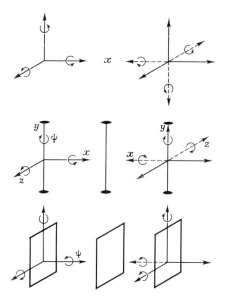

Fig. 1. The changes of signs of translational and rotational displacements under symmetry operations $\bar{1}$, 2, and $M$ (top to bottom).

The case which involves an inversion center is particularly simple, since $\mathbf{R}_h$ is then always equal to $\mathbf{R}_{\bar{h}}$. Therefore, for an arbitrary pair of molecules, we obtain [using (3.1)]

$$\partial^2 U / \partial t^0_{\mu s}\, \partial t^h_{\mu' s'} = S\, \partial^2 U / \partial t^0_{\mu' s'}\, \partial t^h_{\mu s} \qquad (3.4)$$

the value of $S$ being equal to $+1$ if both displacements are linear or angular, and to $-1$ for the mixed case.

*For a second-order axis*, in the general case we have

$$\partial^2 U/\partial t^0_{\mu s}\,\partial t^h_{\mu' s'} = S\,\partial^2 U/\partial t^0_{\mu s}\,\partial t^{h(2)}_{\mu' s'} \tag{3.5}$$

with $S$ equal to $+1$ for $xz, x\psi, z\psi, y\varphi, yh, \varphi\eta$.

We see that

$$\partial^2 U/\partial t^0_{\mu s}\,\partial t^h_{\mu' s'} = S\,\partial^2 U/\partial t^0_{\mu s}\,\partial t^{h(2)}_{\mu' s'} \tag{3.6}$$

if the vector $\mathbf{R}_h$ is parallel to the symmetry axis, and that

$$\partial^2 U/\partial t^0_{\mu s}\,\partial t^h_{\mu' s'} = S\,\partial^2 U/\partial t^0_{\mu' s'}\,\partial t^{h(2)}_{\mu s} \tag{3.7}$$

if the vector $\mathbf{R}_h$ is perpendicular to it.

*For a mirror plane*, in the general case we have

$$\partial^2 U/\partial t^0_{\mu s}\,\partial t^h_{\mu' s'} = S\,\partial^2 U/\partial t^0_{\mu s}\,\partial t^{h(m)}_{\mu' s'} \tag{3.8}$$

with $S$ equal to $+1$ for the same pairs of coordinates as in the case of axis 2. Eq. (3.6) is valid for the $m$ plane parallel to the vector $\mathbf{R}_h$ and Eq. (3.7) holds true for the $m$ plane perpendicular to the vector $\mathbf{R}_h$.

Note that the coupling coefficients, satisfying Eqs. (3.4)–(3.8), vanish at $S = -1$ when $h = 0$, i.e., in the case when both displacements refer to the same molecule.

Thus, for molecules occupying position $\bar{1}$ in a crystal, Eq. (3.4) for $h = 0$ takes the form

$$\partial^2 U/\partial t^0_{\mu s}\,\partial t^0_{\mu' s'} = S\,\partial^2 U/\partial t^0_{\mu' s'}\,\partial t^0_{\mu s}$$

As we know, $S = -1$ if one of the two displacements is angular and the other is linear; on the other hand, the derivatives on the right and the left side of the above equation are identical. Hence all derivatives of the type

$$\partial^2 U/\partial x^0_\mu\,\partial \varphi^0_{\mu'}$$

vanish.

Similarly, in the case of molecules occupying positions 2 and $m$ in a crystal, a number of derivatives with respect to displacements of a particular molecule also vanish.

It is obvious that additional relationships may appear if a lattice has a number of symmetry elements at the same time. As an example, let us consider the crystal of naphthalene, group $P2_1/a$, $Z = 2$.

Equations (3.2) and (3.3) hold, due to the existence of the $2_1$ axis in the lattice:

$$\lambda^h_{\mu s \mu' s'} = \lambda^{h(2_1)}_{\mu' s \mu s'}, \qquad \lambda^h_{\mu s \mu s'} = \lambda^{h(2_1)}_{\mu' s \mu' s'}$$

Equation (3.4) follows, in this case, from the existence of an inversion center which is occupied by the molecule

$$\lambda^h_{\mu s \mu' s'} = S\lambda^h_{\mu' s' \mu s} \quad (\mu = \mu', \ \mu \neq \mu')$$

Combining these formulas we get

$$\lambda^h_{\mu s \mu' s'} = \lambda^{h(2_1)}_{\mu' s \mu s} = S\lambda^{h(2_1)}_{\mu s \mu' s} = S\lambda^h_{\mu' s' \mu s}$$

$$\lambda^h_{\mu s \mu s'} = \lambda^{h(2_1)}_{\mu' s \mu' s} = S\lambda^{h(2_1)}_{\mu' s' \mu' s} = S\lambda^h_{\mu s' \mu s}$$

We shall not concern ourselves with other properties of the coupling co-efficients. It is not the coefficients but rather the elements of the dynamic matrix $T_{\mu s \mu' s'}$ [Eq. (1.6)] that are of particular interest; furthermore, the expressions for the matrix elements are simplified in view of Eqs. (3.2), (3.3)

$$\lambda^h_{\mu s \mu' s'} = \lambda^{h\sigma}_{\mu' s \mu s'}, \quad \lambda^h_{\mu s \mu s'} = \lambda^{h\sigma}_{\mu' s \mu' s'} \tag{3.9}$$

which are valid if the lattice incorporates a symmetry element $\sigma$ coupling the molecules that are not translationally identical, and also by virtue of Eq. (3.4)

$$\lambda^h_{\mu s \mu' s'} = S\lambda^h_{\mu' s' \mu s} \tag{3.10}$$

for the case when molecules occupy a special position. In the latter case the simplification is quite remarkable, if the position of a molecule coincides with a center of inversion.

### 4. The Limiting Frequencies and Their Eigenvectors

The theory of lattice vibration can best be checked by measurements of infrared absorption frequencies or Raman frequencies. Since the wavelengths of light exceed intermolecular distances by approximately three orders of magnitude, the resonance light absorption frequencies correspond to the optical vibration frequencies of the lattice for the case of $K = 0$. The acoustic frequencies are equal to zero. It would be worthwhile at this stage to pay particular attention to the solution of Eq. (1.6) for the limiting case. The dynamic matrix is considerably simplified, since

$$T_{\mu s \mu' s'} = \sum \lambda^h_{\mu s \mu' s'}$$

The matrix of coefficients $T_{\mu s \mu' s'}$ is symmetric. To show this we write

$$T_{\mu s \mu' s'} = \lambda^0_{\mu s \mu' s' i} + \sum (\lambda^h_{\mu s \mu' s'} + \lambda^{\bar h}_{\mu s \mu' s'})$$

and

$$T_{\mu' s' \mu s} = \lambda^0_{\mu' s' \mu s} + \sum (\lambda^h_{\mu' s' \mu s} + \lambda^{\bar h}_{\mu' s' \mu s})$$

But by virtue of Eq. (3.1) every term of $T_{\mu' s' \mu s}$ is equal to the term of $T_{\mu s \mu' s'}$

written directly above, Hence

$$T_{\mu s \mu' s'} = T_{\mu' s' \mu s} \tag{4.1}$$

The existence of relationships between the coupling coefficients $\lambda$ for various space groups and the numbers $Z$ of molecules in a cell, as well as for specific symmetry positions of the molecules, permits further simplifications.

Now we shall prove that in the case of crystals with one independent molecule in a cell, which are described by a simple Bravais group, the matrix of coefficients breaks down into four equal blocks for $Z = 2$, into 16 equal blocks for $Z = 4$, etc.

For this purpose, we write $T_{\mu s \mu' s'}$ as

$$T_{\mu s \mu' s'} = \lambda^0_{\mu s \mu' s'} + \sum (\lambda^h_{\mu s \mu' s'} + \lambda^{h\sigma}_{\mu s \mu' s'})$$

where, depending on the symmetry class of the crystals considered, each part of the $h$ summation is identified with one of the two sums over symmetrically independent regions (molecule zero being in the same position for both terms). We have for $Z = 2$

$$T_{1s1s'} = \lambda^0_{1s1s'} + \sum (\lambda^h_{1s1s'} + \lambda^{h\sigma}_{1s1s'}), \qquad T_{2s2s'} = \lambda^0_{2s2s'} + \sum (\lambda^{h\sigma}_{2s2s'} + \lambda^h_{2s2s'})$$

According to Eq. (3.9) every term in the second equation is equal to the corresponding term of the first equation. We have the same for two other rows of the matrix:

$$T_{1s1s'} = \sum (\lambda^h_{1s2s'} + \lambda^{h\sigma}_{1s2s'}), \qquad T_{2s1s'} = \sum (\lambda^{h\sigma}_{2s1s'} + \lambda^h_{2s1s'})$$

Taking note of Eq. (4.1), we obtain finally

$$T_{1s1s'} = T_{2s2s'}, \qquad T_{1s2s'} = T_{2s1s'} = T_{2s'1s} = T_{1s'2s} \tag{4.2}$$

i.e., the matrix of coefficients breaks down into four equal blocks, which are symmetric submatrices of the original matrix. Now we turn to the case $Z = 4$. For two translationally identical molecules numbered by the index 1, which are displaced by the vectors $t_{1s}$ and $t_{1s'}$, symmetric transformations that produce equivalent positions for the crystal may be defined, with the molecules displaced by the vectors $t_{2s}$ and $t_{2s'}$, $t_{3s}$ and $t_{3s'}$, $t_{4s}$ and $t_{4s'}$. In this case the vector $\mathbf{R}_h$ will transform in accordance with the symmetry of the space group. Thus

$$\lambda^h_{1s1s'} = \lambda^{h\sigma_1}_{2s2s'} = \lambda^{h\sigma_2}_{3s3s'} = \lambda^{h\sigma_3}_{4s4s'}$$

In the special case of $h = 0$ we have $\lambda^0_{\mu s \mu s'} = \lambda^0_{\mu' s \mu' s'}$. Let us consider now two displaced molecules that are not translationally identical. If the indices 1 and 2 are used to number the molecules, a symmetry operation may be specified which permutes the indices, as in $12 \leftrightarrow 21$, or transforms them into

34 or 43. Clearly, there are no other possible transformations. Hence

$$\lambda^h_{1s2s'} = \lambda^{h\sigma_1}_{2s1s'} = \lambda^{h\sigma_2}_{3s4s'} = \lambda^{h\sigma_3}_{4s3s'}$$

Similarly,

$$\lambda^h_{1s3s'} = \lambda^{h\sigma_1}_{3s1s'} = \lambda^{h\sigma_2}_{2s4s'} = \lambda^{h\sigma_3}_{4s2s'}$$

$$\lambda^h_{1s4s'} = \lambda^{h\sigma_1}_{4s1s'} = \lambda^{h\sigma_2}_{2s3s'} = \lambda^{h\sigma_3}_{3s2s'}$$

A technique similar to the one used just now, i.e., breaking down a sum into terms corresponding to symmetrically independent regions, may be applied to prove the validity of the following formulas:

$$T_{1s1s'} = T_{2s2s'} = T_{3s3s'} = T_{4s4s'}$$

$$T_{1s2s'} = T_{2s1s'} = T_{3s4s'} = T_{4s3s'}$$

$$T_{1s3s'} = T_{3s1s'} = T_{2s4s'} = T_{4s2s'}$$

$$T_{1s4s'} = T_{4s1s'} = T_{2s3s'} = T_{3s2s'}$$

Now we see from Eq. (4.1) that the matrix falls into 16 symmetric blocks.

Another essential simplification of the problem is achieved if the molecular position coincides with the crystal's center of inversion. In this case Eq. (3.10) makes it possible to define the elements of the $T_{\mu s'\mu s'}$ matrix blocks, corresponding to two linear or two angular displacements as

$$T_{\mu s\mu's'} = \lambda^0_{\mu s\mu's'} + 2 \sum_{h\neq h(\bar{1})} \lambda^h_{\mu s\mu's'} \qquad \begin{pmatrix} s,s' = 1,2,3 \\ s,s' = 4,5,6 \end{pmatrix}$$

All elements of the matrix blocks for the mixed case vanish:

$$T_{\mu s\mu's'} = 0 \qquad \begin{pmatrix} s = 1,2,3, & s' = 4,5,6 \\ s' = 1,2,3, & s = 4,5,6 \end{pmatrix}$$

We have considered so far two basic simplifications of the dynamic matrix that are imposed by symmetry. Two major types of information—the multiplicity of the general position of the space group, and whether or not the position of the molecule coincides with a center of inversion—may indicate the principal features of the optical spectra of the lattice of molecular crystals.

Having made these general observations we shall now consider some examples.

*The Group* $P\bar{1}$, $Z = 1$. This is probably the simplest case. A number of crystals of this symmetry are known, e.g., hexamethylbenzene, to mention one of the most common compounds. In this case $T_{\mu s\mu's'} = T_{1s1s'} = T_{ss'}$ since $Z = 1$, i.e., the coefficients form a $6 \times 6$ matrix. Due to the existence of the inversion center, the matrix breaks down into two blocks. Both blocks have a diagonal form, since $T_{ss'} = 0$ for the quadrants corresponding to mixed indices.

Equation (1.6) can then be written as two $3 \times 3$ eigenvalue problems:

$$\omega^2 e_s = \sum T_{ss'} e_{s'}, \qquad s = 1, 2, 3$$

$$\omega^2 e_s = \sum T_{ss'} e_{s'}, \qquad s = 4, 5, 6$$

The first equation corresponds to translational vibrations. As we know, the three acoustic frequencies vanish. Translational vibrations are acoustic vibrations, and for this reason all the three roots of the first secular equation should vanish. This follows immediately, in the case of the group $P\bar{1}$, $Z = 1$ under discussion. Indeed, as a corollary of the general theorem, we have that a parallel shift of all the molecules by segments of equal length cannot create a force acting on any separate molecule.

In the present case all molecules in the crystal are parallel. Assume that the crystal is shifted along either the $x$, $y$, or $z$ axis. The generalized force components along the inertial axes of the molecule are $T_{ss'} e_{s'}$, and hence $T_{ss'} = 0$ ($s = 1, 2, 3$). Thus all translational frequencies are seen to vanish. Consequently, the dynamic problem is reduced to just one third power secular equation, which describes the angular rocking motion of the molecule

$$\omega^2 e_s = \sum T_{ss'} e_{s'}, \qquad s, s' = 4, 5, 6$$

By way of illustration, the matrix element $T_{\varphi\psi}$ may be determined from

$$(I_\varphi I_\psi)^{\frac{1}{2}} \cdot T_{\varphi\psi} = \frac{\partial^2 U}{\partial \varphi^0 \partial \psi^0} + 2 \sum_h \frac{\partial^2 U}{\partial \varphi^0 \partial \psi^h} \qquad (h \neq h(\bar{1}))$$

As will be shown below, in practice it is usually sufficient to take into account only the interactions of molecule zero with its nearest neighbors. The molecular coordination number as a rule ranges from 10 to 14, and is usually 12. In our case it will be sufficient to account for 5–7 pair interactions. The reader may recall the meaning of the derivatives that occur in the expressions for the dynamic matrix elements. Thus, $\partial^2 U/\partial \varphi^2$ is the second derivative at $\varphi = 0$ calculated for the curve $U = U(\varphi)$ where $U$ is the lattice energy and $\varphi$ is the displacement of molecule zero about the $x$ axis in the molecular system of inertial axes. $\partial^2 U/\partial \varphi^0 \partial \varphi^h$ is the second derivative at $\varphi^0 = 0$, $\varphi^h = 0$ is calculated for the surface, $U = U(\varphi^0, \varphi^h)$, etc.

As is clearly seen from this argument, the calculation of the matrix elements by the method of atom–atom potentials is a rather cumbersome procedure, and it therefore requires a computer. It is obvious that to evaluate the second derivative at point zero a section of the surface $U(\varphi^0, \varphi^1)$ should be reproduced in at least 24 points. Consequently, to evaluate a single coupling coefficient the lattice energy must be computed 24 times.

Three vibration frequencies $\omega_1$, $\omega_2$, and $\omega_3$ may be found from the appropriate secular equation. The successive substitution of the values of these three frequencies makes it possible to determine the relative amplitude of each frequency. Normalizing the values $e_s$ by means of the equation $e_4{}^2 + e_5{}^2 + e_6{}^2 = 1$ we obtain a three-dimensional unit vector $\mathbf{e}(\omega_1)$ which is the eigenvector of $\omega_1$. A similar procedure is used to determine the eigenvectors corresponding to the frequencies $\omega_2$ and $\omega_3$. Vibrations of molecules along their eigenvectors are harmonic; these vibrations are normal vibrations. To define completely the libration of a molecule in a crystal of the hexamethylbenzene type one should specify the directions of three unit vectors $\mathbf{e}(\omega_1)$, $\mathbf{e}(\omega_2)$, $\mathbf{e}(\omega_3)$ and the frequencies $\omega_1$, $\omega_2$, $\omega_3$ of harmonic vibrations referred to the appropriate axes of a molecule (or a crystal). Vibration about the axes of inertia is clearly nonharmonic. For example, from Eq. (3.2) we have

$$t_\varphi = e_\varphi(\omega_1)e^{-i\omega_1 t} + e_\varphi(\omega_2)e^{-i\omega_2 t} + e_\varphi(\omega_3)e^{-i\omega_3 t}$$

It is obvious that the values of these amplitudes are of no interest, since in the problem as formulated, determination of the phase shifts between different terms is meaningless.

*The Space Group* $P\bar{1}$, $Z = 2$. Now we shall take up the case of molecules occurring in a general position. Although the symmetry of the lattice in this case is the same as in the case discussed above, the relationships between the coupling coefficients are different; the coefficients now satisfy Eqs. (3.9). We have two molecules in a cell, and in this case Eq. (4.2) follows from Eq. (3.9).

Let us write Eqs. (1.6) in the form

$$\omega^2 e_{1s} = \sum L_{ss'} e_{1s'} + \sum M_{ss'} e_{2s'}$$
$$\omega^2 e_{2s} = \sum M_{ss'} e_{1s'} + \sum L_{ss'} e_{2s'}$$

$$(4.3)$$

with the notations

$$L_{ss'} = T_{1s1s'} = T_{2s2s'}, \qquad M_{ss'} = T_{1s2s'} = T_{2s1s'}$$

We shall also introduce new coordinates, which are the sums and the differences of the original coordinates. Vibrations in the new coordinates will have the amplitudes

$$e_s^S = e_{1s} + e_{2s}, \qquad e_s^A = e_{1s} - e_{2s}$$

The addition and subtraction of Eqs. (4.3) will lead to two independent sets of equations

$$\omega^2 e_s^S = \sum T_{ss'}^S e_{s'}^S, \qquad \omega^2 e_s^A = \sum T_{ss'}^A e_{s'}^A \qquad (4.4)$$

in which we use the notations: $T_{ss'}^S = T_{1s1s'} + T_{1s2s'}$ and $T_{ss'}^A = T_{1s1s'} - T_{1s2s'}$.

As an example we give the value of the coefficient $T_{34}^S$:

$$(MI_\varphi)^{1/2} T_{34}^S = \frac{\partial^2 U}{\partial z^0 \partial \varphi^0} + \frac{\partial^2 U}{\partial z^0 \partial \varphi^1} + \frac{\partial^2 U}{\partial z^0 \partial \varphi^{1'}} + \frac{\partial^2 U}{\partial z^0 \partial \varphi^2} + \frac{\partial^2 U}{\partial z^0 \partial \varphi^{2'}} + \cdots$$

and of the coefficient $T_{34}^A$:

$$(MI_\varphi)^{1/2} T_{34}^A = \frac{\partial^2 U}{\partial z^0 \partial \varphi^0} + \frac{\partial^2 U}{\partial z^0 \partial \varphi^1} - \frac{\partial^2 U}{\partial z^0 \partial \varphi^{1'}} + \frac{\partial^2 U}{\partial z^0 \partial \varphi^2} - \frac{\partial^2 U}{\partial z^0 \partial \varphi^{2'}} + \cdots$$

Unprimed indices refer to the molecules that are translationally identical to molecule zero. In practice the sum usually consists of about 10 to 14 terms representing the nearest neighbors' contributions. Equations (4.4) make it possible to calculate nine nonvanishing frequencies and their nine eigenvectors (the total number of vibration branches is 12; at $K = 0$, $\omega = 0$ for the three acoustic branches). Joint variations of the sums or the differences of appropriate molecular coordinates lead to harmonic (normal) vibrations.

The vibrations labeled by S may appropriately be called symmetric vibrations. In this case the symmetry of mutual arrangement of the molecules is maintained when two molecules with a common inversion center approach each other remaining in phase and then move apart. The difference of coordinates vanishes identically; it is the sum of coordinates that executes harmonic vibrations. Conversely, the A vibrations may conveniently be called antisymmetric. The sum of coordinates vanishes identically at all times, and the difference of coordinates vibrates harmonically provided the molecules with a common inversion center always move in the same direction.

Equations (4.4) yield the values of 12 frequencies and their 12 eigenvectors. Thus we see that harmonic (normal) vibrations produce combined changes of six sums or six differences of appropriate molecular coordinates along the directions defined by six-component vectors. Translational vibration of the molecule along a certain axis or molecule libration are examples of a composite form of motion, described by the sum of components of twelve vectors

$$\mathbf{e}^S(\omega_1),\ \mathbf{e}^S(\omega_2), \ldots, \mathbf{e}^S(\omega_6), \qquad \mathbf{e}^A(\omega_1),\ \mathbf{e}^A(\omega_2), \ldots, \mathbf{e}^A(\omega_6)$$

It should be emphasized again that the mode of such vibrations cannot be determined since the phase shifts of normal vibrations are unknown. In contrast to the simplest type of molecular vibrations in a crystal of the hexamethylbenzene type, in the present case there are no directions in the cell which can be defined as directions of pure vibrations or rotations.

The number of frequencies considered in the above discussion is 12. This applies to the case when $K \neq 0$. For $K = 0$ there are three vanishing frequencies (the acoustic frequencies). These frequencies should be sought among the antisymmetric vibrations. The procedure is easily seen to account for the existence of zero solutions.

From Eq. (4.3) the force can be written as

$$T_{1s11}x_1 + T_{1s12}y_1 + T_{1s13}z_1 + T_{1s14}\varphi_1 + T_{1s15}\psi_1 + T_{1s16}\eta_1$$

$$+ T_{1s21}x_2 + T_{1s22}y_2 + T_{1s23}z_2 + T_{1s24}\varphi_2 + T_{1s25}\psi_2 + T_{1s26}\eta_2$$

Let us shift the whole crystal along the $x_1$ axis in the system of inertia axes of the molecules labeled by the index 1. In the case of the group $P\bar{1}$, $Z = 2$ for the axes, the molecules with the subscript 2 will shift by antiparallel vectors of equal length. The requirement that the generalized force be equal to zero is

$$T_{1s11} - T_{1s21} = 0$$

which can be rewritten, using the notations adopted in Eq. (4.4), as

$$T_{s1}^A = 0 \qquad (4.5)$$

Shifting the crystal along the $y$ and $z$ axes, we shall similarly get $T_{s2}^A = T_{s3}^A = 0$. Since this reasoning holds true for a generalized force acting on any coordinate, only the lower diagonal block of the matrix $T_{ss'}^A$ is different from zero. Three frequencies out of the six which occur in the general case for $\mathbf{K} = 0$ vanish. Nonvanishing frequencies belonging to class A correspond to librations.

In the calculation of dynamic elements by the method of atom–atom potentials the accuracy check of Eq. (4.5) in the nearest-neighbor interaction scheme is of special interest.

*The Group* P2$_1$, $Z = 2$. Simplification of the form of the dynamic matrix is the same for all cases of two symmetrically coupled molecules in a cell. Equations (4.4) hold true for $Z = 2$ and for the space groups $P\bar{1}$, P2, P2$_1$, P$m$, P$c$. The vibrations are classified as class S and class A vibrations.

The S and A vibrations may again be called symmetric and antisymmetric, respectively. The symmetry for the groups P2$_1$ and P2 is axial symmetry, while for the groups P$m$ and P$c$ it is symmetry with respect to a plane.

We would have 12 vibration frequencies for $\mathbf{K} = 0$; however, three of these vanish. The conditions of vanishing for the group P2$_1$ (and other groups) are different from those for the group P$\bar{1}$. The feature of the group P$\bar{1}$, $Z = 2$ that is responsible for the difference is the fact that the inertial axes $x$, $y$, $z$ of the first molecule are antiparallel to the inertia axes of the second molecule. This is not the case when the molecules have a common axis or a plane of symmetry.

The requirement that the generalized force acting on the coordinate $t_{1s}$ when the crystal is shifted as a whole, should vanish, has the following form:

$$T_{1s11}x_1 + T_{1s12}y_1 + T_{1s13}z_1 + T_{1s21}x_2 + T_{1s22}y_2 + T_{1s23}z_2 = 0 \quad (4.6)$$

Here $x_1, y_1, z_1$ and $x_2, y_2, z_2$ are the components of the shift along the inertial axes of the first and second molecules, respectively. These expressions cannot be substantially simplified for the group P2$_1$ (and other groups) and for arbitrary mutual positions of the inertial axes and the crystal axes.

Equations (4.5) are not valid in this case. It is also not clear a priori which three frequencies vanish and whether they belong to class A or S, since the result depends on the mutual orientation of the inertial axes and the crystal axes.

Equations of the type (4.6) obviously produce various relationships between the matrix elements. These relationships may conveniently be derived in the crystallographic system of coordinates.

*The Group* P2$_1$/a, $Z = 2$. This symmetry, which may be called the naphthalene class symmetry, is encountered quite often. A remarkable fact is that almost all symmetric aromatic molecules adopt this symmetry when crystallizing. As the crystal contains molecules coupled by the 2$_1$ axis that are not translationally identical, the matrix $T_{\mu s \mu' s'}$ breaks down into four blocks, which are equal in pairs:

$$T_{1s1s'} = T_{2s2s'}, \qquad T_{1s2s'} = T_{2s1s'}$$

But, besides these formulas, we have to take into account the relationships arising from the existence of the center of inversion occupied by the molecules in the crystal, i.e., the situation when Eq. (3.4) applies

$$\lambda^h_{\mu s \mu' s'} = S \lambda^h_{\mu' s' \mu s}$$

Hence, reasoning similar to that for the group P$\bar{1}$, $Z = 1$ which we considered earlier, leads to

$$T_{1s2s'} = \lambda^0_{1s1s'} + 2 \sum_{h \neq h(\bar{1})} \lambda^h_{1s1s'} \qquad \begin{pmatrix} s', s = 1,2,3 \\ s', s = 4,5,6 \end{pmatrix}$$

$$T_{1s1s'} = 0 \qquad \begin{pmatrix} s = 1,2,3; & s' = 4,5,6 \\ s = 4,5,6; & s' = 1,2,3 \end{pmatrix}$$

In an analogous way we get for $T_{1s2s'}$

$$T_{1s1s'} = 2 \sum_{h \neq h(\bar{1})} \lambda^h_{1s2s'} \qquad \begin{pmatrix} s', s = 1,2,3 \\ s', s = 4,5,6 \end{pmatrix}$$

$$T_{1s2s'} = 0 \qquad \begin{pmatrix} s = 1,2,3; & s' = 4,5,6 \\ s = 4,5,6; & s' = 1,2,3 \end{pmatrix}$$

The equations of motion $\omega^2 e_{\mu s} = \sum T_{\mu s \mu' s'} e_{\mu' s'}$ may assume the form

$$\omega^2 e_{1s} = \sum T_{1s1s'} e_{1s'} + \sum T_{1s2s'} e_{2s'} \Big\}$$
$$\omega^2 e_{2s} = \sum T_{2s1s'} e_{1s'} + \sum T_{2s2s'} e_{2s'} \Big\} \qquad (s, s' = 1,2,3)$$

$$\omega^2 e_{1s} = \sum T_{1s1s'} e_{1s'} + \sum T_{1s2s'} e_{2s'} \Big\}$$
$$\omega^2 e_{2s} = \sum T_{2s1s'} e_{1s'} + \sum T_{2s2s'} e_{2s'} \Big\} \qquad (s, s' = 4,5,6)$$

and, applying the technique used to derive Eq. (4.4), we obtain

$$
\left.\begin{aligned}
\omega^2 e_s^{\mathrm{S}} &= \sum T_{ss'}^{\mathrm{S}} e_{s'}^{\mathrm{S}} \\
\omega^2 e_s^{\mathrm{A}} &= \sum T_{ss'}^{\mathrm{A}} e_{s'}^{\mathrm{A}}
\end{aligned}\right\} \quad (s, s' = 1, 2, 3)
$$

$$
\left.\begin{aligned}
\omega^2 e_s^{\mathrm{S}} &= \sum T_{ss'}^{\mathrm{S}} e_{s'}^{\mathrm{S}} \\
\omega^2 e_s^{\mathrm{A}} &= \sum T_{ss'}^{\mathrm{A}} e_{s'}^{\mathrm{A}}
\end{aligned}\right\} \quad (s, s' = 4, 5, 6)
$$

(4.7)

The vibrations in crystals such as that of naphthalene are characterized by twelve frequencies and the eigenvectors corresponding to these frequencies. The eigenvectors fall into four categories: S translation, A translation, S libration and A libration. The labels S and A specify the symmetry of vibrations with respect to the $2_1$-axis. Each category is described by a vector in a three-dimensional vector space, whose orientation relative to the axes of a molecule (or a crystal) may be determined.

When we say that the total number of vibration frequencies is 12 we neglect the fact that at $\mathbf{K} = \mathbf{0}$ these frequencies include the vanishing acoustic frequencies. There can be only nine nonvanishing frequencies at $\mathbf{K} = \mathbf{0}$ for crystals like that of naphthalene. The vanishing frequencies are seen to be defined by the roots of the first two equations in (4.7). Depending on the value of the angle between the inertial axes and the crystal axes, the frequencies belonging to either of the two classes A or S may vanish. This is also true for one of the frequencies A(S) and two frequencies S(A).

By way of example we shall calculate some of the matrix elements. For translation vibrations

$$
MT_{12}^{\mathrm{S}} = M(T_{1112} + T_{1122})
$$

$$
= \frac{\partial^2 U}{\partial x^0 \, \partial y^0} + 2\left(\frac{\partial^2 U}{\partial x^0 \, \partial y^1} + \frac{\partial^2 U}{\partial x^0 \, \partial y^2} + \cdots\right) + 2\left(\frac{\partial^2 U}{\partial x^0 \, \partial y^{1'}} + \frac{\partial^2 U}{\partial x^0 \, \partial y^{2'}} + \cdots\right)
$$

For libration

$$
(I_4 I_\eta)^{1/2} T_{56}^{\mathrm{A}} = (I_4 I_\eta)^{1/2} (T_{1516} - T_{1526})
$$

$$
= \frac{\partial^2 U}{\partial \psi^0 \, \partial \eta^0} + 2\left(\frac{\partial^2 U}{\partial \psi^0 \, \partial \eta^1} + \frac{\partial^2 U}{\partial \psi^0 \, \partial \eta^2} + \cdots\right)
$$

$$
- 2\left(\frac{\partial^2 U}{\partial \psi^0 \, \partial \eta^1} \quad \frac{\partial^2 U}{\partial \psi^0 \, \partial \eta^2} + \cdots\right)
$$

Unprimed coordinates refer to the molecules that are translationally identical to molecule zero. All the terms in parentheses account for the contributions from the half-space, which implies that the contributions of molecules situated at distance $\mathbf{R}_h$ and $\mathbf{R}_{\bar{h}}$ from molecule zero need not be taken into account separately.

*The Group* $P2_1/a$, $Z = 4$. As follows from the detailed discussion above, this symmetry is most convenient for the packing of molecules that do not have a center of symmetry. This situation is quite common and, if only for this reason, it deserves to be considered separately, all the more so because, for the lower systems, and for the case of molecules that are crystallographically equivalent, the relationships that apply to this group are also valid for all groups of symmetry.

The equations of motion (1.6) may be written as

$$\omega^2 e_{1s} = L_{ss'} e_{1s'} + M_{ss'} e_{2s'} + N_{ss'} e_{3s'} + P_{ss'} e_{4s'}$$

$$\omega^2 e_{2s} = M_{ss'} e_{1s'} + L_{ss'} e_{2s'} + P_{ss'} e_{3s'} + N_{ss'} e_{4s'}$$

$$\omega^2 e_{3s} = N_{ss'} e_{1s'} + P_{ss'} e_{2s'} + L_{ss'} e_{3s'} + M_{ss'} e_{4s'}$$

$$\omega^2 e_{4s} = P_{ss'} e_{1s'} + N_{ss'} e_{2s'} + M_{ss'} e_{3s'} + L_{ss'} e_{4s'}$$

where obvious notations have been introduced for the equal coefficients $T$. This set of equations may be arranged into smaller independent sets by the introduction of the following new variables

$$e^{SSS} = e_{1s} + e_{2s} + e_{3s} + e_{4s}$$

$$e^{SAA} = e_{1s} + e_{2s} - e_{3s} - e_{4s}$$

$$e^{ASA} = e_{1s} - e_{2s} + e_{3s} - e_{4s}$$

$$e^{AAS} = e_{1s} - e_{2s} - e_{3s} + e_{4s}$$

Then the equations of motion take the form

$$\omega^2 e_s^{SSS} = \sum T_{ss'}^{SSS} e_{s'}^{SSS}, \qquad \omega^2 t_s^{SAA} = \sum T_{ss'}^{SAA} t_{s'}^{SAA}$$
$$\omega^2 e_s^{ASA} = \sum T_{ss'}^{ASA} e_{s'}^{ASA}, \qquad \omega^2 t_s^{AAS} = \sum T_{ss'}^{AAS} t_{s'}^{AAS}$$
(4.8)

$(s, s' = 1, 2, ..., 6)$ where

$$T^{SSS} = T_{1s1s'} + T_{1s2s'} + T_{1s3s'} + T_{1s4s'}$$

$$T^{SAA} = T_{1s1s'} + T_{1s2s'} - T_{1s3s'} - T_{1s4s'}$$

$$T^{ASA} = T_{1s1s'} - T_{1s2s'} + T_{1s3s'} - T_{1s4s'}$$

$$T^{AAS} = T_{1s1s'} - T_{1s2s'} - T_{1s3s'} + T_{1s4s'}$$

Thus, we see that of the four molecules in the cell, the molecules of each pair having common symmetry elements (as in the case, e.g., for the group $P2_1/a$, $Z = 4$), yield 24 frequencies classified into four groups with different vibration modes. The eigenvectors in the present case have six components, i.e., libration and translation are not separated. There are 21 nonvanishing frequencies. It is not obvious, a priori, to which group the vanishing frequencies belong.

*The Group Pbca, Z* = 4. We shall now proceed to discuss the results for the case of the benzene crystal. The regularities observed in the examples given earlier are well-enough defined to enable us to formulate the results immediately.

Since a cell contains four molecules, we see that Eqs. (4.8) are valid for this case. Since the molecule now is not in a general position, but occupies a center of inversion, each of the equations breaks down into two: one for translation and the other describing libration. The vanishing frequencies are to be looked for in the four translation groups.

Table 1 summarizes the results of the discussion. It includes the lower systems and lists data only for simple cells containing one independent molecule.

### *Table 1*

LIMITING FREQUENCIES FOR THE SPACE GROUPS OF LOWER SYSTEMS

| Number of molecules in a cell | Is the molecule in a special position | Space groups | The number of branches | The number of limiting frequencies | List of frequencies | |
|---|---|---|---|---|---|---|
| 1 | no | P1 | 6 | 3 | $\omega$ | |
| 1 | yes | $P\bar{1}$ <br> $P\bar{1}$    P2 | 6 | 3 | $\omega_{lib}$ | $\omega_{tr}$ |
| 2 | no | $P2_1$    P$m$ <br> P$c$ | 12 | 9 | $\omega^S$ | $\omega^A$ |
| 2 | yes | $P2_1/a$ <br> $P2/m$ | 12 | 9 | $\omega^S_{tr}$ <br> $\omega^S_{lib}$ | $\omega^A_{tr}$ <br> $\omega^A_{lib}$ |
| 4 | no | $P2_1/a$ <br> $P2_12_12_1$* | 24 | 21 | $\omega^{SSS}$ <br> $\omega^{SAA}$ <br> $\omega^{SSS}_{lib}$ | $\omega^{ASA}$ <br> $\omega^{AAS}$ <br> $\omega^{ASA}_{lib}$ |
| 4 | yes | *Pbca* <br> *Pmma*** | 24 | 21 | $\omega^{SAA}_{lib}$ <br> $\omega^{SSS}_{tr}$ <br> $\omega^{SAA}_{tr}$ | $\omega^{AAS}_{lib}$ <br> $\omega^{ASA}_{tr}$ <br> $\omega^{AAS}_{tr}$ |
| 8 | no | *Pbca* <br> *Pmma*** | 48 | 45 | six groups <br> of frequencies | |

* All groups of the V and $C_{2v}$ classes that have a simple cell.
** All P groups of the Class $V_h$.

5. THE DYNAMIC PROBLEM FOR A NAPHTHALENE CRYSTAL

In this section we give a detailed discussion of the equations of lattice dynamics for a crystal of naphthalene, which has the type of packing very often encountered for centrosymmetric molecules.

Let us write the equations of motion (1.6) in a general form specifying, however, that a cell contains two molecules.

$$\omega^2 e'_{1s} = \sum (T'_{1s1s'} e'_{1s'} + T'_{1s2s'} e'_{2s'})$$

$$\omega^2 e'_{2s} = \sum (T'_{2s1s'} e'_{1s'} + T'_{2s2s'} e'_{2s'})$$

There are no simplifications for $\mathbf{K} \neq \mathbf{0}$ and, for arbitrary orientation, the Hermitian matrix $T'_{\mu s \mu' s'}$ has no zero elements. Thus the matrix does not have a block form. We shall now give the values of the dynamic elements (the formulas are considerably simplified by symmetry) and show that the matrix becomes real after the introduction of the vector $\mathbf{e}_{\mu s}$ with components

$$e'_{11},\ e'_{12},\ e'_{13},\qquad ie'_{14},\ ie'_{15},\ ie'_{16},$$

$$e'_{21},\ e'_{22},\ e'_{23},\qquad ie'_{24},\ ie'_{25},\ ie'_{26}$$

instead of the 12-component eigenvector with components

$$e'_{11},\ e'_{12},\ e'_{13},\ e'_{14},\ e'_{15},\ e'_{16},$$

$$e'_{21},\ e'_{22},\ e'_{23},\ e'_{24},\ e'_{25},\ e'_{26}$$

This will also prove that the libration components of the 12-component vector are shifted by $\pi/2$ with respect to the translation components.

Now, the elements of the dynamic matrix have to be calculated

$$T'_{\mu s \mu' s'} = \sum \lambda^h_{\mu s \mu' s'} \exp(i\mathbf{K} \cdot \mathbf{R}_h)$$

We have already discussed the restrictions imposed on the coefficients $\lambda^h_{\mu s \mu' s'}$ by the symmetry of the group $P2_1/a$, $Z = 2$. These restrictions assume the form

$$\lambda^h_{\mu s \mu' s'} = s \lambda^h_{\mu' s' \mu s} \qquad \text{for} \quad \mu = \mu',\ \mu \neq \mu'$$

and also

$$\lambda^h_{\mu s \mu' s'} = \lambda^{h2_1}_{\mu' s \mu s'}, \qquad \lambda^h_{\mu s \mu s'} = \lambda^{h2_1}_{\mu' s \mu' s'}$$

The submatrices $T'_{1s1s'}$ and $T'_{2s2s'}$ have the same structure. If we write the expression for the matrix element in the form

$$T'_{\mu s \mu' s'} = \sum [\lambda^h_{\mu s \mu' s'} \exp(i\mathbf{K} \cdot \mathbf{R}_h) + \lambda^{\bar{h}}_{\mu s \mu' s'} \exp(i\mathbf{K} \cdot \mathbf{R}_{\bar{h}})]$$

and take note of the fact that the molecular position coincides with the center of inversion, we obtain

$$
\begin{array}{ll}
& s \qquad\ \ s' \\[4pt]
T'_{\mu s \mu s'} = \lambda^0_{\mu s \mu s'} + 2 \sum_{h \neq h(\bar{1})} \lambda^h_{\mu s \mu s'} \cos \mathbf{K} \cdot \mathbf{R}_h & \begin{array}{l}(1,2,3)(1,2,3)\\(4,5,6)(4,5,6)\end{array} \\[18pt]
T'_{\mu s \mu s'} = 2i \sum_{h \neq h(\bar{1})} \lambda^h_{\mu s \mu s'} \sin \mathbf{K} \cdot \mathbf{R}_h & \begin{array}{l}(1,2,3)(4,5,6)\\(4,5,6)(1,2,3)\end{array} \\[18pt]
T'_{\mu s \mu' s'} = 2 \sum_{h \neq h(\bar{1})} \lambda^h_{\mu s \mu' s'} \cos \mathbf{K} \cdot \mathbf{R}_h & \begin{array}{l}(1,2,3)(1,2,3)\\(4,5,6)(4,5,6)\end{array} \\[18pt]
T'_{\mu s \mu' s'} = 2i \sum_{h \neq h(\bar{1})} \lambda^h_{\mu s \mu' s'} \sin \mathbf{K} \cdot \mathbf{R}_h & \begin{array}{l}(1,2,3)(4,5,6)\\(4,5,6)(1,2,3)\end{array}
\end{array}
\tag{5.1}
$$

The matrix $T'_{\mu s \mu' s'}$ is Hermitian, but its elements in the present case are either real numbers or pure imaginary numbers. For the real elements we have

$$
\begin{array}{ll}
& s \qquad\ \ s' \\[4pt]
T'_{\mu s \mu' s'} = T'_{\mu' s' \mu s} & \begin{array}{l}(1,2,3)(1,2,3)\\(4,5,6)(4,5,6)\end{array}
\end{array}
$$

and for the pure imaginary elements

$$
\begin{array}{ll}
T'_{\mu s \mu' s'} = -iT'_{\mu' s' \mu s} & \begin{array}{l}(1,2,3)(4,5,6)\\(4,5,6)(1,2,3)\end{array}
\end{array}
$$

These formulas follow from the general treatment and directly from (3.4).

The validity of the statement made at the beginning of the section is now obvious: All elements of the matrix of coefficients will be real for the vectors in which the libration components are shifted by 90°. Let us write the basic equations in the form:

$s = 1, 2, 3$:

$$
\begin{aligned}
\omega^2 e'_{1s} = &\sum_{s'=1,2,3} T'_{1s1s'} e'_{1s'} + \sum_{s'=4,5,6} T'_{1s1s'} e'_{1s'} + \sum_{s'=1,2,3} T'_{1s2s'} e'_{2s'} \\
&+ \sum_{s'=4,5,6} T'_{1s2s'} e'_{2s'}
\end{aligned}
$$

$s = 4, 5, 6$:

$$
\begin{aligned}
\omega^2 e'_{1s} = &\sum_{s'=1,2,3} T'_{1s1s'} e'_{1s'} + \sum_{s'=4,5,6} T'_{1s1s'} e'_{1s'} + \sum_{s'=1,2,3} T'_{1s2s'} e'_{2s'} \\
&+ \sum_{s'=4,5,6} T'_{1s2s'} e'_{2s'}
\end{aligned}
$$

$s = 1, 2, 3$:

$$\omega^2 e'_{2s} = \sum_{s'=1,2,3} T'_{2s1s'} e'_{1s'} + \sum_{s'=4,5,6} T'_{2s1s'} e'_{1s'} + \sum_{s'=1,2,3} T'_{2s2s'} e'_{2s'}$$
$$+ \sum_{s'=4,5,6} T'_{2s2s'} e'_{2s'}$$

$s = 4, 5, 6$:

$$\omega^2 e'_{2s} = \sum_{s'=1,2,3} T'_{2s1s'} e'_{1s'} + \sum_{s'=4,5,6} T'_{2s1s'} e'_{1s'} + \sum_{s'=1,2,3} T'_{2s2s'} e'_{2s'}$$
$$+ \sum_{s'=4,5,6} T'_{2s2s'} e'_{2s'}$$

Introduce a real, symmetric matrix $T_{\mu s \mu' s'}$ (unprimed), which is related to $T'_{\mu s \mu' s'}$ as follows: $T' = iT$ and $T' = -iT$ for the mixed elements of the matrix that are situated to the left and to the right from the principal diagonal, respectively, and $T' = T$ for the nonmixed case.

We now rewrite the equations, multiplying the second and the fourth equation by $i$ and introducing some new unprimed quantities. The resulting equations have the same form as the original ones, i.e.,

$$\omega^2 e_{1s} = \sum (T_{1s1s'} e_{1s'} + T_{1s2s'} e_{2s'}), \qquad \omega^2 e_{2s} = \sum (T_{2s1s'} e_{1s'} + T_{2s2s'} e_{2s'})$$

The expressions for $T_{\mu s \mu' s'}$ are given by Eq. (5.1) with the $i$s dropped. We shall repeatedly refer to these equations later.

The matrix cannot be simplified if a complete solution of the problem is to be obtained, i.e., if the behavior of all the 12 dispersion surfaces $\omega(K)$ is examined. The above equations, which can now be written only formally, do not break down into two sets. The frequencies are given by the roots of a $12 \times 12$ secular equation, and the eigenvectors have 12 components. All the required coupling coefficients should be evaluated prior to the calculation of the dynamic matrix elements, and the matrices $T_{\mu s \mu' s'}$ should be obtained at the mesh points defined by the vector $K$. Twelve frequencies and 144 components of twelve 12-component vectors are to be found.

It will be noted, however, that the solution of the problem is facilitated by the fact that only the nearest-neighbor interactions are needed. For a crystal of naphthalene, an appreciable contribution comes from the following neighbors (Fig. 2): molecules with a common index $\mu$ that occupy the center at [010], i.e., $R = b$, or at [001]. i.e., $R = c$, and molecules with a different $\mu$ and the coordinates $[\frac{1}{2}\frac{1}{2}0]$, $[\frac{\bar{1}}{2}\frac{1}{2}0]$, i.e., $R = \frac{1}{2}(a+b)$, $R = \frac{1}{2}(-a+b)$ or $[\frac{1}{2}\frac{1}{2}1]$, $[\frac{\bar{1}}{2}\frac{1}{2}1]$, i.e., $R = \frac{1}{2}(a+b)+c$, $R = \frac{1}{2}(-a+b)+c$.

The matrix elements have the following form: for the unmixed indices $s$

$$T_{\mu s \mu s'} = \lambda^0_{\mu s \mu s'} + 2\lambda^{001}_{\mu s \mu s'} \cos K \cdot R_{001} + 2\lambda^{010}_{\mu s \mu s'} \cos K \cdot R_{010}$$
$$+ 2\lambda^{\frac{1}{2}\frac{1}{2}0}_{\mu s \mu s'} \cos K \cdot R_{\frac{1}{2}\frac{1}{2}0} + 2\lambda^{\frac{\bar{1}}{2}\frac{1}{2}0}_{\mu s \mu s'} \cos K \cdot R_{\frac{\bar{1}}{2}\frac{1}{2}0}$$
$$+ 2\lambda^{\frac{1}{2}\frac{1}{2}1}_{\mu s \mu s'} \cos K \cdot R_{\frac{1}{2}\frac{1}{2}1} + 2\lambda^{\frac{\bar{1}}{2}\frac{1}{2}1}_{\mu s \mu s'} \cos K \cdot R_{\frac{\bar{1}}{2}\frac{1}{2}1}$$

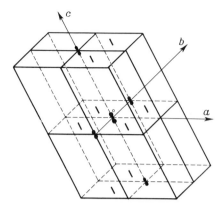

**Fig. 2.** The crystal lattice of naphthalene. The nearest neighbours of molecule zero taken into account are shown.

while $T_{\mu s \mu' s'}$ is given by the same expression with the term $\lambda^0$ dropped; for the mixed indices $s$

$$T_{\mu s \mu s'} = 2\lambda^{001}_{\mu s \mu s'} \sin \mathbf{K} \cdot \mathbf{R}_{001} + 2\lambda^{010}_{\mu s \mu s'} \sin \mathbf{K} \cdot \mathbf{R}_{010}$$
$$+ 2\lambda^{\frac{1}{2}\frac{1}{2}0}_{\mu s \mu s'} \sin \mathbf{K} \cdot \mathbf{R}_{\frac{1}{2}\frac{1}{2}0} + 2\lambda^{\overline{\frac{1}{2}}\frac{1}{2}0}_{\mu s \mu s'} \sin \mathbf{K} \cdot \mathbf{R}_{\overline{\frac{1}{2}}\frac{1}{2}0}$$
$$+ 2\lambda^{\frac{1}{2}\frac{1}{2}1}_{\mu s \mu s'} \sin \mathbf{K} \cdot \mathbf{R}_{\frac{1}{2}\frac{1}{2}1} + 2\lambda^{\overline{\frac{1}{2}}\frac{1}{2}1}_{\mu s \mu s'} \sin \mathbf{K} \cdot \mathbf{R}_{\overline{\frac{1}{2}}\frac{1}{2}1}$$

while $T_{\mu s \mu' s'}$ is defined by the same expression.

There is one special case when the problem is essentially simplified, viz. when we consider the cross section of the dispersion surfaces by the plane $\mathbf{K} \| \mathbf{b}$.

Any sum involved in (5.1) includes the terms related to molecules with the centers of gravity at $m, n, p$ and $\overline{m}, n, \overline{p}$ or $m + \frac{1}{2}, n + \frac{1}{2}, p$ and $\overline{m} + \frac{1}{2}, n + \frac{1}{2}, \overline{p}$. The first sum may be rewritten, as

$$2 \sum_{h \neq h(\overline{1})} (\lambda^{mnp} \cos \mathbf{K} \cdot \mathbf{R}_{mnp} + \lambda^{\overline{m}n\overline{p}} \cos \mathbf{K} \cdot \mathbf{R}_{\overline{m}n\overline{p}})$$

For $\mathbf{K} \| \mathbf{b}$ we have

$$\mathbf{K} \cdot \mathbf{R}_{mnp} = \mathbf{K} \cdot \mathbf{R}_{\overline{m}n\overline{p}} = K R_y$$

and hence both terms have the same factor $\cos K R_y$.

Equation (5.1) has the form (recall that the imaginary unity disappears after the transformation of $T'$ into $T$)

$$\begin{array}{cc} s & s' \\ (1,2,3) & (1,2,3) \\ (4,5,6) & (4,5,6) \end{array} \quad T_{\mu s \mu s'} = \lambda^0_{\mu s \mu s'} + 2 \sum_{h \neq h(\overline{1})} (\lambda^{mnp}_{\mu s \mu s'} + \lambda^{\overline{m}n\overline{p}}_{\mu s \mu s'}) \cos K R_y$$

$(1, 2, 3) (4, 5, 6)$
$(4, 5, 6) (1, 2, 3)$
$$T_{\mu s \mu s'} = 2 \sum_{h \neq h(\bar{1})} (\lambda^{mnp}_{\mu s \mu s'} + \lambda^{\bar{m}n\bar{p}}_{\mu s \mu s'}) \sin K R_y$$

$(1, 2, 3) (1, 2, 3)$
$(4, 5, 6) (4, 5, 6)$
$$T_{\mu s \mu' s'} = 2 \sum_{h \neq h(\bar{1})} (\lambda^{m+\frac{1}{2}n+\frac{1}{2}p}_{\mu s \mu' s'} + \lambda^{\bar{m}+\frac{1}{2}n+\frac{1}{2}\bar{p}}_{\mu s \mu' s'}) \cos K R_y$$

$(1, 2, 3) (4, 5, 6)$
$(4, 5, 6) (1, 2, 3)$
$$T_{\mu s \mu' s'} = 2 \sum_{h \neq h(\bar{1})} (\lambda^{m+\frac{1}{2}n+\frac{1}{2}p}_{\mu s \mu' s'} + \lambda^{\bar{m}+\frac{1}{2}n+\frac{1}{2}\bar{p}}_{\mu s \mu' s'}) \sin K R_y$$

For $0n0$, $m0p$ and similar indices the second term in brackets does not appear. The matrix will consist of four equal blocks. As is easily seen

$$\lambda^0_{1s1s'} = \lambda^0_{2s2s'}, \qquad \lambda^{mnp}_{1s1s'} = \lambda^{\bar{m}n\bar{p}}_{2s2s'}, \qquad \lambda^{\bar{m}n\bar{p}}_{1s1s'} = \lambda^{mnp}_{2s2s'}$$

Hence for the two upper rows we have $T_{1s1s'} = T_{2s2s'}$. Furthermore,

$$\lambda^{m+\frac{1}{2}n+\frac{1}{2}p}_{1s2s'} = \lambda^{\bar{m}+\frac{1}{2}n+\frac{1}{2}\bar{p}}_{2s1s'}, \qquad \lambda^{m+\frac{1}{2}n+\frac{1}{2}p}_{1s2s'} = \lambda^{\bar{m}+\frac{1}{2}n+\frac{1}{2}\bar{p}}_{2s1s'}$$

and hence $T_{1s2s'} = T_{2s1s'}$. The behavior of the equations of motion in this case is known from the preceding section. Vibrations are classified as symmetric or antisymmetric and satisfy two equations that are mutually independent:

$$\omega^2 e_s^S = \sum T_{ss'}^S e_{s'}^S$$
$$\omega^2 e_s^A = \sum T_{ss'}^A e_{s'}^A \qquad (s, s' = 1, 2, ..., 6)$$

The matrix elements for the unmixed indices $s$ are

$$T_{ss'}^S = \lambda^0_{1s1s'} + 2 \sum_{h \neq h(\bar{1})} (\lambda^{mnp}_{1s1s'} + \lambda^{\bar{m}n\bar{p}}_{1s1s'}) \cos K R_y$$
$$+ 2 \sum_{h \neq h(\bar{1})} (\lambda^{m+\frac{1}{2}n+\frac{1}{2}p}_{1s2s'} + \lambda^{\bar{m}+\frac{1}{2}n+\frac{1}{2}\bar{p}}_{1s2s'}) \cos K R_y$$
$$T_{ss'}^A = \lambda^0_{1s1s'} + 2 \sum_{h \neq h(\bar{1})} (\lambda^{mnp}_{1s1s'} + \lambda^{\bar{m}n\bar{p}}_{1s1s'}) \cos K R_y \qquad (5.2)$$
$$- 2 \sum_{h \neq h(\bar{1})} (\lambda^{m+\frac{1}{2}n+\frac{1}{2}p}_{1s2s'} + \lambda^{\bar{m}+\frac{1}{2}n+\frac{1}{2}\bar{p}}_{1s2s'}) \cos K R_y$$

and for the mixed indices $s$

$$T_{ss'}^S = 2 \sum_{h \neq h(\bar{1})} (\lambda^{mnp}_{1s1s'} + \lambda^{\bar{m}n\bar{p}}_{1s1s'}) \sin K R_y$$
$$+ 2 \sum_{h \neq h(\bar{1})} (\lambda^{m+\frac{1}{2}n+\frac{1}{2}p}_{1s2s'} + \lambda^{\bar{m}+\frac{1}{2}n+\frac{1}{2}\bar{p}}_{1s2s'}) \sin K R_y$$
$$T_{ss'}^A = 2 \sum_{h \neq h(\bar{1})} (\lambda^{mnp}_{1s1s'} + \lambda^{\bar{m}n\bar{p}}_{1s1s'}) \sin K R_y$$
$$- 2 \sum_{h \neq h(\bar{1})} (\lambda^{m+\frac{1}{2}n+\frac{1}{2}p}_{1s2s'} + \lambda^{\bar{m}+\frac{1}{2}n+\frac{1}{2}\bar{p}}_{1s2s'}) \sin K R_y$$

Note that $\lambda$ with a negative index is dropped if the position of the molecule coincides with some point on the $b$ axis or in the $ac$ plane.

For $\mathbf{K} = 0$ the matrix elements which have mixed indices vanish, and both matrices break down into two—libration and translation matrices. As a corollary, we obtain the results that were discussed earlier.

The resulting equations lead to 12 curves $\omega(\mathbf{K})$, three of them running from the origin (which particular curves behave in this manner depends on the angle between the molecular axes of inertia and the crystallographic axes, as discussed earlier) and the other nine branches start from their limiting values which may be obtained from optical data.

Within the framework of the nearest-neighbor approximation (the neighboring molecules were listed earlier) we get the following simplified formulas instead of (5.2):

For unmixed indices

$$T_{ss'}^{S} = \lambda_{1s1s'}^{0} + 2\lambda_{1s1s'}^{001} + 2\lambda_{1s1s'}^{010} \cos Kb$$
$$= + 2(\lambda^{\frac{1}{2}\frac{1}{2}0} + \lambda^{\overline{\frac{1}{2}}\frac{1}{2}0} + \lambda^{\frac{1}{2}\frac{1}{2}1} + \lambda^{\overline{\frac{1}{2}}\frac{1}{2}1}) \cos Kb/2$$

$$T_{ss'}^{A} = \lambda_{1s1s'}^{0} + 2\lambda_{1s1s'}^{001} + 2\lambda_{1s1s'}^{010} \cos Kb$$
$$- 2(\lambda^{\frac{1}{2}\frac{1}{2}0} + \lambda^{\overline{\frac{1}{2}}\frac{1}{2}0} + \lambda^{\frac{1}{2}\frac{1}{2}1} + \lambda^{\overline{\frac{1}{2}}\frac{1}{2}1}) \cos Kb/2$$

and for mixed indices

$$T_{ss'}^{S} = 2\lambda_{1s1s'}^{010} \sin Kb + 2(\lambda^{\frac{1}{2}\frac{1}{2}0} + \lambda^{\overline{\frac{1}{2}}\frac{1}{2}0} + \lambda^{\frac{1}{2}\frac{1}{2}1} + \lambda^{\overline{\frac{1}{2}}\frac{1}{2}1}) \sin Kb/2,$$

$$T_{ss'}^{A} = 2\lambda_{1s1s'}^{010} \sin Kb - 2(\lambda^{\frac{1}{2}\frac{1}{2}0} + \lambda^{\overline{\frac{1}{2}}\frac{1}{2}0} + \lambda^{\frac{1}{2}\frac{1}{2}1} + \lambda^{\overline{\frac{1}{2}}\frac{1}{2}1}) \sin Kb/2.$$

At $K = \pi/b$ we have

$$T_{ss'}^{S} = T_{ss'}^{A} = \lambda_{1s1s'}^{0} + 2\lambda_{1s1s'}^{001} - 2\lambda_{1s1s'}^{010}$$

for unmixed indices and

$$T_{ss'}^{S} = -T_{ss'}^{A} = 2(\lambda^{\overline{\frac{1}{2}}\frac{1}{2}0} + \lambda^{\frac{1}{2}\frac{1}{2}0} + \lambda^{\overline{\frac{1}{2}}\frac{1}{2}1} + \lambda^{\frac{1}{2}\frac{1}{2}1})$$

for mixed indices. The signs of $T_{ss'}^{S}$ and $T_{ss'}^{A}$ are seen to be different for the mixed-index matrix elements. This fact may be interpreted in the following way:

At $K = \pi/b$, i.e., at $\lambda = b/2$ the equations for symmetric vibrations and the equations for antisymmetric vibrations become identical if the rotational components of either class are shifted by $\pi/2$.

At $K = \pi/b$ the antisymmetric and symmetric branches always merge pairwise. We have only six frequencies at this value of $K$. Then, with the curves diverging, this behavior of $\omega(\mathbf{K})$ will recur at greater values of $K$.

6. CALCULATION OF CRYSTAL DYNAMICS BY THE METHOD OF
ATOM–ATOM POTENTIALS

The whole problem of comparing the theoretical and experimental data on lattice dynamics is fundamentally changed if the atom–atom potentials scheme is employed for the calculation of the molecular interaction energy. As was pointed out by the author [2], the success that accompanied the application of the scheme for the calculation of the lattice energy gives every reason to believe that the method will also prove valuable for the evaluation of the dynamic coefficients. For this purpose, computer programs should be written, capable of handling calculations of the lattice energy as a function of two displacements of two molecules (or of one molecule if the zero coefficients are to be evaluated).

Variations of the dimensions of an elementary cell that can be programmed into the algorithm for the calculation of the dynamic coefficients will make it possible to examine the temperature behavior of lattice vibration frequencies.

A variety of computer programs may be written, differing in the degree of complexity. A master computer program should be capable of determining the dispersion surfaces for preassigned cell dimensions and for a series of cells extending along the isobaric section of the energy surface.

One may also construct a computer program that will calculate and minimize the free energy. The simplest way to do this is to use the general formula of the quasi-harmonic approximation (Chap. 6).

A simpler program would handle the limiting frequencies, which could also be calculated in the quasi-harmonic approximation as functions of temperature.

Finally, the simplest problem is the calculation of the rms spectrum frequency, which is uniquely determined if the values of as few as six (zero type) dynamic coefficients are available. According to (2.2) we have

$$\overline{\omega^2} = \frac{1}{6}\left[\frac{1}{M}\frac{\partial^2 U}{\partial u_1{}^2} + \frac{1}{M}\frac{\partial^2 U}{\partial u_2{}^2} + \frac{1}{M}\frac{\partial^2 U}{\partial u_3{}^2} + \frac{1}{I_1}\frac{\partial^2 U}{\partial \Omega_1{}^2} + \frac{1}{I_2}\frac{\partial^2 U}{\partial \Omega_2{}^2} + \frac{1}{I_3}\frac{\partial^2 U}{\partial \Omega_3{}^2}\right]$$

Thus, we see that only six coupling coefficients need to be evaluated for a given elementary cell. If the lattice energy is plotted against each of the six displacements of molecule zero in the vicinity of equilibrium, the values of the second derivatives calculated at a lattice point will give all terms of the above expression for $\overline{\omega^2}$.

The first numerical calculation of the dynamic problem by the method of atom–atom potentials was made by the author and his co-workers. For a series of molecular crystals the values of $\overline{\omega^2}$ were obtained as a function of temperature. The results for $\overline{\omega^2}$ cannot be compared with the experimental data on the behavior of limiting frequencies versus temperature, owing to the fact that the averaging in Eq. (1.4) is carried out over the whole spectrum and for the entire range of the wave numbers. However, as is shown on p. 372, the value of $\overline{\omega^2}$ is, but for a constant factor, equal to the characteristic temperature $\theta$, which can be determined from experimental thermodynamic data. The calculated values of $\overline{\omega^2}$ are discussed and compared with $\theta$ in Chapter 6, Table 6.

The calculation scheme suggested by the author was used by Oliver [3], who calculated the limiting frequencies and their eigenvectors for the benzene crystal. His results are summarized in Table 2. Similar calculations for the naphthalene and anthracene crystals are reported by Pawley [4] and Weulersse [5]. Mukhtarov [6] and the author have calculated the intermolecular vibration frequencies for the crystals of naphthalene, anthracene, diphenyl, and also obtained the $K \| b$ sections of the dispersion surfaces.

We shall describe the results of this paper, since the calculations are quite thorough and comprehensive, giving frequency versus temperature data (the calculations were made for a series of cells measured at different temperatures). The symmetry group for all of these crystals is $P2_1/a$, $Z = 2$. The dynamics of such lattices was discussed in detail in the preceding sections. The quasi-harmonic model and atom–atom potentials were used for the calculation of the matrix of dynamic coefficients; the values of the potential parameters involved are given in Ch. 2 on p. 170. Numerical results on the libration of the naphthalene molecule [7] (frequencies and eigenvectors) at

*Table 2*

FREQUENCIES AND EIGENVECTORS OF THE BENZENE CRYSTAL

| | SSS | | | | ASA | | |
|---|---|---|---|---|---|---|---|
| $\omega$ | $U$ | $V$ | $W$ | $\omega$ | $U$ | $V$ | $W$ |
| 100 | 0.978 | −0.015 | 0.002 | 96 | 0.899 | 0.2374 | 0.035 |
| 85 | 0.013 | 0.950 | 0.050 | 82 | 0.196 | 0.985 | −0.020 |
| 69 | −0.001 | 0.045 | 0.915 | 65 | 0.030 | 0.016 | 0.875 |

| | SAA | | | | AAS | | |
|---|---|---|---|---|---|---|---|
| $\omega$ | $U$ | $V$ | $W$ | $\omega$ | $U$ | $V$ | $W$ |
| 89 | 0.898 | −0.063 | 0.005 | 83 | 0.956 | 0.078 | −0.099 |
| 55 | 0.058 | 0.916 | 0.015 | 51 | 0.074 | 0.986 | 0.013 |
| 46 | 0.004 | −0.013 | 0.959 | 40 | 0.089 | −0.013 | 0.975 |

## Table 3

FREQUENCIES AND EIGENVECTORS OF MOLECULAR LIBRATIONS
FOR THE NAPHTHALENE CRYSTAL

| $\omega^S_{exp}$ | $\omega^S_{calc}$ | $U$ | $V$ | $W$ | $\omega^A_{exp}$ | $\omega^A_{calc}$ | $U$ | $V$ | $W$ |
|---|---|---|---|---|---|---|---|---|---|
| 126 | 109 | 0.947 | −0.298 | −0.113 | 125 | 116 | 0.997 | −0.077 | 0.012 |
| 79 | 74 | 0.298 | 0.954 | −0.017 | 71 | 65 | 0.075 | 0.896 | 0.438 |
| 51 | 51 | 0.113 | −0.018 | 0.993 | 46 | 40 | 0.022 | 0.437 | 0.899 |

$T = 293°K$ are summarized in Table 3. The table also lists the frequency values obtained experimentally [8]. The values of $\omega$ are given in inverse centimeters.

The relative orientation of the axes of symmetric and antisymmetric librations and of the inertial axes of the molecule is shown in Fig. 3. As can be seen from the figures and the table, close as the vibrational axes are to the inertial axes of the molecule, they do not coincide. It is appropriate to note, nevertheless, that until now all studies of intermolecular vibrations by means of Raman spectra have assumed that the axes of normal vibrations should coincide with the inertial axes. The origin of this error, resulting in mis-assignments of the frequencies, is difficult to trace. The results of our calculations suggest the following conclusions.

The angle between the inertial axes and the eigenvectors of the symmetric and antisymmetric vibrations corresponding to the frequency of 116 cm$^{-1}$ does not exceed 10°; however, there are two modes of low-frequency asymmetric vibrations whose eigenvectors form an angle of about 30° with the inertia axes $V$ and $W$.

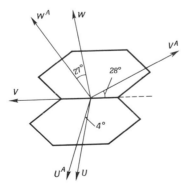

*Fig. 3.* The orientation of vibrational axes relative to the inertial axes of the molecule for the naphthalene crystal.

## Table 4

FREQUENCIES AND EIGENVECTORS OF MOLECULAR TRANSLATIONS
FOR THE NAPHTHALENE CRYSTAL

| $\omega_{exp}$ | $\omega_{calc}^S$ | $a$ | $b$ | $C$ | $\omega_{exp}$ | $\omega_{calc}^S$ | $a$ | $b$ | $C$ |
|---|---|---|---|---|---|---|---|---|---|
| 0 | 0 | — | — | — | 98 | 85 | 0.857 | 0.04 | 0.515 |
| 73 | 51 | 0.012 | 0.993 | 0.017 | 0 | 0 | — | — | — |
| 0 | 0 | — | — | — | 39 | 41 | −0.513 | −0.028 | 0.857 |

Table 4 compares the calculated values of translation vibration frequencies for naphthalene with the experimental results [9]. The temperature is $T = 293°K$, the same value as in the case of librations.

The table demonstrates that the directions of translation vibrations are close to the crystal axes (also cf. Fig. 3), and therefore, terminologically, we may refer to the vibrations of types $a$, $b$, and $c$. Of the three curves corresponding to translation vibrations, which vanish at $\mathbf{K} = 0$ (acoustic branches), two ($a$ and $c$ types) belong to the asymmetric mode and one ($b$ type) to the symmetric mode. The $C$ axis[1] in Table 4 is the third axis orthogonal to the crystallographic $a$ and $b$ axes, which are mutually orthogonal.

## Table 5

ANTHRACENE: LIBRATIONS OF THE MOLECULE

| $\omega_{exp}^S$ | $\omega_{calc}^S$ | $U$ | $V$ | $W$ | $\omega_{exp}^A$ | $\omega_{calc}^A$ | $U$ | $V$ | $W$ |
|---|---|---|---|---|---|---|---|---|---|
| 121 | 155 | 0.969 | −0.156 | −0.193 | 125 | 135 | 0.995 | −0.074 | −0.067 |
| 70 | 85 | 0.197 | 0.956 | 0.217 | 65 | 62 | −0.038 | 0.899 | −0.437 |
| 39 | 34 | 0.150 | −0.248 | 0.957 | 45 | 47 | 0.093 | 0.432 | 0.897 |

Table 5 gives the results of our calculations of the libration frequencies and eigenvectors for anthracene [10], in comparison with the frequencies available from experimental measurements [8]. The temperature is $T = 293°K$. Here, as in the case of naphthalene, deviations of the axes of normal librations from the principal inertia axes of the molecule are insignificant.

The results of calculations of the translation vibrations at $T = 293°K$ are

---

[1] There is no assignment of frequencies to the symmetric or antisymmetric modes in the paper referred to earlier.

## Table 6

ANTHRACENE: TRANSLATIONS OF THE MOLECULE

| $\omega^S_{calc}$ | $a$ | $b$ | $c$ | $\omega^A_{calc}$ | $a$ | $b$ | $c$ |
|---|---|---|---|---|---|---|---|
| 0 | — | — | — | 101 | 0.893 | −0.051 | 0.582 |
| 59 | 0.065 | 0.982 | 0.035 | 0 | — | — | — |
| 0 | — | — | — | 42 | −0.579 | −0.060 | 0.810 |

summarized in Table 6. There are no published data on experimental measurements of the translation vibrations spectrum of anthracene. Experimental data on the spectrum of translations and librations of diphenyl are given in Ref. [11]. However, some frequencies of the spectrum are missing. Table 7 contains the experimentally measured libration frequencies and the frequencies

## Table 7

DIPHENYL: VIBRATIONS

| $\omega_{exp}$ | $\omega^S_{calc}$ | $U$ | $V$ | $W$ | $\omega_{exp}$ | $\omega^A_{calc}$ | $U$ | $V$ | $W$ |
|---|---|---|---|---|---|---|---|---|---|
| — | 94 | 0.956 | 0.121 | 0.266 | — | 50 | 0.970 | 0.242 | 0.022 |
| 43 | 36 | −0.256 | 0.784 | 0.566 | — | 82 | −0.232 | 0.951 | −0.205 |
| 94 | 119 | −0.140 | −0.609 | 0.780 | 59 | 88 | −0.070 | 0.194 | 0.978 |

and eigenvectors calculated at $T = 220°K$. In this case deviations of the inertial axes from the vibration axes for the $V^s$ and $W^s$ normal vibrations are so large that assignment of either type to any inertia axis is no longer feasible.

Calculated and measured values of the translation vibration frequencies at $T = 293°K$ are given in Table 8. The assignment of the frequencies of translations in paper [11] is apparently erroneous.

The calculated results for the spectra of the naphthalene, anthracene, and diphenyl crystals that are given in this section agree quite well with the

## Table 8

DIPHENYL: TRANSLATIONS

| $\omega_{exp}$ | $\omega^S_{calc}$ | $a$ | $b$ | $c$ | $\omega_{exp}$ | $\omega^A_{calc}$ | $a$ | $b$ | $c$ |
|---|---|---|---|---|---|---|---|---|---|
| — | 0 | — | — | — | — | 80 | 0.892 | −0.515 | −0.275 |
| 56 | 57 | 0.323 | 0.735 | −0.020 | — | 0 | — | — | — |
| — | 0 | — | — | — | 116 | 92 | 0.576 | −0.211 | 0.805 |

experimental data. Noteworthy in this respect is the remarkable agreement between experimental and calculated values of frequencies for the naphthalene and anthracene crystals, reported in a paper of Suzuky, based on comprehensive research of a high experimental standard.

Discrepancies between our figures and the experimental data, which are observed at high frequencies, appear to be largely due to failure to include the interaction between the intramolecular and crystal vibrations.

The temperature dependence of the frequencies of the optical branches in the quasi-harmonic approximation is discussed by Mukhtarov and the author in the paper referred to earlier. As an example, we give the numerical results for the spectrum of the naphthalene crystal; the calculation was carried out for the temperature range $0°$–$300°K$. The calculated and experimental frequency versus temperature curves for the naphthalene crystal are plotted in Fig. 4a for librations and in Fig. 4b for translations. The dotted lines are experimental curves, while the solid lines give the theoretical results. The letters A and S refer to the antisymmetric and symmetric vibration modes respectively, $u, v, w, a, b, c$ are the types of vibrations. As can be seen from Fig. 4, the experimental and calculated curves agree quite well for the low-frequency branches (discrepancies are less than 10%), approaching each other at low temperatures. The fact that the calculated curves fall more steeply at high temperatures is apparently due to the anharmonic character of vibrations, which cannot be fully accounted for within the framework of the quasi-harmonic approximation. We also note that there is considerable disagreement between the calculated and the experimental curve for a $b$-type translation of comparatively low frequency. Apparently the accuracy of the experimental curve should be improved.

Considerable disagreement between the calculated and experimental curves, which tends to increase (up to 20%) in the low-temperature range, is observed for the high-frequency branches of $u^A$, $u^S$, $v^S$ librations. This situation may well be due to the interaction between the intramolecular and lattice vibrations. There are two low-frequency bands of 176 and 195 $cm^{-1}$ in the vibrational spectra of the naphthalene molecule (the frequency next in magnitude is $359 cm^{-1}$). The existence of low-frequency vibrations of this kind, naturally, renders incorrect the approximation in which a molecule in the lattice is treated as rigid. Moreover, relatively small frequency increments of low-frequency intramolecular vibrations and high-frequency lattice vibrations, as is well known from the theory of vibrations in classical mechanics or from perturbation theory in quantum mechanics, result in substantial changes of the frequencies of interacting vibration systems even if such systems are engaged in weak kinematic or potential interaction. Both types of interactions play a part in the present case.

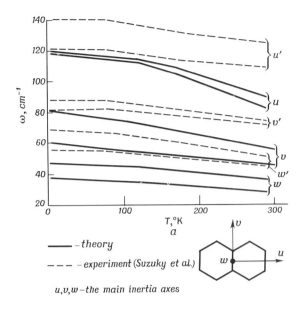

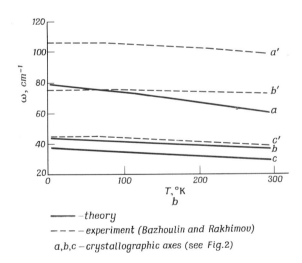

Fig. 4. The temperature behavior of libration (a) and translation (b) frequencies of molecular vibrations for the naphthalene crystal.

In the high-temperature range the experimental and calculated curves come closer together. Insofar as the assumption concerning the interaction between intramolecular and lattice vibrations is valid, the fact that the curves approach each other at higher temperatures may naturally be explained by the anharmonic character of the lattice vibrations, all the more so that this factor, as the low-frequency branches show, reduces the frequencies with the growth of vibration amplitudes, while the interaction between the intramolecular and lattice vibrations must clearly lead to greater disagreement between the frequency of a true lattice vibration and the result calculated in the quasi-harmonic approximation.

An analysis of the behavior of the frequency versus temperature curves for the anthracene and diphenyl crystals also indicates good agreement with the experiment. Such an analysis is in many respects similar to the treatment of the naphthalene crystal that we have discussed.

The same paper gives the sections of the dispersion surfaces $\omega(\mathbf{K})$ for $\mathbf{K} \| \mathbf{b}$ for the naphthalene, anthracene, and diphenyl crystals. The plots for naphthalene ($T = 293°\mathrm{K}$) are given in Fig. 5.

To calculate the sections of the dispersion surfaces the matrix of dynamic coefficients was diagonalized for every value of $K$. It proved sufficient to take the $K$ increment equal to 0.05 in the cases studied; to make sure that the assignment of frequencies to vibration modes was correct, the step value was reduced to 0.01 in the regions where different branches came too close together or in the vicinity of the points of intersection. Comparison of the calculated and experimental frequencies, as well as the study of frequency versus temper-

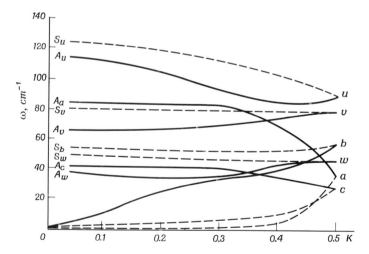

Fig. 5. A section of the dispersion surfaces for the naphthalene crystal. S.A.: symmetric, antisymmetric; u, v, w: libration; $a, b, c$: translation frequencies.

ature curves, reveals a good fit to the experimental data. However, it is quite obvious that the calculated results can be substantially improved if a better choice of the atom–atom potential curves is made. Besides, it is quite possible that the formulation of the problem should be refined by the introduction of intramolecular vibrations. It is not unlikely, for example, that the reversed order and excessively high values of the calculated S and A frequencies for the strongest pair in the naphthalene crystal spectrum are due precisely to the interaction between the intermolecular and intramolecular vibrations.

7. THE MEAN VIBRATION AMPLITUDE

According to the fundamental principles of statistical physics, the frequency of a normal vibration determines its mean energy

$$\varepsilon = \frac{1}{2}\hbar\omega + \frac{\hbar\omega}{\exp(\hbar\omega/kT) - 1}$$

On the other hand, the translation mean energy is equal to $\frac{1}{2}m\omega^2 a^2$. We can thus find the vibration amplitude. However, we are concerned with the mean displacement along a certain axis. The solution of the dynamic problem yields, for every frequency, normalized eigenvectors with components $e_i$ ($i = 1, 2, \ldots, 6z$). To calculate the quadratic mean displacement along the $X$ axis, the quadratic mean displacement produced by a normal vibration with a frequency $\omega$, i.e., $\varepsilon/m\omega^2$, should be multiplied by $e_1{}^2$ (which, of course, is the square of the direction cosine). Thus, in the system of molecular coordinates, the quadratic mean displacements along the $x$, $y$, and $z$ axes are, respectively, given by

$$e_1{}^2\varepsilon/m\omega^2, \qquad e_2{}^2\varepsilon/m\omega^2, \qquad e_3{}^2\varepsilon/m\omega^2.$$

To find the contribution made to a displacement along the unit vector $\mathbf{e}$ by the vibration frequency $\omega$ with the eigenvector $\mathbf{e}$ we have to calculate

$$(\varepsilon/m\omega^2)(\mathbf{e}\cdot\mathbf{1})^2 = (\varepsilon/m\omega^2)e_i e_j l_i l_j$$

It is seen that the pattern of a translation displacement is determined by the tensor

$$(\varepsilon/m\omega^2)e_i e_j$$

This contribution comes from a single normal vibration. There are $6Z$ vibrations per cell. Hence, the tensor sought is given by

$$T_{ij} = \sum_{6Z} (\varepsilon/m\omega^2)e_i e_j$$

If the data available for the calculation include not only limiting frequencies but also dispersion surfaces, the averaging for $T_{ij}$ should be carried out over a reciprocal space unit (i.e., over the Brillouin zone).

We have discussed the calculation of the molecular translation displacements. It is remarkable that the tensor elements are expressed in terms of the translation components of eigenvectors.

In very much the same way, for rotations about the $x$, $y$, and $z$ axes, the mean angular displacements from equilibrium positions are given by

$$\theta_1 = e_4\,\varepsilon/I_1\,\omega^2, \qquad \theta_2 = e_5\,\varepsilon/I_2\,\omega^2, \qquad \theta_3 = e_6\,\varepsilon/I_3\,\omega^2$$

where $I_1, I_2, I_3$ are the corresponding moments of inertia.

The contribution to the angular rotation about the axis $\mathbf{l}$ that comes from the vibration $\omega$ with the eigenvector $\mathbf{e}$ is:

$$(\boldsymbol{\theta}\cdot\mathbf{l})^2 = \theta_i\,\theta_j\,l_i\,l_j$$

The tensor $\omega_{ij} = \theta_i\,\theta_j$ is calculated as follows:

$$T_{ij} = \sum_{6Z} e_i\,e_j\,\varepsilon(\omega)/(I_i\,I_j)^{1/2}\,\omega^2$$

where $e$ and $e_j$ are the libration components of eigenvectors.

The only calculation that has been carried out so far in accordance with this scheme is described by Pawley in the paper referred to above. He studied the naphthalene and anthracene crystals at room temperature.

For naphthalene the tensors have the numerical values

$$T = \begin{vmatrix} 4.14 & 0.08 & -0.05 \\ & 4.14 & -0.18 \\ & & 3.44 \end{vmatrix} \times 10^{-2}\,\text{Å}^2$$

$$\omega = \begin{vmatrix} 21.9 & 2.0 & 0.6 \\ & 15.7 & 2.5 \\ & & 17.2 \end{vmatrix} \text{degree}^2$$

Referring as they do to an organic molecular crystal, these figures are quite representative: the mean displacement of the center of gravity is of the order of 0.02 Å, the mean libration angle is 4°.

The $T$ and $\omega$ tensors may be calculated from an analysis of the X-ray crystal data, since the structural amplitude of X-ray scattering, given by

$$F = \sum f_j\,e^{-M_j}\exp(hx_j + ky_j + lz_j)$$

includes the so-called Debye–Waller temperature factor $e^{-M_j}$. The quantity $M_j$ is directly related to the quadratic mean displacement of the atom

$$M_j = 8\pi^2\overline{u_j^2}(\sin^2\theta)/\lambda^2$$

where $\theta$ is the X-ray scattering angle (the Bragg angle) and $\lambda$ is the X-ray wavelength (cf. p. 244). The value of $\overline{u_j^2}$ is largely dependent on intermolecular vibrations. The amplitudes of intramolecular vibrations may be estimated in various ways, e.g., by electron diffraction from gases. Their rms amplitudes are at least one order of magnitude smaller than those of the intermolecular vibrations.

The calculation of $\overline{u_j^2}$ by means of the tensors $T_{ij}$ and $\omega_{ij}$ is therefore of particular interest. It is clear enough that, when engaged in all normal vibrations, an atom vibrates anisotropically. One may put

$$\overline{u^2} = \sum_{i=1}^{3} \sum_{j=1}^{3} U_{ij} l_i l_j$$

where $U_{ij}$ is a symmetric tensor having six independent components and $l$ is the unit direction vector. Using a technique which is now quite common in X-ray studies, we can determine the $U_{ij}$ tensors by the least-squares method. An oriented ellipsoid is assigned to every atom. A typical pattern, which is obtained by using X-ray diffraction data from a sucrose crystal, is presented in Fig. 6. Patterns of this kind can be produced as visual displays by modern computers.

It is obvious that

$$\overline{u^2} = \sum_{i=1}^{3} \sum_{j=1}^{3} U_{ij} l_i l_j = \sum_{i=1}^{3} \sum_{j=1}^{3} [T_{ij} l_i l_j + \omega_{ij} (\mathbf{l} \times \mathbf{r})_i \cdot (\mathbf{l} \times \mathbf{r})_j]$$

The formula assumes that the molecule vibrates as a rigid body. The vector $\mathbf{r}$ is directed from the center of gravity of the molecule to a constituent atom and its length gives the distance. Trueblood and Shoemaker [12] have shown that

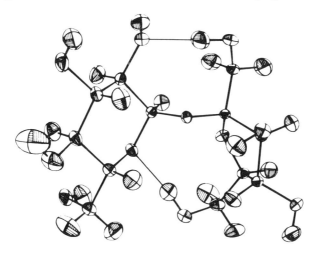

*Fig. 6.* An X-ray analysis pattern for the sucrose crystal.

this formula is valid for molecules occupying a symmetry position in the crystal. One more tensor should be introduced in the general case. The atomic tensors $U_{ij}$ are determined, as a rule, with insufficient accuracy, and in many cases the conclusions concerning the character of the atomic vibrations based on the values of these atomic tensors may be somewhat risky. The computation accuracy for the tensors $T_{ij}$ and $\omega_{ij}$ is considerably better, since measurements of all the atomic tensors are utilized. For a molecule having, say, 20 atoms, the atomic vibrations are characterized by 60 rough experimental measurements. The least-squares method is then applied to those 60 numbers and 12 new numbers (six components of $T_{ij}$, six components of $\omega_{ij}$) are obtained. This will clearly improve the computation accuracy. Yet even here it would be unrealistic to hope for an improvement of several percent.

One should therefore be satisfied if the values of the diagonal elements of the tensors $T_{ij}$ and $\omega_{ij}$ calculated by the method of atom–atom potentials and measured by X-ray diffraction techniques agree by about 20–30%, while the off-diagonal elements agree in their order of magnitude.

8. REORIENTATION OF MOLECULES

So far we have discussed the vibrational motion of molecules. However, this is not the only type of motion possible. Reorientational motion of molecules in a crystal, which was predicted earlier, has been confirmed by NMR techniques. This type of motion may occur if reorientation barriers are not much higher than $kT$. This implies in its turn that reorientation is possible only for appropriately shaped molecules, e.g., for molecules shaped something like disks, cones, spheres, or cylinders. Reorientation of molecules is quite compatible with rigorous crystallic ordering. A distinct electron density X-ray map shows the average positions of atoms. If reorientation occurs in such a way that atoms retain their average positions, the X-ray experiment will not be affected in the least.

This is exactly the situation in the benzene and hexamethylbenzene crystals. Reorientation of each molecule is effected by a rotation about the sixth-order axis belonging to the molecule.

The frequency of reorientation should satisfy the equation

$$\omega_c = \omega_0 \, e^{-U/kT}$$

where $U$ is the activation energy of reorientational motion. When the temperature rises the frequency of reorientation gradually increases, starting from zero. At a fairly typical height of the barrier of about 3 kcal/mole the number of reorientational steps per second is approximately $10^5$ at 85°K, while it is of the order of $10^8$ at 170°K. The vibration frequency of the benzene molecule is of the order of $10^{12}$. It is thus clear that the molecule is in a state of vibration for most of the time. Even at temperatures close to the melting point, one step will occur in about a thousand vibrations. Hence it follows that for ordered crystals thermodynamic quantities are insensitive to molecular reorientation.

As pointed out above, data on molecular reorientation in a crystal are obtained from the studies of NMR line widths. Typically it is the resonance of protons that is observed. At low temperatures the test substance produces a broad line. At a certain temperature the reorientation frequency reaches the value corresponding to the line width. According to theory, at this temperature the line width will rapidly shrink.

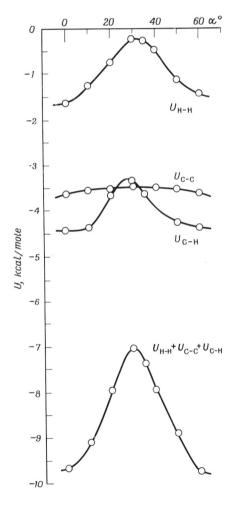

*Fig. 7.* The energy of interaction between an isolated rotating molecule and neighboring molecules which are at rest. Not only the integral curve is shown in the figure, but also curves illustrating the behavior of the individual terms of the total energy which arise due to the interaction between atoms of different kinds.

It is not infrequently suggested that the instant of NMR line narrowing corresponds to a phase transition. The suggestion is erroneous, and the line-narrowing temperature is actually the temperature at which the method becomes sensitive to molecular reorientation.

Considering that reorientation is a rather rare phenomenon, its discussion in terms of some geometrical or physical model should proceed from the assumption that the neighbors of a rotating molecule are at rest. A rough estimate (without any calculations) of the possibility of reorientation can be obtained by examining the situations brought about by various rotations of the molecule between different positions of equilibrium. Rotation is impossible if at any instant the distance between the nearest atoms belonging to different molecules is reduced, say, by half.

The energy curve corresponding to reorientational rotation may readily be calculated by the method of atom–atom potentials.

We have made such calculations for the benzene crystal. The results are shown in Fig. 7. The 4.9 kcal/mole barrier appears when the molecule swings away from the equilibrium position by 35°. The values seem to be somewhat greater than the experimental figures. There is no point in seeking a model that would give better agreement with the experiment, since the experimental data are rather conflicting. It may be appropriate to point out that the rotation barrier becomes substantially lower if one assumes that reorientational rotation involves two or three neighboring molecules simultaneously. The rotation barrier is found to be close to 3 kcal/mole, if reorientation of a molecule is assumed to involve its neighbors belonging to the first coordination sphere. It is noteworthy that the use of the method of atom–atom potentials may lead to a better understanding of the mechanism of such delicate phenomena as the reorientational steps of molecules in a crystal.

REFERENCES

1. L. D. Landau and E. M. Lifschitz, "Mechanics" (in Russian). State Publ. House of Phys. and Math. Lit., Moscow, 1958.
2. A. I. Kitaigorodsky, *J. Chim. Phys. Physiochim. Biol.* **63**, 9 (1966).
3. D. A. Oliver, Ph.D. Thesis, London Univ., London, 1967.
4. G. S. Pawley, *Phys. Status Solidi* **20**, 347 (1967).
5. P. Weulersse, *C. R. Acad. Sci. Ser. B* **264**, 327 (1967).
6. E. I. Mukhtarov, Thesis, Univ. of Baku, 1969.
7. A. I. Kitaigorodsky and E. I. Mukhtarov, *Kristallografiya* **14**, 784 (1969).
8. M. Suzuki and M. Ito, *Spectrochim. Acta Part A* **24**, 1091 (1968).
9. P. A. Bazhoulin and A. A. Rakhimov *Fiz. Tverd. Tela* **8**, 2163 (1966).
10. A. I. Kitaigorodsky and E. I. Mukhtarov, *Opt. Spektrosk.* To be published.
11. A. A. Rachimov, Thesis, Moscow, 1966.
12. K. N. Trueblood and D. P. Shoemaker, *Acta Crystallogr, Sect. A* **24**, 209 (1968).

# Chapter IV
# Methods of Investigating Structure and Molecular Movement

## A. Diffraction Methods

### 1. METHODS OF STRUCTURE DETERMINATION, THEIR ACCURACY, AND OBJECTIVITY

The properties of a substance are determined unambiguously by its structure. For this reason, structural data form the basis of any classification of substances and serve as a test of theories used for predicting properties.

The structure of a molecule can be described by the mutual arrangement of its constituent atoms, meaning their time-average positions. In addition, to understand a number of the properties of matter requires information on the nature of the vibrations; i.e., their amplitudes and frequencies need to be known.

Despite the fact that a molecule is an electron-nuclear system, the conception of the atom as a structural unit retains its full sense. The electron density of a molecule can be broken up, with high accuracy, into spherical distributions about the corresponding nuclei. Hence, it is usually permissible not to distinguish between the location of the center of the atom and the atomic nucleus.

Thus, one of the expedient descriptions of a molecule consists in stating

233

the coordinates of the atomic centers and indicating the number of electrons associated with each atom. The number of electrons associated with an atom differs little from the electrical charge on its nucleus. In other words, the atoms of a molecule are more or less electrically neutral particles. However, if these differences are substantial or are of fundamental interest, one may speak of atomic charge, meaning the corresponding difference.

The model of a molecule as a point system in which each point is characterized by a chemical species of atom and by an atomic charge located at that point is the principal model in problems connected with the structures of organic substances.

Of course, a molecule can be described more accurately by imposing its electron density onto a system of atomic nuclei. But there are unfortunately no objective methods of determining electron density with any considerable accuracy, whether by experiment or by calculation. Theoretical calculations made by the approximate methods of quantum chemistry hardly give an idea of the accuracy and authenticity of the results obtained. Since the possibility of comparing these calculations with experiment is problematic, the use of the results of these calculations evokes little confidence.

Even the determination of atomic charges from experiment is a very complex problem. There are a number of fine methods for determining the properties of an atom in a molecule. The most important of these are the nuclear magnetic and nuclear quadrupole resonance methods. However, the chemical shifts of frequencies of these resonances cannot be interpreted unambiguously. The former measures the screening of the atomic nucleus by the electron cloud of the atom, and the latter measures the gradient of the electric field at the site of the nucleus. Recalculations to atomic charges are difficult primarily because the frequency shifts of these resonances may be caused not only by changes in atomic charge, but also by changes in energy level distribution of the electrons.

As to determination of electron density or at least atomic charges by X-ray diffraction methods, the accuracy of these methods is simply not high enough. In the very best investigations even the number of electrons per atom can be determined only to a few percent.

Assessments of atomic charges from measurement of the dipole moments of the molecule are naturally highly indefinite, because the measured value is the dipole moment of the whole molecule. The distribution of this moment over the separate parts of the molecule, and, even more so, the localization of the dipole centers of individual parts of the molecule, or of the whole molecule, is arbitrary.

Only the determination of the fraction of the electron density due to an unpaired electron stands somewhat apart. In this case, i.e., when the molecule is a radical, the distribution of such an electronic charge over all the atoms of the molecule can be judged successfully from the fine structure of the electron magnetic resonance spectra.

In contrast to the unsatisfactory situation regarding experimental methods of determining the electron density of a molecule and even atomic charges, experiment offers great possibilities in determining the structure geometry. There are also fairly satisfactory methods of judging the dynamics of a molecule.

The structure of a free molecule, i.e., of a substance in the gaseous state, is determined by gas electron diffraction methods and by molecular rotation spectra. The latter are measured both by optical spectroscopic procedures and by microwave spectroscopy. Radio spectroscopic measurements are very precise. But the number of independent parameters that can be determined by such a method (any method of rotational spectroscopy) is very small. For this reason, interesting results are obtained only for simple molecules or for molecules in which only a small number of structural details have to be established.

Gas electron diffraction methods can also be recommended only for molecules consisting of small numbers of atoms, due to the relatively poor experimental data. Information on gas scattering is contained in the location and intensities of about a dozen diffraction rings. It is clear that in this case problems involving only a few parameters (not more than three or four with complete confidence) can be tackled.

The principal methods by means of which practically all the material on molecular geometry and crystal geometry has been accumulated are the diffraction methods of studying single crystals. X-ray diffraction studies are much more popular for this purpose than neutron diffraction (owing to the complexity and cumbersomeness of the method) and electron diffraction (owing to methodological difficulties).

Rather complex instruments yield a great deal of experimental material constituting data on several thousand diffracted rays. The diffraction angle and intensity of a ray are the experimental data which, after mathematical processing, carried out at present almost exclusively by electronic computers, gives information on the coordinates of the atoms in a crystal unit cell.

To judge the possibilities and precision of the method it must be remembered that the higher the atomic number of the atom in question, the stronger are the X rays scattered by it. For this reason, light atoms are difficult to detect and are found with less precision in the presence of heavy ones. On the other hand, molecules that contain a small number of heavy atoms (one or two) are incomparably easier objects of study because in a rough approximation such a structure scatters as if it were built up only of heavy atoms.

If the molecule contains very many atoms, determination of its structure is practically possible only if it contains one or two heavy atoms. In many cases heavy atoms are specially introduced into the molecule under the assumption that this causes no change in the part of the structure not associated directly with this "extra" atom (the isomorphic substitution method employed in biological crystallography).

It is fortunate that the introduction of one heavy atom for a large number of light ones decreases only slightly the accuracy of determination of the light-atom coordinates.

The situation is different if we are interested in the coordinates of the carbon atom in tetraiodoethylene or those of the hydrogen atom in any hydrocarbon. In this case the ratio of heavy and light atoms is such that the experimental determination of the coordinates of the light atoms becomes practically senseless. This means that on the basis of general crystal-chemical and structural chemical principles the coordinates of the light atoms can be established beforehand to a much greater accuracy than experiment is capable of doing.

The problem of determining the coordinates of hydrogen atoms in organic substances is of substantial interest. Until now the only direct experiment by which this problem could be solved was that of neutron diffraction. Neutron scattering is not related directly to atomic number, and hydrogen atoms scatter neutrons better than many other atoms. By this method the coordinates of hydrogen atoms can be determined quite objectively to a good accuracy.

A good accuracy in a diffraction experiment is 0.01 Å. A number of workers claim to have determined interatomic differences to a better accuracy. But comparison of investigations carried out by different workers, or comparison of structural data obtained independently with respect to distances which should unquestionably be identical, show that in estimating the accuracy of experiment it is injudicious to rely on general theoretical formulas.

There are a number of errors in X-ray diffraction studies that cannot be estimated practically. That is why the only solid basis for judging the possibilities of the method is not theoretical calculation of the error, but simple comparison of different measurements of the same substance. Such a comparison leads to the figure 0.01 Å, which, incidentally, is not at all bad, taking into account that it amounts to less than one percent of the magnitude of any interatomic distance. The inevitable errors just mentioned, that cannot be estimated properly in practice, are related to the difference between the crystal model for which the theory of X-ray diffraction intensities is built (and by means of which the computer calculates the final result) and the real crystal. In diffraction theory it is impossible to take into account the specific dislocational structure of a concrete object. It is also impossible to deal with absolutely pure preparations. Furthermore, even an insignificant impurity may seriously distort the average unit cell structure determined by diffraction methods.

It goes without saying that structure determination methods involving single crystals are much more effective than those involving investigations of the gaseous state (not to mention the study of liquids, since determination of the structure of a molecular liquid by diffraction methods is practically an impossible task). Apart from the fact that the structure of the molecule is determined much more accurately by crystallographic investigation, one obtains, in addition, valuable information on the relative placement of the molecules in the crystal.

A great deal of experimental material has been accumulated to date, enabling us to make numerous generalizations. A large number of regularities can be formulated on the basis of structural experiments alone, concerning bond lengths, valence angles, optimal molecular conformations, the nature of their mutual arrangement, etc.

Reverting to the determination of the coordinates of hydrogen atoms, note that this problem can be solved not only by neutron diffraction studies. A very sensitive method of determining the coordinates of protons of a single crystal is low-resolution nuclear magnetic resonance. It can be shown (see below) that the anisotropy of breadth of NMR lines is related unambiguously to the proton coordinates in the unit cell. This new method is now under development.

In a number of cases nuclear quadrupole resonance (NQR) is of great help in determining the structure of crystals. As will be discussed below, if the molecule has nuclei with quadrupole moments, this method enables unambiguous determination of the number of independent molecules in the unit cell, and when studying a single crystal it makes it possible to determine the orientation of the moecule in the cell.

Both radiospectroscopic methods are also valuable supplements to optical methods of investigation in studying lattice dynamics.

Nuclear magnetic resonance (and NQR) methods positively establish the presence or absence of molecular reorientation in the unit cell. The temperature dependence of NQR frequencies can be put in direct relation to the vibration amplitudes of the molecules in the cell.

Determination of the vibration frequencies of molecules still remains mainly a task to be solved by infrared absorption and Raman spectrum methods.

As to the determination of the type of the so-called dispersion surfaces (dependence of the vibration frequency on the wave vector), the main methods are those involving measurement of diffuse neutron scattering; measurement of diffuse X-ray scattering is also helpful.

2. PRINCIPLES OF THE DIFFRACTION METHOD OF STUDYING CRYSTAL STRUCTURE

Crystal structure is determined by diffraction studies. These include X-ray, electron, and neutron diffraction. The first of these is by far the most popular. The physical principles of the methods do not differ essentially, and in the following, for the sake of definiteness, we shall discuss only X-ray diffraction.

X-rays are scattered by atoms in all directions. But the scattered waves travel in the same phase and enhance one another only in certain selected directions. The directions of the diffracted beams can easily be determined with the aid of Bragg's law, which reduces essentially to the following.

If the unit cell is represented by a point (node, as crystallographers sometimes call it), the crystal will be a space lattice of such nodes. Planes can be drawn through the lattice points in innumerable ways. If one plane can be drawn, there will infallibly be other planes parallel to it: the "nodal" planes of a space lattice fall into equidistant families. Each family is characterized by the direction of the normal to the crystal axes and by the interplanar distance. Figure 1 illustrates the division of a lattice into systems of planes with the indices (230) for a two-dimensional case.

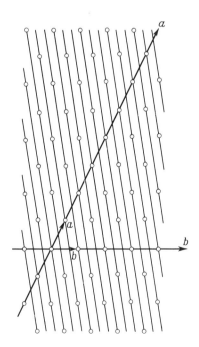

*Fig. 1.* Set of nodal planes (230).

Evidently a strong diffracted beam can arise only if it is "reflected" by some family of planes. However, this condition is only a necessary one. For the resulting beam to be perceptible, it is required in addition that the different planes of the same family give "reflected" beams with a path difference of a whole number of wavelengths.

Thus we come to the Bragg equation

$$2d \sin \theta = n\lambda$$

The left-hand side of this equation, as is evident from Fig. 2, represents the path difference between beams as a function of the interplanar distance $d$ and the glancing angle $\theta$.

To obtain all the possible diffracted beams, the crystal must be set at all possible angles to the incident beam. Then different families of planes will fall into the "reflecting" position.

The number of diffracted beams the crystals give rise to may be very large. Evidently reflections will be given by all the families of planes with interplanar distances ranging from the maximum distance for the crystal in question to a minimum value, which we find from the condition $\sin \theta \leqslant 1$; we get $d_{min} = \lambda/2$.

The geometry of the diffraction pattern unambiguously determines the geometry of the space lattice. Determining the angles $\theta$ and the directions of the reflected beams relative to the crystal axes, one establishes (to the order of 0.0001 Å in the best experiments) the edge lengths and the angles between the edges of the crystal's unit cell. As to the structure of the crystal, it is derived by analyzing the intensities of the "reflected" beams. This relation is demonstrated below.

The fundamentals are illustrated by Fig. 3, which is a schematic projection of the structure of hexamethylbenzene. The "nodes" of the unit cell are selected at the centers of the molecules, a certain system of planes is drawn

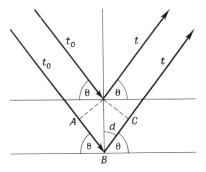

*Fig. 2.* Graphical interpretation of Bragg equation.

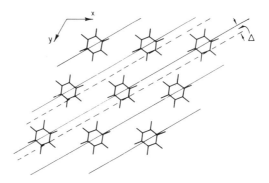

**Fig. 3.** Interplanar distances in hexamethylbenzene.

(their traces are visible, and the "nodal" planes should be imagined perpendicular to the plane of the drawing), on which an X-ray beam is incident at the required angle, i.e., that which satisfies Bragg's law.

The plane drawn through the nodes (solid lines) serves as the origin for reckoning the phases of the scattered waves. To make the phase relations between the waves scattered by atoms visible, planes are drawn through the atoms (broken line).

Consider any atom of the unit cell, the position of which is characterized by the radius vector $\mathbf{r}$. The path difference between the wave scattered by this atom and the wave that would be scattered by an atom situated at the node is calculated in exactly the same manner as in deriving Bragg's law. This path difference is equal to $2\Delta \sin \theta$ where $\Delta = \mathbf{u} \cdot \mathbf{r}$ ($\mathbf{u}$ being a unit vector). Thus, the phase of a wave scattered by any atom with respect to the coordinate origin of the unit cell (i.e., the path difference multiplied by $2\pi/\lambda$), equals $\alpha = \mathbf{s} \cdot \mathbf{r}$, where $\mathbf{s}$ is a vector perpendicular to the reflecting family of planes, and $|\mathbf{s}| = (4\pi \sin \theta)/\lambda$.

It can be shown that the vector $\mathbf{s}$ introduced in this way is equal to $2\pi\mathbf{H}$ for a crystal in a reflecting position, where $\mathbf{H}$ is the reciprocal lattice vector

$$\mathbf{H} = h_1 \mathbf{a}_1{}^* + h_2 \mathbf{a}_2{}^* + h_3 \mathbf{a}_3{}^*$$

where the $h_i$ are whole numbers, and the unit cell edges of the reciprocal lattice are determined by the equations

$$\mathbf{a}_i \cdot \mathbf{a}_j{}^* = \delta_{ij}$$

($\delta_{ij}$ is the Kronecker symbol).

The intensity of the diffracted beam depends on the geometry of the experiment and the structure of the crystal. If $F$ is the amplitude of scattering of one unit cell (in the general case $F$ is a complex quantity), the intensity

$$I(hkl) = k|F|^2$$

where $k$ depends on the size, shape, and absorption of the crystal, on the intensity and wavelength of the incident beam, and on the geometry of the experiment. This coefficient is, of course, a function of the angle of scattering. Formulas for $k$ for different experimental conditions can be found in handbooks. The square of the amplitude modulus $F$ is determined by the crystal structure, and for this reason $F$ is called the structure amplitude.

The sensitivity of the beam intensity to structure is due to the fact that the atoms of the unit cell scatter in various phases depending on their mutual arrangement. The phase for the $k$th atom, as we have just seen, is equal to $2\pi\mathbf{H}\cdot\mathbf{r}_k$. Hence, the structure amplitude

$$F = \sum_1^N f_k \exp 2\pi i\mathbf{H}\cdot\mathbf{r}_k$$

where $f_k$ is the amplitude of scattering by the $k$th atom, which will be discussed below.

For the centrosymmetric case

$$F = 2\sum_1^{N/2} f_k \cos 2\pi\mathbf{H}\cdot\mathbf{r}_k$$

If $f_k$ and $\mathbf{r}_k$, i.e., the structure, are known, the intensity of each reflected beam, i.e., for any $\mathbf{H}$, can be calculated. It equals $n^2|F|^2$, where $n$ is the number of unit cells in the crystal volume involved in the scattering.

X-ray diffraction studies solve the reverse problem, namely, the structure is determined from many hundreds and thousands of measurements of $|F|^2$ values.

The first stage of the work consists in searching for the crude structure of the crystal. One judges the correctness of the direction of search by comparing the structure amplitude values measured experimentally with those calculated for the model accepted by the formula just given. The most difficult are the first steps. After there is confidence that the model selected is correct, at least with respect to the majority of structure amplitude signs (phases), the calculated and measured structure amplitudes are brought closer together by the least-squares method, by minimizing the expression

$$R = \frac{\sum ||F_{obs}| - |F_{calc}||}{\sum |F_{obs}|}$$

Selection of a structure model involves primarily selection of the correct positions of atoms, of course. However, besides this, it is necessary to select the correct atomic amplitude values $f$.

Another method, generally equatable with the least-squares method, is the so-called "Fourier" method.

For a centrosymmetric crystal, the electron density inside the unit cell has the form

$$\rho(\mathbf{r}) = 2 \sum F_{\mathbf{H}} \cos 2\pi \mathbf{H} \cdot \mathbf{r}$$

Thus, if the structure has been "guessed" and the signs of $F_{\mathbf{H}}$ are mainly correct, $\rho(\mathbf{r})$ can be plotted, its maxima found, the signs of $F_{\mathbf{H}}$ recalculated, a new $\rho(\mathbf{r})$ plotted, and the maxima of $\rho(\mathbf{r})$, i.e., the coordinates of the maxima, can be finally determined.

Now let us consider in greater detail what the quantity $f$ in the expression for the structure amplitude is. Each electron is characterized by its own "distribution function" $\rho_i(x, y, z)$, which indicates the density of the "electron cloud" that appears when an infinite number of its paths are imposed upon one another. The total electron density in the atom (in numbers of electrons per unit volume) is given by

$$\rho(x, y, z) = \sum_{i=1}^{Z} \rho_i(x, y, z)$$

The quantity $\rho \, dv$, where $dv$ is an element of volume, is the average number of electrons per volume $dv$.

Scattering by the volume element $dv$ is obviously expressed as follows: $\rho \, dv \, e^{i(\mathbf{s} \cdot \mathbf{r})}$, where $\mathbf{r}$ is the radius vector connecting the coordinate origin with the element $dv$. In accordance with the definition of the atomic scattering factor we have

$$f(\mathbf{s}) = \int_v \rho \, e^{(i\mathbf{s} \cdot \mathbf{r})} \, dv \tag{2.1}$$

Knowledge of the function $\rho(x, y, z)$ determines the atomic factor. In the general case the atomic factor is a function of the vector $\mathbf{s}$; this means that atomic scattering depends on the argument $(\sin \theta)/\lambda$ and on the orientation of the atom in the scattering position. We introduce the main simplification always introduced when computing atomic scattering, namely, the assumption that the electron distribution is spherically symmetrical. It is quite obvious that in this case the value of $f$ is independent of the orientation of the atom, and its scattering becomes a function of the scalar argument $f = f((\sin \theta)/\lambda)$. Obviously, this assumption conforms to the electron distribution of a free atom quite well.

In the case of spherical symmetry the expression (2.1) of the atomic factor can conveniently be transformed by introducing the function

$$U(r) = 4\pi r^2 \rho(r) \tag{2.2}$$

The origin of coordinates is taken at the center of the atom.

Obviously, $U(r)$ is the number of electrons enclosed between spheres of radii $r$ and $r+dr$. The function $U(r)$ is called the radial distribution and, as we shall see in the following, characterizes the atom very vividly.

As is known, an element of volume in spherical coordinates equals

$$dv = r^2 \sin \alpha \, d\alpha \, dr \, d\varphi$$

Selecting the direction of $s$ as the sphere radius from which $\alpha$ is measured, one obtains $\angle sr = \alpha$. Substituting the value of $\rho$ from (2.2) and making use of the above expression for $dv$, one gets

$$f(s) = (1/4\pi) \int_0^\infty \int_0^\pi \int_0^{2\pi} U(r) \, e^{isr \cos \alpha} \sin \alpha \, dr \, d\alpha \, d\varphi$$

Integrating over the angles, there is no difficulty in finding

$$f(s) = \int_0^\infty U(r) \, [(\sin sr)/sr] \, dr$$

Thus, the atomic factor is a function of $(\sin \theta)/\lambda$; the form of this function is determined by the radial distribution of electrons in the spherically symmetrical atom.

The dependence of $f$ on the scattering angle is shown in Fig. 4.

Let $f_0$ be the scattering amplitude of an atom in an equilibrium position. Suppose the thermal shift of the atom is $\Delta$. Then the scattering amplitude will be $f_0 \, e^{i2\pi \cdot \mathbf{H} \cdot \mathbf{\Delta}}$

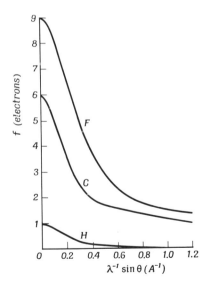

*Fig. 4.* Dependence of $f$ on scattering angle for H, C, and F atoms.

Let the projection of vector $\mathbf{\Delta}$ on the vector $\mathbf{H}$ be denoted by $z$.

It is just this projection that determines the scattering at the given angle. Then the instantaneous scattering amplitude can be written as

$$f_0 e^{i2\pi Hz}$$

Let the probability of an atom shifting a distance of $z$ be $w(z)$. We accept a Gaussian distribution for the shift of the atom, i.e., we assume that

$$w(z) = (2\pi D)^{-\frac{1}{2}} e^{-Z^2/2D}$$

where the standard deviation of the distribution $D$ designates the mean quadratic deviation $\overline{u_H}^2$ in the direction of the reciprocal vector.

The average amplitude of interest equals

$$f_0 \int_{-\infty}^{\infty} w(z) e^{i2\pi Hz} \, dz = f_0 e^{-M}$$

where

$$M = 8\pi^2 \overline{u_H}^2 \left(\frac{\sin\theta}{\lambda}\right)^2 = 2\pi^2 \overline{(\mathbf{\Delta}\cdot\mathbf{H})^2}$$

The latter expression can also be rewritten as

$$M = 2\pi^2 \overline{(\Delta_x H_x + \Delta_y H_y + \Delta_z H_z)^2}$$

where $\Delta_x$, $\Delta_y$, and $\Delta_z$ are constants having the meaning of projections of the vibration on the axes of the reciprocal lattice. This is the expression for a crystal of orthorhombic symmetry.

It is not difficult to grasp that in the general case the expression for $M$ can be reduced to

$$M = (h^2\alpha + k^2\beta + l^2\gamma + 2hk\delta + 2kl\varepsilon + 2hl\eta)$$

thus characterizing the anisotropy by six constants.

Of great importance is the following circumstance, well known to practical workers using X-ray diffraction. Quite exact coordinates of atoms can be found by assuming the atom-temperature factor to be isotropic. In this way we first find the correct structure with respect to atomic arrangement, establishing the insignificant anisotropy of the atomic scattering factor independently (and then only in very precise experiments).

### 3. SPHERICITY OF ATOMS

By direct computation (namely, by evaluation of the Fourier integral) one can obtain the scattering function of the electrons of a given atom from the dependence of the atomic factor on the scattering angle and direction. The absence of any substantial dependence of the atomic factor on the scattering direction is proof of spherical symmetry of the electron density of the atom. Thus it follows from X-ray diffraction practice that to a fair accuracy the electron density of a molecule and of a crystal is the superposition of spherical atoms. Of course, this refers to time-and-space-averaged electron density, since only such an electron distribution can be found by experiment.

This result is rather surprising. It might have been thought that a substantial part of the electrons belong to the molecule as a whole and participate in the chemical binding. However, it turns out that the establishment of the chemical binding consists mainly in the juxtaposition of atoms, and in the additive overlapping of the electron clouds of adjacent atoms.

Atoms owe their principal deviations from sphericity to their heat motion and, as experiment shows, mainly to the vibration of the molecules as a whole. A substantial role in the vibrations of atoms in crystals is played by the swinging of molecules about their centers of gravity. For this reason, the vibration amplitudes of atoms located near the center of gravity of a molecule should be considerably smaller than that of "outer" atoms. It appears, accordingly, that the electron density of atoms remote from the center is always more diffuse than that of the central atom. Naturally, due to the swinging of the molecules as mentioned above, the electron density of the outer atoms is "smeared" anisotropically, being more diffuse in the direction perpendicular to the radius connecting the atom in question with the center of gravity of the molecule. An indirect idea of the magnitude of this anisotropy can be obtained from the fact that the angular swing amplitudes are usually between 4° and 5°.

The deviations of the electron density of an atom from spherical symmetry due to intramolecular vibrations are evidently substantially smaller. Reliable determination of this anisotropy is very difficult.

It is quite obviously very complicated to separate the deviations from spherical symmetry due to movement of the atom from deformation of the electron cloud due to chemical bonding. More or less strict ideas of their respective contributions were obtained from a study of diamond crystals. It may be considered established that an upper estimate of the "bonding charge" located midway between two adjacent carbon atoms, disturbing the additivity of the spherical electron densities is 0.2 electron. In other words, the electron density of diamond should be imagined as spherical atoms which yield about 0.4 electron for the formation of a condensed electron density in the centers of the bond lines.

Thus, the deviation of the carbon atom from spherical symmetry is insignificant. It follows from other studies that similar percentage deviations occur in other light atoms, in particular, in a valency-bonded hydrogen atom having only one electron. As to heavy atoms, their deviation from sphericity probably cannot be detected by present-day experimental techniques.

The scheme of atom–atom potentials widely used in this book is based on the following assumption: To within a few percent the electron densities of the atoms in a molecule and in a molecular crystal can be represented as the sum of the spherically symmetrical densities of the individual atoms.

We see no grounds to make exception from this rule for aromatic systems and for multiple bonds. The structure of the anthracene crystal has been studied very accurately; to the accuracy of the experiment the electron density of the molecule can be represented by the sum of the densities of spherical atoms (Fig. 5). Thus, neither the movements of electrons along aromatic rings, nor the presence of bonds formed by $p$-electrons, perceptibly disturbs the sphericity of atoms.

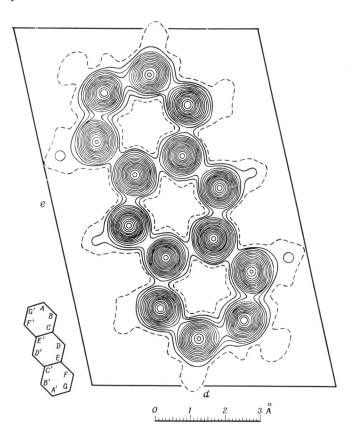

*Fig. 5.* Anthracene. Section of three-dimensional electron density series, coinciding with plane of molecule.

4. ACCURACY OF STRUCTURAL DETERMINATIONS

Two main circumstances determine the accuracy to which interatomic distances can be determined from the electron density distribution of a substance. The first is the limited resemblance of a real crystal to the ideal crystal model for which the principal formula relating the structure to the intensities of the scattered X-rays (or other radiations) is valid, i.e., the limited accuracy of computing the structural amplitudes $F_{calc}$. The second circumstance is the difficulty of measuring and taking into account various experimental factors, primarily beam absorption, i.e., the limited accuracy to which the observed structural amplitude values $F_{obs}$ are known.

The experimental difficulties can be minimized by preparing crystal samples of spherical shape, exposure to strictly monochromatic radiation, and recording the integrated intensities by ionization methods. By employing these, and other, measures the amplitudes can be measured, generally speaking, to about 1 or 2%. Note that in most structural studies, however, the accuracy is much worse, probably not better than 10 or 20%. The practice of X-ray diffraction analysis has shown that improvement of the accuracy of measurement of structural amplitudes very often affects the accuracy of the results obtained only to a very insignificant extent. This is due primarily to the fact that it is almost always possible to accumulate such extensive experimental material that there are several score structural amplitude values for each geometrical parameter of the structure. For example, in the study of the naphthalene crystal the intensities of 644 reflections were measured. The molecule occupies a position with a center of symmetry in the crystal. Therefore five radius vectors of carbon atoms have to be determined. This makes 43 reflections for each coordinate.

In view of the fact that the intensities of reflections may differ from one another by thousands of times (and that these differences can be established), a measurement error of the order of 10% does not seem so large.

Theoretical calculations of the probable accuracy of measurement of the electron density of atoms and of interatomic distances have shown that with very roughly measured reflection intensities fairly accurate values of the quantities interesting the crystallographer can be obtained by increasing the amount of experimental data.

However, there is another reason why exact intensity measurements may turn out to be inadequate. The formula for structural amplitude given above is valid for an ideal crystal of small size—of the order of 1000–10,000 Å. Only in this case is it possible to neglect secondary phenomena, i.e., scattering by one atom of a wave already scattered by another (the so-called dynamic scattering effects).

It is very fortunate for X-ray diffraction practice that crystals possess block structures as a rule. The size of the blocks is just about that necessary to allow one to neglect the dynamic effects. However, it is not simple to make sure that the crystal blocks are of the required size. The usual practice is to do no more than attempt to enhance the block structure of the crystal by quenching in liquid air, or by other such means.

The mosaic (block) structure of a crystal arises as a result of the presence in the crystal of a large number of linear and helical dislocations.

Dislocation displacements of one part of the lattice relative to another may result in a great variety of shifts and rotations of neighboring regions. There is no good way of taking into account completely the nature of the distortions, and of estimating their effect on reflection intensity. Furthermore, if one block is turned but slightly with respect to its neighbor, a beam reflected by one of them will be partly reflected by the other in the reverse direction, and this will make the intensity lower than it would be if the blocks were greatly disoriented.

It is clear from the above that it is practically impossible to take this pheno-menon into account. Since dislocational movements are crystallographically regular, block disorientation may have different effects on beams reflected from different systems of planes.

Finally, another consequence of crystal mosaic structure is that the unit cells at the block boundaries are distorted. The structure we measure is that of an average unit cell, to which deformed cells also make their contribution. The number of boundary cells with a block about 1000 Å in size will equal approximately 1% of the number of cells in the block. Impurities, which are always present, may increase this percentage.

The inevitability of the error we make in calculating structural amplitude by means of the main formula relating it to structure renders attempts at gaining maximum accuracy in estimating reflection intensities not always justified. Experience shows that a measurement accuracy of more than 10% is often useless for the goal in mind.

What then is the accuracy to which interatomic distances and electronic densities are established in the best investigations, i.e., in those where many score reflections are measured for each unknown structural parameter? Numerous attempts have been made to calculate the probable errors, but the most convincing estimates can be made from the data of X-ray diffraction studies, since a great deal of experimental material has been accumulated which therefore allows the comparison of many independent estimates of the same values.

Above we emphasized an important feature of organic crystals, namely, that in them intermolecular interactions are much weaker than intramolecular. The bond energy of atoms not connected by valency bonds is two orders lower than that of those that are. The properties of a free molecule differ only in-significantly from those of a molecule that is part of a crystal. As to the most rigid parameter, the bond length, it may be considered firmly established that bond lengths in a free molecule and in a crystalline molecule coincide within limits better than the experimental accuracy of a crystal-structure investigation (cf. Chapter I).

Accordingly, we can get a good idea of accuracy by comparing chemically equivalent bond lengths occupying different positions in the crystal and therefore determined independently of one another. An example of such a comparison for naphthalene, anthracene, and acenaphthalene is shown in Fig. 16, Chapter II. Comparison of bond length of the same molecule in different polymorphic modifications is also fairly indicative. For example, Litaka gives the following table for three crystalline forms of glycine [1]:

| | | | |
|---|---|---|---|
| N–C | 1.491 | 1.484 | 1.474 |
| C–C | 1.527 | 1.521 | 1.524 |
| $[C-O]_1$ | 1.254 | 1.233 | 1.252 |
| $[C-O]_2$ | 1.237 | 1.257 | 1.255 |
| N–O | 2.687 | 2.701 | 2.690 |

On the basis of a large number of structural investigations (including investigations of the same objects by different authors) it can be stated that the probable error of interatomic distance determination is about 0.01 Å in the best studies.

The accuracy of determining electron density has been discussed above: The number of electrons per atom can be determined to 0.2–0.3 electron even for such atoms as C, N, and O.

## 5. COMPARISON OF X-RAY, ELECTRON, AND NEUTRON DIFFRACTION ANALYSES

The procedures of X-ray, electron, and neutron diffraction techniques differ considerably. But the methods of treating the results of observations are almost identical because of the full analogy between the formulas relating scattered intensity distribution to structure.

Similar methods can be used to calculate, from intensity measurements, the electron density of a substance from X-ray diffraction data, the electrostatic potential from electron diffraction data, and the arrangement of atomic nuclei from neutron diffraction data.

It will be readily appreciated that all the diffraction methods supplement each other and their joint use for solving the structure of one and the same object may be of great interest. Let us consider the specific features of neutron diffraction and of electron diffraction studies. The theory of structure determination by these methods is, on the whole, universal. The difference regards mainly the atomic scattering factor $f$.

*a. Electron Diffraction*

The motion of electrons is described by Schrödinger's wave equation

$$\nabla^2 \Psi + (8\pi^2 m/h^2)(E-V)\Psi = 0 \qquad (5.1)$$

where $\Psi(x, y, z)$ is a wave function, the squared modulus of which gives the probability of an electron being at a given point. The total energy of an electron beam $E$ is given by the accelerating voltage $p$: $E = ep$ ($e$ being the charge of an electron) and determines the wavelength $\lambda$ of the primary mono-chromatic wave incident on the object

$$\Psi_0 = Ae^{ik_0 x} \qquad (5.2)$$

so that $\lambda^{-1} = k_0/2\pi = (2mE/h)^{1/2}$; $\Psi_0$ is the solution of Eq. (5.1) in the absence of the term $V$ (potential energy), which begins to play a part only when the wave falls upon some object—an atom, a molecule, or a crystal, in the potential of which $\varphi(\mathbf{r})$ the electron acquires the potential energy $V(\mathbf{r}) = \varphi(\mathbf{r})e$. This is what causes scattering. Thus, the "scattering matter" for electron diffraction is the electrostatic potential $\varphi(\mathbf{r})$, which plays the same role here as electron density $\rho(\mathbf{r})$ in X-ray diffraction. This analogy, as will be seen below, is complete and applies also the main formulas of the theory of scattering.

Schrödinger's equation is solved in the kinematic approximation, i.e., under the condition that the secondary beams are weak, by representing the function sought $\Psi$ as the sum of the primary wave $\Psi_0$ and a small scattered one $\Psi'$:

$$\Psi = \Psi_0 + \Psi'$$

Substituting this expression into (5.1) and bearing in mind that $\Psi_0$ is the solution of (5.1) without the term $V$, one gets the following equation for $\Psi'$:

$$\nabla^2 \Psi' + k_0{}^2 \Psi' = U(\mathbf{r})\Psi(\mathbf{r}); \qquad U(\mathbf{r}) = (8\pi^2 me/h^2)\varphi(\mathbf{r})$$

In mathematical form this is the so-called generalized Poisson equation describing, in particular, the propagation of an electromagnetic disturbance in a medium containing charges, which is further evidence of the analogy indicated above. Its solution

$$\Psi'(\mathbf{r}) = (1/4\pi)\int U(\mathbf{r}_1)\Psi(\mathbf{r}_1) R^{-1}e^{ikR} dv_1$$

where $R = |\mathbf{r} - \mathbf{r}_1|$ ($\mathbf{r}_1$ being the vector inside the scattering volume). Here the integrand contains $\Psi = \Psi_0 + \Psi'$. Making use of the basic condition of the kinematic theory of scattering, namely, the weakness of secondary waves $\Psi' \ll \Psi_0$, i.e., substituting $\Psi_0$ for $\Psi$, which signifies that further scattering $\Psi'$ of once-formed secondary waves is neglected, one obtains

$$\Psi' = (1/4\pi)\int U(\mathbf{r}) Ae^{i\mathbf{k}_0 \mathbf{r}}R^{-1}e^{ikR} dv \qquad (5.3)$$

A similar approximation is made in the kinematic theory of X-ray scattering. There, its admissibility is still better grounded since the absolute amplitude values of the secondary X-ray waves are much smaller than those of electrons, which interact much more strongly with matter. The removal of this approximation is the main feature of the dynamic theory of X-ray, as well as of electron scattering. Exchanging $R$ for $r - (\mathbf{n} \cdot \mathbf{r}_1)$ in (5.3) ($\mathbf{n}$ being the unit vector in the direction $\mathbf{k}$), so that $e^{ikR} = e^{ikr} e^{-i\mathbf{k} \cdot \mathbf{r}_1}$, and bearing in mind that $\mathbf{k}_0 - \mathbf{k} = 2\pi \mathbf{H}$, i.e., introducing the reciprocal space vector, one finds

$$\Psi' = \frac{2\pi m e}{h^2} \frac{e^{i\mathbf{k} \cdot \mathbf{r}}}{r} A F_{el}(\mathbf{H})$$

This means that at a large distance from the object $\Psi'$ is a spherical wave, the amplitude of which is proportional to the initial amplitude $A$ and to the Fourier integral $F(\mathbf{H})$ (over the potential of the object)

$$F(\mathbf{H}) = \int \varphi(\mathbf{r}) e^{2\pi i \mathbf{r} \cdot \mathbf{H}} \, dv \tag{5.4}$$

The factor $2\pi m e/h^2$ is analogous to $e/mc^2$ in X-ray diffraction studies, and it may be regarded as the scattering of a certain unit. It is convenient to express scattering in these arbitrary units, as scattering is similarly expressed in electron units in X-ray diffraction studies.

Considering the separate atom as the scattering object and transferring (5.4) to spherical coordinates, one obtains the following expression for the atomic scattering amplitude

$$f_{el}(H) = \int \varphi(r) r^2 \frac{\sin 2\pi H r}{2\pi H r} \, dr$$

which is similar to the expression for the X-ray atomic factor.

Representing the unit cell potential as the sum of the potentials of its constituent atoms

$$\varphi(\mathbf{r}) = \sum_i \varphi(\mathbf{r} - \mathbf{r}_i)$$

the structural amplitude can be written in the usual form as the sum of atomic amplitudes with phase factors

$$F_{el}(\mathbf{H}) = \sum f_{el} e^{2\pi i (\mathbf{r} \cdot \mathbf{H})} \tag{5.5}$$

By analogy with the formula for electron density we get the following representation of the crystal potential $\varphi(x, y, z)$ as a Fourier series

$$\varphi(x, y, z) = (1/V) \sum \sum \sum F_{el} e^{-2\pi i (\mathbf{r} \cdot \mathbf{H})}$$

As in X-ray diffraction studies, the moduli $|F_{el}|$ are found from the reflection intensities on the electron diffraction pattern. Thus, processing experimental electron diffraction data by the Fourier method yields the potential pattern of the crystal lattice $\varphi(x, y, z)$.

The full analogy of the mathematical apparatus makes the theory of crystal structure determination and the concrete methods of interpretation (e.g., $F_{el}^2$ series, etc.) the same for X-ray, electron, and neutron diffraction studies. However, the differences in the physical nature and character of the function determining scattering give rise to some physical distinctions between the methods.

Note the following main features of the potential function $\varphi(x, y, z)$.

The potential of a crystal consists of the positive potential of its nuclei $\varphi_+ = Ze/r$ and the negative potential of its electron shells $\varphi_-$

$$\varphi = \varphi_+ + \varphi_-$$

Owing to the concentrated nature of the nuclear charge and the diffuseness of the electron shells, $\varphi_+ > \varphi_-$, i.e., $\varphi > 0$. The role of the electron shells reduces to a screening the nuclear potential, and the course of the atomic potential depends not only on the number, but also on the nature of the electron distribution in the shell. A crystal potential formed by superposition of atomic potentials is a positive continuous periodic function, the maxima of which correspond to atoms.

The potential $\varphi(x, y, z)$ is a more diffuse function than the electron density $\rho(x, y, z)$; this follows from the well-known relation of Thomas–Fermi-statistic atomic theory

$$\rho \sim \varphi^{3/2}; \qquad \varphi \sim \rho^{2/3} \tag{5.6}$$

This is illustrated by Fig. 6. Hence it follows that the atomic scattering curves for electrons $f_{el}$ decline faster than the corresponding curves for X-rays, $f_x$. Furthermore, from expression (5.6) and Fig. 6 it will be seen that the ratio of the peak heights of light and heavy atoms in X-ray diffraction differs from that in electron diffraction, and is more favorable to the light atoms in the latter case.

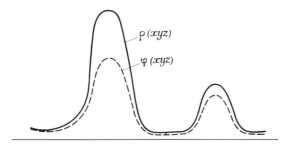

*Fig. 6.* Comparative diagram of electron density ($\rho(x, y, z)$) and potential ($\varphi(x, y, z)$).

The distinctive feature of $F_{el}$ and $F_{el}^2$ series in electron diffraction studies is that the number of terms is from two to four times smaller than in X-ray diffraction studies, which follows from the faster decline of the atomic scattering curve. For the same reason, electron diffraction series converge faster than in the case of X-ray diffraction; the break-off error in them is smaller, but the resulting function is less sharp. It may be noted in conclusion that since the potential and charge distributions in the crystal are related by the Poisson equation

$$\nabla^2 \varphi = -4\pi(\rho_+ - \rho_-)$$

the structural amplitudes for electrons $F_{el}$ and for X rays $F_X$ are also connected by the relation

$$F_{el} = (F' - F_X)/\pi H^2$$

where $F' = \sum_j Z_j e^{2\pi i(r_j \cdot \mathbf{H})}$ ($Z$ is the nuclear charge).

*b. Neutron Diffraction*

Schrödinger's equation (5.1) is valid not only for electrons, but for microparticles in general; in particular, it can be used to examine neutron scattering. Let us first consider the scattering of a monochromatic neutron beam by a separate nucleus. Here the primary wave $\Psi_0$ has the same form (5.2). The wavelength of the slow (thermal) neutrons used in structural neutron diffraction analysis is approximately 1 Å. When it hits a nucleus, the neutron is scattered to an extent depending, according to (5.1), on the shape and nature of the potential function $V(r)$ of the nucleus, the initial energy $E$ of the neutron, and the reduced mass $m$ of the system. Like electron scattering, which depends on the electrostatic potential of the object, neutron scattering depends on the nuclear force potential. The specific feature of these forces is that they diminish very rapidly with distance. The nuclear field is concentrated in the immediate vicinity of the nucleus and has a radius of $r_0 \sim 10^{-13}$ cm. The solution again has the form of the sum of the initial and scattered waves, which is spherically symmetrical, and is quite analogous to the case of electron diffraction:

$$\Psi = \Psi_0 + f_n(H) e^{ikr}/r \qquad (5.7)$$

[cf. (5.2)]. Strict solution of Eq. (5.1) for neutrons after substitution of (5.7) gives, as is known, an important and very simple result, which we will now obtain qualitatively from general reasoning. Both in the theory of X-ray scattering and in that of electron scattering the dependence of the atomic scattering amplitude on the angle $\theta$ (or on $\mathbf{H}$) comes as a result of taking into account the phase relations between the secondary waves in the Fourier integral (5.2). The scattering volume is of the same order of magnitude as

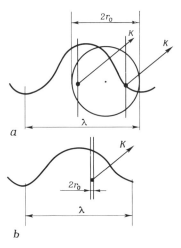

*Fig. 7.* Illustration of specific nature of atomic scattering factor of neutrons.

the length of the primary wave (Fig. 7a), and the secondary waves emitted from different points of it have different phases. In the case of slow neutron scattering $\lambda \gg r_0$ (Fig. 7b), and the phase shift is also very small, i.e., all the secondary waves emitted from the nucleus are in the same phase when scattered in any direction, and therefore neutron scattering by the nucleus is spherically symmetrical, and the amplitude $f_n(\theta)$ is independent of the angle of scattering $\theta$; it is a constant value. Hence, the $f_n$ "curve" does not decline. Thus, neutron scattering by nuclei is the ideal case of scattering from "point" atoms. The structural amplitude of coherent scattering by the crystal will have a form similar to (5.5)

$$F_n = \sum_i f_n e^{2\pi i(\mathbf{r}_j \cdot \mathbf{H})}$$

and here, in contrast to $f_X$ and $f_{el}$, the nuclear amplitude $f_n$ has a constant value, as we have just established. Formally, the Fourier integral of the nuclear scattering power of the entire unit cell could be introduced in this case also, presenting this power as the sum of $f_n$-weighted $\delta$ functions, each of which describes a point scattering center

$$n(\mathbf{r}) = \sum f_{nj} \delta_j(\mathbf{r} - \mathbf{r}_j)$$

$$F_n = \int n(\mathbf{r}) e^{2\pi i(\mathbf{r} \cdot \mathbf{H})} dv \tag{5.8}$$

However, this idea acquires real physical sense only if the heat motion of the atomic nuclei in the crystal is taken into account. Then the point nucleus spreads out into a certain region $\chi(\mathbf{r})$, in accordance with which the constant

$f_n$ should be multiplied in the usual way by a temperature factor and as a result turns out to decline in the case of nuclei in crystals, and depends on $(\sin\theta)/\lambda$ according to the law of this temperature factor. For example, in the case of spherically symmetrical vibration

$$f_n = \text{const}\, e^{-B[(\sin\theta)/\lambda]^2}$$

The superposition of nuclei spread out by heat motion is a certain continuous function

$$n(\mathbf{r}) = \sum_j f_n \chi_i(\mathbf{r} - \mathbf{r}_j)$$

known as the nuclear scattering power function, or nuclear density function for short (but not literally because this function has nothing to do with actual density). This function is also periodic, like $\rho(\mathbf{r})$ or $\varphi(\mathbf{r})$. By reversing the Fourier integral (5.8), we get the Fourier-series representation of $n(\mathbf{r})$

$$n(\mathbf{r}) = (1/V) \sum \sum \sum F_n\, e^{-2\pi i(\mathbf{r}\cdot\mathbf{H})}$$

where $|F_n|$ is found from a neutron diffraction experiment. Thus, the experimental data obtained by studying crystals by neutron diffraction techniques can also be treated by the Fourier method, the mathematical apparatus of all three diffraction methods being identical.

However, as in the case of electron diffraction, the difference in the physical nature of the scattering gives rise to a number of specific features of Fourier series in neutron diffraction studies. The most important of them is that some nuclei have positive scattering amplitudes $f_n$, and some negative (most $f_n$ are positive). For example, for potassium $f_{nK} = 0.35 \times 10^{-12}$ cm, for carbon $f_{nC} = 0.64 \times 10^{-12}$ cm, and for hydrogen $f_{nH} = -0.38 \times 10^{-12}$ cm. If the crystal contains nuclei with negative $f_n$, they are represented in the "nuclear density" pattern by dips, rather than peaks. In these cases the general condition of nonnegativity of scattering power, valid for X rays and electrons, does not hold.

$f_n$ values have no systematic dependence on atomic number. This makes it possible, for instance, to determine the hydrogen atom positions very well by neutron diffraction, even in the presence of the very heaviest (in the usual sense) atoms, to investigate structures consisting of atoms that cannot be distinguished by X rays and electrons owing to the closeness of their atomic numbers (e.g., alloys of the type of Fe–Co, etc), to study phenomena related to differences in isotopic composition. The presence of a magnetic moment in the neutron makes it also possible to study the magnetic structure of crystals. An essential feature of the Fourier series in neutron diffraction studies is the sharpness of the peaks in the "nuclear density" function $n(\mathbf{r})$. However, owing to the slow decline of amplitudes, which depend only on the temperature factor, the break-off effects affect neutron diffraction series very strongly.

The figures below will give an idea of the different scattering power of matter with respect to X rays, electrons, and neutrons.

The scattering by electrons, X rays, and neutrons, respectively, for zero $\theta$ angle, is characterized in units of $10^{-12}$ cm as follows:

| | | | |
|---|---|---|---|
| H atom: | 5.3 | 0.28 | $-0.38$ |
| C atom: | 24.5 | 1.69 | 0.66 |
| Cl atom: | 46.0 | 4.79 | 0.99 |

These are the scattering amplitude values $a$. $f$ values are recalculated to $a$ by the formula

$$a = (e^2/mc^2)f = 0.28 \times 10^{-12}f$$

### 6. FINDING AND ELUCIDATING THE STRUCTURES OF MOLECULAR CRYSTALS

*a. General Approaches. Heavy Atom Method*

The determination of the structure of any crystal falls into two stages: Determination of the phases (signs in the case of centrosymmetry) of the structural amplitudes, and the elucidation of the structure.

The second stage is always automated and carried out on electronic computers by standard programs. The first stage still requires creative intervention on the part of the researcher, but has also already been developed sufficiently to be entrusted mainly to the computer.

By far the most popular method is that of building Patterson series, known also as $F^2$ series. Denoting electron density by $\rho(\mathbf{r})$, we calculate the integral

$$A(\mathbf{u}) = (1/V) \int \rho(\mathbf{r})\rho(\mathbf{r}+\mathbf{u})\,dv \tag{6.1}$$

bearing in mind that $\rho(\mathbf{r})$ can be represented as a Fourier series (discarding the coefficient $1/V$)

$$\rho(\mathbf{r}) = \sum_{h,k,l=-\infty}^{+\infty} F_{hkl}\,e^{-2\pi i(\mathbf{r}\cdot\mathbf{H})}$$

Substitution into (6.1) gives an integral of the form

$$\int_0^a e^{2\pi i(n-m)x/a}\,dx = \begin{cases} 0, & n \neq m \\ 1, & n = m \end{cases} \tag{6.2}$$

where $n$ and $m$ are integers. Hence,

$$A(\mathbf{u}) = \sum_{h,k,l=-\infty}^{+\infty} F_{hkl}^2\,e^{-2\pi i(\mathbf{u}\cdot\mathbf{H})}$$

$$= F_{000}^2 + 2\sum_{h,k,l=1}^{+\infty} F_{hkl}^2\cos 2\pi(\mathbf{u}\cdot\mathbf{H}) \tag{6.3}$$

The function $A(\mathbf{u})$ may be called an $F^2$ series. This series $A(\mathbf{u})$ can be built on the basis of experiment (since knowledge of the signs or phases of the series' terms is not necessary). For this reason we must try and see what structural data can be extracted from the function $A(\mathbf{u})$.

We make use of the conception of electron density as the sum of $N$ atomic functions

$$\rho(\mathbf{r}) = \sum_{p=1}^{N} \rho_p(\mathbf{r} - \mathbf{r}_p)$$

Each of the atomic functions can be expanded in a Fourier series

$$\rho_p(\mathbf{r} - \mathbf{r}_p) = \sum_{h,k,l=-\infty}^{+\infty} f_p \exp[-2\pi i(\mathbf{r} - \mathbf{r}_p) \cdot \mathbf{H}] \qquad (6.4)$$

where $f_p$ is the atomic factor of the $p$th atom, and thus the electron density can be represented as

$$\rho(\mathbf{r}) = \sum_{p=1}^{N} \sum_{h,k,l=-\infty}^{+\infty} f_p \exp[-2\pi i(\mathbf{r} - \mathbf{r}_p) \cdot \mathbf{H}] \qquad (6.5)$$

Substituting this formula for $\rho(\mathbf{r})$ into the integral of interest $A(\mathbf{u})$, one gets

$$A(\mathbf{u}) = (1/V) \int_V \sum_{p=1}^{N} \sum_{h,k,l=-\infty}^{+\infty} f_p \exp[-2\pi i(\mathbf{r} - \mathbf{r}_p) \cdot \mathbf{H}]$$

$$\times \sum_{s=1}^{N} \sum_{h,k,l=-\infty}^{+\infty} f_s \exp[-2\pi i(\mathbf{r} - \mathbf{r}_s + \mathbf{u}) \cdot \mathbf{H}] \, dV \qquad (6.6)$$

The integral $A(\mathbf{u})$ breaks down into separate terms, which, owing to the condition (6.2) of orthogonality of exponential functions, equals zero whenever functions with different $\mathbf{H}$ are multiplied. In all other cases the integral results in the expression

$$\sum_{s=1}^{N} \sum_{p=1}^{N} \sum_{h,k,l=-\infty}^{+\infty} f_p f_s \exp\{-2\pi i[\mathbf{u} - (\mathbf{r}_s - \mathbf{r}_p)] \cdot \mathbf{H}\} \qquad (6.7)$$

which need only be summed with respect to $p$ and $s$. Separating the expressions $p = s$ out of this sum, we obtain

$$A(\mathbf{u}) = \sum_{p=1}^{N} \sum_{h,k,l=-\infty}^{+\infty} f_p^2 \exp(-2\pi i \mathbf{u} \cdot \mathbf{H})$$

$$+ \sum_{p=1}^{N} \sum_{s=1}^{N} \sum_{h,k,l=-\infty}^{+\infty} f_p f_s \exp\{-2\pi i[\mathbf{u} - (\mathbf{r}_s - \mathbf{r}_p)] \cdot \mathbf{H}\} \qquad (6.8)$$

Each sum (6.7) resembles an atomic function in structure, coinciding with it formally if the scattering power of the "atom" $f_p f_s$ is at the point $\mathbf{r}_s - \mathbf{r}_p$

of the $\mathbf{u}$ space. The function (6.7) may be called the interatomic vector function, because its form depends on the atomic factors of the two atoms $s$ and $p$, and on the vectorial distance $\mathbf{r}_s - \mathbf{r}_p$ between them.

This circumstance determines the properties of interest of the function $A(\mathbf{u})$, i.e., of the $F^2$ series. Obviously, the $F^2$ series can be represented as the sum of interatomic functions for all possible pairs of atoms in the crystal unit cell. This means that maxima should form in the $F^2$ series for $\mathbf{u}$ values encountered in the crystal as interatomic vectors.

It follows from the last formula of the series $A(\mathbf{u})$ for a crystal containing $N$ atoms in its unit cell, that it consists of $N^2$ maxima, of which $N$ are trivial (distance between an atom and itself, i.e., the interatomic vector equals zero). This maximum at the point [000] is called the zero maximum. All the other maxima of the $F^2$ series should, in principle, make it possible to facilitate the solution somewhat, and sometimes to solve the structural problem entirely.

If the unit cell contains 20 atoms, the $F^2$ series consists of 380 nontrivial interatomic functions. Obviously, these maxima will overlap in complex structures, and analysis of the $F^2$ series will be of hardly any use.

However, the situation is different if one of the 20 atoms is a heavy one (say, Cl, Br, or I in an organic molecule). Since the interatomic functions have a height proportional to the product of the atomic numbers, the 19 interatomic vectors connecting light atoms with the heavy one will stand out sharply among the rest (not to mention maxima of the type Cl–Cl).

It can easily be seen that in the presence of one or two heavy atoms per molecule interpretation of the $F^2$ series is a quite feasible task. If the crystal symmetry is known the interatomic vectors will mostly result in a uniform scheme and in any case in an insignificant number of possible schemes of placement of these heavy atoms in the unit cell.

Since the heavy atoms give the predominant contribution to the structural amplitude, it may be assumed, to a first approximation at least, that the phases of the structural amplitudes are determined by these heavy atoms. Then, the electronic computer calculates the electron density series $\rho(\mathbf{r}) = \sum F_H \exp(-2\pi i \mathbf{H} \cdot \mathbf{r})$ where the moduli $F_H$ are derived from experiment, and the phases from calculation for a structure composed of heavy atoms only.

Among the maxima of this series there will appear peaks corresponding to the light atoms. In some favorable cases with fortunate structure geometry and an advantageous ratio of heavy to light atoms, the structure will become manifest immediately. In other cases, only part of the structure may be disclosed since a large number of false maxima may appear. Then the structure has to be "guessed" on the basis of crystal chemical experience. The phases of the structural amplitudes are determined again and a new electron density series is built. This work continues until the false peaks vanish and each $\rho(\mathbf{r})$

maximum acquires unambiguous interpretation. It is this crystal chemical experience that is difficult to program.

The heavy-atom method is the main technique in structural analysis. If the molecule does not contain heavy atoms and is more or less complex, the investigator prefers to solve the same structural problem on another object containing a heavy atom.

The introduction of a heavy atom does not solve the problem in all cases. If it is necessary to determine the structure of a crystal made up of atoms of the same weight, the general approach consists in making use of the relationships between structural amplitudes for different H of the same crystal.

We shall not dwell upon this question here, since it cannot be dealt with in brief; so we shall restrict ourselves to general remarks, referring the reader to a monograph by the present author [2].

There are two types of relations between amplitudes, namely, authentic and probability relations. The basic equation of the relation between amplitudes arises because the structural amplitudes of certain indices can be represented as elements of the Gramm determinant, which possesses the property of nonnegativity. As to the probability relations, they arise, because the phases of the trigonometrical functions contained in the expressions for the structural amplitudes are distributed uniformly in a circle.

On the basis of the theory of the relations between structural amplitudes, programs can be drawn up for computers, by means of which the signs and even the phases of all the structural amplitudes can be found, provided their moduli are known (these are given by experiment).

A number of successful cases of the use of this method for rather complex structures consisting of atoms of one species have been published [3]. Still, this method has not gained very great popularity. In the author's monograph cited above it is shown that the possibilities of determining the signs of structural amplitudes from measured intensities decrease rapidly with increasing number of atoms in the unit cell. The limit is about 50–100 atoms per unit cell.

*b. Special Methods for Molecular Crystals. Structure Seeking in the Absence of Heavy Atoms*

A number of new possibilities in searching for a crystal structure arise if a crude structure of the molecule is known and the mutual arrangement of molecules in the crystal is to be determined.

Of considerable help, first, is the geometrical model of a molecular crystal, described in detail in Chapter I, i.e., the fact that in the first approximation an organic crystal may be regarded as a close packing of rigid molecules.

With the aid of a computer one seeks such a mutual arrangement of the molecules that all the determining contacts between them are the same for identical atom pairs. Of course, there must not be any molecules that penetrate into others or "hang in midair."

Of great help in carrying out such geometrical analysis is work with models [4].

When the molecule has only three degrees of freedom, geometrical analysis is, as a rule, unambigous. Many examples are given in the author's book "Organic Chemical Crystallography" [4a].

By way of illustration of the possibilities of geometrical analysis, the structure of 2,6-dimethylnaphthalene was discussed above in Chapter I.

If the number of degrees of freedom of the molecule is not large, work with a computer will not require much time either. It is quite possible to scan all the molecular orientations. (To begin with it is sufficient to calculate the distances between the atoms of neighboring molecules for points in the space of $\varphi$, $\psi$, and $\eta$, every 5 or even every 10° apart, immediately discarding variants in which, say, $R_{HH}$ is less than 2.0 Å, $R_{CH}$ is less than 2.7 Å, etc.) This work usually results in a few (and perhaps even only one) allowed regions. Then, continuing work with the geometrical crystal model, one can find a molecular placement which gives an ideal packing.

The search for ideal packing in the case of a molecule in a general position takes much more time. If we work blindly and divide the unit cell edge into 30 or 50 parts, the number of points in the six-dimensional parameter space for which distances have to be figured out is of the order of $10^6$–$10^8$.

Apparently, such a "frontal" attack on the solution of the problem is not always expedient with the present-day capabilities of computers. A group of Soviet authors [5] have employed the gradient descent method successfully in solving a similar problem. The program is compiled so as to move from an arbitrary point of parameter space toward a decreasing number of forbidden contacts.

Quite obviously, the geometrical model can be replaced by the physical one. Then the intermolecular distances will be only intermediate data. They are used to calculate the energy surface, among the deepest minima of which the correct structure is to be sought. These methods of work are quite clear from the contents of Chapter II.

Finally, it is of great interest to combine the geometrical (energy) approach with an analysis of a diffraction experiment. One may, say, as has already been done in several studies, minimize the sum of the $R$-factor and the potential energy. The difficulty lies in the arbitrariness of selecting the weight factor.

The structural amplitudes of a molecular crystal can always be regarded as a function of the degrees of freedom of the molecule (three in the case of the type $P2_1/a$, $Z = 2$; six for one molecule in a general position, etc.).

Thus, the search for the structure consists in calculating either

(a)   the structural amplitudes, or
(b)   the intermolecular distances, or
(c)   the molecular interaction energies

as functions of these three or six degrees of freedom of the molecule.

Various versions of combined programs can be made up to carry out these calculations in various sequences. We believe that the most expedient method of searching for a structure is to calculate a small number (30–50) of structural amplitudes and simultaneously to analyze the molecule's geometrical placement. It is always worthwhile to begin the work by examining the models of possible packings on a structure-finder (see Chapter I).

Other special methods for finding the structures of molecular crystals are isomorphous substitution, diffraction by optical masks, and the method of molecular transforms, which is the mathematical analogue of the latter. These methods have long been among the tools used by X-ray crystallographers. The structural investigation of protein molecules containing many thousands of atoms, an epoch-making event in biological physics, became possible only with the aid of specially synthesized protein derivatives containing heavy atoms.

In this case the study is made possible by the isomorphism of the structure of a protein crystal with that of its derivative. Suppose these two structures are identical in all respects except for substitution at the X position of a light atom by a heavy atom. We can write the Patterson function for these two crystals as

$$P(\mathrm{A}) = f_\mathrm{A}^2 \delta(\mathbf{r}) + \sum_i f_\mathrm{A} f_i \delta(\mathbf{r} - \mathbf{r}_{Xi}) + \sum_i f_\mathrm{A} f_i \delta(\mathbf{r} - \mathbf{r}_{iX}) + \sum_{i,j} f_i f_j \delta(\mathbf{r} - \mathbf{r}_{ij})$$

$$P(\mathrm{B}) = f_\mathrm{B}^2 \delta(\mathbf{r}) + \sum_i f_\mathrm{B} f_i \delta(\mathbf{r} - \mathbf{r}_{Xi}) + \sum_i f_\mathrm{B} f_i \delta(\mathbf{r} - \mathbf{r}_{iX}) + \sum_{i,j} f_i f_j \delta(\mathbf{r} - \mathbf{r}_{ij})$$

Here A is a light atom, substituted in the other crystal by the heavy atom B. The subscripts $i$ and $j$ indicate all the other atoms of equal weight and position in both crystals.

Reducing to a common scale and subtracting, one can build the series

$$\Delta P = (f_\mathrm{A}^2 - f_\mathrm{B}^2) \delta(\mathbf{r}) + \sum_i (f_\mathrm{A} - f_\mathrm{B}) f_i [\delta(\mathbf{r} - \mathbf{r}_{Xi}) + \delta(\mathbf{r} - \mathbf{r}_{iX})]$$

With the exception of the zero peak, this series gives the structure and its inversion through point X as seen from the position of, i.e., "visible" from, the substituted atom, i.e., the totality of the vectors $\mathbf{r}_{iX}$ and $\mathbf{r}_{Xi}$. Of course, the scheme of isomorphous substitution just described may be found useful not only in the case of protein structures, but for other organic substances as well.

The optical mask method, due to Taylor and Lipson, is well known to the English reader and has been described in detail in a number of monographs (see, e.g., [6]). Holes arranged like the atoms of a molecule are marked on a black screen to the scale 1 Å = 0.65 mm. Such optical masks can be prepared for different projections of the molecule onto the screen plane. By observing the Fraunhofer diffraction from such a screen, one gets the Fourier transform of the molecule.

It is not difficult to show the validity of the following statement: At a point of reciprocal space where the Fourier transform of the molecule equals zero, the structural amplitude should also equal zero.

With the aid of this theorem alone this or that mutual arrangement of the molecules can be discarded by comparing the experimental intensities of the reciprocal lattice points with the Fourier-transform pattern of the molecule. Comparison of experiments with diffraction on optical masks is also possible, and this can even be carried out by machine.

The molecular transform method unquestionably speeds up the trial and error method for molecular crystals. A total program can be set up, combining all the methods mentioned above for the trial and error method, for use in a single computer program.

## 7. HEAT WAVE SCATTERING

Lattice dynamics have a substantial influence on diffractional effects. Moreover, the study of X-ray and neutron scattering by matter is essentially the only experimental method of establishing the shapes of the surfaces $w(\mathbf{k})$ (vibration frequency versus wave vector) over the entire range of wave vectors.

The effect of thermal motion on X-ray diffraction reduces to two phenomena: a decrease of the intensity of the reflected beams, which we discussed above; and the generation of a diffuse pattern of its own.

Owing to thermal motion, the placement of the crystal atoms no longer corresponds to an ideal lattice at each instant. Suppose the atom in the $k$th position of the $L$th unit cell has shifted by the vector $\mathbf{\Delta}_k^L$ from its ideal position. This atom gives an elementary wave of the same amplitude $f_k$. However, the phase of the wave will change relative to that which existed in the ideal lattice by the value $(\mathbf{k} - \mathbf{k}_0) \cdot \mathbf{\Delta}_k^L = \mathbf{s} \cdot \mathbf{\Delta}_k^L$. Thus, the amplitude of the wave sent by an atom displaced from its equilibrium position should be represented by

$$g_k^L = f_k \exp(i\mathbf{s} \cdot \mathbf{\Delta}_k^L)$$

At the same moment of time the shifts of the $k$th atom in the different unit cells may have a great variety of values. Accordingly, the values of $g_k{}^L$ will differ for the $k$th atoms of different unit cells. The average scattering for $k$th atoms will be

$$\overline{g_k{}^L} = (1/N) \sum_L g_k{}^L = f_k(1/N) \sum \exp(is\cdot\Delta_k{}^L) = f_k\overline{\exp(is\cdot\Delta_k{}^L)}$$

This average value can be presented in the form $e^{-M_k}$ where $M_k$ is a pure number. The quantity $e^{-M_k}$ is known as the temperature factor.

The calculation of the quantity $M$ from lattice dynamics data was discussed in Chapter III. It was shown that the temperature factor of X-ray coherent scattering takes lattice dynamics into account in purely integral form and is of little use in studying molecular movements.

On the other hand, as will now be shown, a great deal of information is contained in the diffuse scattering of rays, determined not by the average structural amplitude

$$\bar{F} = \sum \overline{g_k{}^L}\exp(i2\pi\mathbf{H}\cdot\mathbf{r}_k)$$

but by the instantaneous values

$$F_L = \sum g_k{}^L \exp(is\cdot\mathbf{r}_k)$$

The experimentally determined intensities are, of course, related to the average values of the structural factor, since the time of measurement is fairly large compared to the period of vibration of the atom ($\sim 10^{-13}$ sec); and this vibration period is very large compared to the period of an electromagnetic X-ray wave ($10^{-18}$ sec). Hence, we are indeed justified in considering the electromagnetic field of the instantaneous distribution of matter in the lattice (the crystal matter is immobile relative to the incident wave).

This makes it possible to regard scattering by vibrating molecules as diffraction by standing waves.

Let us consider a simple example, namely the diffraction of X rays by a cubic crystal in a state of stationary transverse vibration along the [100] axis.

Essentially, the standing wave of thermal vibrations imposes a superlattice on the static pattern, with a period equal to the length of the thermal wave. In Fig. 8 the length of the thermal wave is about $3a$.

The letters denote the points of the new superlattice. This lattice will contain, for example, sets of planes of the type $AB'$, slightly inclined with respect to the basic plane $AA'$ and possessing almost the same period. Only these superlattice planes, those closest to the basic plane, need be taken into consideration, for only they give reflections and these only in the first order. This being so, the main reflection should be accompanied by two "diffuse reflections" occurring at angles of $\theta\pm\alpha$ where $\alpha$ is a small quantity.

Each of the waves has its own superlattice. Thus, the diffracted beam will be surrounded by numerous "diffuse" beams. A hazy spot of rays "reflected

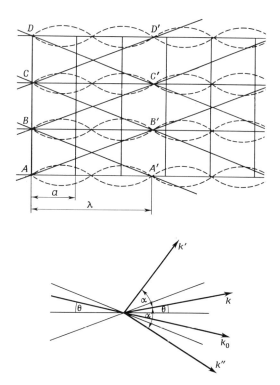

*Fig. 8.* Thermal vibration of wavelength $AA'$ traveling along $AA'$ gives rise to additional nodal planes $AB'$, $BC'$, $CD'$, and $BA'$, $CB'$, $DC'$.

from heat waves" will appear around the diffraction spot on the film. This intensity distribution (of diffuse rays around the diffracted one) is not necessarily symmetrical. Symmetry will occur in very rare cases (theoretically, only upon complete isotropy of vibration).

We are dealing with reflections of only the first order, and only from $AB'$. The reasons for this are obvious: Reflections from planes drawn through $A$ and $D'$ and other such planes come into the vicinity of a different basic reflection. As to the reflections of the second, third, etc., orders from a thermal wave of given length, they are equivalent to the first-order reflections from thermal waves of one-half, one-third, etc., the length.

The condition of diffraction from heat waves should have the usual form

$$2\pi \mathbf{H}' = \mathbf{k} - \mathbf{k}_0$$

where $\mathbf{H}'$ is a vector with the direction of the normal to the reflecting planes and a magnitude of $1/d$. It is evident from Fig. 9 that $\mathbf{H}' = \mathbf{H} + \boldsymbol{\tau}$, where $\boldsymbol{\tau}$ is the thermal wave vector. Indeed, it is obvious from the figure that $ON/H =$

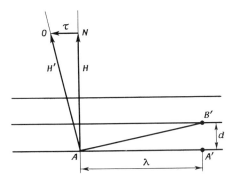

*Fig. 9.* Condition of diffraction due to heat waves.

$d/\lambda$ and hence, $ON = Hd/\lambda = \tau$. Thus the condition of diffraction has the form

$$2\pi(\mathbf{H}+\tau) = \mathbf{k} - \mathbf{k}_0$$

Figure 10 illustrates the geometrical sense of the condition that diffraction from the thermal waves of a crystal in a given direction occurs if the thermal wave vector joins the end of the vector $\mathbf{k}$ with the nearest reciprocal lattice point.

None of the $\tau$ vectors extends beyond a parallelohedral zone built around the reciprocal lattice point. Indeed,

$$\tau = \tau_1\mathbf{a}^* + \tau_2\mathbf{b}^* + \tau_3\mathbf{c}^*$$

Suppose there is a vector $\tau'$ which extends beyond this zone. There will always be a vector $\tau$ which ends at the point lying within the zone, and

$$\tau' = \tau + \mathbf{H}$$

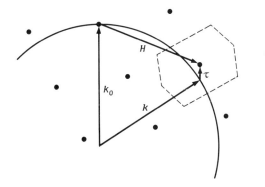

*Fig. 10.* Interpretation of diffraction due to heat waves by means of reciprocal lattice.

where $\mathbf{H}$ is the reciprocal lattice vector. From the wave formula

$$\Delta_k{}^L = U_k \exp[i2\pi(\tau \cdot \mathbf{r}_k{}^L)] \exp(i\omega t)$$

it follows that the vector $\tau'$ gives no new solutions owing to the periodicity of the lattice because

$$\mathbf{H} \cdot \mathbf{r}_k{}^L = \mathbf{H} \cdot \mathbf{r}_k + \mathbf{H} \cdot \mathbf{R}_L = \mathbf{H} \cdot \mathbf{r}_k + m$$

where $m$ is an integer and $\mathbf{r}_k{}^L = \mathbf{R}_L + \mathbf{r}_k$. Thus, $\tau$ and $\tau'$ give the same waves (differing only in initial phase, but not essentially).

This limitation is easily understood. The vector can be of any order of smallness, i.e., the length of the standing waves can be, generally speaking, of any order of largeness for an infinite crystal, but the vector $\tau$ cannot be greater than the size of the elementary reciprocal unit cell. This means that the wavelength cannot be smaller than the size of a direct lattice unit cell.

The maximum length obviously depends on the size of the crystal as a whole. As with any proper vibrations of a limited body, half of the maximum length of the standing wave equals the size of the body. The assortment of wavelengths cannot be continuous owing to boundary conditions: The boundaries of the crystal must be nodes of the standing wave.

Thus, only one order of reflection from the thermal waves falls into the zone of each given reciprocal lattice node.

Translating the condition of diffraction $2\pi(\mathbf{H}+\tau) = \mathbf{k}-\mathbf{k}_0$ "into the language of the true crystal lattice," gives the following meaning: By varying the angle of incidence $\theta$ over a wide range we can get a diffuse "reflection" from a given region $hkl$ at an angle of $\theta + \phi \approx 2\theta_B$. The angle $\theta + \phi$ varies only within 1–2°. As long as the angle $\theta$ varies within 30°, the diffuse spot will be visible on the film.

The condition $2\pi(\mathbf{H}+\tau) = \mathbf{k}-\mathbf{k}_0$ can be derived more rigorously by investigating the expression for the diffuse intensity $I_2$, which has the form

$$I_2 = I_e \left| \sum_L (F_L - \bar{F}) \exp(i\mathbf{s} \cdot \mathbf{R}_L) \right|^2$$

From the same equation also follow conclusions concerning the intensity distribution of the diffuse scattering.

The zone of each of the reciprocal lattice points can be characterized by the scattering intensity as a function of $\mathbf{H}+\tau$. The surfaces of the function $I(\mathbf{H}+\tau)$, called equal scattering (isodiffusion) surfaces, can be plotted in reciprocal space. Here we use the reciprocal space for representing the intensity of the thermal scattering $I_2(\mathbf{s})$, which, like $I_1(\mathbf{s})$, can be presented as a function of the reciprocal lattice coordinates $\varepsilon, \eta, \zeta$. The function $I_2(\mathbf{s})$ declines from its maximum value very slowly; the corresponding regions around the reciprocal lattice points are therefore considerable in size and may even merge.

Measurement of diffuse scattering is a rather troublesome task. Such measurements have been carried out for only an insignificant number of molecular crystals. By way of example, Fig. 11 shows the isolines of an anthracene crystal for a unit cell of its reciprocal lattice (plane section only), formed by the points $\bar{5}\cdot 0\cdot 10$, $\bar{4}\cdot 0\cdot 10$, $\bar{5}09$, and $\bar{4}09$.

The above discussion of the conditions of the diffraction geometry of rays on thermal waves shows that waves with wave vectors $\tau$ contribute intensity to the reciprocal space at points distant by the vectors $\tau$ from the lattice points.

The expression for the structural amplitude

$$F_L = \sum f_j \exp[i s \cdot (\mathbf{r}_j + \mathbf{\Delta}_j{}^l)] = \sum f_j \exp[i 2\pi (\mathbf{H} + \tau) \cdot (\mathbf{r}_j + \mathbf{\Delta}_j{}^l)]$$

can be considered for each mode (each frequency) and wave vector $\tau$.

Expanding in a series of shift powers, one obtains

$$F(\mathbf{H} + \tau) = \sum f_j \mathbf{U}_{yj}^{\tau} \cdot 2\pi (\mathbf{H} + \tau) \exp[i 2\pi (\mathbf{H} + \tau) \cdot \mathbf{r}_j]$$

Here $\mathbf{U}_{yj}$ is the shift the $j$th atom receives on participating in the $y$th mode of vibration with the wave vector $\tau$.

All the vibration modes make their contribution to the scattering intensity at the same reciprocal space point. Nevertheless, it is quite possible to determine all the dispersion surfaces, since scattering intensity values can be analyzed at points displaced by the value of the vector $\tau$ relative to different reciprocal lattice points. The contributions of different modes will be different for different points. This enables a large number of equations to be written,

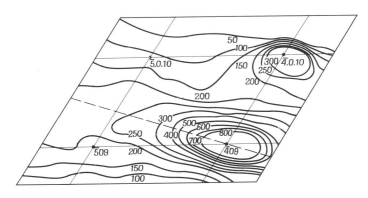

*Fig. 11.* Anthracene. Experimental isodiffuse lines.

from which the displacements can be determined and then recalculated to vibration frequencies by means of expressions relating the displacement, energy, and frequency.

To make clear the possibilities of the method, we write the formula for the intensity of X-ray scattering by the thermal waves of a simple single-atom lattice:

$$I(\tau) = \frac{|H+\tau|^2 f^2 e^{-2M}}{m} \sum_{\gamma=1}^{3} \frac{E_\gamma \cos^2[H+\tau] \cdot e_\gamma}{v_\gamma{}^2(\tau)}$$

In this case the intensity is created by three acoustic waves. If, for instance, the wave vector coincides with the crystal symmetry axis and is parallel to the diffraction vector, only longitudinal acoustic waves will contribute to the intensity, because for transverse waves

$$\cos^2[H+\tau] \cdot e_\gamma = 0$$

For high temperatures, the intensity has a very simple expression

$$I(\tau) = \frac{(H+\tau)^2 f^2 e^{-2M} kT}{m} \frac{1}{v_e{}^2(\tau)}$$

Studies of thermal diffuse X-ray scattering—theory, experiment, and exhaustive references—can be found in a monograph [7]. Similar information in the field of neutron diffraction studies can be found in another useful collection [8].

A complete investigation of the dispersion surfaces for a naphthalene crystal has been undertaken by British workers (Pawley, private communication). For molecular crystals of this type, formulas can be compiled relating experiment and all the twelve dispersion surfaces (frequencies and proper vectors) calculated by means of the atom–atom potential scheme.

$$I(\tau) = \sum_{s=1}^{12} \frac{[(H+\tau) \cdot \eta_s]^2}{m\omega_s} \left( \frac{1}{e^{\hbar\omega_s(\tau)/kT} - 1} + \frac{1}{2} \right) \hbar$$

where

$$\eta_s = \sum_n e_{sn}(\tau) f_n e^{i(H+\tau) \cdot r_n}$$

and the $e_s$ are proper vectors, which can be calculated by the atom–atom potential method, as was described in Chapter III.

In the formulas written above the $r_n$ are the coordinates of the atoms in the unit cell, $m$ is the mass of the unit cell, and H is the reciprocal lattice vector of the lattice point about which scattering is being studied.

## B. Nuclear Magnetic Resonance

1. THEORETICAL FUNDAMENTALS OF NUCLEAR MAGNETIC RESONANCE IN
A SOLID

NMR lines in the solid are very broad, hundreds of thousands of times broader than is natural for an isolated spin. The main reason for their distention is the local magnetic field set up on a given nucleus by the precessional magnetic moments of the surrounding nuclei. As a result of this precession, the local field $\mathbf{h}$ at the point $\mathbf{R}$ is determined by the component $\mu_n$ of the magnetic moment in the direction of the constant magnetic field $\mathbf{H}$

$$\mathbf{h} = \frac{3(\mu_n \cdot \mathbf{r})\mathbf{r} - \mu_n}{R^3} \qquad (1.1)$$

where $\mathbf{r} = \mathbf{R}/R$. By means of not very complicated transformations and taking into account that

$$\mu_n = \mu\mathbf{n} \qquad (1.2)$$

where $\mathbf{n} = \mathbf{H}_0/H_0$, we write out the components of the vector $\mathbf{h}$ in the following form

$$h_x = \mu_n R^{-3}[(3r_x{}^2 - 1)n_x + 3r_y r_x n_y + 3r_z r_x n_z]$$
$$h_y = \mu_n R^{-3}[(3r_x r_y n_x + (3r_y{}^2 - 1)n_y + 3r_z r_y n_z] \qquad (1.3)$$
$$h_z = \mu_n R^{-3}[3r_x r_z n_x + 3r_y r_z n_y + (3r_z{}^2 - 1)n_z]$$

It is evident from (1.3) that the vector $\mathbf{h}$ can also be represented as the product of a certain symmetrical matrix $\|A_{kl}\|$ by the vector $\mathbf{n}$:

$$\mathbf{h} = \mu_n \|A_{kl}\| \mathbf{n} \qquad (1.4)$$

The matrix

$$\|A_{kl}\| = R^{-3}(3r_k r_l - \delta_{kl}) \qquad (1.5)$$

is called the dipole–dipole interaction tensor, and $\delta_{kl}$ is the Kronecker symbol.
With the aid of either of the above representations of the vector $\mathbf{h}$, (1.1) or (1.4), the value of the local field can easily be obtained by scalar multiplication of $\mathbf{h}$ by the vector $\mathbf{n}$. In the case of (1.4) one has

$$h_{\text{loc}} = \mathbf{h} \cdot \mathbf{n} = \mu_n A_{kl} h_k h_l$$

Directing the field $\mathbf{H}$ along the $z$ axis and passing over to spherical coordinates, we get the quantity $h_{\text{loc}}$ as a function of the angle between the vectors $\mathbf{H}$ and $\mathbf{R}$

$$h_{\text{loc}} = \mu_z R^{-3}(3\cos^2\gamma - 1)$$

Allowance for quantum-mechanical interaction between nuclei of the same species results in corresponding additions to the value of $h_{loc}$ on the given nucleus. For example, for an isolated pair of protons an elementary quantum calculation carried out by Pake showed that one-third of the measured value of $h_{loc}$ is due to the resonance exchange of energy between protons.

In the general case, with random configurations of the nuclei, the absorption line is a broad, shapeless hump.

The shape of NMR lines in solids can be described theoretically by the moment method. In this method the shape of the line $g(h)$ is characterized by a set of numbers (moments) which describe the behavior of the function $g$ more and more remote from the point $h = 0$ as their ordinal numbers increase. We remind the reader that the $n$th-order moment of the function $g$ is the quantity

$$S_n = \langle h^n \rangle = \int_{-\infty}^{+\infty} h^n g(h)\, dh \bigg/ \int_{-\infty}^{\infty} g(h)\, dh$$

NMR lines are symmetrical about the point $h = 0$. This means that all the odd moments become zero (the number of negative terms in the integrand equals the number of positive terms). Thus, only the second, fourth, and higher even moments need be calculated. The second moment can be estimated as the average square of the local field

$$S_2 = \int_{-\infty}^{\infty} h^2 g(h)\, dh \bigg/ \int_{-\infty}^{\infty} g(h)\, dh$$

Van Vleck has shown [9] that the second moment of a single crystal consisting of resonating nuclei with the spin quantum number $I$ and the gyromagnetic relation $\Gamma$ at a distance $r$ from one other is given by

$$S_2 = (3/8\pi^2) I(I+1) N^{-1} \Gamma^2 h^2 \sum_{j>k} \sum_k (3\cos^2 \gamma_{jk} - 1)^2 r_{jk}^{-6}$$

$$+ (N^{-1} h^2/12\pi^2) \sum_j \sum_f I_f (I_f + 1) \Gamma_f^2 (3\cos^2 \gamma_{jf} - 1)^2 r_{jf}^{-6} + \cdots \quad (1.6)$$

here $\Gamma_f$ and $I_f$ refer to nuclei of different species, $r_{jk}$ is the length of the vector connecting the $j$th and $k$th nuclei, $\gamma_{jk}$ is the angle between $\mathbf{H}_0$ and the vector $\mathbf{r}_{jk}$, and $N$ is the total number of resonating nuclei. If nuclei of a third species are present in the crystal, the formula becomes one term longer, etc. In the case of a polycrystalline sample this expression should be averaged over all $\gamma$ values. Van Vleck also obtained a formula for the fourth moment.

The ratio of the second moment to the fourth gives information on the shape of the line. For a Gaussian shape, this ratio is 1.32, for a rectangular signal it is 1.16. The totality of all the even moments should, in principle, enable complete characterization of the line contours. However, owing to computational difficulties and insufficient experimental accuracy one is obliged in most cases to restrict oneself to calculating only the second moment. In its initial form the Van Vleck formula is suitable for use in structural analysis by the NMR method. Obviously, the trial-and-error method can be used for this purpose. The second moment is calculated for a certain trial structure. The result is compared with experiment. If necessary, the trial structure is then modified and the calculation is repeated until satisfactory agreement between the experimental and calculated second moment is achieved. Owing to the very sharp dependence of the calculated results on distance, they can very often be completed successfully without resorting to electronic computers. The predominant contribution to the second moment is that of the immediate vicinity of the proton (or other nucleus) in question, within a radius of the order of 5 Å. The contribution of a nucleus situated on the surface of this sphere to the second moment is three orders smaller than that of its closest neighbor. For this reason, allowance for angles and distances is made only inside this sphere. To allow for the more remote interactions in the calculation, the following integral is used

$$\int 4\pi r^2 \, dr \, n_0 r^{-6} = n_0 \int_{r_o}^{\infty} 4\pi r^{-4} \, dr = n_0 \tfrac{4}{3}\pi r_0^{-3}$$

where $n = N/r^3$ is the number of protons per unit volume. It can be seen that this integral contains neither proton–proton line orientation nor $\mathcal{C}$–H bond length. In investigating the Van Vleck formula many authors have made use of the formal separation of the moment into its intra- and intermolecular parts. This technique is especially popular when using NMR for analysis of polymer crystallinity.

The next major advance in the development of the theory of structural analysis of single crystals by the NMR method was due to McCall and Hamming [10]. They were the first to examine in the general case the problem of the experimental volume required and the maximum number of parameters obtainable from the complete experiment. The line of reasoning followed by these authors is simple. It is necessary to express a direction in the Van Vleck formula (1.6) by the polar angles $\theta$ and $\varphi$. Making use of the relation

$$\cos \gamma_{kl} = \cos \theta \cos \theta_{kl} + \sin \theta \sin \theta_{kl} \cos(\varphi - \varphi_{kl})$$

they found that

$$
\begin{aligned}
(3\cos^2 \gamma_{kl} - 1)^2 = \; & 9a^4 + 9b^4c^4 + 9b^4d^4 + 54a^2b^2c^2 + 54a^2b^2d^2 \\
& + 54b^2c^2d^2 + 108ab^3cd^2 + ab^3c^2d + 108a^2b^2cd \\
& + 36b^4cd^3 + 36b^4c^3d + 36ab^3d^3 + 36a^3bc \\
& + 36a^3bd - 6a^2 - 6b^2c^2 - 6b^2d^2 - 12abc \\
& - 12abd - 12b^2cd + 1
\end{aligned}
$$

where

$$a = \cos\theta\cos\theta_{kl}, \qquad c = \cos\varphi\cos\varphi_{kl}$$

$$b = \sin\theta\sin\theta_{kl}, \qquad d = \sin\varphi\sin\varphi_{kl}$$

After cumbersome transformations the second moment can be expressed as a function of the angles $\theta$ and $\varphi$:

$$\Delta H^2(\theta,\varphi) = \Gamma \sum_{k=1}^{22} R_k F_k(\theta,\varphi)$$

The authors divided the total second moment into two parts: $R_k$—the coordinate part, and $F_k$—the angular part; they are given by

$$R_k = w \sum_i \sum_j x_{ij}^r y_{ij}^s z_{ij}^t r_{ij}^{-(6+r+s+t)}$$

$$F_k(\theta,\varphi) = \sin^a\theta\cos^b\theta\sin^c\varphi\cos^d\theta \tag{1.7}$$

The constants $w$, $r$, $s$, and $t$ and $a$, $b$, $c$, and $d$ are listed in the original work [10] for values of $k$ from 1 to 22.

McCall and Hamming called the functions $R_k$ lattice sums. Only 15 of the 22 lattice sums can be independent in the case of lowest crystal symmetry. The other seven are related to the 15 independent sums by

$$R_{20} = -9R_7 - 3R_{10} - 3R_{14}$$

$$R_{21} = -9R_6 - 3R_{15} - 3R_{12}$$

$$R_{22} = -9R_5 - 3R_{11} - 3R_{13}$$

$$R_{19} = -27R_1 - 9R_4 - 3R_{10} + 3R_8 - 9R_9 - 9R_3$$

$$R_{17} = 27R_1 + 9R_4 + 3R_{10} - 3R_8 - 3R_9$$

$$R_{18} = 27R_1 - 3R_{10} - 3R_8 + 3R_9 + 9R_3$$

$$R_2 = -R_4 + 6R_1 - R_3$$

This means that a complete NMR experiment enables the values of not more than 15 structure parameters to be found.

A very essential question is that of the necessary and sufficient volume of experiment. Fifteen independent equations with respect to the 15 unknown $R_k$ can be obtained, according to McCall, from five settings of the crystal, for each of which a $\varphi$-dependence is registered at an assigned small pitch and constant $\theta$. $\varphi$-dependence is preferable to $\theta$-dependence for experimental reasons: Satisfactory sensitivity is usually attained only with cylindrical samples compactly filling the coils of conventional NMR spectrometers. Dereppe and co-authors [11] have demonstrated that structural analysis by the NMR method encounters a number of limitations:

1. owing to insufficient sensitivity the accuracy of the experimental second moment values is low in many cases.

2. the effect of movements in the lattice may distort observed $\theta$-dependencies in a complex manner.

3. 15 structural parameters can be determined only for crystals of the triclinic system.

With increasing crystal symmetry $n$ lattice sums cease to be linearly independent of the $15-n$ others. This decreases the number of determinable parameters to $15-n$, and the maximum number of atoms whose coordinates can be obtained by NMR turns out to be:

$$\frac{15}{3} + 1 = 6, \quad \text{for triclinic crystals}$$
$$\frac{9}{3} = 3, \quad \text{for monoclinic crystals, and}$$
$$\frac{6}{3} = 2, \quad \text{for orthorhombic and cubic crystals.}$$

Thus, the possibilities of the NMR method may seem rather low. But taking into account the difficulties of the neutron diffraction method and the very small capabilities of X-ray diffraction analysis in determining the coordinates of hydrogen atoms, the aid afforded by the NMR method appears very valuable at the concluding stages of structural analysis of organic crystals.

On the other hand, the high sensitivity of the NMR method to various kinds of molecular movements (vibrations, reorientations, rotations, diffusion) makes it a fairly independent method of analyzing lattice structure and dynamics.

In the presence of internal mobility at a frequency of $\gtrsim 10^4$ Hz, the nucleus in question is subjected to the action of the time-averaged value of the local field $\langle h(t) \rangle$ rather than its instantaneous value $h(t)$, which results in a change in breadth and shape of the absorption line. Analysis of the temperature and angular dependence of the spectrum characteristics makes it possible, in principle, to calculate nuclear movement, given the structure geometry, or to solve the reverse problem. This problem is practically impossible to solve in the general form, but is simplified considerably if the spin system consists of isolated groups of nuclei. In such a case it becomes unnecessary to calculate the function $g(h)$, and the problem is solved by calculating the value and angular dependence of the splitting appearing in the NMR spectra.

Such calculations were carried out for various concrete "movement figures" and the corresponding angular dependencies of the value $\langle h \rangle$ were obtained. In the general form the problem can be examined for an isolated proton pair. All imaginable "movement figures" of a $p$-$p$ vector, independent of the physical nature of the mobility, are contained in the set of all the trajectories which can be assigned on a sphere. This gives a foundation on which the calculation formulas can be systematized, and by which the general elements in concrete calculations can be expressed. It can be shown that, in the general case, the value of the average local field can be represented in a quadratic form with respect to the variables $n_x$, $n_y$, and $n_z$ ($\mathbf{n} = \mathbf{H}_0/H_0$) by

$$\langle h \rangle = \pm \tfrac{3}{2}\mu R^{-3}(\bar{a}_{xx}n_x^2 + \bar{a}_{yy}n_y^2 + \bar{a}_{zz}n_z^2 + \cdots) \tag{1.8}$$

with the coefficients

$$\bar{a}_{kl} = \sum_{i=1}^{m} p_i a_{kl}^i, \qquad k,l = x,y,z \tag{1.9}$$

for discrete movement figures with $m$ positions of the $p$-$p$ vector, and by

$$\bar{a}_{kl} = \int\int p(\mathbf{r})a_{kl}\,ds \tag{1.10}$$

for continuous movement figures, with the probability density distribution function of $p$-$p$ vector orientations equal to $p(\mathbf{r})$; the coefficients $a_{kl}$ are the components of the unit vector $\mathbf{r}^i$:

$$a_{kl}^i = 3r_k^i r_l^i - \delta_{kl}$$

All possible angular dependencies of the value $\langle h \rangle$ differ from one another in the elements of the matrix $\|\bar{a}_{kl}\|$ of coefficients of the form (1.8). There is a correspondence between the class of symmetry of the movement figure and the form of the matrix $\|\bar{a}_{kl}\|$. The particular simplicity of the NMR spectra of organic substances containing water (since one $p$-$p$ vector is much shorter than all the others) gives rise to special possibilities. For example, one can determine the symmetry of the orientations of the water molecules in fibrous proteins, study diffusional and vibrational mobility in hydrates, etc.

It should be pointed out that the capabilities of NMR for studying angular vibrations of molecules are limited. The change in the splitting value $2\,\Delta h$ is related linearly to the average square amplitude

$$2\,\Delta h \sim \langle \alpha^2 \rangle$$

Therefore, if $2\,\Delta h$ has the quite perceptible value of $1$–$1.5\,\mathrm{G}$ for a deviation angle of about $10°$ (e.g., the water molecule), the effect is practically unnoticeable in the case of heavy molecules with $\alpha \sim 1°$.

2. INVESTIGATION OF MOLECULAR MOVEMENT IN A CRYSTAL BY THE
   NMR METHOD

Gutowsky and Pake [12] have studied the temperature dependence of the NMR line breadth for nuclei of spin $\tfrac{1}{2}$ in some molecular crystals. The temperature was varied between $90°K$ and the melting point of the substance. In some cases it was found possible to compare the observed line structure and the breadth transitions of these lines with the existence and frequency of definite types of hindered rotation in the solid state. An examination was made of the quantitative effect of this type of movement on the structure and the second moment of the absorption line. It was found that comparatively slow movements (at a frequency of $\sim 10^5$ Hz) already caused perceptible line narrowing compared to their breadth in the absence of movements.

In 1,2-dichloroethane, 1,1,1-trichloroethane, and perfluorethane the breadth jumps of the lines are close to phase transitions and heat capacity anomalies. In 1,2-dichloroethane these jumps correspond to rotation about the longitudinal axis of the molecule. Many molecules (acetonitrile, methyl iodide, nitromethane, dimethylmercury, ammonia) give an NMR line at 90°K corresponding to proton rotation about a third-order axis. Data have been obtained on the movements of the molecules of 2,2-dimethylpropane, ethanol, methanol, acetone, methylamine, and ethyl halides.

Powles and Gutowsky [13] have studied $CH_3$-group reorientation in a number of organic crystals at low temperatures: in methylchloroform, 2,2-dinitropropane, and 2-chloro-2-nitropropane. They studied the molecular movements in the solid by observing the temperature dependence of the shape of the NMR lines. The authors point out two aspects of these observations. First, it can be established whether the $CH_3$ group is actually reoriented about the corresponding third-order axis. If such reorientation occurs, the breadth of the NMR line will decrease with increasing temperature within a certain temperature interval. They succeeded in showing that these 'movements are thermally activated hindered rotation. Powles and Gutowsky derived an expression relating the line breadth to the frequency of reorientation, which in its turn gives the rate of reorientation as a function of temperature; by means of this function the height of the reorientation barrier can be estimated. Secondly, studying the dependence of the shape of the NMR line on the sample temperature gave the authors a number of ideas concerning the mechanism of reorientation. The simplest reorientation model is the classical passage of the particle over the potential barrier $E_0$. But if only protons participate in the movement, a tunnel effect is quite possible owing to the small mass and low barrier height.

Since the shape of the line may vary simultaneously with its breadth, the most satisfactory measure of the temperature effects is the second moment. Using the general reasoning given at the beginning of the previous section, there is no difficulty in writing formula (1.8) and the corresponding formula for the second moment in the case of an isolated triplet of protons situated at the vertices of an equilateral triangle.

If a given state of nuclear spin has a lifetime long enough for the local fields responsible for the line breadth to fluctuate, the resulting line broadening will be the average over this lifetime. In its turn, in accordance with the principle of inderminacy of lifetime the $\delta t$ of the given state is related to the spectral line breadth by the expression $(h\delta v)\delta t \approx h/2\pi$. For this reason, the interval $\frac{1}{2}\pi\delta v$ will be a suitable time for averaging. For an immobile $CH_3$ group, $\delta v$ is approximately $2 \times 10^4$ Hz. Actually this interval is itself a mean value, because the lifetimes of nuclear spin states have a distribution pattern. Movement decreases line breadth only when the frequency of reorientation begins to exceed the resonance line breadth $\delta v$.

Powles and Gutowsky [13] have shown that the experimental results for the $CH_3$ group can be described equally well by three different models: (1) random reorientation of the Brownian movement type, (2) classical movement across a threefold potential barrier, (3) tunnel effect. In their work, Powles and Gutowsky are inclined to favor the assumption that the tunnel effect plays the predominant part in methylchloroform, 2,2-dinitropropane, and 2-chloro-2-nitropropane crystals.

A study of the $CH_3$ group rotation in hexamethylbenzene within the temperature interval from 1°K to room temperature gave the following results. Above 95°K, $S_2 = 13$ G$^2$, and in the temperature interval 1–38°K, $S_2 = 19$ G$^2$. Calculation gives 32.7 G$^2$ for a rigid lattice. Thus, methyl group rotation does not cease even at 1°K.

The work of Powles and Gutowsky was developed further by Waugh and Fedin [14], who showed that the barriers hindering the rotation of molecules or groups in the system can be estimated on the basis of a very simple experiment. If $T$ is the temperature at which the line starts to narrow, the potential barrier

$$V_0 = kT(2.5)\log{(n/\Delta)}(kT/2I)^{1/2} \qquad (2.1)$$

where $I$ is the moment of inertia, $n$ is the barrier multiplicity factor, and $\Delta$ is the line breadth in a rigid lattice.

The fact that the potential barrier depends on $n$ and $\Delta$ logarithmically means that a still simpler approximate formula may often give quite satisfactory results. Calculations show that, for the molecules usually encountered, the right-hand side of Eq. (2.1) $\log{n/\Delta}(kT/2I)^{1/2}$ varies from substance to substance by not more than 10%. If not very high accuracy is required, one may write

$$V_0 \ \text{(kcal/mole)} \approx 37T_c \ \text{(°K)}$$

This linear dependence was first established for three concrete molecules, but it is clear from what has been said above that it is of much more general significance.

It is interesting to note that formula (2.1) gives estimates of the potential barrier height in good agreement with those obtained by measuring the spin–lattice relaxation time. Since (2.1) is derived from a purely classical model, the agreement indicated is evidence that the tunnel effect plays a minor part in the mechanism of hindered rotation. An additional thorough analysis will have to be made before final conclusions can be drawn on this point.

Reorientation of a molecule as a whole was first detected by Andrew [15] in solid benzene. The second moment of the NMR signal in benzene decreases one and a half times when the temperature is raised from 77 to 110°K. This jump was identified with hindered rotations about the sixth-order axis.

McCall and Douglass [16] studied the temperature dependence of $S_2$ in an adamantane crystal and established that above $-130°C$, $S_2 = 0.9\,G^2$, and below that point $S_2 = 40\,G^2$. Theoretical estimates of the intramolecular contribution showed $31.5\,G^2$, and of the intermolecular contribution (in a rigid lattice) $7\,G^2$. The authors considered that the agreement between theory and experiment was good. The pronounced line narrowing on passing into the rotational crystalline state was explained quantitatively by assuming isotropic rotation of the adamantane molecules; the calculated value was $S_2 = 0.95\,G^2$. However, careful calculations of the intramolecular contribution to the second moment of adamantane showed that it could not be greater than $16\,G^2$. The intermolecular contribution was found to equal $6.63\,G^2$, and the total second moment, $22.4\,G^2$. After finding out about these results, McCall recalculated the intramolecular contribution and "reduced" it to $20.16\,G^2$. At the conclusion of the subsequent discussion it was decided to repeat the experiment. The low-temperature second moment in adamantane was actually found to equal $22\,G^2$. This is an instructive example of the magnitude of the errors which may arise if the second moment of NMR signals in molecular crystals are measured'without sufficient care.

3. DETERMINING PROTON COORDINATES IN ORGANIC CRYSTALS

A large number of studies of this type have been published. However, the value of some of them is reduced considerably, it seems to us, by the fact that dynamic effects are superimposed on geometrical effects. The first study in this field was made by Andrew and Eades [17] in 1953. They were the first to combine X-ray diffraction and NMR of broad lines for more exact location of hydrogen nuclei in benzene. The authors solved this problem in a non-trivial way by making use of a preliminary study of 1,3,5-deuterobenzene. A simplified scheme of their reasoning was approximatcly as follows: Let $S_2$ be the observed second moment in an ordinary benzene molecule; it consists of an intramolecular part $S_2'$ and an intermolecular part $S_2''$ due to the interaction of all the other nuclei in the crystal

$$S_2 = S_2' + S_2''$$

The change in $S_2'$ and $S_2''$ due to deuterium substitution was calculated theoretically. To a first approximation the deuterium nuclei may be considered nonmagnetic. Therefore the main term, corresponding to the interaction of the protons with their closest neighbors, is eliminated from the intramolecular part of the second moment. The term characterizing interaction with the diametrically opposite nucleus is also eliminated. Then only interaction with two protons is taken into account in $S_2'$. As to $S_2''$, this quantity decreases to half and the following system of equations appears:

$$S_2 = S_2' + S_2''$$

$$S_2^D = aS_2' + \tfrac{1}{2}S_2''$$

where $a \ll \tfrac{1}{2}$. Andrew and Eades had to do with a more complex system of equations, since they made allowance for the fact that the magnetic moment of deuterium differs from zero if taken very exactly. $S_2'$ and $S_2''$ were found by solving this system, and from these $r_{CH}$ was determined, i.e., the length of the C–H bond in benzene. The determination was accurate to 0.02 Å.

In 1955 Andrew and Hyndman [18] studied the structure of urea $(NH_2)_2CO$ by the NMR method. It had been known from X-ray diffraction data that the four heavy atoms were in the same plane. Figure 12 shows the arrangement of the heavy atoms; it possesses $C_{2v}$ symmetry. The interatomic distances and angles were found by X-ray diffraction methods. What was not known was whether all four hydrogen atoms were in the same plane as the heavy nuclei (a) or arranged symmetrically above and below this plane (b) or whether the molecular configuration was of lower symmetry than $C_{2v}$.

Urea crystals possess tetragonal symmetry and crystallize as needles of square cross section, elongated along the tetragonal axis. At first the crystal was mounted with its tetragonal axis (001) vertical. Then the (110) axis was set vertical. Since the crystal was elongated in the (001) direction it had to be cut up into four pieces and these pieces put together. Then the angles between the direction of the field and the (001) axis were varied. Several records were taken for each setting. The second moment was calculated for each record, and the average was found for each setting. The probable error was $0.3\,G^2$. The minimum value of the second moment was found to be $13\,G^2$, and the maximum $29\,G^2$.

The theoretical values of the second moment were calculated for planar and nonplanar models by the Van Vleck formula.

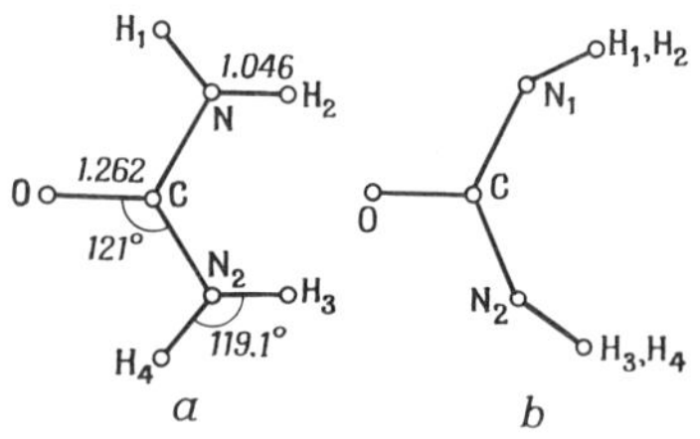

Fig. 12. The arrangement of the atoms in area.

To find the parameters more exactly, the theoretical value of the second moment was expanded in a Taylor series. The comparative influences on the second moment of the various structural parameters were estimated. The problem was reduced to three equations with three unknowns. The final result was: length of N–H bond 1.046 Å, $H^1N^1H^2$ angle 119.1°, OC $N^1$ angle 120.5°. Bond length determinations were accurate to ±0.01 Å, and angle determinations to ±2°. The proton–proton distance in both amino groups was 1.803 Å. The protons are arranged as in Fig. 12a.

Emsley and Smith [19] studied single thiourea crystals and repeated part of the results of Andrew and Hyndman for urea.

Dereppe, Touillaux, and Van Meersche determined the structure of oxalic acid dihydrate $C_2H_2O_4 \cdot 2H_2O$, paying most of their attention to precision proton placement.

The proton resonance spectra were taken at room temperature with a Varian spectrometer (56.6 MHz) under the following conditions: modulation frequency 0.99 Hz, modulation amplitude 0.99 G, time constant 3 sec, rate of passage, 3.78 G/min. The crystal was set in a Teflon goniometer. Figure 13 shows typical $S_2$ versus $\theta, \varphi$ dependencies.

When a polycrystalline sample was cooled, $S_2$ increased to a value 1.391 times greater than the second moment at 25°C. The authors assumed that the temperature dependence of the second moment would be the same with any orientation of a single crystal. According to this hypothesis, the lattice sum (1.7) for a rigid structure would be 1.391 times higher than the experimental values. Then a "coarse" structure was selected and its lattice sums were calculated. The linear divergence factor for this structure was found to equal

$$I = \sum_k |R_{k_0} - R_{k_e}| \Big/ \sum_k |R_{k_0}| = 0.22$$

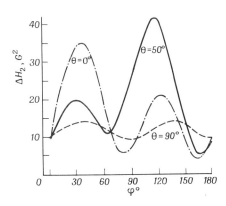

Fig. 13. The typical $S_2$ (or $\Delta H_2$) curves. Three curves $S_2(\varphi)$ for $\theta = 0°, 50°, 90°$ are shown.

After this the proton structure was refined. The authors had at their disposal nine lattice sums (differing from zero) for determining the values of nine coordinates. In principle, nine equations with nine unknowns of the form

$$\Delta R_k = \sum_i (\delta R_{k_o}/\delta u_i)\Delta x_i = \sum_i a_{ki}\Delta x_i$$

$$k = 1,...,9, \quad i = 1,...,9, \quad \Delta R_k = R_{k_0} - R_k$$

can be written. The structure-refining corrections to the trial coordinates can be found by solving this system of equations.

Figure 14 represents the change in the divergence factor during successive refining of the coordinates by the least-squares method. It is evident that with $I_0 = 0.22$ already $I_3 = 0.12$ and $I_{13} = 0.07$, and the last six steps practically did not change the structure.

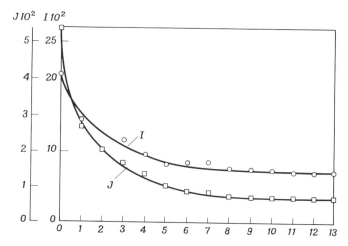

**Fig. 14.** The change in the linear (I) and quadratic (J) divergence factor during refining.

This study can hardly be called very convincing because the "trial" and final structures differ very little. Besides, the final structure does not coincide very well with that obtained by neutron-diffraction methods. In any case, the difference between the neutron-diffraction structure and the final NMR structure is of the same order as that between the final and trial structures.

Gorskaya and Fedin [20] studied the dependence of the NMR signal in naphthalene on the orientation of the single crystal in a magnetic field. The sample was in the shape of a cylinder cut out of a large single crystal. Figure 15 compares the experimental $\varphi$-dependence of the second moment with three calculated dependencies. Curve 1 was obtained under the assumption that the C–H bonds in the precisely established structure of naphthalene are 1.08 Å

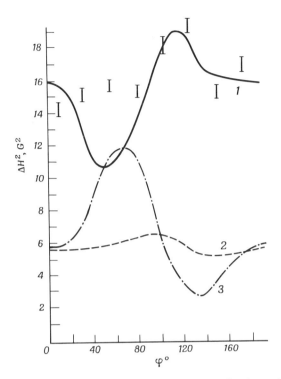

**Fig. 15.** The theoretical $\varphi$-dependence of the second moment for the naphthalene crystal. Experimental data are shown also.

long. Curve 2 corresponds to a C–H distance of 1.03 Å, and curve 3 to 1.13 Å. 1.08 Å is the average of a large number of data on C–H bond length accumulated in organic crystal chemistry. Figure 15 shows distinctly that NMR data are, in the main, in agreement with this value, but the noticeable difference between curve 1 and the experimental $\varphi$-dependence indicates that NMR may enable determination of minor deviations of the molecular geometry from the ideal model.

An NMR spectromer designed for determining the placement of protons in the crystal structure should meet a number of specific requirements. First, the spectrometer must be furnished with goniometric devices for exact orientation of the crystal in the magnetic field. Secondly, special measures should be taken to ensure sufficient sensitivity when observing NMR in single crystals of small volume. The sensitivity requirements are expecially high when molecular crystals are to be studied; owing to large relaxation times, high radio-frequency field intensities $H_1$ cannot be used. Thirdly, the magnetic field and reproducibility of its change must be highly stable. This requirement is due to the fact that in some cases sufficient sensitivity is ensured only if the method of signal accumulation is used. Fourthly, the instrument should give a zero line of extreme stability, because stability of the zero guarantees the accuracy of measurement of the second moments.

The first to achieve orientation of a crystal in the magnetic field of an NMR spectrometer was Pake [21]. The goniometer device he used was a very simple one. A radio-frequency coil was set vertically in the magnet gap; the upper face of the single crystal placed in the coil was glued to a cylindrical Teflon rod. A pointer at the upper end of the rod swinging against a large-diameter circular scale made it possible to establish the crystal orientation to 0.5°. The short-coming of this system is that the crystal must be cylindrical in shape because arbitrary shape greatly decreases the filling factor.

The shape of the radio-frequency coil can be adapted to that of the available crystal by winding it tightly on the crystal. The coil then becomes rectangular in cross section. A brass box with a radio-frequency bridge is placed on the upper end of a vertical brass tube, the lower end of which is set at the center of the magnet gap; the coil is fixed rigidly to this part of the tube.

There is no need to prove the great possibilities of the NMR method in investigating lattice dynamics. On the other hand, it is still difficult to say how good this method can be in studying the geometry of organic molecules.

It seems important to us to investigate further the usefulness of the NMR method for studying proton coordinates and in cases where no reorientation of molecules, or parts of molecules, and no proton exchange occur.

4. THEORY OF NUCLEAR QUADRUPOLE RESONANCE

Nuclear quadrupole resonance was first observed by Dehmelt and Krüger [22] in 1949. Nuclear quadrupole resonance is a division of spectroscopy that studies the energy levels in solids, when the distances between these levels correspond to emission or absorption of electromagnetic vibrations with wavelengths in the decimeter and meter ranges (frequencies from 1 to 1,000 MHz). What is the nature of these energy levels?

The nuclei of many elements have shapes other than spherical. The measure of deviation of the shape of a nucleus from spherical is its quadrupole moment

$$Q = (1/e) \int (3z_m{}^2 - r^2) \, dx \, dy \, dz \qquad (4.1)$$

where $e$ is the charge of an electron, $\rho$ is the density of charge distribution in the nucleus, $r^2 = x^2 + y^2 + z^2$, and $z_m$ is an axis coinciding with the direction of nuclear spin $I$.

It is evident from (4.1) that $Q$ is positive if the nucleus is drawn out

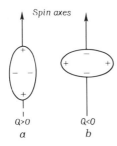

*Fig. 16.* Atomic quadrupole nuclei.

along the spin vector (a), and negative if it is compressed in this direction (b) (Fig. 16).

The quadrupole moment is related to the nuclear spin $I$: If $I = 0$ or $\frac{1}{2}$, $Q = 0$; $Q$ can differ from zero only if $I \geqslant 1$.

Consider an atom with a nucleus having a quadrupole moment different from zero, contained in a molecule located in a crystal. In Fig. 17, A is an atom containing a nonspherical nucleus and $A_1$ is an atom linked to A with a covalent bond. The electric field is nonuniform along the direction of the chemical bond, and the measure of this nonuniformity is the intensity gradient $q_{zz} = \partial^2 V/\partial z^2$. Here $V$ is the electrostatic potential at the center of the nucleus, set up by all the surrounding charges. Strictly speaking, $q$ is a tensor value with nine components, and the mathematical description of the interaction energy of such a nonuniform electric field with a nucleus in which the symmetry of charge distribution differs from spherical is very complicated. It is usually assumed that $q$ is axially symmetrical relative to the $z$ axis, i.e., $\partial^2 V/\partial x^2 = \partial^2 V/\partial y^2$ and coordinate axes are selected in which the tensor is diagonal in form. Then it is shown by quantum-mechanical methods that the energy of a nucleus with the quadrupole moment $Q$ situated in an electric field with a

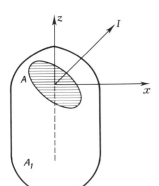

*Fig. 17.* Quadrupole nucleus in a nonuniform electric field set up by the electron cloud of a molecule; $z$ axis along direction of bond.

gradient $q$ axially symmetrical relative to the direction of the chemical bond
has the form

$$E_m = eq_{zz}Q[3m^2 - I(I+1)]/4I(2I-1) \qquad (4.2)$$

In Eq. (4.2) $m$ is the magnetic quantum number at the given nuclear spin $I$;
it varies a unit at a time from $I$ to $-I$, i.e., has $2I+1$ values. It is evident from
(4.2) that the quadrupole energy levels are twice degenerate with respect to $m$,
since $E_m$ depends only on the absolute value of $m$ and not on its sign. For
instance, there are only two different energy levels for N and Cl nuclei. It is
clear that passages between them can result in the appearance of only one
absorption or emission line. To stimulate transitions between the levels (4.2),
an electromagnetic field with the quanta

$$hv = E_{(m)} - E_{(m-1)}$$

has to be applied to the crystal.

The formula for the frequencies at which nuclear quadrupole resonance
may occur has the form

$$v_{res} = \frac{3eQq_{zz}}{4hI(2I-1)}(2|m_I| - 1)$$

where $m_I$ is the larger in absolute value of the two magnetic quantum numbers
involved in the transition. In the radiospectroscopic literature the term
"quadrupole interaction constant" is usually used to denote the values of
$eQq/h$ expressed in millions of hertz, the Planck constant being omitted from
the formula. Then we get $v_{res} = \frac{3}{4}eQq$ for $I = 1$, $v_{res} = \frac{1}{2}eQq$ for $I = \frac{3}{2}$, etc.

Very often the electric field gradient $q$ has no axial symmetry. Then the
asymmetry parameter $\eta = (q_{xx} - q_{yy})/q_{zz}$ is taken into consideration.

The quadrupole interaction constant $eQq_{zz}$ and the asymmetry parameter $\eta$
are the information nuclear quadrupole resonance experiments give about a
chemical compound.

The characteristic feature of quadrupole resonance is that the chemical
effects on $eQq_{zz}$ and $\eta$ are not small. Thus, the quadrupole frequencies of
$^{35}$Cl in tri- and dichloromethane are 38.2809 and 35.9912 MHz, respectively;
i.e., differ in the second significant digit, whereas the accuracy of measurement
of the frequency may be as good as six or seven correct significant digits. The
values of $\eta$ obtained from experiment vary from 0.05 to 0.8.

Nonequivalent placement of the molecules in a crystal's unit cell gives a well-
resolved line structure. For example, $CHCl_3$ shows a two-component quadrupole resonance spectrum with the frequencies 38.3081 and 38.2537 MHz,
which indicates that there are two crystallographically different species of
molecules in the crystal. The magnitude of these crystalline effects exceeds the
line breadth by one or two orders.

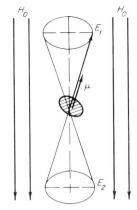

*Fig. 18.* Precession of a nucleus in the presence of a constant magnetic field $H_0$.

If an external constant magnetic field $H_0$ is applied to the crystal, the state (Fig. 18) characterized by the precession angle $\theta$ will have a higher energy than that characterized by the angle $180° - \theta$. Since the energy difference $E_2 - E_1 = \Delta E \ll kT$ between the states where $\mu_I$ has the same direction as the constant magnetic field and opposite to the field, these new levels will be populated as a result of thermal motion, and degeneration with respect to $m$ will be eliminated. Thus, simple qualitative reasoning shows that if a constant magnetic field is applied to the crystal the quadrupole resonance lines from each nucleus will be split (Fig. 19).

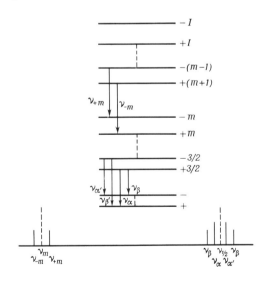

*Fig. 19.* Zeeman splitting of quadrupole energy levels ($I = \frac{3}{2}$).

The overall effect of this splitting in a polycrystalline sample will differ from that in a single crystal. In a single crystal the signals from all the nuclei add up and give a split multicomponent quadrupole resonance spectrum. In a polycrystal the application of the magnetic field causes substantial broadening of the lines owing to the unordered crystal grain orientation which accordingly lowers the signal height. This property of quadrupole resonance lines is utilized for identifying the signal: If the line broadens when a weak magnetic field is applied, the phenomenon observed is quadrupole resonance; if it does not, the signal is of a different origin.

In the simplest case the magnitude of level splitting on application of a field $H_0$ is equal to

$$\Delta E = \pm m \mu_I H_0 \cos \theta / I$$

where $\theta$ is the angle between the direction of $H_0$ and the symmetry axis of $q$, and $\mu_I/I$ is the gyromagnetic ratio, equal to 417.214 Hz/G for the $^{35}$Cl nucleus. It follows from this formula that no splitting will occur if the external magnetic field $H_0$ is at right angles to the symmetry axis of $q$. A more complex case is that where $H_0$ removes degeneration from the level $E_{1/2}$. The states with $m = +\frac{1}{2}$ and with $m = -\frac{1}{2}$, being neighbors $\Delta m = 1$ apart, interact with respect to the field $H_0$. Calculations show that in this case the perturbing field $H_0$ gives rise to mixed levels described by the formula

$$E_\pm = \frac{eQq_{zz}}{4I(2I-1)} \left[\tfrac{3}{4} - I(I+1)\right] \pm \left[\cos^2 \theta + (I+\tfrac{1}{2})^2 \sin^2 \theta\right] \frac{\mu_I}{2I} H_0 \quad (4.3)$$

It follows from (4.3) that zero splitting is impossible at any $\theta$ angle. Considering Fig. 19 we see that at a nuclear spin of $I = \frac{3}{2}$ Zeeman splitting of quadrupole resonance results in the appearance of four observable lines. With the fields $H_0$ usually used no $E_+ \rightleftarrows E_-$ transition is observed in the quadrupole resonance spectrum: The frequency corresponding to it is two orders lower than $\nu_{res}$. Making use of Fig. 19 and of Eqs. (4.2) and (4.3), we find that the internal lines of the Zeeman quadruplet are separated by the frequency interval

$$\nu_{\alpha\alpha'} = 3(\mu_I/I) H_0 \cos \theta + (\mu_I/I) H_0 \left[\cos^2 \theta + (I+\tfrac{1}{2})^2 \cos^2 \theta\right]^{1/2}$$

It follows from this formula that the internal, most intense lines of the Zeeman multiplet coincide at a certain orientation of the magnetic field, falling on the frequency $\nu_{res}$. The angle $\theta$ at which this occurs for the spin $I = \frac{3}{2}$ can be determined from the condition $\tan \theta = 2^{1/2}$, i.e., $\theta = 54°45'$. This is utilized for structural studies of single crystals: By varying the orientation of the crystal in a constant magnetic field, one finds the directions of $H_0$ which do not split the signal; these directions form a cone, the axis of which indicates the orientation of the chemical bond (see below).

Nuclear quadrupole resonance lines are often greatly broadened, which decreases their amplitude accordingly, and this sharply raises equipment sensitivity requirements. The nuclei surrounding the one under study may have magnetic moments. If so, the nucleus under investigation is in an environment of local magnetic fields. Depending on the magnitude of these local magnetic fields and their orderliness in the crystal, such an internal Zeeman effect may either result in the appearance of fine structure in the quadrupole resonance line or, more often, in broadening it.

Important static factors affecting line breadth are defects and stresses in the crystal. Crystal lattice defects cause unordered spreading of the electric field gradient values $q$, thus drawing the corresponding nuclei out of the main resonance region; this broadens the line. A similar effect is caused by stresses arising in the crystal upon rapid crystallization or cooling, local heating, etc. The quantitative theory of line broadening due to stresses and defects in the crystal can be built only when the law of intermolecular interaction and the effect of the interaction on the electric field gradient in the crystal become known. This interesting problem has still to be solved.

The shape of quadrupole resonance lines may be affected by vibration of the molecules in the crystal. Thermal vibrations limit the lifetime of the quadrupole energy levels and thus determine the "natural" line breadth. It is clear from the above that this contribution is much smaller than the effect of static factors. However, it is qualitatively quite clear that, besides their effect on the lifetime of energy levels, thermal vibrations may also exert an influence similar to defects and stresses, causing additional spread of the $q$ values from nucleus to nucleus, i.e., broadening the line.

The lifetime of the levels is characterized by the spin–lattice relaxation time $T_1$. Interaction of the radio-frequency field quanta with the system of quadrupole energy levels evens out the populations of these levels. If the radio-frequency field is removed from the sample, thermal vibrations will cause the difference of populations $n$ created by the action of that field to tend to the initial difference of populations $n_0$ (depending on the experiment temperature) according to the law

$$\Delta n = \Delta n_0 (1 - e^{t/T_1})$$

where $t$ is the time elapsed from the moment of removal of the radio-frequency field, and $T_1$ is the spin–lattice relaxation time.

If $T_1$ is large, the lifetime in the excited state will also be large. Under such conditions prolonged action of a radio-frequency field may prevent the thermal vibrations from restoring equilibrium in the numbers of transitions from the upper level to the lower one and vice versa; there comes a moment when the populations of the upper and the lower levels become equal. After this the amount of energy absorbed from the radio-frequency field per unit time will be exactly equal to the amount of energy emitted by the system of nuclear spins, i.e., the presence of the sample in the circuit stops being perceptible. This is known as resonance saturation, which has to be prevented in NMR experiments by using minimal energy densities in the circuit. In nuclear quadrupole resonance experiments saturation is much more difficult to reach. For $p$ dichlorobenzene, $T_1$ equals $2 \times 10^{-2}$ sec. Nevertheless, at a certain energy density in the circuit this phenomenon may result in substantial line broadening.

For more direct characterization of line shape, the concept of spin–spin relaxation time $T_2$ is introduced into NQR. Let the function $g(v)$ describe the resonance frequency distribution resulting from the action of all static and dynamic factors. This function is normalized by the usual condition $\int_{-\infty}^{\infty} g(v) \, dv = 1$. Then, by definition, the spin–spin relaxation time is

$$T_2 = 2g(v_{\text{res}}) \qquad (4.4)$$

Therefore, observation of line shape makes it possible to judge relaxation processes. The values of $T_2$ for $p$-dichlorobenzene vary from $5 \times 10^{-4}$ to $7 \times 10^{-5}$ sec when impurities are added to the crystal, whereas $T_1$ does not change.


5. Use of NQR in Studying the Structure of Molecular Crystals

*a. Determining Molecular Orientation in the Unit Cell*

A consistent investigation of the Zeeman splitting of the NQR spectrum was carried out by Dean and co-workers [23] for 1,2,4,5-tetrachlorobenzene. In this work a specially designed semiautomatic apparatus was used, by means of which the crystal structure of the object could be studied in full. The authors concluded that quadrupole resonance may be of substantial help in X-ray diffraction studies of certain objects. The accuracy of bond direction determination was slightly higher than in the X-ray method, the error not exceeding $\pm 1°$.

Splitting of a line into its components depends on the orientation of the crystal in the magnetic field. Splitting is the same for all translationally identical atoms but differs, generally speaking, for atoms bound by symmetry elements. However, there is a special direction which remains unchanged or is inverted by symmetrical operations peculiar to the crystal. In this case the splitting pattern is the same for both crystallographically nonequivalent atoms. Note that any direction satisfies this requirement in a centrosymmetric crystal.

Zeeman splitting is utilized to find the directions of the electric field gradient, by searching, in the case of each line of the zero field (i.e., for each nucleus), for the direction of the magnetic field relative to the crystal axes, where splitting does not occur. It can be shown that, for a spin of $\frac{3}{2}$, zero splitting occurs with the field directions lying on the surface of a double cone with the direction of the field gradient as its axis. The directions of the zero fields form an angle of 54°44′ with the cone axis. The structural investigation carried out by this method consists in searching for zero splitting cones. A very important circumstance is the possibility of measuring the electric field asymmetry at the atomic nucleus. The cone is circular only in the ideal case of absence of asymmetry.

### b. Studying Lattice Dynamics

Bayer [24] examined the effect of torsional vibrations of molecules on $q_{zz}$, i.e., on the frequency of quadrupole resonance. Infrared and Raman spectra showed that the lowest frequencies of these vibrations are not less than two orders higher than the possible frequencies of quadrupole resonance. Bayer showed that in reality the nuclei are in an electric field, the parameter $q$ of which differs from what it would be if the molecules were completely immobile, and that the effective value of $q$ depends on the temperature. Calculation gave the following temperature dependence of the quadrupole resonance frequency for $I = \frac{3}{2}$:

$$\frac{1}{v_{res}}\frac{dv}{dT} = -\frac{3h^2}{8\pi^2 k T^2}\left[\frac{e^{hv_x/kT}}{A_x(e^{hv_x/kT}-1)^2} + \frac{e^{hv_y/kT}}{A_y(e^{hv_y/kT}-1)^2}\right] \quad (5.1)$$

where $v_{res}$ is the quadrupole resonance frequency at $\eta = 0$, $v$ is the frequency under study, $T$ is the temperature, $v_x$ and $v_y$ are the frequencies of torsional vibrations relative to the $x$ and $y$ axes of the tensor $q$, and $A_x$ and $A_y$ are the corresponding moments of inertia. The torsional vibrations relative to the $z$ axis do not contribute to $q$, but change only $\eta$. Figure 20 is the frequency versus temperature curve calculated from (5.1). It is evident that the resonance frequency grows substantially with declining temperature. The temperature coefficient of the quadrupole resonance frequency equals 18.7 kHz/deg for $p$-dibromobenzene, and 2.7 kHz/deg for $p$-dichlorobenzene. Owing to this strong dependence on temperature the specimen has to be thermostatically controlled. Even a slight temperature drop (2–3°) over the specimen may broaden the line manyfold.

The temperature variation of the quadrupole resonance frequency predicted by Bayer holds well for most of the compounds studied. Detailed investigations of the temperature dependence of NQR frequencies and verification of the applicability of Bayer's theory and its more complicated versions were carried out by Babushkina [25] in the author's laboratory.

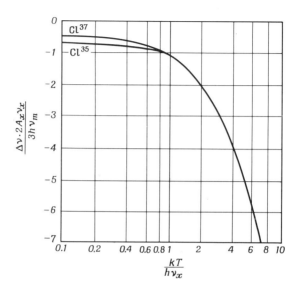

**Fig. 20.** Temperature dependence of quadrupole resonance frequencies of $^{35}Cl$ and $^{37}Cl$ according to Bayer theory.

## c. Solid Solutions

Of some significance among the applications of quadrupole resonance are investigations of solid solutions. Monfils [26] was the first to suggest characterizing the influence of an impurity molecule on the NQR signal in a solid solution by the number $N$ of adjacent molecules of the matrix, brought out of resonance by the impurity molecule. At low impurity concentrations it may be supposed that $N$ is a property of the impurity molecule in the given lattice and is independent of concentration. Then it can easily be shown that the signal intensity decreases with increasing impurity concentration according to the law

$$A = A_0 e^{-Nc}$$

where $A_0$ is the amplitude of the signal in the pure sample. Three types of $A$ versus $c$ dependencies are observed in experiments (Fig. 21). Curve 1 is characteristic of unlimited solubility within the given interval of concentration $c$; curve 2 indicates limited solubility; and curve 3 corresponds to the case where the shape and size of the impurity molecule entirely prevents the formation of a solid solution.

Fedin and the present author [27] showed that the matrix molecules in solid solutions of a number of compounds in $p$-dichlorobenzene displayed no perceptible shifts of NQR frequency. They also attempted to predict the value of $N$ for several systems on the basis of organic crystal chemical data, proceeding from the assumption that the lowering of $A$ in a solid solution is due

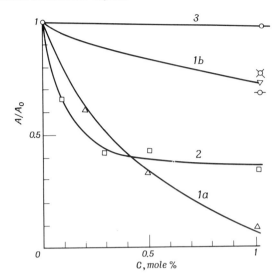

**Fig. 21.** Dependence of ampltiude of NQR signal on concentration of impurity in *p*-dichlorobenzene. Curves: (1) unlimited solubility; (a) impurity molecule larger than matrix molecule; (b) impurity molecule smaller than matrix molecule; (2) limited solubility; (3) no solubility.

to static lattice distortions. This assumption was based on spin–lattice relaxation times $T_1$ measured in a solid solution, which indicated that $T_1$ was independent of $c$. At the same time, according to the data of Hirai, $T_1$ depends very greatly on temperature (Fig. 22). It may therefore be considered that at low $c$ the lattice dynamics remain unchanged to a first approximation, and conclusions may be drawn as to the static effect of the impurity on the quadrupole resonance signal.

*d. Polymorphism*

The main regularities of this phenomenon should be displayed in NQR spectra. The spectral manifestations may depend on whether an enantiotropic or monotropic transformation is occurring.

Investigation of phase transitions in a number of molecular crystals, namely: *o*-dichlorobenzene, chloranil, trichloroacetamide, dihalomethanes, dichloro-dinitromethane, trichlorogermane, iodobenzene, etc., discloses an increase in multiplicity of the NQR spectrum in the low-temperature phase. This establishes unambiguously a transition of the first kind involving a jumpwise change in symmetry (NQR registers changes in the number of independent molecules in the unit cell).

When the multiplicity of the NQR spectrum does not change, phase

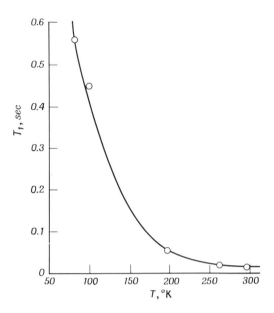

*Fig. 22.* Temperature dependence of $T_1$ in $p$-dichlorobenzene.

transitions in NQR may still be accompanied by a frequency jump, as well as by a change in slope of the frequency versus temperature curve. Examples of such behavior are the $\alpha \rightarrow \beta$ transition in $p$-dichlorobenzene, where the NQR frequency of the $^{35}Cl$ of the $\beta$-phase is 24 kHz lower than in the $\alpha$-phase at a phase transition temperature of 32°C, and the $\alpha \rightarrow \beta$ transition in tetrabromobenzene. The phase transition at 32°C in tetrabromobenzene is characterized not only by a 0.15% jump in the NQR frequency, but also by a change in slope of the frequency versus temperature dependence before and after the phase transition, $(1/\nu)(d\nu/dT)$ equaling $3.9 \times 10^{-5}$ $deg^{-1}$ above the phase transition and $5.9 \times 10^{-5}$ $deg^{-1}$ below it.

When slow phase transitions are observed by the NQR method, the peculiarities of the mutual arrangement and movement of the same molecules in different crystal lattices at the same temperature can be studied.

Reviewing the collection of all available experimental data, it may be concluded that the difference between the average NQR frequencies of two modifications at the same temperature does not, as a rule, exceed $\sim 1.5\%$ of the average signal frequency. These frequency shifts do not differ in value from the usual differences for chemically identical, but crystallographically nonequivalent atoms.

Investigation of the effect of crystal fields on NQR frequency encounters major difficulties. There are practically no special studies in this field.

REFERENCES

1. Y. Iitaka, *Acta Crystallogr.* **14**, 1 (1961).
2. A. I. Kitaigorodsky, "The Theory of Crystal Structure Analysis." Consultants Bureau, New York, 1961.
3. See for instance I. L. Kare, K. Britts, and S. Brenner, *Acta Cystallogr.* **17**, 506 (1964).
4. A. I. Kitaigorodsky, *Izv. Akad. Nauk SSSR Ser. Khim.* **No. 6**, 587 (1946).
4a. A. I. Kitaigorodsky, "Organic Chemical Crystallography." Consultants Bureau, New York, 1961.
5. I. M. Gelfand, E. B. Vull, S. L. Ginsburg, and Yu. G. Fyodorov, "Ravine Method in X-ray Diffraction Analysis Problems" (in Russian). Nauka Publ. House Moscow, 1966.
6. H. Lipson, C. A. Taylor, "Fourier Transforms and X-ray Diffraction." Bell, London, 1958.
7. J. L. Amoros and M. Amoros, "Molecular Crystals—Their Transforms and Diffuse Scattering," Wiley, New York, 1967.
8. P. A. Egelstaff, ed., "Thermal Neutron Scattering," Academic Press, New York, 1965.
9. J. F. Van Vleck, *Phys. Rev.* **74**, 1168 (1948).
10. G. McCall and R. Hamming, *Acta Crystallogr.* **12**, 81 (1959).
11. J. M. Dereppe, R. Touillaux, and M. Van Meersche, *J. Chim. Phys. Physiochim. Biol.* **63**, 9, 1265 (1966).
12. H. S. Gutowsky and G. E. Pake, *J. Chem. Phys.* **18**, 162 (1950).
13. G. J. Powles and H. S. Gutowsky, *J. Chem. Phys.* **23**, 1692 (1955).
14. J. Waugh and E. I. Fedin, *Fiz. Tverd. Tela* **4**, 2233 (1962).
15. E. R. Andrew, *J. Chem. Phys.* **18**, 607 (1950).
16. D. W. McCall and D. C. Douglass, *J. Chem. Phys.* **33**, 777 (1960).
17. E. R. Andrew and R. Eades, *Proc. Roy. Soc. Ser. A* **218**, 537 (1953).
18. E. R. Andrew and D. Hyndman, *Discuss. Faraday Soc.* **35**, 19, 195 (1955).
19. J. W. Emsley and J. A. Smith, *Proc. Chem. Soc. London* **1**, 53 (1958).
20. N. V. Gorskaya and E. I. Fedin, *Zh. Strukt. Khim.* **9**, 560 (1968).
21. G. S. Pake, *J. Chem. Phys.* **16**, 327 (1948).
22. H. Dehmelt and H. Krüger, *Naturwissenschaften* **37**, 11 (1950).
23. C. Dean, M. Pollak, B. Craven, and G. Jeffrey, *Acta Crystallogr.* **11**, 710 (1958).
24. H. Bayer, *Z. Phys.* **130**, 227 (1951).
25. T. A. Babushkina, Thesis, Inst. of Chem. Phys. of Acad. Sci. of USSR, Moscow, 1968.
26. A. Monfils and D. Grosjean, *Physica (Utrecht)* **12**, 30 (1956).
27. E. I. Fedin and A. I. Kitaigorodsky, *Kristallografiya* **6**, 406 (1961).

# Chapter V
# Thermodynamic Experiments

1. Measuring Thermal Expansion

The most precise way of finding the expansion tensor or the coefficient of volume expansion is the X-ray diffraction method of measuring the unit-cell parameters of a crystal.

For exact measurements, one selects several interplanar distance. $d_{hkl}$, to which correspond large Bragg angles $\theta_{hkl}$, and measures their values as a function of temperature. There is no difficulty in obtaining X-ray reflections of the unknown and of a standard (say, common salt) on the same X-ray diagram.

In the monoclinic system the unit cell is described by four parameters. Therefore, it is generally sufficient to follow the behavior of four reflections. An alternative method is to measure a large number of reflections and treat the results by the least squares method.

Thermal expansion is usually described by means of the principal coefficients of expansion $\alpha_{11}$, $\alpha_{22}$, and $\alpha_{33}$. These values are numerically equal to the change in unit length of the crystal in the direction of the principal axes of the thermal expansion ellipsoid, caused by a 1° rise in the temperature of the crystal.

The thermal ellipsoid appears as a result of deformation of a sphere of unit radius. Hence, each radius vector of the ellipsoid has the value $1 + \Delta$, where

294

$\Delta$ is the coefficient of linear expansion in the direction of the radius vector in question, equal to

$$\Delta = \alpha_{11} x^2 + \alpha_{22} y^2 + \alpha_{33} z^2$$

In this equation the thermal expansion is referred to the ellipsoid's main axes.
The thermal expansion of a crystal may be expressed relative to its lattice axes as

$$\Delta = A_1 \alpha^2 + A_2 \beta^2 + A_3 \gamma^2 + A_4 \beta\gamma + A_5 \gamma\alpha + A_6 \alpha\beta$$

where $A_1$, $A_2$, $A_3$, $A_4$, $A_5$, and $A_6$ are constants characterizing the expansion, and $\alpha$, $\beta$, and $\gamma$ are the direction cosines of the given direction relative to a rectangular coordinate system fixed to the crystal.

For a monoclinic crystal, owing to symmetry requirements, the directions $\alpha,\beta,\gamma$; $\alpha,\bar\beta,\gamma$; $\bar\alpha,\beta,\bar\gamma$, and $\bar\alpha,\bar\beta,\bar\gamma$ are equivalent. Hence $A_4 = A_6 = 0$, and the formula becomes

$$\Delta = A_1 \alpha^2 + A_2 \beta^2 + A_3 \gamma^2 + A_5 \gamma\alpha$$

To find the constants $A_1$, $A_2$, $A_3$, and $A_5$ we consider the directions lying in the crystal planes $ac$ and $bc'$, where $c'$ is the normal to the plane $ab$.

For plane $ac$ we have $(\beta = 0)$

$$\Delta = A_1 \alpha^2 + A_3 \gamma^2 + A_5 \gamma\alpha$$

Instead of presenting $\Delta$ as a function of the two related values $\alpha$ and $\gamma$, we introduce the angle $\psi$ between the direction in question and that of the $a$ axis, then

$$\Delta = A_1 \cos^2\psi + A_3 \sin^2\psi + A_5 \cos\psi \sin\psi$$

Trigonometric transformation gives

$$\Delta = \tfrac12\{(A_1+A_3) + (A_1-A_3)\cos 2\psi + A_5 \sin 2\psi\}$$

or

$$\Delta = m + n\cos 2\psi + p\sin 2\psi$$

where $m = \tfrac12(A_1+A_3)$; $n = \tfrac12(A_1-A_3)$, and $p = \tfrac12 A_5$.
Determining from experiment a series of $\Delta_i$ values for different $\psi_i$, one finds the optimum values of $m$, $m$, and $p$ by the least-squares method. For this purpose, one seeks the minimum of the expression

$$V_{\mathrm{I}} = \sum_i (\Delta_i - m - n\cos 2\psi_i - p\sin 2\psi_i)^2$$

Equating $\partial V_{\mathrm{I}}/\partial \Delta_i$, $\partial V_{\mathrm{I}}/\partial m$, etc., to zero, one obtains

$$m\sum i + n\sum \cos 2\psi_i + p\sum \sin 2\psi_i = \sum \Delta_i$$
$$m\sum \cos 2\psi_i + n\sum \cos^2 2\psi_i + p\sum \sin 2\psi_i \cos 2\psi_i = \sum \Delta_i \cos 2\psi_i$$
$$m\sum \sin 2\psi_i + n\sum \sin 2\psi_i \cos 2\psi_i + p\sum \sin^2 2\psi_i = \sum \Delta_i \sin 2\psi_i$$

Solving three equations with three unknowns, one gets the best values of $A_1$, $A_3$, and $A_5$.

For plane $bc'$, we have ($\alpha = 0$)

$$\Delta_2 = A_2 \beta^2 + A_3 \gamma^2$$

Denoting by $\varphi_i$ the angle between one of the directions of expansion and the $b$ axis, one obtains

$$\Delta_i = A_2 \cos^2 \varphi_i + A_3 \sin^2 \varphi_i$$

The optimum values of $A_2$ and $A_3$ are found as above by minimizing the expression

$$V_{II} = \sum (\Delta_i - A_2 \cos^2 \varphi_i - A_3 \sin^2 \varphi_i)^2$$

Equating $\partial V_{II}/\partial A_2$ and $\partial V_{II}/\partial A_3$ to zero, one obtains

$$A_2 \sum \cos^4 \varphi_i + A_3 \sum \cos^2 \varphi_i \sin^2 \varphi_i = \sum \Delta_i \cos^2 \varphi_i$$

$$A_2 \sum \cos^2 \varphi_i \sin^2 \varphi_i + A_3 \sum \sin^4 \varphi_i = \sum \Delta_i \sin^2 \varphi_i$$

In this way the best values of the expansion constants $A_1$, $A_2$, $A_3$, and $A_5$ can be found. The constant $A_3$ is found independently from two measurements made in the $ac$ and $bc$ planes.

The principal coefficients of thermal expansion $\alpha_{11}$, $\alpha_{22}$, and $\alpha_{33}$ and the angle of slope $\psi_0$ of the ellipsoid axis of length $\alpha_{11}$, relative to the crystal's $a$ axis is found from the equations

$$A_1 + A_3 = \alpha_{11} + \alpha_{33}$$

$$A_1 - A_3 = (\alpha_{11} - \alpha_{33}) \cos 2\psi_0$$

$$A_5 = (\alpha_{11} - \alpha_{33}) \sin 2\psi_0$$

$$A_2 = \alpha_{22}$$

In monoclinic crystals the ellipsoid axis $\alpha_{22}$ coincides with the $b$ axis.

The expansion tensors of organic crystals have been determined over a wide range of temperatures only in a small number of cases.

Typical data for several cases are illustrated by parameter versus temperature graphs, by temperature curves of expansion tensor elements, and by expansion patterns and their cross sections.

Direct comparison of the expansion tensor with crystal structure cannot, of course, be expected to be very successful. To explain expansion one has to resort to calculations. These show, quite naturally, that crystal structure is determined by the joint behavior of the expansion and elasticity tensors and the heat capacity.

Nevertheless, some simple relationships strike the eye when examining the above experimental results. It is clear, for instance, that from purely geometrical considerations expansion will always be minimal in the directions for which the contribution of the intermolecular distances is small. It is also clear that expansion will be smaller in the direction of the long axes of the molecules. In crystals of aromatic compounds the direction normal to the planes of the carbon networks often becomes important; in such directions the contribution of intermolecular distances is large and expansion will be maximal.

Expansion patterns can be coordinated with structures relatively simply when there is only one molecule per unit cell. But if there are two, four, or more, and they are turned relative to one another in a complex manner, the relationship between the expansion tensor and the structure is no longer obvious.

The almost complete absence of thermal expansion in the directions in which molecules are linked by hydrogen bonds is an expected fact. This effect is especially prominent in pentaerythrite crystals.

Thus, in themselves, investigations of the expansion tensor are of limited value for understanding the interactions of molecules. However, without information on the expansion tensor, the search for and testing of models of solids is impossible.

Cell parameter versus temperature curves, expansion tensor element curves, and the appearance of expansion patterns are shown for several crystals in Figs. 1–20. These data were obtained by Ryzhenkov and are reproduced from his thesis [1].

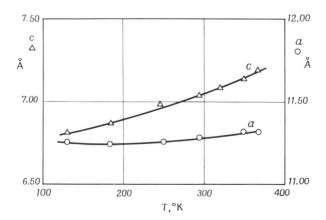

*Fig. 1.* Temperature dependence of unit cell parameters of tetraphenylsilicane.

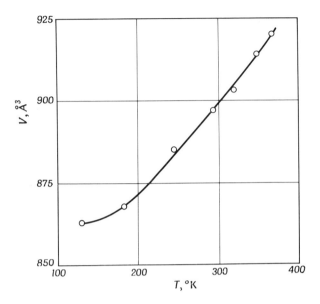

**Fig. 2.** Temperature dependence of unit cell volume of tetraphenylsilicane.

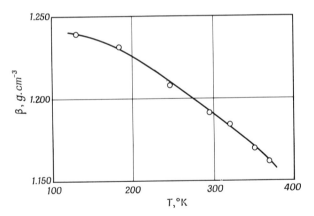

**Fig. 3.** Density versus temperature for tetraphenylsilicane.

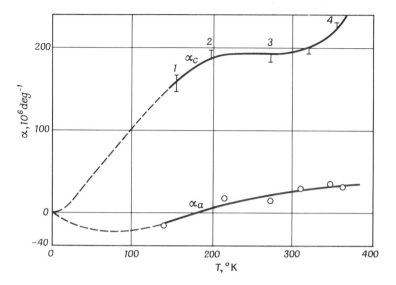

***Fig. 4.*** Temperature dependence of linear expansion coefficients of tetraphenylsilicane.

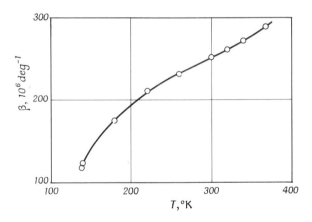

***Fig. 5.*** Temperature dependence of volume expansion coefficient of tetraphenylsilicane.

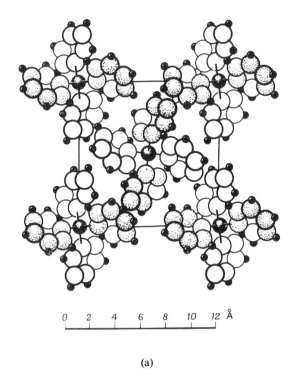

0    2    4    6    8    10    12  Å

(a)

$$\alpha_i = \alpha_2 c_2^2 + \alpha_3 c_3^2$$

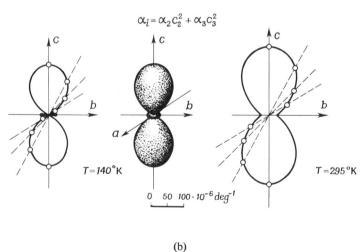

$T = 140°K$

$T = 295°K$

0    50    100 · $10^{-6}$ deg$^{-1}$

(b)

*Fig. 6.* (a) Structure of tetraphenylsilicane and (b) plane and three-dimensional expansion figures at 140 and 295°K (black area, region of negative expansion).

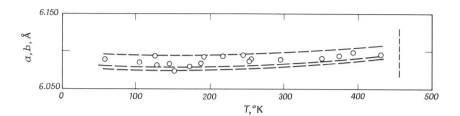

**Fig. 7.** Temperature dependence of parameters $a$ and $b$ of pentaerythrite.

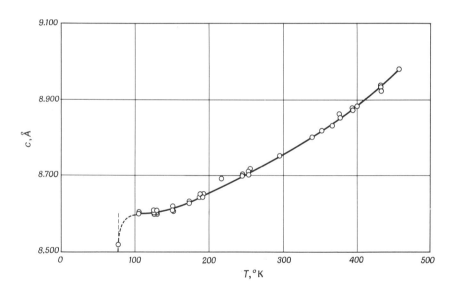

**Fig. 8.** Temperature dependence of parameter $c$ of pentaerythrite.

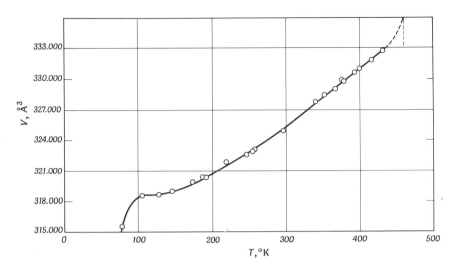

*Fig. 9.* Temperature dependence of unit cell volume of pentaerythrite.

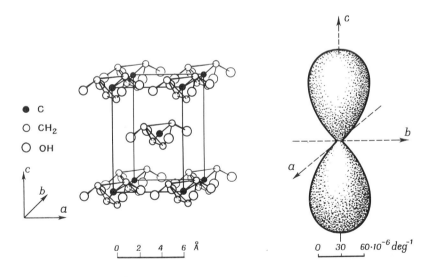

*Fig. 11.* Structure of pentaerythrite and three-dimensional expansion figure at 300°K.

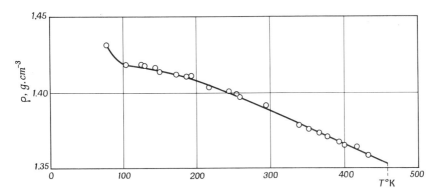

*Fig. 10.* Density versus temperature for pentaerythrite.

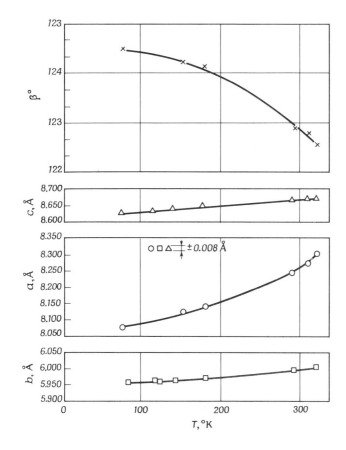

*Fig. 12.* Temperature dependence of unit cell parameters of naphthalene crystal.

304          5 Thermodynamic Experiments

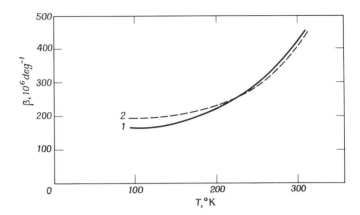

**Fig. 13.** Temperature dependence of volume expansion coefficient of naphthalene crystal (broken line, measurements with quartz dilatometer).

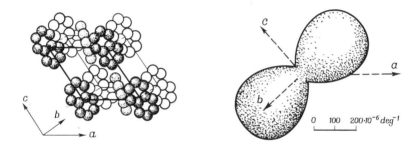

**Fig. 14.** Structure and three-dimensional expansion figure of naphthalene at 300°.K

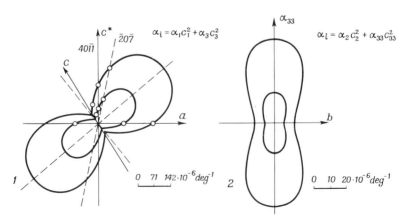

**Fig. 15.** Expansion figures of naphthalene: (1) in plane ($h0l$); (2) in plane perpendicular to main $\alpha_{\parallel}$-axis. External figures constructed for 308°K, internal figures for 90°K.

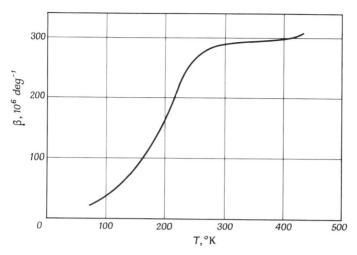

*Fig. 16.* Temperature dependence of volume expansion coefficient of anthracene.

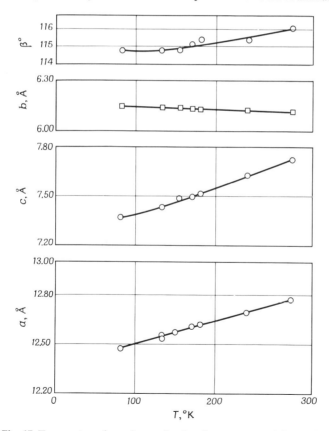

*Fig. 17.* Temperature dependence of unit cell parameters of dibenzyl.

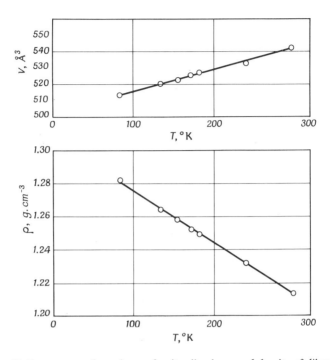

**Fig. 18.** Temperature dependence of unit cell volume and density of dibenzyl.

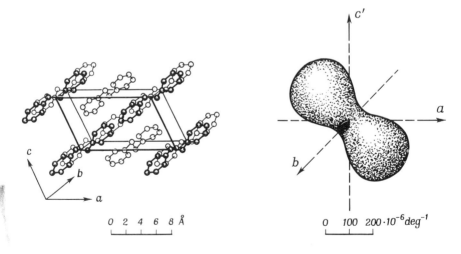

**Fig. 19.** Structure of dibenzyl and three-dimensional expansion figure at 295°K. Shaded area is negative expansion along *b*.

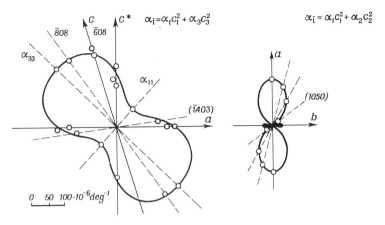

*Fig. 20.* Polar expansion patterns of dibenzyl in planes ($h0l$) and ($hk0$).

## 2. MEASURING THE ELASTICITY TENSOR OF A SINGLE CRYSTAL

The elasticity tensors of molecular crystals of low symmetry have been studied very little. These measurements can be made in practice only by determining the velocity of sound traveling in different crystallographic directions.

A fairly satisfactory method of measuring coefficients of adiabatic elasticity is the ultrasonic pulse method. The measurements are made so as to enable examination of the crystal without reference to its crystal faces.

The basic relationships are corollaries of the equation of motion of plane waves, which can be written (neglecting gravitational forces) as

$$\partial \sigma_{ik}/\partial x_k = \rho\, \partial^2 U_i/\partial t^2 \qquad (2.1)$$

here $\rho$ is the density of the medium, and $U_i$ is the $i$th component of shift along the $x_i$ axis.

If the elastic properties of the medium obey Hooke's law, the system of equations (2.1) may have a solution in the form of a plane wave

$$U_i = A_i \exp\{i(2\pi/\lambda)(lx+my+nz-v_n t)\}$$

where $A_i$ is the amplitude of shift of particles along the coordinate axes $x$, $y$, and $z$; $l$, $m$, and $n$ are the direction cosines of the wave propagation direction; $v_n$ is the rate of propagation of the wave in the direction $[l,m,n]$, and $\lambda$ is the wavelength.

After mathematical transformations, the system (2.1) can be rewritten as [2]

$$(\Gamma_{11}-\rho v_i{}^2)A_x + \Gamma_{12}A_y + \Gamma_{13}A_z = 0$$
$$\Gamma_{12}A_x + (\Gamma_{22}-\rho v_i{}^2)A_y + \Gamma_{23}A_z = 0 \qquad (2.2)$$
$$\Gamma_{13}A_x + \Gamma_{23}A_y + (\Gamma_{33}-\rho v_i{}^2)A_z = 0$$

where $\Gamma_{ik}$ is a function of the elastic constants and direction cosines

$$\Gamma_{11} = l^2 c_{11} + m^2 c_{66} + n^2 c_{55} + 2mn c_{56} + 2nl c_{15} + 2lm c_{16}$$

$$\Gamma_{22} = l^2 c_{66} + m^2 c_{22} + n^2 c_{44} + 2mn c_{24} + 2nl c_{46} + 2lm c_{26}$$

$$\Gamma_{33} = l^2 c_{55} + m^2 c_{44} + n^2 c_{33} + 2mn c_{34} + 2nl c_{35} + 2lm c_{46}$$

$$\Gamma_{12} = l^2 c_{56} + m^2 c_{24} + n^2 c_{34} + mn(c_{23}+c_{44}) + nl(c_{36}+c_{45}) + lm(c_{25}+c_{46})$$

$$\Gamma_{13} = l^2 c_{15} + m^2 c_{46} + n^2 c_{35} + mn(c_{35}+c_{46}) + nl(c_{13}+c_{55}) + lm(c_{14}+c_{56})$$

$$\Gamma_{23} = l^2 c_{16} + m^2 c_{26} + n^2 c_{45} + mn(c_{25}+c_{46}) + nl(c_{14}+c_{56}) + lm(c_{12}+c_{66})$$

It is known that (2.2) can have nonzero solutions only if the determinant of this system equals zero,

$$\begin{vmatrix} \Gamma_{11}-\rho v_k{}^2 & \Gamma_{12} & \Gamma_{13} \\ \Gamma_{12} & \Gamma_{22}-\rho v_k{}^2 & \Gamma_{23} \\ \Gamma_{13} & \Gamma_{23} & \Gamma_{33}-\rho v_k{}^2 \end{vmatrix} = 0 \qquad (2.3)$$

The roots of this equation correspond to the three velocities of plane elastic waves propagating in a random direction in an anisotropic medium. One of these waves is longitudinal and two of them are shear waves. But the waves will be purely longitudinal or purely shear only if they are traveling in singular directions [3]. Then the polarization vector of one of the waves coincides with the wave normal, and those of the other two waves are at right angles to it owing to their mutual orthogonality. However, in the general case the polarization vectors of all three waves, while remaining orthogonal to each other, will be oriented randomly to the direction of propagation of the wave front. Such waves are known as quasi longitudinal and quasi-shear waves. An intermediate case is also possible, where the polarization vector of only one of the waves is at right angles to the wave normal. The polarization vectors of the other two waves are then at random angles to the direction of wave travel. Then the first wave is purely shear, and the other two are quasi longitudinal and quasi shear, respectively.

The vectorial characteristic equation (2.3) known in the theory of elasticity as Christoffel's equation, is usually used in solving the problem of finding, given the elastic constants of the crystal, the phase velocities of the plane waves, and the polarization directions in each of these waves for any selected direction in the crystal.

We are interested in the reverse problem, namely, in determining the elasticity tensor from the elastic wave velocities measured experimentally in various crystallographic directions.

The uniform stress $\sigma_{kl}$ and uniform strain $\varepsilon_{ij}$ are tensors of rank two. Hooke's law, generalized for a single crystal, has the form

$$\varepsilon_{ij} = S_{ijkl}\sigma_{kl} \qquad (2.4)$$

where the $S_{ijkl}$ tensor of rank four is called the compliance tensor. For any deformation component $\varepsilon_{ij}$, the right-hand side of the equation consists of nine terms. There are altogether also nine equations, and therefore 81 $S_{ijkl}$ coefficients.

Equation (2.4) can be reversed, i.e., written in the form

$$\sigma_{ij} = C_{ijkl}\varepsilon_{kl} \qquad (2.5)$$

$C_{ijkl}$ is called the elasticity tensor. It can easily be shown that

$$S_{ijkl} = S_{jikl} \quad \text{and} \quad C_{ijkl} = C_{jikl}$$

Of the 81 components, 36 are independent; the same is true of the elasticity tensor.

To avoid cumbersome equations it is customary to use so-called matrix notation. The transition is accomplished according to the schemes

$$\begin{vmatrix} \sigma_{11} & \sigma_{12} & \sigma_{31} \\ \sigma_{12} & \sigma_{22} & \sigma_{23} \\ \sigma_{31} & \sigma_{23} & \sigma_{33} \end{vmatrix} \rightarrow \begin{vmatrix} \sigma_1 & \sigma_6 & \sigma_5 \\ \sigma_6 & \sigma_2 & \sigma_4 \\ \sigma_5 & \sigma_4 & \sigma_3 \end{vmatrix}$$

and

$$\begin{vmatrix} \varepsilon_{11} & \varepsilon_{12} & \varepsilon_{31} \\ \varepsilon_{12} & \varepsilon_{22} & \varepsilon_{23} \\ \varepsilon_{31} & \varepsilon_{23} & \varepsilon_{33} \end{vmatrix} \rightarrow \begin{vmatrix} \varepsilon_1 & \tfrac{1}{2}\varepsilon_6 & \tfrac{1}{2}\varepsilon_5 \\ \tfrac{1}{2}\varepsilon_6 & \varepsilon_2 & \tfrac{1}{2}\varepsilon_4 \\ \tfrac{1}{2}\varepsilon_5 & \tfrac{1}{2}\varepsilon_4 & \varepsilon_3 \end{vmatrix}$$

Thus, the subscripts in the rank-four tensor are substituted according to the scheme

$$\begin{array}{cccccc} 11 & 22 & 33 & 23,32 & 31,13 & 12,21 \\ 1 & 2 & 3 & 4 & 5 & 6 \end{array}$$

Here

$$S_{ijkl} = S_{mn} \qquad \text{when } m \text{ and } n \text{ equal } 1, 2, \text{ or } 3$$

$$2S_{ijkl} = S_{mn} \qquad \text{when either } m \text{ or } n \text{ equal } 4, 5, \text{ or } 6$$

$$4S_{ijkl} = S_{mn} \qquad \text{when both } m \text{ and } n \text{ equal } 4, 5, \text{ or } 6$$

Hooke's law acquires the form

$$\varepsilon_i = S_{ij}\sigma_j \quad \text{or} \quad \sigma_i = C_{ij}\varepsilon_j$$

The factors 2 and 4 are introduced in the $S_{ij}$ term, and hence need not be introduced in $C_{ij}$.

Thus, compliance and elasticity are characterized by square $6 \times 6$ matrices. But the matrices are symmetric,

$$C_{ij} = C_{ji} \quad \text{and} \quad S_{ij} = S_{ji}$$

This follows, in particular, from formulas (2.4) and (2.5), where the elasticity coefficients can be expressed as second derivatives with respect to energy. The order of differentiation is immaterial. Hence the symmetry. Thus, in a triclinic crystal elasticity and compliance have 21 independent elements each. The transition from $C_{ij}$ to $S_{ij}$ is accomplished by inversion of the matrix

$$C_{ij} = S_{ij}^{-1}$$

The number of independent elements of the tensor decreases further with increasing crystal symmetry.

Monoclinic crystals have 13 independent elements. The matrix (axis $2\|Y$) has the form

$$\begin{bmatrix} C_{11} & C_{12} & C_{13} & 0 & C_{15} & 0 \\ & C_{22} & C_{23} & 0 & C_{25} & 0 \\ & & C_{33} & 0 & C_{35} & 0 \\ & & & C_{44} & 0 & C_{46} \\ & & & & C_{55} & 0 \\ & & & & & C_{66} \end{bmatrix}$$

Orthorhombic crystals have nine, and cubic crystals three independent elements.

To obtain an expression for compressibility, both volume and linear, it is necessary to invert the elasticity tensor.

### 3. CALCULATING ELASTIC CONSTANTS OF SINGLE CRYSTALS FROM EXPERIMENTALLY MEASURED ELASTIC WAVE VELOCITIES

In studying the elastic properties of crystals one usually uses an orthogonal coordinate system fixed to the crystal. We shall deal with monoclinic crystals whose crystallographic [b] axis is a second-order axis of symmetry and in which the form of the elasticity tensor $|C_{ij}|$ does not change on rotating about it. Therefore the $y$-axis is superposed on the crystallographic [b] axis. The choice of $x$ and $z$ axes is arbitrary in the crystallographic plane [$ac$]. For convenience in orienting the crystal, the $x$ and $z$ axes are selected so that the $x$ axis coincides with the crystallographic [a] axis, and the $z$ axis, with the normal to the [ab] plane. (This is especially convenient if the crystals under study have distinct cleavage along [ab]).

To determine all 13 independent elastic moduli of a monoclinic crystal by the pulse ultrasonic method, one measures the velocities of propagation of elastic waves in at least six arbitrary nonequivalent crystallographic directions [4]. As has already been said, three waves propagate in each of these directions, one of them being quasi longitudinal and the other two quasi shear.

Measuring all the 18 velocities of these waves and substituting into Christoffel's equation (2.3) one gets a system of 18 equations with 13 unknowns. Since the velocities are found experimentally with a certain error (due to the apparatus used), the equations will be somewhat incompatible. It is necessary, therefore, to find the probabilities of the constants, and the accuracies of the resulting constants will differ.

Consider the solution of this system for a concrete choice of directions of elastic wave propagation.

The problem of determining the elastic constants from the experimentally measured velocities $v_k$ is greatly simplified if at least one of the three waves propagating in the prescribed direction is a pure one.

For convenience in orienting and preparing samples, it is good practice with monoclinic crystals to measure sonic velocities in the direction of the following edges of a cube constructed on the $x$, $y$, and $z$ axes:

$$[001], \ [110], \ [010], \ [101], \ [100], \ [011]$$

The [010] direction is singular, and all three waves propagating in it are pure waves. Of the three waves propagating in each of the directions [001], [101], and [100], one is a purely shear wave. It has a polarization vector extending along the singular direction [010]. Finally, only quasi-longitudinal and quasi-shear waves propagate in the last two directions.

Table 1 lists the symbols of all the 18 measured velocities of propagation of the elastic waves together with the polarization for each of the waves, and shows the orientation of the samples.

Having selected the directions indicated for measuring the propagation velocities of plane elastic waves in the crystals, Christoffel's general equation (2.3) can be reduced to a maximally simple form, convenient for determining the tensor components $|C_{ij}|$.

For example, for propagation of the elastic waves in the direction [001], the direction cosines are $l = m = 0$, and $n = 1$, and hence

$$\begin{vmatrix} C_{55} - \rho v_k^2 & 0 & C_{35} \\ 0 & C_{44} - \rho v_k^2 & 0 \\ C_{35} & 0 & C_{33} - \rho v_k^2 \end{vmatrix} = 0$$

## Table 1

DESIGNATION OF VELOCITIES OF ELASTIC WAVES DEPENDING ON DIRECTION
OF PROPAGATION AND DISPLACEMENT DIRECTIONS.
ORIENTATION OF SPECIMENS IN SELECTED ORTHOGONAL COORDINATE SYSTEM

| $v_i$ | Type of wave | Direction of propagation | Polarization | Orientation of specimen in coordinate axes |
|---|---|---|---|---|
| $v_1$ | Quasi longitudinal | 001 | 001 | |
| $v_2$ | Quasi shear | 001 | 100 | |
| $v_3$ | Shear | 001 | 010 | |
| $v_4$ | Quasi longitudinal | 110 | 110 | |
| $v_5$ | Quasi shear | 110 | 110 | |
| $v_6$ | Quasi shear | 110 | 001 | |
| $v_7$ | Longitudinal | 010 | 010 | |
| $v_8$ | Quasi shear | 010 | 001 | |
| $v_9$ | Quasi shear | 010 | 100 | |
| $v_{10}$ | Quasi longitudinal | 101 | 101 | |
| $v_{11}$ | Quasi shear | 101 | 101 | |
| $v_{12}$ | Shear | 101 | 010 | |
| $v_{13}$ | Quasi longitudinal | 100 | 100 | |
| $v_{14}$ | Shear | 100 | 010 | |
| $v_{15}$ | Quasi shear | 100 | 001 | |
| $v_{16}$ | Quasi longitudinal | 011 | 011 | |
| $v_{17}$ | Quasi shear | 011 | 011 | |
| $v_{18}$ | Quasi shear | 011 | 100 | |

The roots of this equation are the velocities $v_1$, $v_2$, and $v_3$ (in the notation of Table 1).

For propagation of the elastic waves in the direction [110] the direction cosines equal $l = m = \sqrt{2}/2$, and $n = 0$. Equation (2.3) has the form

$$
\begin{vmatrix}
C_{11} + C_{66} - 2\rho v_k{}^2 & C_{12} + C_{66} & C_{15} + C_{46} \\
C_{12} + C_{66} & C_{22} + C_{66} - 2\rho v_k{}^2 & C_{25} + C_{46} \\
C_{15} + C_{46} & C_{25} + C_{46} & C_{55} + C_{44} - 2\rho v_k{}^2
\end{vmatrix} = 0
$$

Here $k = 4, 5, 6$, i.e., the roots of this equation are the velocities $v_4$, $v_5$, and $v_6$.

We have altogether six secular equations of the third order with respect to the velocity of propagation. All 18 velocities are connected with the 13 $C_{ik}$ coefficients.

The calculation of $C_{ik}$ from the velocities of sound propagation is a very cumbersome task and has to be done with computers. Owing to the limited accuracy of the experiment, the equations may turn out to be incompatible. Evidently the best way is to reduce the problem to the reverse. In other words, the $C_{ik}$ are determined very roughly at first, and then refined by the least-squares method to yield the sonic velocities that coincide best with experiment.

Algebraic transformations yield several simple relations between the sonic velocities and the elements of the elasticity tensor. For example,

$$
C_{22} = \rho v_7{}^2, \qquad C_{33} - C_{11} = \rho v_1{}^2 + \rho v_2{}^2 - \rho v_{13}^2 - \rho v_{15}^2
$$

If the problem at hand is to test the theory by means of which the $C_{ik}$ are calculated, it is best to check relations similar to those written against experiment.

4. Elasticity Tensors of Naphthalene, Stilbene, Tolan, and Dibenzyl Single Crystals at Room Temperature and Normal Pressure

Table 2 gives the velocities of propagation of longitudinal and shear waves measured by Teslenko [5] in single crystals of naphthalene, stilbene, tolan, and dibenzyl at room temperature and normal pressure. The measurements were accurate to about 1%. The wave propagation velocities were measured several times for each of the directions enumerated and for different samples. Some of the velocity values in a few of the samples sometimes displayed a spread exceeding the tolerable error of measurement. This may evidently be attributed to internal sample defects: heterogeneity, block structure, cracks, etc. The data given in the table are averages of several measurements. The table also indicates the absolute error involved in measuring each velocity.

*Table 2*

VELOCITY OF SOUND IN NAPHTHALENE, DIBENZYL, TOLAN,
AND STILBENE CRYSTALS AT ROOM TEMPERATURE AND NORMAL PRESSURE

| $V_i$ | Velocity of sound in multiples of $10^5$ cm/sec | | | |
|---|---|---|---|---|
| | Naphthalene | Dibenzyl | Tolan | Stilbene |
| $V_1$ | $3.34 \pm 0.02$ | $2.56 \pm 0.02$ | $2.48 \pm 0.02$ | $2.64 \pm 0.02$ |
| $V_2$ | $1.06 \pm 0.01$ | $1.50 \pm 0.01$ | $2.09 \pm 0.02$ | $2.32 \pm 0.02$ |
| $V_3$ | $1.70 \pm 0.02$ | $1.67 \pm 0.02$ | $1.60 \pm 0.02$ | $1.68 \pm 0.02$ |
| $V_4$ | $3.13 \pm 0.02$ | $2.84 \pm 0.02$ | $2.62 \pm 0.02$ | $2.86 \pm 0.02$ |
| $V_5$ | $1.25 \pm 0.02$ | $1.26 \pm 0.02$ | $1.38 \pm 0.02$ | $1.40 \pm 0.02$ |
| $V_6$ | $1.44 \pm 0.02$ | $1.60 \pm 0.02$ | $1.85 \pm 0.02$ | $2.03 \pm 0.02$ |
| $V_7$ | $2.94 \pm 0.02$ | $2.49 \pm 0.02$ | $3.10 \pm 0.02$ | $2.89 \pm 0.02$ |
| $V_8$ | $1.64 \pm 0.02$ | $1.83 \pm 0.02$ | $1.66 \pm 0.02$ | $1.81 \pm 0.02$ |
| $V_9$ | $1.96 \pm 0.02$ | $1.40 \pm 0.02$ | $1.27 \pm 0.02$ | $1.40 \pm 0.02$ |
| $V_{10}$ | $3.22 \pm 0.02$ | $2.29 \pm 0.02$ | $3.10 \pm 0.02$ | $3.32 \pm 0.02$ |
| $V_{11}$ | $1.46 \pm 0.02$ | $1.38 \pm 0.01$ | $1.07 \pm 0.01$ | $1.12 \pm 1.01$ |
| $V_{12}$ | $1.63 \pm 0.01$ | $1.82 \pm 0.02$ | $1.45 \pm 0.02$ | $1.70 \pm 0.02$ |
| $V_{13}$ | $2.63 \pm 0.03$ | $3.04 \pm 0.02$ | $2.64 \pm 0.02$ | $2.84 \pm 0.02$ |
| $V_{14}$ | $1.91 \pm 0.02$ | $1.54 \pm 0.01$ | $1.28 \pm 0.01$ | $1.45 \pm 0.01$ |
| $V_{15}$ | $1.33 \pm 0.02$ | $1.28 \pm 0.01$ | $2.20 \pm 0.02$ | $2.34 \pm 0.02$ |
| $V_{16}$ | $2.96 \pm 0.03$ | $2.68 \pm 0.02$ | $2.82 \pm 0.02$ | $3.06 \pm 0.03$ |
| $V_{17}$ | $2.26 \pm 0.02$ | $1.08 \pm 0.01$ | $1.20 \pm 0.02$ | $1.20 \pm 0.02$ |
| $V_{18}$ | $1.20 \pm 0.02$ | $1.60 \pm 0.02$ | $1.80 \pm 0.02$ | $1.85 \pm 0.02$ |

It is evident from the data of Table 2 that the velocities of elastic waves in the crystals under study are considerably anisotropic. A complete picture of the velocity anisotropy in a given crystal can be obtained by constructing the normal velocity surface.

In the case of monoclinic crystals the section of the normal velocity surface cut by the plane $xz$ is especially important since it is a symmetry plane.

Solving Christoffel's equation for a wave propagating in this plane, one gets a triplet of numbers for each direction, corresponding to the three velocities of propagation of elastic waves in that direction.

Figures 21–24 show the sections of the normal velocity surface cut by the plane $(ac)$ for naphthalene, stilbene, tolan, and dibenzyl.

It can be seen from these sections that the anisotropy of velocities is very considerable. Therefore, any slight error in orientation will give an incorrect velocity value.

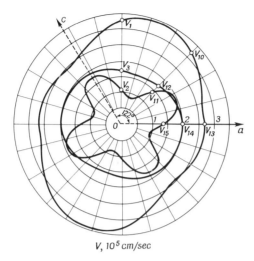

$V, 10^5 cm/sec$

**Fig. 21.** Sections of normal velocity surfaces in naphthalene single crystal cut by plane (*ac*). Here and in the following (in Figs. 22–24) circles indicate experimental points of velocities $V_k$ of propagation in the direction of [001], [101], and [100]. Meaning of velocities is given in Table 1.

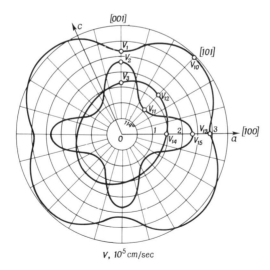

$V, 10^5 cm/sec$

**Fig. 22.** Sections of normal velocity surfaces in stilbene single crystal cut by plane (*ac*).

316

5 Thermodynamic Experiments

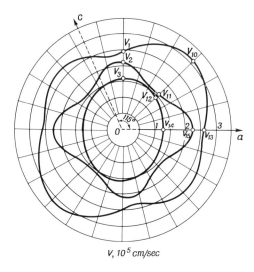

V, 10⁵ cm/sec

*Fig. 23.* Sections of normal velocity surfaces in tolan single crystal cut by plane (*ac*).

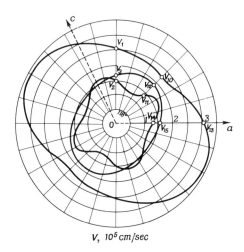

V, 10⁵ cm/sec

*Fig. 24.* Section of normal velocity surfaces in dibenzyl single crystal cut by plane (*ac*).

The results of recalculation of the velocities to find the elastic constants for naphthalene, dibenzyl, tolan, and stilbene at room temperature and normal pressure are listed in Table 3. The experimental error is about 2% for $C_{22}$, $C_{44}$, and $C_{66}$, and 3–4% for $C_{11}$, $C_{33}$, $C_{55}$, and $C_{13}$. For $C_{23}$ and $C_{12}$, the error is about 10%, and the values of the constants $C_{15}$, $C_{25}$, $C_{35}$, and $C_{46}$ can only be considered tentative.

## Table 3

ELASTIC CONSTANTS AND COMPLIANCE MODULI
OF NAPHTHALENE, DIBENZYL, SILBENE, AND TOLAN AT ROOM TEMPERATURE

| | Elasticity constants, units of $10^{10}$ dyne/cm² | | | | Compliance moduli, units of $10^{-11}$ cm²/dyne | | | |
|---|---|---|---|---|---|---|---|---|
| | Naphtha-lene | Dibenz-ene | Stilbene | Tolan | Naphtha-lene | Dibenz-ene | Stilbene | Tolan |
| 11 | 7.80 | 9.45 | 9.30 | 7.85 | 1.90 | 2.06 | 2.26 | 1.58 |
| 22 | 9.90 | 6.80 | 9.20 | 8.55 | 5.63 | 2.15 | 1.90 | 2.06 |
| 33 | 11.90 | 7.20 | 7.90 | 6.45 | 4.72 | 2.09 | 2.44 | 2.01 |
| 44 | 3.30 | 3.10 | 3.25 | 2.90 | 3.09 | 3.54 | 3.15 | 3.46 |
| 55 | 2.10 | 2.55 | 6.40 | 5.45 | 37.85 | 5.25 | 1.57 | 2.14 |
| 66 | 4.15 | 2.60 | 2.45 | 1.85 | 2.45 | 4.16 | 4.22 | 5.41 |
| 12 | 2.30 | 3.95 | 5.70 | 3.50 | −0.46 | −0.65 | −0.81 | −0.75 |
| 13 | 3.40 | 4.15 | 5.70 | 1.15 | −0.67 | −0.74 | −1.13 | 0.09 |
| 23 | 4.45 | 3.35 | 4.85 | 3.50 | −4.05 | −0.64 | −0.58 | −0.88 |
| 15 | −0.6 | −2.4 | −0.3 | 0.3 | 0.88 | 1.56 | −0.55 | 0.25 |
| 25 | −2.7 | −0.8 | −0.5 | 2.5 | 12.72 | −0.14 | 0.63 | −0.74 |
| 35 | 2.9 | −0.7 | −0.5 | 0.9 | −11.92 | −0.35 | 1.07 | 0.07 |
| 46 | −0.5 | 0.8 | 0.5 | 0.1 | 0.37 | −1.10 | −0.61 | −0.13 |

Table 3 also contains the compliance moduli $S_{ij}$ for the crystals studied. The matrix of elastic compliances is the reciprocal to that of $C_{ij}$. In the $C_{ij}$ matrix a number of nondiagonal elements are measured with a large (up to 30–50%) error. On passing from $C_{ij}$ to $S_{ij}$, as shown by a direct estimate, the error of the initial matrix elements is distributed somewhat over all the elements of the $S_{ij}$ matrix. For this reason, great care must be exercised in dealing with questions of accuracy of the $S_{ij}$.

It is interesting to use the compliance tensor for characterizing the anisotropy of strains in the crystals, studied upon uniform compression. We are able to calculate compressibility in any direction under the action of unit hydrostatic pressure. Linear compressibility depends upon the direction. We can characterize it by a vector of definite length, which describes a certain surface representing uniform compression.

Linear compressibility $\beta$ is related to the compliance tensor $S_{ij}$ by the expression

$$\beta = S_{ijkl} l_i l_j$$

where the $l$'s are the direction cosines.

For the monoclinic system, this pattern can be presented in the form

$$\beta = A_1 l_1{}^2 + A_2 l_2{}^2 + A_3 l_3{}^2 + A_5 l_1 l_3$$

## Table 4

TENSOR COMPONENTS $A_i$ FOR DIBENZYL,
STILBENE, TOLAN, AND NAPHTHALENE

| | $A_i$ $(10^{-11}$ cm$^2$/dyne) | | | |
|---|---|---|---|---|
| | Dibenzyl | Stilbene | Tolan | Naphthalene |
| $A_1$ | 0.67 | 0.32 | 0.92 | 0.78 |
| $A_2$ | 0.86 | 0.51 | 0.42 | 1.12 |
| $A_3$ | 0.71 | 0.73 | 1.21 | −0.01 |
| $A_5$ | 1.08 | 1.15 | −0.43 | 1.67 |

here $l_{1-3}$ are the direction cosines of the given direction relative to the orthogonal coordinate system we have chosen. The tensor components $A$ are expressed as follows through the compliance tensor (in matrix notation):

$$A_1 = S_{11} + S_{12} + S_{13}, \qquad A_2 = S_{12} + S_{22} + S_{23}$$

$$A_3 = S_{13} + S_{23} + S_{33}, \qquad A_5 = S_{15} + S_{25} + S_{35}$$

Table 4 gives the values of $A_1$, $A_2$, $A_3$, and $A_5$ calculated from the compliance moduli $S_{ij}$ for dibenzyl, stilbene, tolan, and naphthalene.

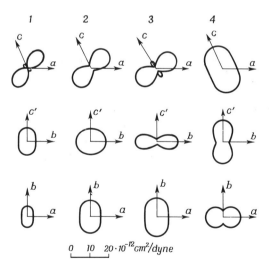

$$0 \quad 10 \quad 20 \cdot 10^{-12} \text{cm}^2/\text{dyne}$$

**Fig. 25.** Section of uniform compression figure of stilbene (1), dibenzyl (2), naphthalene (3), and tolan (4) cut by coordinate planes.

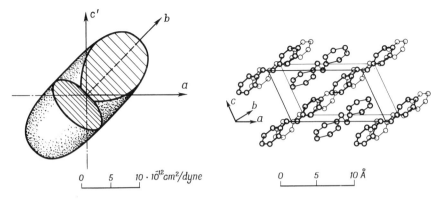

**Fig. 26.** Uniform compression figure of dibenzyl and arrangement of dibenzyl molecules in unit cell.

An idea of the uniform compression can be obtained by examining the figure sections cut by the coordinate planes. Such sections are shown in Fig. 25. Figures 26–29 present the three-dimensional uniform compression figure and the arrangement of molecules in the unit cell for the crystals under consideration.

In dibenzyl the direction of greatest compression coincides with that of the major axis of the molecule. This result seems fairly natural. Indeed, the inter-molecular vectors pointing in the direction of the major axis of the molecule are larger than the corresponding transverse vectors.

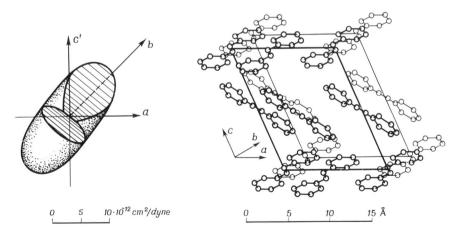

**Fig. 27.** Uniform compression figure of stilbene and arrangement of stilbene molecules in unit cell.

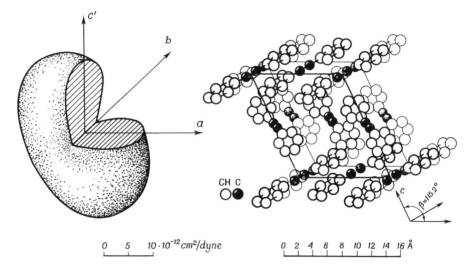

*Fig. 28.* Uniform compression figure of tolan and arrangement of tolan molecules in unit cell.

Attention is also drawn to the fact that the direction of best compressibility is not the normal to the cleavage plane (*ab*) of the crystal. Hence, the maximum shift of molecules upon hydrostatic compression should be interpreted as approachment of the closely packed layers $2_1 \, a$ with simultaneous slip of the surfaces relative to one another. In the case of stilbene and tolan, the direction of greatest compression is more difficult to relate to molecular arrangement, because their structures consist of two symmetrically independent layers of

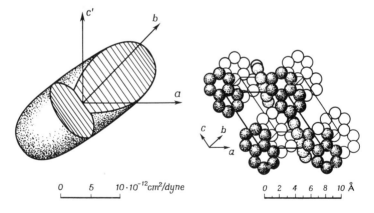

*Fig. 29.* Uniform compression figure of naphthalene and arrangement of naphthalene molecules in unit cell.

identical packing. Nevertheless, in these cases, as in that of dibenzyl, the uniform compression figures obtained suggest that the maximum molecular shift due to hydrostatic pressure will occur on approachment of the closely packed layers $2_1 a$ with simultaneous displacement of one relative to the other.

The flat naphthalene molecules are arranged in the unit cell in such a way that the direction of best compression corresponds to approachment of the molecular planes with slip along these planes.

In the cases under consideration the uniform compression figures have the same appearance and differ only in size and orientation with respect to the coordinate axes.

*Table 5*

PACKING FACTORS OF STILBENE,
DIBENZYL, NAPHTHALENE, AND TOLAN

| Crystal | $K$ |
| --- | --- |
| Stilbene | 0.720 |
| Dibenzyl | 0.705 |
| Naphthalene | 0.702 |
| Tolan | 0.685 |

The fact that the absolute compressibility values are symbatic with the packing coefficients, as follows from the data of Table 5, is of great help in giving the "geometrical" interpretation of these results.

We believe that symbasis of compressibility and looseness of molecular packing in organic crystals is a general rule, corresponding to the undirected nature of intermolecular interaction.

5. INVESTIGATION OF THE ELASTIC PROPERTIES OF POLYCRYSTALLINE SAMPLES

A large number of papers have been devoted to the elastic properties of polycrystals. One of the crucial problems dealt with in them is that of determining the elastic constants of a polycrystalline aggregate from the elastic properties of its component crystallites. Of some interest also is the reverse problem: How far is it possible to determine the elastic characteristics of a crystal from measurements made on polycrystalline samples?

The simplest case is that of random orientation of the small crystallites composing the polycrystal.

322 5 Thermodynamic Experiments

5 Thermodynamic Experiments

Such an elastically isotropic medium is characterized by two independent elastic constants. The matrix of elasticity coefficients in this case has the form

$$\begin{vmatrix} C_{11} & C_{12} & C_{12} & 0 & 0 & 0 \\ C_{12} & C_{11} & C_{12} & 0 & 0 & 0 \\ C_{12} & C_{12} & C_{11} & 0 & 0 & 0 \\ 0 & 0 & 0 & C_{44} & 0 & 0 \\ 0 & 0 & 0 & 0 & C_{44} & 0 \\ 0 & 0 & 0 & 0 & 0 & C_{44} \end{vmatrix} \tag{5.1}$$

The following condition of isotropy must be fulfilled

$$C_{44} = \tfrac{1}{2}(C_{11} - C_{12}) \tag{5.2}$$

But even in this simple case no strict way of calculating the elastic properties of a polycrystal from known properties of a single crystal has as yet been found. One of the main difficulties is due to grain boundary phenomena, consisting of variation of stresses and strains from grain to grain when an external stress is applied to the crystal. There is no exact way of taking these phenomena into account.

Methods of calculating the elastic characteristics of polycrystals from the corresponding elastic properties of crystallites (averaging methods) are based on the assumption that there are no internal stresses in the unstrained aggregate.

Let us consider briefly some of the methods of averaging. The first and simplest averaging methods were suggested by Voigt [6] and Reuss [7]. In both methods interaction between the crystallites composing the polycrystal is neglected. This makes it possible to find directly the elastic constants $\bar{C}_{ijkl}$ (in Voigt's method) and of the compliance moduli $\bar{S}_{ijkl}$ (in Reuss's method) of the polycrystal as the averages of the corresponding values of $C'_{ijkl}$ and $S'_{ijkl}$ of the individual crystallites.

Voigt's method is based on the assumption that the strains are uniform (continuous) over the entire polycrystal. In Reuss's method all the components of the stress tensor at the grain boundaries are assumed to be continuous.

For real crystals, these assumptions are not true, and Voigt and Reuss averages can be suitable only for special models. For example, Voigt averaging is fulfilled for a stack of single-crystal anisotropic plates, randomly oriented relative to one another, and Reuss averaging, for a pack of anisotropic single-crystal rods with their axes at random directions [8]. In the cases indicated the conditions of equality of strains and stresses are fulfilled only if loading is in one definite direction. In other words, such models are only ideal one-dimensional analogues of aggregates that do not exist in reality.

If a polycrystal is regarded as a nonuniform elastoanisotropic medium, the expressions for the elasticity and compliance tensors at any point of a randomly selected crystal will have the form

$$C'_{ijkl} = l_{ip} l_{jq} l_{kr} l_{ls} C_{pqrs} \qquad (5.3)$$

$$S'_{ijkl} = l_{ip} l_{jq} l_{kr} l_{ls} S_{pqrs} \qquad (5.4)$$

Here the quantities with primes denote crystallite constants in a coordinate system common to the entire sample, and those without primes, constants of the crystallite in its principal coordinate system. The symbols $l_{ik}$ denote the cosines of the angles between the principal axes of the crystallite in question and the coordinate axes. The elastic constants and compliance moduli of a polycrystal are obtained by integrating (5.3) and (5.4) over the entire orientation region defined by the Euler angles

$$\bar{C}_{ijkl} = (1/8\pi^2) \int C'_{ijkl} \sin\theta \, d\theta \, d\varphi \, d\psi$$

$$\bar{S}_{ijkl} = (1/8\pi^2) \int S'_{ijkl} \sin\theta \, d\theta \, d\varphi \, d\psi$$

After transformations, Voigt averaging gives [9]

$$\bar{C}_{11} = \tfrac{1}{5}(3A+2B+4C) = \lambda + 2\mu$$

$$\bar{C}_{12} = \tfrac{1}{5}(A+4B-2C) = \lambda \qquad (5.5)$$

$$\bar{C}_{44} = \tfrac{1}{5}(A-B+3C) = \mu$$

Here $\lambda$ and $\mu$ are Lamé's constants.

The condition of isotropy,

$$\bar{C}_{44} = \tfrac{1}{2}(\bar{C}_{11} - \bar{C}_{12}) \qquad (5.6)$$

should be fulfilled.

The following notation is used in (5.5) and (5.6)

$$3A = C_{11} + C_{22} + C_{33}$$

$$3B = C_{12} + C_{13} + C_{23}$$

$$3C = C_{44} + C_{55} + C_{66}$$

Given below are the expressions for some important elasticity characteristics of a polycrystal, obtained as a result of Voigt averaging, which we shall require in the following.

The modulus of volume contraction

$$K_V = \tfrac{1}{3}(A+2B)$$

The shear modulus                                                        (5.7)

$$G_V = \tfrac{1}{5}(A - B + 3C)$$

Quite similarly, Reuss averaging gives

$$\bar{S}_{11} = \tfrac{1}{5}(3A' + 2B' + C')$$

$$\bar{S}_{44} = \tfrac{1}{5}(4A' - 4B' + 3C') = 1/G_R \qquad (5.8)$$

$$1/K_R = 3A' + 6B'$$

Here

$$3A' = S_{11} + S_{22} + S_{33}$$

$$3B' = S_{23} + S_{31} + S_{12}$$

$$3C' = S_{44} + S_{55} + S_{66}$$

It should be emphasized again that Voigt averaging presumes uniformity of the deformation components at the crystallite interfaces. Actually, on passing across crystallite interfaces only the stress components normal to the interface and the shear strain components remain constant.

Since the grains are arranged at random in a polycrystal aggregate and their elastic properties are anisotropic, the stresses in different grains should be different, and additional forces must arise to preserve equilibrium at the grain boundaries. In real crystals such conditions do not generally exist.

It is clear from what has been said that Voight results will agree satisfactorily with experiment only if the anisotropy is very small. With increasing crystallite anisotropy, Voigt averaging departs from the real values, giving figures that are too high.

Reuss averaging presumes that all the stress components are continuous at the interfaces of differently oriented crystallites, though this is true only of the stress components normal to the crystallite interfaces. As a result of this, Reuss averaging gives too low a value of the quantity measured, and the greater the anisotropy of the crystallites constituting the aggregate, the larger will be the discrepancy.

These circumstances, namely, the excessive values obtained with Voigt averaging and the deficient values with Reuss averaging, were studied by Hill [10], who showed strictly that the real macroscopic values of the elasticity moduli lie between the average values obtained by Voigt and by Reuss averaging. The most probable values of the elasticity moduli of a polycrystal are the arithmetical or geometrical means of the elastic constants determined by these methods. The following formulas should be approximately correct:

$$K = \tfrac{1}{2}(K_V + K_R), \qquad G = \tfrac{1}{2}(G_V + G_R) \qquad (5.9)$$

If the measurements are carried out properly on the polycrystal (effects of porosity and texture eliminated) the experimental values of the constants do not differ from those calculated after Voigt, Reuss, Hill by more than 1–2%, even in the case of such strongly anisotropic materials as copper [11], corundum, etc. [12, 13]. There is reason to expect that this method can also be successfully applied to organic substances.

## 6. MEASURING AND CALCULATING ELASTIC PROPERTIES OF POLYCRYSTALS

In the simplest case, where the crystallites are uniformly distributed, the matrix of elasticity coefficients has the form of (5.1).

The elastic constants of such a sample (there are altogether two of them) can easily be found from the rate of propagation of longitudinal and shear waves in it.

The elastic constants are determined by the following commonly known formulas

$$\rho v_{long}^2 = C_{11}, \qquad \rho v_{sh}^2 = C_{44}$$

The constant $C_{12}$ is found from the condition of isotropy (5.2).

If the samples are prepared from a thoroughly ground powder compressed into a pellet, compression under a piston will result in a texture of hexagonal symmetry [14]. Such a texturized sample is characterized by five independent elastic constants. The elasticity coefficient tensor in this case will be written as

$$
\begin{vmatrix}
C'_{11} & C'_{12} & C'_{13} & 0 & 0 & 0 \\
C'_{12} & C'_{11} & C'_{13} & 0 & 0 & 0 \\
C'_{13} & C'_{13} & C'_{33} & 0 & 0 & 0 \\
0 & 0 & 0 & C'_{44} & 0 & 0 \\
0 & 0 & 0 & 0 & C'_{44} & 0 \\
0 & 0 & 0 & 0 & 0 & C'_{66}
\end{vmatrix}
\qquad (6.1)
$$

Here

$$C'_{66} = \tfrac{1}{2}(C'_{11} - C'_{12}) \qquad (6.2)$$

(The elastic moduli of the texturized samples are denoted with a prime.) In this notation the direction of compression of the powder in the cylindrical mold is taken as the $z$ axis.

The elastic constants are determined by measuring the velocity of sound in three nonequivalent directions: along the compression axis (along the $z$ axis), perpendicular to the compression axis (perpendicular to the $z$ axis) and at

45° to the compression axis (45° to the $z$ axis). The symbols used to denote the velocities depending on the direction of wave propagation and polarization are given in Table 6.

*Table 6*

DESIGNATION OF ELASTIC WAVE VELOCITIES IN
GRAIN-ORIENTED SPECIMENS DEPENDING ON DIRECTION OF
PROPAGATION AND POLARIZATION

| Type of wave | Direction of propagation | Polarization | Orientation of specimen | Velocity designation |
|---|---|---|---|---|
| Longitudinal | Along $z$ axis | Along $z$ axis | | $v_1$ |
| Shear | Along $z$ axis | $\perp 0z$ | | $v_2$ |
| Longitudinal | Along $y$ axis | Along $y$ axis | | $v_3$ |
| Shear | Along $y$ axis | Along $z$ axis | | $v_4$ |
| Shear | Along $y$ axis | Along $x$ axis | | $v_5$ |
| Quasi longitudinal | Along $ab$ | Along $ab$ | | $v_6$ |
| Quasi shear | Along $ab$ | Along $cd$ | | $v_7$ |

The elastic constants of the resulting samples of hexagonal symmetry are connected with the above velocities by the relations

$$\rho v_1^2 = C_{33}', \qquad \rho v_2^2 = C_{11}', \qquad \rho v_3^2 = C_{44}'$$

$$\rho v_4^2 = C_{44}', \qquad \rho v_5^2 = C_{66}'$$

where $\rho$ is the density.
From (6.2) we find

$$C_{12}' = C_{11}' - 2C_{66}'$$

To determine the constants $C_{13}'$, it is necessary to consider propagation of the waves in the sample with the 45° $xz$ cut (see last line of Table 6).
Solving the Christoffel determinant for this cut, one gets the following expressions for $C_{13}'$:

$$C_{13}' = (-C_{44}') \pm \{[(C_{11}'+C_{44}') - 2\rho v_k^2][(C_{44}'-C_{33}') - 2\rho v_k^2]\}^{1/2} \quad (6.3)$$

or

$$C_{13}' = (-C_{44}') \pm \{(C_{11}'+C_{44}')(C_{44}'+C_{33}') - 2\rho v_7^2 \cdot 2\rho v_6^2\}^{1/2}$$

By means of Eq. (6.3), $C_{13}'$ can be obtained from measurements of the velocity $v_7$ or $v_6$.

Using reduced formulas, all the elasticity moduli (6.1) of a texturized sample with hexagonal symmetry can be determined from the experimentally measured velocities $v_1$–$v_7$.

To get an idea of the elastic properties of an isotropic polycrystal, the result should be averaged by the Voigt–Reuss–Hill method. Note that this method should give very accurate results in this case, because the anisotropy of elastic properties of the pellet is not great. The modulus of volume compression $k$ and the shear modulus $G$ are calculated from formulas (5.7). To find the same values in the Reuss approximation one first calculates the compliance matrix $S_{ij}'$ which is the inverse of the matrix $C_{ij}'$ (6.1), and then applies formula (5.8). The resultant values are found from (5.9). The elasticity of polycrystalline naphthalene was studied by Teslenko. The naphthalene pellets for this experiment were compressed under a load of 30 tons. The density of the pellet, determined pycnometrically, was $1.16_7$ gm$^7$/cm$^3$. The results of measurements of the velocity of sound in the corresponding directions are listed in Table 7. The components of the elasticity tensor $C_{ij}'$ and the compliance tensor $S_{ij}'$ of the texturized sample, calculated from the velocities of wave propagation in this sample, are given in Table 8. The results of averaging of the compression and shear moduli ($K$ and $G$, respectively) are shown in Table 9.

## Table 7

SONIC VELOCITIES IN GRAIN-ORIENTED NAPHTHALENE
PELLET AT ROOM TEMPERATURE ($10^5$ cm/sec)

| $v_1$ | $v_2$ | $v_3$ | $v_4$ | $v_5$ | $v_6$ | $v_7$ |
|------|------|------|------|------|------|------|
| 2.82 | 2.84 | 1.52 | 1.53 | 1.76 | 2.88 | 1.45 |

## Table 8

ELASTICITY AND COMPLIANCE MODULI OF
TEXTURIZED NAPHTHALENE SPECIMEN AT 20°C[a]

| $ij$ | 11 | 33 | 44 | 66 | 12 | 13 |
|------|------|------|------|------|------|------|
| $C'$ | 9.44 | 9.30 | 2.72 | 3.63 | 2.18 | 4.48 |
| $S'$ | 1.38 | 1.71 | 3.68 | 2.75 | 0.00 | −0.66 |

[a] $C'_{ij}$ units $10^{10}$ dyne/cm²; $S'_{ij}$ units $10^{-11}$ cm²/dyne.

## Table 9

VOLUME COMPRESSION MODULUS $k$ AND SHEAR
MODULUS $G$ IN VOIGT–REUSS–HILL APPROXIMATION
AT ROOM TEMPERATURE ($10^{10}$ dyne/cm²)

| $k_V'$ | $k_R'$ | $k_H'$ | $G_V$ | $G_R'$ | $G_H'$ |
|------|------|------|------|------|------|
| 5.61 | 5.52 | 5.56 | 2.96 | 2.80 | 2.88 |

## 7. CALORIMETRY

This classical method, of fundamental importance for investigating thermo-dynamic regularities, is described in a large number of books and reviews. In application to organic solids calorimetry is described in an extensive review by Westrum and McCullough [15]. For this reason, we shall restrict ourselves to a few remarks.

To study thermodynamic functions it is necessary to have a $C_p$ versus $T$ curve at one's disposal. As a rule, the heat capacity is measured at the pressure of the saturated vapor. This value is practically equal to $C_p$.

Since the internal pressure of a crystal is a value of the order of hundreds of thousands of atmospheres, it is quite legitimate to consider the experimentally measured $C_p$ referred to zero external pressure. Thermodynamic potentials are determined from the calorimetric experiment by integrating the function $C_p(T)$. The enthalpy equals $H = \int_0^T C_p \, dT$ (to the accuracy of the term at $T = 0$), and the entropy $S = \int_0^T (C_p/T) \, dT$. The thermodynamic potential is also calculated from information about only the $C_p$ versus $T$ curve. At $p = 0$, the thermodynamic potential equals the free energy, and the enthalpy is equal to the internal energy.

Thus, for normal (i.e. zero) pressure, all the thermodynamic potentials can be found from calorimetric experiments.

The derivatives of the thermodynamic functions with respect to volume can be determined only by measuring the expansion and elasticity (cf. Chapter VI).

Since the $C_p$ versus $T$ curve has to be integrated to obtain correct values of the thermodynamic functions, it is essential to make measurements at low temperatures. This is because the possibilities of extrapolating $C_p$ to zero by means of the $T^3$ law are limited by the low values of the characteristic temperature. The low-temperature approximation can be carried out with sufficient accuracy only below the temperature of liquid helium.

In the experiment the average heat capacity is measured for some temperature interval $\Delta T$. As a rule, this gives no errors provided $\Delta T$ is less than $0.1T$.

The main difficulty in designing a precise measuring instrument is creating adiabatic conditions. The design should be such that the only heat delivered to the sample is that from a special heater. There should be no heat exchange with the medium when the equilibrium temperature is established.

As the calorimeter temperature rises during the measurements, the calorimeter shell is also heated to minimize heat exchange. When the influx of energy to the sample ceases, a new temperature equilibrium is established. This requires a time period of the order of tens of minutes; in special cases (say, near a phase transition point) the time period may increase even to days.

Apart from the adiabatic calorimeter, heat measurements are sometimes made in an isothermal calorimeter.

Calorimetry experts believe that the adiabatic calorimeter has substantial advantages over the isothermal, in application to organic substances, at any rate.

Heat capacity is a property which cannot be measured with high accuracy. Most authors give heat capacity values to four significant figures. However, the spread of heat capacity values in carrying out measurements with the same instruments and with the same samples is within $\pm 0.2\%$. The measurements of different authors very often coincide to only $1\%$. Besides the inaccuracies of measurement, of great importance in this respect is the evidently faulty control of sample purity. Unless special precautions are taken, an organic substance obtained from a synthesist usually contains up to $1\%$ impurities. Of course, this may tell strongly on the heat capacity values.

The unbiased errors of heat capacity measurements are more or less obliterated during calculation of the thermodynamic functions. For this reason, the entropy values may be given to four significant figures. In reality, the entropy (order of value 50–100 cal/deg mole) can be determined to 0.1 unit.

An important element of thermodynamic measurement is the calorimetric measurement of phase transitions. The foremost requirement in such studies is that highly pure substances be used. A great variety of types of behavior of substances on passing through transition points has been described in the literature. McCullough [15] suggests dividing transitions into seven types according to the nature of the $C_p$ versus $T$ curve in the transition region.

It seems to us that many contradictions and misunderstandings would be eliminated if measurements of the heat of transition could be made simultaneously with microscopic observation of the shift of the transition boundary as it moves along a single crystal.

## 8. ISOTHERMAL COMPRESSIBILITY

Compressibility measurements at constant temperatures are carried out directly by determining the decrease in volume caused by exerting a given pressure on a piston. A large number of substances have been studied. The results and methods have been described exhaustively by Bridgman [16].

Compressibility measurements of organic substances are simplified by the fact that their compressibilities are at least an order higher than that of crystals in which the atoms are linked by valence bonds. Müller showed in his time that the linear compressibility of paraffin at right angles to the valence chain is from 3 to $12 \times 10^{-12}$ cm$^2$/dyne whereas along the chain it is less than $3 \times 10^{-13}$.

To obtain the compressibility values corresponding to changes in size of the unit cell of a crystal, one either has to resort to X-ray measurements under pressure or to squeeze the samples thoroughly before measurements to eliminate possible voids (cf. Section 5).

The direct result of the measurement is the volume contraction $\Delta V$ relative to the initial volume $V_0$. In published data one usually finds plots of the ratio $\Delta V/V$ against pressure.

From these data, generally speaking, one can calculate directly the volume compressibility in the form $-(1/V_0)(\partial V/\partial p)_T$. The true compressibility as found in thermodynamic formulas $-(1/V)(\partial V/\partial p)_T$ can also be calculated from experiments. As might have been expected, the volume compressibility always decreases with increasing pressure.

Direct measurements of volume contraction under pressure at constant temperature are especially interesting when studying the dependence of compressibility on volume. Ultrasonic measurements are both more accurate and incomparably simpler in application to the elasticity tensor. However, it is not easy to make them at high pressures. For this reason, adiabatic compressibilities and elasticity tensors are known only for zero pressure.

At the same time, there is no need to prove the great importance of measurements at variable pressure for the study of molecular interaction. Only in this way can one establish the equation of state of a body and with its aid calculate the derivatives of internal energy of the crystal with respect to volume at constant pressure and at constant volume.

One of the main thermodynamic relations is

$$(\partial E / \partial V)_T = T(\alpha / \beta) - P$$

At a temperature of absolute zero, the internal energy is equal to the lattice energy and we get a direct method of determining the dependence of lattice energy on pressure, and hence on volume.

Unfortunately, no such studies have been carried out owing to great experimental difficulties. Measurements at $T > 0$ give the dependence of energy on volume but do not enable separation of the lattice energy from the vibrational components unless additional assumptions are made (see also Chapter VI).

*Table 10*

ISOTHERMAL COMPRESSIBILITY $\beta$ OF SOME CRYSTALS [$(cm^2/dyne) \times 10^{-11}$]

| Substance | $P$ (kbar) 0 | 2 | 4 | 6 | 8 | 10 | 12 | 14 | 16 | 18 | 20 | 22 |
|---|---|---|---|---|---|---|---|---|---|---|---|---|
| Naphthalene | 1.88 | 1.28 | 1.05 | 0.89 | 0.74 | 0.60 | 0.50 | 0.45 | 0.41 | 0.38 | 0.36 | 0.36 |
| Anthracene | 1.73 | 1.06 | 0.88 | 0.76 | 0.66 | 0.58 | 0.52 | 0.45 | 0.41 | 0.36 | 0.34 | 0.33 |
| Urea | 1.44 | 0.79 | 0.64 | 0.73 | 0.59 | 0.52 | 0.47 | 0.43 | 0.40 | 0.37 | 0.36 | 0.34 |
| Hexamethylene- tetramine | 1.22 | 0.88 | 0.76 | 0.68 | 0.61 | 0.55 | 0.50 | 0.45 | 0.43 | 0.39 | 0.35 | 0.33 0.29 |
|  |  |  |  |  |  |  |  |  |  |  |  | 0.28 0.28 |
| Anthraquinone | 1.18 | 0.95 | 0.76 | 0.64 | 0.58 | 0.52 | 0.48 | 0.44 | 0.40 | 0.36 | 0.34 | 0.34 |

Some of Bridgman's data on measurements of the volume isothermal compressibility of organic substances are given in Table 10. Comparison of the data for naphthalene with the value of $\beta_S$, measured ultrasonically and recalculated to $\beta_T$ shows that these values agree satisfactorily.

## 9. MEASURING THE HEAT OF SUBLIMATION

Comparatively few results of direct measurements of the sublimation heat of organic crystals have been published. Sublimation heats can be determined, as a matter of principle, by direct measurement of the amount of energy required to evaporate a given amount of substance in calorimeters similar to those used for determining heat capacity. In these measurements sublimation should be effected in such a manner as to maintain thermodynamic equilibrium. Delivery of electric energy and removal of the vapor should be so balanced that the calorimeter is always approximately at a constant temperature. Besides, the system should operate so that there is no dead space in it; the volume initially occupied by the solid specimen should not be filled with vapor at the end of the process and no changes in temperature should occur. The problems involved in direct determination of sublimation heat still remain unsolved. Usually sublimation heats are determined from the temperature dependence of vapor pressure above the crystalline phase, meaning the pressure of the gaseous phase in equilibrium with the condensed phase. To distinguish between vapor pressure in equilibrium with a solid and equilibrium pressure above a liquid one uses the term "sublimation pressure." The most popular of the existing methods of measuring saturated vapor pressures are:

(1)   Static methods, which reduce directly to measurement of the saturated vapor pressure at a given temperature, there being practically a constant equilibrium between the condensed phase and the vapor. The equilibrium pressure is determined directly by means of a mercury, ionization, spiral, membrane, or other pressure gauge.

(2)   Method of vapor transfer by a flow of inert gas (streaming method). A flow of an inert gas is passed over the substance, preheated to a definite temperature, which carries the vapors of the substance into a condenser. The velocity of the stream is adjusted so that the gas becomes saturated in the vapors of the substance under study. Since complete saturation of the gas is difficult to achieve, this method does not give very accurate results.

(3)   Method of evaporation from an open Langmuir surface. This method is based on determining the rate of evaporation of the substance from an open surface under vacuum. The mass loss per unit surface area of the substance per unit time is given by the relation

$$G = \alpha p (M/2\pi RT)^{\frac{1}{2}}$$

where $\alpha$ is the coefficient of accommodation on evaporation (Langmuir coefficient). A disadvantage of this method is that the coefficient $\alpha$ must be known.

(4)   Knudsen effusion method, based on measuring the rate of efflux of the vapor through a small opening from the space containing the saturated vapor. The disadvantage is the difficulty of constructing a cell with definite

parameters. If possible errors are taken into account, the Knudsen method is the most reliable one for low vapor pressures.

The initial equation for calculating the vapor elasticity at any temperature is the Clapeyron–Clausius equation based on the second law of thermodynamics. This equation gives the relation between the pressure and temperature along the phase equilibrium curve. From the condition of phase equilibrium in a two-phase, one-component system follows the equation

$$dP/dT = (S_1 - S_2)/(V_1 - V_2)$$

where $S_1$, $S_2$, $V_1$, and $V_2$ are the molecular entropies and volumes of the two phases. In this formula the difference $S_1 - S_2$ can conveniently be expressed through the heat of transition from one phase to the other. This transition occurs at constant pressure and temperature.

Substituting $H = T(S_2 - S_1)$, we find the Clapeyron–Clausius formula

$$dP/dT = H/T(V_2 - V_1), \qquad dP/dT = H/(T\Delta V)$$

In the case of vaporization (evaporation or sublimation) this formula can be simplified. Consider the case of equilibrium between a solid or liquid and its vapor. Since the volume of gas $V_2$ is usually much larger than that of the condensed phase $V_1$, the latter may be neglected, i.e.,

$$dP/dT = H/TV_2$$

Further, considering the vapor an ideal gas, one may put $V_2 = RT/P$, which results in

$$dP/dT = HP/RT^2$$

or

$$\frac{dP}{P} = \frac{H}{R}\frac{dT}{T^2}, \qquad \ln P = -\frac{H}{R}\frac{1}{T} + B$$

In temperature intervals wherein the heat of transition may be considered constant, the quantity $H/R$ can be denoted by $A$, and then the formula

$$\ln P = -(A/T) + B$$

determines the change in pressure of the saturated vapor with temperature. This formula is employed in experimental determinations of sublimation heats.

The experimental determination of the saturated vapor pressure dependence on temperature from the rate of efflux of the vapor through a small orifice reduces to the following. It is supposed that the amount of gas passing out of a small orifice in the vessel wall is equal to the amount of gas exerting pressure on an equal area. If the number of molecules contained in unit volume is $n$ and their mean velocity is $\bar{v}$, the number of impacts of the molecules per unit area of the wall can be calculated by integrating the Maxwell distribution [17]. The total number of impacts per unit wall area per unit time equals

$$\gamma = P/(2\pi MT)^{\frac{1}{2}}$$

where $M$ is the molecular weight and $T$ is the temperature. The corresponding mass of molecules (mass of gas) is obtained by multiplying by the mass of one molecule. The amount of substance in grams will be

$$G = 5.833 \times 10^{-2} P(M/T)^{\frac{1}{2}}$$

The pressure in this formula should be in millimeters of mercury.

If the orifice allows all the molecules incident on its area to pass through it, then $G$ in the previous formula is equal to the mass of substance passing through the orifice. Evidently, the smaller the wall thickness compared to the diameter of the orifice, the better will this condition be fulfilled. At the same time, the presence of an orifice in the vessel containing the substance disturbs the condensed-phase–vapor equilibrium. The deviation from equilibrium, i.e., the difference between the saturated vapor pressure and the pressure found by the effusion method, depends on the diameter of the effusion orifice. To reduce or entirely eliminate the deviation from equilibrium, it is necessary for the flow through the orifice to be "molecular," i.e., for the free path of a molecule $L$ to be equal to or greater than, in order of magnitude, the orifice radius: $L \geqslant r$. Under such conditions the rate of flow of the gas is no longer determined by collision of the molecules with each other (as at ordinary pressures) but by their collision with the tube walls. But if the area of the orifice is so large and the pressure so high that the flow of the gas ceases to be a molecular flow, Knudsen's formula cannot be used.

Thus the orifice is made small and a correction is introduced to make allowance for part of the molecules, those obliquely incident on the orifice, being reflected back from the inner side wall.

The correction coefficient can easily be calculated by assuming elastic collision. For $0 \leqslant l/r \leqslant 1.50$ ($l$ being the wall thickness and $r$ the orifice radius) it is equal to

$$K = [1+(l/2r)]^{-1}$$

But the impact is not elastic and complex absorptional interactions are possible. This is the weak spot of the Knudsen method, which makes it necessary to introduce one more coefficient into the formula, with no hope of determining it theoretically. Most authors assume it equal to one.

## *Table 11*

SUBLIMATION DATA FOR ORGANIC CRYSTALS

| Compound | Formula | T (°K) | $H_s$ (kcal/mole) |
|---|---|---|---|
| | Normal paraffins | | |
| Methane | $CH_4$ | 90 | 2.20 |
| Ethane | $C_2H_6$ | 90 | 4.90 |
| Propane | $C_3H_8$ | 86 | 6.81 |
| Butane | $C_4H_{10}$ | 107 | 8.57 |
| Pentane | $C_5H_{12}$ | 143 | 10.03 |
| Hexane | $C_6H_{14}$ | 178 | 12.15 |
| Heptane | $C_7H_{16}$ | 183 | 13.83 |
| Octane | $C_8H_{18}$ | 216 | 16.27 |
| Nonane | $C_9H_{20}$ | 219 | 17.82 |
| Decane | $C_{10}H_{22}$ | 243 | 20.26 |
| Undecane | $C_{11}H_{24}$ | 236.6 | 21.82 |
| Dodecane | $C_{12}H_{26}$ | 263 | 24.30 |
| Hexadecane | $C_{16}H_{34}$ | 291 | 32.24 |
| | Cycloparaffins | | |
| Cyclopropane | $C_3H_6$ | 145 | 6.99 |
| Cyclobutane | $C_4H_8$ | 145 | 8.71 |
| Cyclopentane | $C_5H_{10}$ | 122 | 10.19 |
| Cyclohexane | $C_6H_{12}$ | 186 | 11.13 |
| Cycloheptane | $C_7H_{14}$ | 134 | 12.78 |
| Cyclooctane | $C_8H_{16}$ | 166 | 14.05 |
| Cyclotetradecane | $C_{14}H_{30}$ | 320 | 32.0; 21.0 |
| | Polycyclic aromatic compounds | | |
| Benzene | $C_6H_6$ | | 10.7 |
| Azulene | $C_{10}H_8$ | | 16.2 |
| Naphthalene | $C_{10}H_8$ | | 17.3; 15.7; 18.3; 19.6 |
| Anthracene | $C_{14}H_{10}$ | | 24.4; 23.3; 22.8 |
| Phenanthrene | $C_{14}H_{10}$ | | 20.7 |
| 1,2-Benzanthracene | $C_{18}H_{12}$ | | 28.8 |
| Tetracene | $C_{18}H_{12}$ | | 29.8; 28.1 |
| Triphenylene | $C_{18}H_{12}$ | | 25.6 |
| Pentacene | $C_{22}H_{14}$ | | 37.7; 33.5 |
| Coronene | $C_{24}H_{12}$ | | 30.7 |
| Tribenzonaphthalene | $C_{26}H_{16}$ | | 33.9 |
| Tetrabenzoperylene | $C_{34}H_{18}$ | | 28.2 |

Thus, the pressure of interest in millimeters of mercury is found from the loss in weight of the sample $G$ by means of the formula

$$P = 17.14(G/\alpha KAT)(T/M)^{1/2}$$

where $A$ is the area of the orifice in square centimeters; $G$ is the change in weight, in grams; $t$ is the time of effusion; and $\alpha$ is a coefficient depending on adsorption of molecules on inner walls of orifice.

By varying the temperature of diffusion we get the experimental dependence of saturated vapor pressure on temperature. The experimental points are treated by the least-squares method, bearing in mind the equation $\ln P = -(A/T) + B$, given above. Here $A = H/R$, and hence the sublimation heat in calories

$$H = 4.575A$$

The heat of sublimation thus found is in the following relation to lattice energy:

$$-L = \Delta H_T + 2RT + \tfrac{9}{8}R\Theta$$

where $\tfrac{9}{8}R\Theta$ is the energy of the zero-point vibrations in the Debye approximation; $2RT$ is the difference between $6RT$, the energy per molecule of solid; and $4RT$, the energy per molecule of gas.

In concluding this section we give in Table 11 data on the sublimation of some groups of organic crystals.

REFERENCES

1. A. P. Ryzhenkov, Thesis, Univ. of Moscow, 1969.
2. M. J. P. Musgrave, *Proc. Roy. Soc. Ser. A* **226**, 339–356 (1954).
3. K. S. Alexandrov, Thesis, Univ. of Moscow, 1957.
4  K S. Alexandrov, *Soviet Physics–Crystallograph* **3**, 623–626 (1958).
5. V. F. Teslenko, Thesis, Univ. of Moscow, 1966.
6. W. Voigt, "Lehrbuch der Kristallphysik." Leipzig, 1928.
7. A. Z. Reuss, *Z. Angew. Math. Mech. B.* **9**, 49 (1929).
8. S. D. Volkov and N. A. Klinskikh, *Dokl. Akad. Nauk SSSR* **146**, 565–568 (1963).
9. R. F. S. Hearmon, *Advan. Phys.* **6**, 19, 323–361 (1956).
10. R. Hill, *Proc. Phys. Soc. London Sect. A* **65**, 349–354 (1952).
11. M. Markham, Mater. *Res. Stand.* **1**, 2, 107, (1962).
12. D. H. Chung, *Phil. Mag.* **8**, 89, 833 (1963).
13. W. Grandal, D. Chung, and T. Gray, "Mechanical Properties of Engineering Ceramics," Wiley (Interscience), New York, 1961.
14. I. P. Galashkevich, Thesis, Moscow, 1966.
15. E. E. Westrum and J. P. McCullough, *Phys. Chem. Org. Solid State* **1**, 87 (1963).
16. P. W. Bridgman, "The Physics of High Pressure." Bell, London, 1931: second edition, 1949.
17. L. D. Landau and E. M. Lifshitz, "Statistical Physics." Pergamon, Oxford, 1968.

# Chapter VI
# The Theory of Thermodynamics

1. GENERAL RELATIONSHIPS

*a. Scalar Form*

The thermodynamic properties of a substance, i.e., the behavior of a body when energy in the form of heat or work is transferred to it, are fully described if the directly measurable quantities are known, such as volume $V$ and any of heat capacities, say $C_p$, both these quantities being functions of two parameters of state, for example, $p$ and $T$.

As is known from statistical physics, the functions $V(p, T)$ and $C_p(p, T)$ are not independent. For a complete description of the behavior of a body, it is sufficient to have information about one of the thermodynamic potentials given as a function of two variables of state. It is most convenient to start by considering the so-called free energy of a body $F$ specified as a function of volume (per mole or molecule) and temperature. The function $F(V, T)$ is convenient because it is associated with the structural model of a body in the simplest way. Namely,

$$F = -kT \ln Z$$

where the sum over states $Z$ can be calculated if the values of the energy levels of the system $E_m$ and also the dependence of all the $E_m$ on volume are known:

$$Z = \sum_m e^{-E_m(V)/kT}$$

If $F(V, T)$ is specified, then the values of all the thermodynamic quantities measured can be obtained by differentiation. As is known, these relations are arrived at in the following way.

The first law of thermodynamics states that a change in the internal energy of a body is equal to the sum of the work done on it and the heat supplied to it:

$$dE = dQ + dA$$

The amount of heat absorbed by a body is an inexact differential, but $dS = dQ/T$ is an exact differential according to the second law of thermodynamics; the state function $S$ is called the entropy. Introducing the free energy $F = E - TS$, we get

$$dF = -S\,dT - p\,dV$$

From this we have the equalities

$$S = -(\partial F/\partial T)_V, \qquad p = -(\partial F/\partial V)_T, \qquad \text{and} \qquad (\partial S/\partial V)_T = (\partial p/\partial T)_V \tag{1.1}$$

If $F = F(V, T)$ is known, the second relation in (1.1) will give the state equation for a body $p = f(V, T)$, and from the first equation we can find any heat capacity by way of further differentiation

$$C_p = T(\partial S/\partial T)_p, \qquad C_V = T(\partial S/\partial T)_V \tag{1.2}$$

etc.

An experimental investigation of the state equation for a body is a difficult undertaking. At best we have at our disposal data on the linear cross sections of $V(T)$ and $V(p)$. These functions are generally characterized by their derivatives: the isobaric coefficient of volume expansion

$$\alpha = V^{-1}(\partial V/\partial T)_p$$

and the isothermal compressibility

$$\beta_T = -V^{-1}(\partial V/\partial p)_T \tag{1.3}$$

Among the heat capacities, we choose $C_p$ for measurement. Here, again because of practical difficulties, we usually have information on these quantities as functions of one variable. The quantities $\alpha$ and $C_p$ are known, as a rule, as functions of temperature. Isothermal compressibility measured on a press is determined as a function of pressure, and adiabatic compressibility

$$\beta_S = -V^{-1}(\partial V/\partial p)_S$$

measured from the velocity of sound is known to be a function of temperature.
  Having a model of a body, we shall work out an expression for the free
energy $F$ as a function of volume and temperature. By direct differentiation we
obtain the values of the isothermal compressibility, expansion coefficient, and
heat capacity $C_p$ as functions of $V$ and $T$, after which we can compare theory
with experiment. Thus, using the second equation from (1.1), we get

$$1/\beta_T = V(\partial^2 F/\partial V^2)_T \tag{1.4}$$

The quantities $F$, $V$, and heat capacities are convenient to determine per mole
of a substance. The first of Eqs. (1.1) yields

$$C_V = -T(\partial^2 F/\partial T^2)_V \tag{1.5}$$

And, finally, the third equation from (1.1) is transformed, with the help of the
relation

$$(\partial V/\partial p)_T (\partial p/\partial T)_V (\partial T/\partial V)_p = -1$$

which is valid for any function $V(p, T)$, into the form

$$(\partial S/\partial V)_T = -\partial^2 F/\partial V\,\partial T = \alpha/\beta_T \tag{1.6}$$

We thus have

$$\alpha = -\frac{1}{V}\frac{\partial^2 F/\partial V\,\partial T}{\partial^2 F/\partial V^2} \tag{1.7}$$

  If the values of $\beta_S$ and $C_p$ are needed for comparing with experiment, we
shall resort to the auxiliary equations

$$C_p - C_V = T(\alpha^2/\beta_T)V \tag{1.8}$$

$$\beta_S/\beta_T = C_V/C_p \tag{1.9}$$

$$\beta_S = \beta_T - (T\alpha^2 V/C_p) \tag{1.10}$$

To find a relation between $C_p$ and $C_V$, we pass over from the variables $p$
and $T$ to $V$ and $T$. Let $V = f(p, T)$. We write the function $S(V, T)$ in the form
$S = S[f(p, T), T]$. We have

$$(\partial S/\partial T)_p = (\partial S/\partial f)_T (\partial f/\partial T)_p + (\partial S/\partial T)_f$$

$$= (\partial S/\partial V)_T (\partial V/\partial T)_p + (\partial S/\partial T)_V$$

Substituting (1.6) into the last expression yields (1.8). Similarly, writing

$V = V(p, T)$ as $V = V[p, f(S, p)]$, we get

$$(\partial V/\partial p)_S = (\partial V/\partial f)_p (\partial f/\partial p)_S + (\partial V/\partial p)_f$$
$$= (\partial V/\partial T)_p (\partial T/\partial p)_S + (\partial V/\partial p)_T$$

Considering that

$$(\partial T/\partial p)_S (\partial p/\partial S)_T (\partial S/\partial T)_p = -1$$

and, taking into account (1.6),

$$(\partial S/\partial p)_T = (\partial S/\partial V)_T (\partial V/\partial p)_T = -(\partial V/\partial T)_p$$

we obtain

$$\left(\frac{\partial T}{\partial p}\right)_S = \frac{(\partial V/\partial T)_p}{(\partial S/\partial T)_p}$$

that is

$$\left(\frac{\partial V}{\partial p}\right)_S = \frac{(\partial V/\partial T)_p^2}{(\partial S/\partial T)_p} + \left(\frac{\partial V}{\partial p}\right)_T$$

which coincides with (1.10), as follows from the definition (1.3). Equation (1.9) results from (1.8) and (1.10).

In this section we summarize all the principal scalar relations that will be needed to verify theory or to calculate the quantities that have not been measured directly.

Indeed, if the proposed model gives the free energy as a function of $V$ and $T$, we can compute, by means of Eqs. (1.4), (1.5), and (1.7), the isothermal compressibility, the heat capacity $C_V$, and the expansion coefficient. By using these quantities and Eqs. (1.8) and (1.10), we can also determine adiabatic compressibility and heat capacity $C_p$.

We can solve the reverse problem, i.e., find any thermodynamic function from the measured data, with an accuracy of up to the value of the energy at the absolute zero of temperature. For this purpose, the quantities $C_p$, $\alpha$, and $\beta$ must be measured over the entire interval of temperatures and pressures.

For processes taking place at normal pressure,

$$S = \int_0^T (C_p/T)\, dt, \qquad E - E_0 = \int_0^T C_p\, dT$$

The values of $C_p - C_V$ obtained by experiment according to (1.8) permit us to find the work of expansion. Indeed, for $p = 0$

$$dQ = dE = C_V\, dT + (\partial E/\partial V)\, dV$$

It is the second term that represents the work of expansion. It is given by

$$(\partial E/\partial V)\, dV = (C_p - C_V)\, dT$$

When a crystal is heated from zero to $T$, this work is equal to

$$\int (T\alpha^2 V/\beta_T)\, dT$$

If we assume that the crystal energy may be represented by the sum of $E^{\mathrm{vib}}$ and the lattice energy $U(V)$, the measurement of $C_p - C_V$ as a function of $T$ will make it possible to compute the increment of the lattice energy (provided the variation of the vibration frequencies with temperature is not great) since

$$\int (T\alpha^2 V/\beta_T)\, dT = U - U_0 + \Delta V E^{\mathrm{vib}} \tag{1.11}$$

### b. Tensor Form

When a substance in the form of a single crystal is used, we can carry out a considerably more fruitful experiment than the one described above. The expansion tensor can be measured relatively easily at constant (zero) pressure:

$$\alpha_{kl} = (\partial \varepsilon_{kl}/\partial T)_p$$

as well as the adiabatic tensor of elasticity $c_{ijkl}^s$ (also at $p = 0$) by measuring the velocity of sound.

In the anisotropic case the work per unit volume is determined from the expression $\sum \sigma_{ij} d\varepsilon_{ij}$. The internal-energy and free-energy differentials therefore have the form

$$dE = V \sum \sigma_{ij}\, d\varepsilon_{ij} + T\, dS$$

$$dF = V \sum \sigma_{ij}\, d\varepsilon_{ij} - S\, dT$$

The following thermodynamic relations arise in this case:

$$\sigma_{ij} = V^{-1}(\partial F/\partial \varepsilon_{ij})_T, \qquad (\partial \varepsilon_{ij}/\partial T)_\sigma = V^{-1}(\partial S/\partial \sigma_{ij})_T$$

$$V^{-1}(\partial S/\partial \varepsilon_{ij})_T = -(\partial \sigma_{ij}/\partial T)_\varepsilon \tag{1.12}$$

Since the isothermal elasticity

$$c_{ijkl}^T = (\partial \sigma_{ij}/\partial \varepsilon_{kl})_T$$

we have

$$c_{ijkl}^T = V^{-1}(\partial^2 F/\partial \varepsilon_{ij}\, \partial \varepsilon_{kl})_T$$

On the other hand,

$$\sigma_{ij} = V^{-1}(\partial E/\partial \varepsilon_{kl})_S$$

and, consequently, the adiabatic elasticity

$$c_{ijkl}^S = V^{-1}(\partial^2 E/\partial \varepsilon_{ij}\, \partial \varepsilon_{kl})_S$$

A relation analogous to (1.6) is obtained by substituting the expression for Hooke's law into (1.12) for $(\partial S/\partial \varepsilon_{ij})_T$. We get

$$\partial S/\partial \varepsilon_{ij} = -V \sum c_{ijkl}^T \alpha_{kl} \qquad (1.13)$$

Thus, knowing, as earlier, how the free energy varies as a function of temperature as well as the shape and size of the unit cell, we can compute all the properties of the crystal.

The isothermal elasticity and entropy are obtained by direct differentiation. By differentiating the entropy, we arrive at relation (1.13) from which the expansion tensor is found. The heat capacity is also determined by direct differentiation of $F$:

$$C_\varepsilon = -T(\partial^2 F/\partial T^2)_\varepsilon$$

Here knowledge of the relations between the heat capacities $c_\sigma$ and $c_\varepsilon$ (heat capacities at all constant stresses and at constant volume and shape, respectively) as well as between the adiabatic elasticity $c^S$ and the isothermal elasticity $c^T$ is also required.

The derivation of these relations is quite analogous to that of Eqs. (1.8) and (1.10). For convenience, not only the elasticity tensor $c_{ijkl}$ but the inverse tensor

$$s_{ijkl} = c_{ijkl}^{-1}$$

should be used.

After appropriate calculations we get

$$c_\sigma - c_\varepsilon = VT \sum c_{ijkl}^T \alpha_{ij} \alpha_{kl} \qquad (1.14)$$

$$s_{ijkl}^S - s_{ijkl}^T = -\alpha_{ij}\alpha_{kl}(TV/C_p) \qquad (1.15)$$

Equation (1.10) can readily be arrived at through the summation of (1.15). To do this, use should be made of a formula where $\beta$ is represented in terms of the elements of the tensor of compliances. The derivation of such a formula presents no difficulty.

For triaxial compression, Hooke's law assumes the form

$$\varepsilon_{ij} = -p\varepsilon_{ijkl}\delta_{kl} = -ps_{ijkl}$$

But volume compression is equal to $\varepsilon_{ii}$, i.e.,

$$\varepsilon_{ii} = -ps_{iikk}$$

whence

$$\beta = s_{1111} + s_{2222} + s_{3333} + 2(s_{1122} + s_{2233} + s_{3311})$$

2. SPECIFIC FEATURES OF THE THERMODYNAMICS OF MOLECULAR CRYSTALS.
INTRODUCTION OF THE CHARACTERISTIC TEMPERATURE

*a. Splitting up the Free Energy into Intermolecular and Intramolecular Parts*

To predict the behavior of an organic molecular crystal, it is expedient to split up its thermodynamic functions into intermolecular and intramolecular components.

The free energy of a crystal may be written as

$$f = U + kT \sum_\alpha \ln(1 - e^{-\hbar\omega_\alpha/kT}) \qquad (2.1)$$

This is the so-called lattice component of free energy. For molecular crystals, however, the total free energy is reduced to the lattice component of free energy $f$ because of the absence of electron exchange. The lattice energy $U$ is equal to the potential energy of interaction between molecules. This quantity was discussed in detail in Chapter 2. The lattice energy is a function of the parameters of the unit cell of a crystal (at a given symmetry of the arrangement of molecules).

The summation in (2.1) is carried out over all the frequencies $\omega$ of normal vibration. We can always easily separate intramolecular frequencies from intermolecular ones and write

$$f = F + f_{\text{mol}}$$

where $F$ is the intermolecular or crystal part of free energy; and $f_{\text{mol}}$ is the intramolecular part which can successfully be calculated, as is known, from the oscillator formula.

Evidently, to the crystal part of the free energy belong 6NZ frequencies ($N$ = number of cells and $Z$ = number of molecules in a cell), and to calculate $F$ it is necessary to know the path of the $6Z$ dispersion surfaces $\omega(\mathbf{k})$.

Neglecting the zero-point energy of the vibrations, which is much lower than the lattice energy, we may write $F$ as

$$F = U_{\text{lat}} + kT \sum \ln(1 - e^{-\hbar\omega(\mathbf{k})/kT})$$
$$= U_{\text{lat}} + kT \sum_{\mathbf{k},s} \ln[2\sinh(\hbar\omega_s(\mathbf{k})/2kT)] \qquad (2.2)$$

where $s$ stands for the enumeration of the branches of the dispersion surfaces. In practice, in order to compute the vibrational part of $F$, its contribution to $F$ must be calculated for a relatively small number of points of reciprocal space and an average value taken.

A theoretical calculation of the dispersion surfaces is a time-consuming procedure; the same also refers to experimental research (e.g., by the method of neutron scattering). Investigation of the possibilities of simpler models therefore acquires importance.

## b. Debye's Expressions for the Thermodynamic Functions

It is well known that the so-called Debye approximation is very useful in the physics of crystals. The essence of Debye's approximation consists in replacing the seemingly quite individual spectrum of lattice vibrations by a simple one-parameter dependence. The free energy (2.1) may be written in the form

$$F = U_{\text{lat}} + kT \int f(\omega) \ln(1 - e^{-\hbar\omega/kT}) \, d\omega$$

In this equation $f(\omega) \, d\omega$ signifies the number of frequencies lying in the interval from $\omega$ to $\omega + d\omega$. Debye showed that at very low temperatures $f(\omega) \sim \omega^2$. At high temperatures all kinds of waves are excited. An attempt may therefore be made to replace a real spectrum by a parabola $A\omega^2$ which is cut off at a certain frequency $\omega_{\text{max}}$. In this case

$$A \int_0^{\omega_{\text{max}}} \omega^2 \, d\omega$$

is an integral with respect to the spectrum which must be equal to the total number of normal vibrations. Debye considered atomic lattices and set this integral equal to three times the number of atoms. The crystal part of the energy of a molecular crystal refers to molecules each of which has six degrees of freedom instead of three. Consequently, in our case

$$A \int_0^{\omega_{\text{max}}} \omega^2 \, d\omega = 6N$$

where $N$ is the number of molecules in the crystal (in one mole). Hence the constant $A = 18N/\omega_{\text{max}}^3$ and

$$F = U_{\text{lat}} + 18RT/\omega_{\text{max}}^3 \int_0^{\omega_{\text{max}}} \omega^2 \ln(1 - e^{-\hbar\omega/kT}) \, d\omega$$

The only spectrum parameter is the maximum frequency $\omega_{\text{max}}$ or, what is more convenient, the characteristic temperature

$$\Theta = \hbar\omega_{\text{max}}/k$$

Integrating by parts and introducing the Debye function

$$D(x) = \frac{3}{x^3} \int_0^x \frac{z^3}{e^z - 1} \, dz$$

we get for the free energy of a crystal, composed of particles with six degrees of freedom,

$$F = U_{\text{lat}} + 2RT[3\ln(1 - e^{-\Theta/T}) - D(\Theta/T)]$$

After differentiation we obtain

$$E = U_{\text{lat}} + 6RT D(\Theta/T)$$

$$S = 6R\ln(1 - e^{-\Theta/T}) + 8R\, D(\Theta/T) \qquad (2.3)$$

$$C_V = 6R[D(\Theta/T) - (\Theta/T)\, D'(\Theta/T)]$$

The behavior of the Debye functions at high temperatures ($\Theta/T \ll 1$) and at low temperatures ($\Theta/T \gg 1$) is of interest.

The characteristic temperatures of many organic crystals are close to 100°K. Thus "high" temperatures are, strictly speaking, unattainable—the crystal begins to melt. The high-temperature tendencies of the quasi-harmonic functions, however, show up quite distinctly in experiments, as will be shown below. At $\Theta/T \ll 1$

$$D(\Theta/T) \cong 1 \quad \text{and} \quad \ln(1 - e^{-\Theta/T}) \cong \ln(\Theta/T)$$

Consequently,

$$F = U_{\text{lat}} + 6RT\ln(\Theta/T) - 2RT, \qquad C_V = 6R$$

$$E = U_{\text{lat}} + 6RT, \qquad\qquad\qquad S = -6R\ln(\Theta/T) + 8R$$

The approximation for low temperatures becomes valid only at temperatures lower than the boiling point of liquid hydrogen. At $\Theta/T \gg 1$

$$D(x) \approx \pi^4/5x^3 \quad \text{but} \quad \ln(1 - e^{-\Theta/T}) \approx 0$$

Consequently,

$$F = U_{\text{lat}} - \frac{2\pi^4 R}{5} \frac{T^4}{\Theta^3}, \qquad E = U_{\text{lat}} + \frac{6\pi^4 R}{5} \frac{T^4}{\Theta^3}$$

$$S = \frac{8\pi^4 R}{5} \frac{T^3}{\Theta^3}, \qquad\qquad C = \frac{24\pi^4 RT^3}{5\Theta^3}$$

It is important to note that the low-temperature formulas must be fulfilled rigorously. Unfortunately, the exact fulfillment of these formulas takes place for temperatures considerably lower than $\Theta/4$, i.e., for temperatures lower than the boiling point of liquid helium.

The high-temperature approximation would be better in the absence of anharmonicity.

If $\Theta$ is assumed to be a constant, it will mean that the vibrational spectrum is invariable with respect to the state parameters. Such a crystal model may be described as harmonic. This approximation is hardly of any interest. If $\Theta = \text{const}$, then $C_p = C_V$ and thermal expansion does not occur. It is, however, not difficult to dispense with this restriction.

*c. The Quasi-Harmonic Model. Scalar Form*

We have said that the crystal spectrum is replaced by a curve with one parameter. This is what was done by Debye. There is no need, however, to consider $\Theta$ a constant. The energy levels and hence the frequencies of the normal vibrations, and, consequently, the characteristic temperature, may be regarded as functions of the size of a unit cell. This model is called quasi harmonic. It may be described in two variants. First, in scalar form: The characteristic temperature and the lattice energy are treated as functions of volume per molecule. And second, in tensor form: The quantities $\Theta$ and $U$ are functions of the parameters of the crystal cell.

Naturally, we shall start with the simpler model. Since the entropy and heat capacity $C_V$ are derivatives of the free energy at constant volume, the quasi-harmonic functions for $E$, $S$, and $C$ coincide with Debye's expressions (2.3), namely,

$$E = U + 6RTD(\Theta/T)$$

$$S = -6R\ln(1-e^{-\Theta/T}) + 8RD(\Theta/T) \qquad (2.3')$$

$$C_V = 6R[D(\Theta/T) - (\Theta/T)D'(\Theta/T)]$$

(We remind the reader that these expressions are written for the crystal parts of the thermodynamic functions.)

Let us now find the bulk modulus (the modulus of uniform compression) $1/\beta_T = V(\partial^2 F/\partial V^2)_T$. We are to seek the first derivative of the free energy

$$\left(\frac{\partial F}{\partial V}\right)_T = 6R\frac{e^{-\Theta/T}}{1-e^{-\Theta/T}}\frac{\partial\Theta}{\partial V} - 2RD'\left(\frac{\Theta}{T}\right)\frac{\partial\Theta}{\partial V} + \frac{\partial U}{\partial V}$$

For Debye's function $D(x)$, however, the following relation holds true:

$$D(x) = \frac{x}{e^x - 1} - \frac{x}{3}D'(x) \qquad (2.4)$$

Using (2.4), we get

$$\left(\frac{\partial F}{\partial V}\right)_T = \frac{\partial U}{\partial V} + \frac{1}{\Theta}\frac{\Theta}{V}6RTD\left(\frac{\Theta}{T}\right) = \frac{\partial U}{\partial V} + 6RT\frac{\partial\ln\Theta}{\partial V}D\left(\frac{\Theta}{T}\right)$$

We now introduce the Grüneisen constant

$$\gamma = -(V/\Theta)(d\Theta/dV) = -V\partial\ln\Theta/\partial V$$

Then

$$\left(\frac{\partial F}{\partial V}\right)_T = \frac{\partial U}{\partial V} - \frac{6RT\gamma}{V}D\left(\frac{\Theta}{T}\right) \qquad (2.5)$$

Differentiating again with respect to volume and multiplying by $V$, we obtain the isothermal bulk modulus of elasticity:

$$\frac{1}{\beta_T} = V\left(\frac{\partial^2 F}{\partial V^2}\right)_T$$

$$= V\left[\frac{\partial^2 U}{\partial V^2} + 6RT\frac{\partial^2 \ln\Theta}{\partial V^2}D\left(\frac{\Theta}{T}\right) + 6R\Theta\left(\frac{\partial \ln\Theta}{\partial V}\right)^2 D'\left(\frac{\Theta}{T}\right)\right] \quad (2.6)$$

Let us now calculate the volume expansion coefficient. According to (1.6),

$$\alpha = \beta_T(\partial S/\partial V)_T$$

But

$$(\partial S/\partial V)_T = (\partial S/\partial\Theta)(\partial\Theta/\partial V)$$

The entropy $S$, like a number of other thermodynamic functions, depends only on the relation $x = \Theta/T$. For such functions, the following equation holds:

$$\frac{\partial f}{\partial\Theta} = -\frac{T}{\Theta}\frac{\partial f}{\partial T} \quad (2.7)$$

Using this relation for evaluating the entropy, we get

$$\frac{\partial S}{\partial V} = -\frac{T}{\Theta}\frac{\partial S}{\partial T}\frac{\partial\Theta}{\partial V}$$

That is,

$$\alpha = V^{-1}\beta_T C_V \gamma = -\beta_T C_V \partial\ln\Theta/\partial V$$

Finally, we have

$$\alpha = -\frac{6R}{V}\frac{\partial\ln\Theta}{\partial V}\left[\frac{\partial^2 U}{\partial V^2} + 6RT\frac{\partial^2\ln\Theta}{\partial V^2}D\left(\frac{\Theta}{T}\right) + 6R\Theta\left(\frac{\partial\ln\Theta}{\partial V}\right)^2 D'\left(\frac{\Theta}{T}\right)\right]^{-1}$$

$$\times\left[D\left(\frac{\Theta}{T}\right) - \frac{\Theta}{T}D'\left(\frac{\Theta}{T}\right)\right]$$

Heat capacities at constant pressure can be expressed in terms of $C_V$ and $\alpha$. Indeed,

$$C_p = T\left(\frac{\partial S}{\partial T}\right)_p = T\frac{\partial S}{\partial\Theta}\frac{\partial\Theta}{\partial V}\left(\frac{\partial V}{\partial T}\right)_p = -C_V\frac{\partial\ln\Theta}{\partial V}TV\alpha$$

It follows, incidentally, from this equation that $\Theta$ falls with increase of volume ($\gamma > 0$). This constitutes an indispensable feature of the quasi-harmonic model.

The heat capacity $C_p$ can, of course, be expressed in terms of the functions $\Theta(V)$ and $U(V)$ with the aid of the thermodynamic relation (1.8) as well. The adiabatic compressibility is defined as

$$\beta_S = \beta_T - \alpha V / C_V \gamma \tag{2.8}$$

One can also deduce directly an expression for $(1/\beta_S) - (1/\beta_T)$ (see below).

Thus, knowing the dependence of the lattice energy on the volume per molecule $U(V)$ and the volume dependence of the characteristic temperature $\Theta(V)$, we can compute in the quasi-harmonic approximation all the thermodynamic functions of a crystal.

The functions $\Theta(V)$ and $U(V)$ are not, however, independent. In fact, it is the state equation (2.5) for a crystal

$$-p = \frac{\partial U}{\partial V} + 6RT \frac{\partial \ln \Theta}{\partial V} D\left(\frac{\Theta}{T}\right)$$

that gives this relation.

Most experiments are conducted at atmospheric pressure; since $\partial U / \partial V \gg 1$ atm, the following relationship between $\Theta(V)$ and $U(V)$ holds true:

$$\frac{\partial U}{\partial V} = -6RT \frac{\partial \ln \Theta}{\partial V} D\left(\frac{\Theta}{T}\right) = -\frac{\partial \ln \Theta}{\partial V} E^{D} \tag{2.9}$$

where $E^{D}$ is the Debye energy for particles with six degrees of freedom.

In spite of the fact that room temperatures are not quite "high" for the majority of organic crystals, this relationship is still vaiid with an accuracy of up to 10–20% (for $T/\Theta \approx 2$–4). It is therefore useful to write down simplified relations for the high-temperature quasi-harmonic approximation. For $D(\Theta/T) \approx 1$, $\ln(1 - e^{-\Theta/T}) \approx \ln(\Theta/T)$ we have

$$F = U + 6RT \ln(\Theta/T) - 2RT, \qquad S = -6R \ln(\Theta/T) + 8R$$

$$E = U + 6RT, \qquad C_V = 6R$$

$$\frac{1}{\beta_T} = V\left[\frac{\partial^2 U}{\partial V^2} + 6RT \frac{\partial^2 \ln \Theta}{\partial V^2}\right] \tag{2.10}$$

$$\alpha = -\frac{6R}{V} \frac{\partial \ln \Theta}{\partial V} \left[\frac{\partial^2 U}{\partial V^2} + 6RT \frac{\partial^2 \ln \Theta}{\partial V^2}\right]^{-1}$$

The state equation now assumes the form

$$\frac{\partial U}{\partial V} = -6RT \frac{\partial \ln \Theta}{\partial V}$$

i.e. $U = -6RT \ln(\Theta/\Theta_0) + U_0$.

Taking a naphthalene crystal as an example, we will show below the degree of validity of the quasi-harmonic approximation and check the usefulness of the formulas given in this section.

Their successful application for the deduction of the thermodynamic functions of crystals by the atom–atom potential method will be possible only for cubic crystals. Only in this case can the functions $U(V)$ and $\Theta(V)$ be unambiguously computed from the law of interaction of molecules. Knowledge of these two functions enables us to introduce the temperature, for example, with the aid of Eq. (2.9). And this gives us a knowledge of all the thermodynamic relationships.

### d. The Quasi-Harmonic Model: Tensor Variant

An investigation of the model in which the expressions for the free energy and its temperature derivatives are in the same Debye form, while the lattice energy $U$ and the characteristic temperature $\Theta$ are treated as functions of the parameters of a crystal unit cell, is of interest. It is common practice to specify the size of the cell at absolute zero and to regard $U$ and $\Theta$ as functions of the elements of the cell strain tensor $\varepsilon_{ik}$. Thus, $U$ and $\Theta$ will be functions of six variables in the case of a triclinic crystal, of four variables for a monoclinic crystal, of three variables for an orthorhombic one, etc.

We are to find expressions for the tensors of expansion and elasticity (isothermal as well as adiabatic). This problem is solved by methods similar to those just described.

Using property (2.4) of Debye's function, we obtain

$$\left(\frac{\partial F}{\partial \varepsilon_{ik}}\right)_T = \frac{\partial U}{\partial \varepsilon_{ik}} + 6RT\frac{\partial \ln \Theta}{\partial \varepsilon_{ik}} D\left(\frac{\Theta}{T}\right)$$

$$\left(\frac{dE}{\partial \varepsilon_{ik}}\right)_T = \frac{\partial U}{\partial \varepsilon_{ik}} + 6R\Theta\frac{\partial \ln \Theta}{\partial \varepsilon_{ik}} D'\left(\frac{\Theta}{T}\right)$$

For the isothermal tensor of elasticity, we get

$$C^T_{ijkl} = \left(\frac{\partial^2 F}{\partial \varepsilon_{ij}\,\partial \varepsilon_{kl}}\right)_T$$

$$= \frac{\partial^2 U}{\partial \varepsilon_{ij}\,\partial \varepsilon_{kl}} + \frac{\partial^2 \ln \Theta}{\partial \varepsilon_{ij}\,\partial \varepsilon_{kl}} 6RT\, D\left(\frac{\Theta}{T}\right) + \frac{\partial \ln \Theta}{\partial \varepsilon_{ij}} \frac{\partial \ln \Theta}{\partial \varepsilon_{kl}} 6R\Theta\, D'\left(\frac{\Theta}{T}\right)$$

We now pass to the computation of the adiabatic elasticity. For the difference $E_{\text{vib}} = E - U$, the following relation holds:

$$\left(\frac{\partial E_{\text{vib}}}{\partial \Theta}\right)_S = \frac{\partial E^{\text{vib}}}{\partial \Theta} + \frac{\partial E^{\text{vib}}}{\partial T}\left(\frac{\partial T}{\partial \Theta}\right)_S$$

If the process is isentropic, we have

$$dS = (\partial S/\partial \Theta)\,d\Theta + (\partial S/\partial T)\,dT = 0$$

from which it follows that

$$(\partial T/\partial \Theta)_S = T/\Theta$$

By differentiating $E_{\text{vib}} = 6RT\,D(\Theta/T)$, we get

$$(\partial E_{\text{vib}}/\partial \Theta)_S = E_{\text{vib}}/\Theta$$

Now we have

$$\frac{\partial E_{\text{vib}}}{\partial \varepsilon_{ik}} = \left(\frac{\partial E_{\text{vib}}}{\partial \Theta}\right)_S \frac{\partial \Theta}{\partial \varepsilon_{ik}} = E_{\text{vib}}\frac{\partial \ln \Theta}{\partial \varepsilon_{ik}}$$

$$\frac{\partial^2 E_{\text{vib}}}{\partial \varepsilon_{ik}\,\partial \varepsilon_{jl}} = E_{\text{vib}}\left(\frac{\partial^2 \ln \Theta}{\partial \varepsilon_{ik}\,\partial \varepsilon_{jl}} + \frac{\partial \ln \Theta}{\partial \varepsilon_{ik}}\frac{\partial \ln \Theta}{\partial \varepsilon_{jl}}\right)$$

Hence

$$C^S_{ijkl} = \frac{\partial^2 U}{\partial \varepsilon_{ik}\,\partial \varepsilon_{jl}} + \left(\frac{\partial^2 \ln \Theta}{\partial \varepsilon_{ik}\,\partial \varepsilon_{jl}} + \frac{\partial \ln \Theta}{\partial \varepsilon_{ik}}\frac{\partial \ln \Theta}{\partial \varepsilon_{jl}}\right)6RT\,D\left(\frac{\Theta}{T}\right)$$

The difference between the adiabatic and isothermal moduli of elasticity is written in the form

$$C^S_{ijkl} - C^T_{ijkl} = \frac{\partial \ln \Theta}{\partial \varepsilon_{ik}}\frac{\partial \ln \Theta}{\partial \varepsilon_{jl}}T\,C_V$$

By using the above reasoning one can, of course, obtain a formula for the adiabatic bulk modulus of elasticity:

$$\frac{1}{\beta_S} = V\frac{\partial^2 U}{\partial V^2} + V\left[\frac{\partial^2 \ln \Theta}{\partial V^2} + \left(\frac{\partial \ln \Theta}{\partial V}\right)^2\right]6RT\,D\left(\frac{\Theta}{T}\right)$$

and a formula for the differences in the moduli:

$$\frac{1}{\beta_S} - \frac{1}{\beta_T} = V\left(\frac{\partial \ln \Theta}{\partial V}\right)^2 T\,C_V$$

This formula is of course equivalent to (2.8).

Expression (1.13) for the quasiharmonic model becomes

$$\frac{\partial S}{\partial \Theta}\frac{\partial \Theta}{\partial \varepsilon_{ij}} = -V\sum C^T_{ijkl}\alpha_{kl}$$

i.e.

$$C_V\frac{\partial \ln \Theta}{\partial \varepsilon_{ij}} = V\sum C^T_{ijkl}\alpha_{kl}$$

The last expression may be used for comparing the tensor $\partial \ln \Theta/\partial \varepsilon_{ij}$ calculated by the atom–atom potential method with the experimental data.

## 3. EXPERIMENTAL CHARACTERISTIC TEMPERATURE

The concept of the characteristic temperature of a solid body appears when we introduce the Debye model. There is of course always the possibility of representing the actual spectrum of a crystal by a parabola cut off at a certain maximum frequency, so that the areas of the actual and Debye spectra are equal. The characteristic temperature $\Theta = \hbar\omega/k$ obtained in this way will be a state function, like any thermodynamic function.

Determination of $\Theta$ by experiment through the measurement of the actual spectrum of a crystal, however, is a very inexpedient procedure since the evaluation of a spectrum is generally a highly complicated task.

At the same time an investigation into the behavior of the characteristic function is of unquestionable interest, because it clearly shows the degree of applicability of the harmonic and quasi-harmonic models.

With this purpose in mind it is useful to introduce $\Theta$ as an experimental quantity with the aid of a thermodynamic function for which the Debye expression remains true.

The experimental characteristic temperature $\Theta^{\text{exp}}$ may be introduced, without any restriction of generality, either as a function of $V$, $T$, or as a function of the cell parameters and $T$. It is necessary only to agree as to which of the thermodynamic functions will be selected for determination of $\Theta^{\text{exp}}(V, T)$.

An investigation of $\Theta^{\text{exp}}(V, T)$ determined by means of the Debye formula for the entropy

$$S = -6R\ln(1 - e^{-\frac{\Theta^{\text{exp}}}{T}}) + 8RD(\Theta^{\text{exp}/T})$$

has been carried out by the author in collaboration with Koreshkov [1].

In determining $\Theta^{\text{exp}}$ as a function of two variables from the expressions for $S$, formulas (2.3) for $E$ and $C_V$ are no longer valid. Indeed, if $S = S[\Theta^{\text{exp}}(V, T)]$, the formulas for other thermodynamic functions change their form. The relation between free energy and entropy will now take the form

$$S = -\frac{\partial F}{\partial \Theta^{\text{exp}}}\frac{\partial \Theta^{\text{exp}}}{\partial T} + \frac{\partial F}{\partial T}$$

The heat capacities will be given by

$$C_V = T\frac{\partial S}{\partial T} + T\frac{\partial S}{\partial \Theta^{\text{exp}}}\left(\frac{\partial \Theta^{\text{exp}}}{\partial T}\right)_V$$

$$C_p = T\frac{\partial S}{\partial T} + T\frac{\partial S}{\partial \Theta^{\text{exp}}}\left(\frac{\partial \Theta^{\text{exp}}}{\partial T}\right)_p$$

The last formula (for $C_p$) coincides with the quasi-harmonic model.

Let us now consider the behavior of the function $\Theta^{exp}(V, T)$ introduced through the Debye expression for the entropy (but for particles with six degrees of freedom).

Detailed investigation of this physical quantity [2] is unfortunately possible only for a small number of cases. The substances that are really suitable are those for which $C_p$ has been measured over the entire temperature interval (so that $S = \int_0^T (C_p/T)\,dt$ may be computed), and for which the frequencies of the normal intramolecular vibrations are known so that $S_{mol}$ can be calculated and $S_{cr} = S - S_{mol}$ found. These data, necessary as a minimum, allow us to investigate the isobaric cross section (for $p = 0$) of the characteristic temperature. A more complete description of the function $\Theta^{exp}(V, T)$ requires a knowledge of the coefficients of thermal expansion and the compressibility factor as functions of temperature. Crystals that have been examined as thoroughly as this are few in number. We will therefore first consider more extensive data on the isobaric cross sections of $\Theta^{exp}$ for various crystals.

Figure 1 illustrates the behavior of the characteristic temperature for one of the crystals with rigid molecules. A common regularity for these as well as for other crystals is the fall of $\Theta^{exp}$ with rise of temperature. This refers to temperatures higher than the temperature of liquid nitrogen. At low temperatures a slight increase of $\Theta^{exp}$ with temperature is observed. Here we should recall the results (see Chapter III) of measurement of the vibrational intermolecular frequencies obtained by the Japanese authors. They have found that the fall of the frequencies begins from about the temperatures of liquid nitrogen. At low temperatures the frequencies do not change, within the accuracy of their experiment. It is noteworthy that the rapidity of the fall of $\Theta^{exp}$ and lattice frequencies with temperature is of the same order (15–20% within the range from nitrogen temperature to room temperature). This coincidence of the results and, in general, the character of this temperature

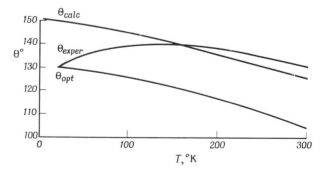

*Fig. 1.* Theoretical and experimental values of the characteristic temperature for naphthalene. $\theta^{oa}$, see Table 5.

change await interpretation. It is believed, however, that it predicts the success of the quasi-harmonic model, because at temperatures in the last hundred degrees down to absolute zero the cell dimensions change but very slightly.

The range within which lies the temperature $\Theta^{exp}$, and the nature of its variation in the isobaric process, are typical of rigid molecules of the naphthalene type.

The frequencies of the vibrations along intermolecular hydrogen bonds are somewhat greater than those of intermolecular vibrations. Therefore, if the vibrations along hydrogen bonds are assigned to the crystal vibrations, it is natural to expect only slight differences in the behavior of $\Theta^{exp}$. The calculation of $\Theta^{exp}$ has been made for crystals of urea, glycine, and hydrazine (see Fig. 2).

Comparing the behavior of $\Theta^{exp}$ of these crystals and of crystals without hydrogen bonding, we see that the picture has changed. $\Theta^{exp}$ has become greater, which is quite natural, because the hydrogen bond is highly rigid. At high temperatures $\Theta^{exp}$ falls more abruptly, and at low temperatures a sharp rise of $\Theta^{exp}$ is observed. It is still not clear to what extent these results can be interpreted within the framework of the quasi-harmonic model.

In the work cited above similar graphs have been plotted for about 20 substances. The isobaric cross sections of $\Theta^{exp}(V, T)$ shown in the figures have proved to be typical. In all molecular crystals without exception the

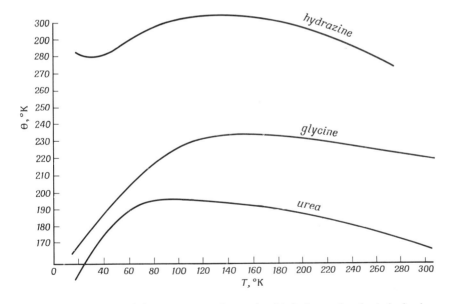

*Fig. 2.* The characteristic temperature of crystals with hydrogen bonds. 1, hydrazine; 2, glycine; 3, urea.

$\Theta^{\mathrm{exp}}(T)$ curves at $p = 0$ are curves with a maximum lying in the region of 100°K (i.e., in the region of the values of the characteristic temperature). The rate of decrease of $\Theta^{\mathrm{exp}}$ in the region of fall varies within narrow limits for very different substances ($\Delta\Theta/\Delta T \sim 0.07\text{--}0.15$).

Large differences are observed in the low-temperature region of rise of $\Theta$. This rise, as has been indicated above, is less distinct in the case of rigid molecules. For flexible molecules the derivatives $d\Theta/dT$ are approximately identical on the left and on the right of the maximum.

The hydrogen bonds, as we see, play a part, but the presence of dipole moments (in full agreement with the results of Chapter II on the interaction of dipole molecules) does not affect the behavior of the isobaric cross section of the characteristic temperature.

We shall now discuss a method for a more complete description of the behavior of $\Theta^{\mathrm{exp}}$. Let us consider the procedure of analysis for scalar and tensor cases.

*a. Scalar Form*

In the harmonic approximation the characteristic temperature is a constant; while in the quasi-harmonic approximation, it is a function of volume only. The experimental characteristic temperature is a function of two variables: $\Theta^{\mathrm{exp}} = \Theta(V, T)$. An explicit dependence of $\Theta^{\mathrm{exp}}$ on temperature is a consequence of anharmonicity. To evaluate the magnitude of the anharmonicity, it is necessary to know the dependence $\Theta^{\mathrm{exp}}(T)$ in the isochoric process. If such a dependence does not exist, then the quasi-harmonic approximation does hold.

Let us consider the derivatives of the characteristic temperature for various parameters in various processes. By differentiating the expression for the crystal part of the entropy with respect to temperature at constant pressure, we find, after certain transformations, that

$$\left(\frac{\partial \Theta^{\mathrm{exp}}}{\partial T}\right)_p = \frac{C_p - C_{\mathrm{D}} - C_{\mathrm{mol}}}{T(\partial S/\partial \Theta)}$$

This formula may be conveniently written in the form

$$\left(\frac{\partial \Theta^{\mathrm{exp}}}{\partial T}\right)_p = \frac{\Theta^{\mathrm{exp}}}{T}\left(1 - \frac{C_p - C_{\mathrm{mol}}}{C_{\mathrm{D}}}\right) \tag{3.1}$$

where $C_{\mathrm{D}} = T(\partial S/\partial T)$ is given by Eq. (1.2).

The determination of $(\partial \Theta^{\mathrm{exp}}/\partial T)_p$ with the help of Eq. (3.1) must, of course, lead to results that agree with the isobaric cross section of $\Theta^{\mathrm{exp}}$ found from the entropy. Equation (3.1) is more expedient to use, as will be seen below.

Differentiating the entropy with respect to volume at constant temperature and taking into account that the molecular contribution to the entropy depends on temperature alone, i.e., $(\partial S_{mol}/\partial V)_T = 0$, we get

$$\left(\frac{\partial \Theta^{exp}}{\partial V}\right)_T = \frac{\alpha}{\beta_T(\partial S/\partial \Theta)}$$

Here $\alpha$ is the coefficient of thermal expansion and $\beta_T$ the isothermal compressibility.

Using Eq. (1.8), we also obtain

$$\left(\frac{\partial \Theta^{exp}}{\partial V}\right)_T \frac{\partial V}{\partial T} = \alpha V \left(\frac{\partial \Theta^{exp}}{\partial V}\right)_T = -\frac{C_p - C_V}{C_D} \frac{\Theta^{exp}}{T} \tag{3.2}$$

Since $\Theta^{exp}$ is a function of two variables, only two derivatives are independent, no matter which, e.g., those written above. All the remaining derivatives are expressed in terms of these two and $\alpha$, $\beta_T$:

$$\left(\frac{\partial \Theta^{exp}}{\partial V}\right)_p = \frac{1}{\alpha V}\left(\frac{\partial \Theta^{exp}}{\partial T_1}\right)_p$$

$$\left(\frac{\partial \Theta^{exp}}{\partial T}\right)_V = \left(\frac{\partial \Theta^{exp}}{\partial T}\right)_p - \alpha V \left(\frac{\partial \Theta^{exp}}{\partial V}\right)_T \tag{3.3}$$

Taking into account Eqs. (3.2) and (3.1), we have

$$\left(\frac{\partial \Theta^{exp}}{\partial T}\right)_V = \frac{\Theta^{exp}}{T}\left(1 - \frac{C_V - C_{mol}}{C_D}\right)$$

$$\left(\frac{\partial \Theta^{exp}}{\partial P}\right)_T = \beta_T V\left(\frac{\partial \Theta^{exp}}{\partial V}\right)_T, \qquad \left(\frac{\partial \Theta^{exp}}{\partial P}\right)_V = \frac{\beta_T}{\alpha}\left(\frac{\partial \Theta^{exp}}{\partial T}\right)_V \tag{3.4}$$

It is clear that if the characteristic temperature $\Theta^{exp}$ is regarded as a function of two variables, we must distinguish two Grüneisen "constants"

$$\gamma_T = -(V/\Theta)(\partial \Theta/\partial V)_T, \qquad \gamma_p = -(V/\Theta)(\partial \Theta/\partial V)_p$$

The thermodynamic behavior of a crystal is determined by these two "constants."

We can, of course, express all the requisite thermodynamic quantities with the aid of $\Theta^{exp}(V, T)$. Equations (3.1) and (3.2) are the most significant. The expression for the heat capacity at constant volume can easily be calculated:

$$C_V = T\left(\frac{\partial S}{\partial T}\right)_V = T\left(\frac{\partial S}{\partial T}\right)_{\Theta exp} + T\left(\frac{\partial S}{\partial \Theta^{exp}}\right)_T\left(\frac{\partial \Theta^{exp}}{\partial T}\right)_V$$

or, taking into account (2.6), which remains valid for the entropy not only in the quasi-harmonic case,

$$C_V = C_D\left[1 - \frac{T}{\Theta^{exp}}\left(\frac{\partial \Theta^{exp}}{\partial T}\right)_V\right] \tag{3.5}$$

356                                                6 The Theory of Thermodynamics

This result as well as the expressions for other thermodynamic functions can however be obtained from (3.1) and (3.2).

For high temperatures, Eq. (3.5) is transformed into

$$C_V = 6R\left[1 - \frac{T}{\Theta^{\text{exp}}}\left(\frac{\partial\Theta^{\text{exp}}}{\partial T}\right)_V\right]$$

In the quasi-harmonic approximation we have

$$\left(\frac{\partial\Theta^{\text{exp}}}{\partial T}\right)_V = 0 \quad \text{and} \quad C_V = 6R$$

Our attention should of course be focused on Eq. (3.4), since it is this equation that determines the difference between reality and the quasi-harmonic model. It is expedient to determine the degree of quasi harmonicity by the formula

$$\eta = -\frac{T}{\Theta^{\text{exp}}}\left(\frac{\partial\Theta^{\text{exp}}}{\partial T}\right)_V \tag{3.6}$$

that is,

$$\eta = \frac{C_V - C_{\text{mol}}}{C_{\text{D}}}$$

b. Tensor Form

In a general case the characteristic temperature is a function of the unit cell parameters and temperature. It is convenient to assume that $\Theta^{\text{exp}} = \Theta(\varepsilon_i, T)$, where $\varepsilon_i$ is the relative change of $a_i$. Let us again assume that $\Theta^{\text{exp}}$ is determined from the experimental values of the crystal part of the entropy. Evidently, the entropy as well as $\Theta^{\text{exp}}$ may be represented by surfaces in the multidimensional space $\varepsilon_i$, $T$. As in the isotropic case, the data of the isobaric experiment make it possible to investigate one of the cross sections of this surface.

Two experiments are of interest: thermal deformation at $p = 0$ and triaxial isothermal compression. In the first experiment the behavior of 0 is established with the aid of the derivatives $(\partial\Theta^{\text{exp}}/\partial\varepsilon_{ij})_p$, and in the second by $(\partial\Theta^{\text{exp}}/\partial\varepsilon_{ij})_T$. We can now introduce the corresponding Grüneisen tensors

$$\gamma_T = -\frac{1}{\Theta}\left(\frac{\partial\Theta^{\text{exp}}}{\partial\varepsilon_{ij}}\right)_T \quad \text{and} \quad \gamma_p = -\frac{1}{\Theta}\left(\frac{\partial\Theta^{\text{exp}}}{\partial\varepsilon_{ij}}\right)_p$$

In the quasi-harmonic model $\gamma_T$ is equal to $\gamma_p$. Determination of the tensors $\gamma_T$ and $\gamma_p$ from directly measurable quantities can be made with the help of the following relations.

From (1.13) we obtain

$$\left(\frac{\partial\Theta^{\text{exp}}}{\partial\varepsilon_{ij}}\right)_T = \frac{V}{(\partial S/\partial\Theta)_T}\sum C_{ijkl}^T \alpha_{kl} \tag{3.7}$$

The necessity to know the isothermal elasticity is a vexing complication. It is the adiabatic elasticities that are usually measured in experiments, and recalculation of $C^S$ into $C^T$ is somewhat difficult.

For estimating $\gamma_p$, we have no formula similar to (3.7). The relation with experiment is provided by the formula

$$(\partial\Theta/\partial T)_p = \sum (\partial\Theta/\partial\varepsilon_{ij})_p \alpha_{ij}$$

Equation (3.1) remains valid for $(\partial\Theta/\partial T)_p$. For this quantity we have, by analogy with (3.3),

$$(\partial\Theta^{\text{exp}}/\partial T)_p = (\partial\Theta^{\text{exp}}/\partial T)_V + \sum (\partial\Theta^{\text{exp}}/\partial\varepsilon_{ij})_T \alpha_{ij} \qquad (3.8)$$

Having determined the sum in (3.8) on the basis of (3.7), we find the experimental value of $(\partial\Theta^{\text{exp}}/\partial T)_V$ which characterizes the degree of departure of a crystal from the quasi-harmonic model. The degree of quasi harmonicity can still be computed with the help of (3.6), but the value of $\eta$ determined from (3.8) will not exactly coincide with (3.6), because the derivatives $(\partial\Theta^{\text{exp}}/\partial T)_V$ in formulas (2.8) and (3.3) have different meanings: we have supposed above that $\Theta$ depends only on the unit cell volume, whereas in calculations carried out with the aid of (3.8) the quantity $\Theta$ is considered to be a function of the cell parameters. In other words, $(\partial\Theta/\partial T)_V$ in (3.6) is a derivative with retention of volume, and in (3.8) it is a derivative with retention of volume and shape.

## 4. THERMODYNAMIC FUNCTIONS OF A NAPHTHALENE CRYSTAL

### a. Scalar Quantities

The crystal of naphthalene plays the role of rock salt in the physics of molecular crystals. It is a typical representative as regards both structure and properties. We have more than once noted both explicitly and implicitly that molecular crystals having no intermolecular hydrogen bonding are very similar to one another. A careful investigation of the thermodynamic functions of a naphthalene crystal will provide us with a clear understanding of the behavior of a large class of substances. Our choice in this respect is very limited. For an exhaustive analysis, we must have a crystal whose heat capacity has been measured from the temperature of liquid helium. Such substances are perhaps several tens in number. Furthermore, it is necsesary to know the expansion tensor and the tensor of elasticity, again from a very low temperature to the melting point. And such substances are only a few in number. Even in the case of naphthalene, the measurement of elasticity and expansion has not yet been extended to the very low temperatures. Besides, we need a crystal, for the molecules of which the spectrum of intramolecular vibrations is known; not to mention that we should have at our disposal experimental data concerning the dynamics of the lattice in order to check the models under study. Intermolecular frequencies of vibrations have been measured for a very limited number of cases, and a complete study of dispersion surfaces (in the case of naphthalene) has been started only recently by English investigators.

The principal data for a naphthalene crystal are collected in 39 columns of Table 1. The values of the physical quantities are given as a function of temperature at atmospheric pressure. The data on the heat capacity $C_p$ have been reported in the literature [3]; the heat capacity $C_{mol}$ has been computed from the values of the frequencies of the intramolecular vibrations [4]; the cell dimensions have been measured and the thermal expansion tensor computed by Ryzhenkov [5]; the tensor of adiabatic elasticity has been obtained by measuring the velocity of sound by Afanassieva [6]; the compliances have been computed by inversion of the matrix $c_{ijkl}$ and from them the values of the adiabatic compressibility $\beta_S$ have been obtained. The thermal expansion coefficient has been measured up to the temperature of liquid nitrogen, and the compressibility factor, up to 100°K.

It is not always easy to estimate the accuracy of experimental data. The low-temperature data available are incomplete. Interpolation and extrapolation procedures therefore acquire great importance. The temperature dependence for the greater part of the physical quantities presented in Table 1 may be of a rather complicated nature and cannot be expressed in terms of simple temperature polynomials. Nevertheless, it is hardly probable that the curves would have a large number of frequently occurring points of inflection. Where possible, the curve was smoothed out by the displacement of one or two points, thereby changing the values of the physical quantities by 2 or 3%.

An especially crucial step is extrapolation to zero. This procedure is additionally complicated by the fact that a number of peculiarities may be expected at temperatures lower than $\Theta/4$ (i.e., lower than 20–30°K). The low-temperature Debye behavior of the thermodynamic functions may manifest itself, as has been shown by a number of authors, at temperatures $T < \Theta/50$. Thus, Debye's formulas may sometimes prove inapplicable for extrapolation to zero.

The introduction of the function $\Theta^{exp}(V, T)$ may be of substantial use in extrapolation and interpolation of experimental data.

We can probably rely on the following features in the behavior of $\Theta^{exp}(V, T)$. At very low temperatures, as follows from the theory of Born and von Kármán that has been confirmed for many simple crystals,

$$\Theta = \Theta_0(1 - bT^2)$$

The initial fall of $\Theta$ may not continue for a long time. In many cases for simple crystals (and also in our case) the curve passes through a minimum, bends, runs through a maximum, and then falls almost linearly with rise in temperature. The linear (or almost linear) fall of $\Theta$ (in an isobaric experiment) usually begins at temperatures close to the characteristic temperature.

The derivative $(\partial\Theta/\partial T)_V$ determines the departure of a crystal from quasi-harmonic behavior. It is natural to assume that $(\partial\Theta/\partial T)_V$ should fall smoothly to zero as $T \to 0$. In the case of the quasi-harmonic model $(\partial\Theta/\partial V)_T = (\partial\Theta/\partial V)_p$. It may be supposed that these derivatives become equal to each other at temperatures lower than, say, 20°K.

These specific features in the behavior of the derivatives have been used to extrapolate the expansion coefficient to zero. As seen from the graph for a naphthalene crystal given in the preceding chapter, extrapolation to zero is extremely difficult—at temperatures of the order of 70–100°K the $\alpha(T)$ curve becomes horizontal, and it is completely impossible to predict where its fall and transition to the $\alpha \sim T^3$ law will begin.

Setting $(\partial\Theta/\partial V)_T$ and $(\partial\Theta/\partial V)_p$ equal to each other at low temperatures, we can find $\alpha$, since extrapolation to zero of the other quantities entering into the expressions for these derivatives presents no difficulty.

Some remarks concerning Table 1 remain to be made.

We resort to integration and use the theorem of the mean to match the integral quantities with the corresponding functions. If this is not done, then, for example, the $\Theta(T)$ curve determined with the aid of the table of Debye's functions from the values of the entropy $S$ will be found unmatched with the values of the derivatives $(\partial\Theta/\partial T)_p$. Column 7 in Table 1 has been computed from the formula $\Delta S = (\overline{C_p - C_{mol}}) \ln(T_2/T_1)$. The data in columns 9, 10, and 11 are based on the table of Debye's functions. The derivative $(\partial\Theta/\partial T)_p$ has been calculated from the formula

$$\left(\frac{\partial\Theta}{\partial T}\right)_p = \frac{\Theta}{T}\left(1 - \frac{C_p - C_{mol}}{C_D}\right)$$

Column 15 and the succeeding columns have been compiled first for temperatures 100°K and higher. The value of $\beta_S$ can readily be extrapolated to zero, and the values of $\alpha$, as was indicated above, for 20 and 60°K were written later, by equating $(\partial\Theta/\partial V)_T$ and $(\partial\Theta/\partial V)_p$ for the lowest temperatures. Column 18 contains the data on the difference $\beta_S - \beta_T$. In column 20 the following is written:

$$(\partial S/\partial V)_T = \alpha/\beta_T$$

From this value and the figure of column 11 we calculate

$$\frac{1}{\Theta}\left(\frac{\partial\Theta}{\partial V}\right)_T = \left(\frac{\partial\ln\Theta}{\partial V}\right)_T = -\frac{\alpha}{\beta_T C_D}$$

**Table 1** Thermodynamic Functions of Naphthalene

| 1 | 2 | 3 | 4 | 5 | 6 | 7 | 8 | 9 | 10 | 11 | 12 | 13 |
|---|---|---|---|---|---|---|---|---|---|---|---|---|
| $T(°K)$ | $C_p$ (cal/mole deg) | $C_{mol}$ | $C_p - C_{mol}$ | $\overline{C_p - C_{mol}}$ | $\ln(T_1/T_2)$ | $\Delta S$ | $S$ | $\Theta(°K)$ | $\Theta/T$ | $C_D$ | $\dfrac{C_p - C_{mol}}{C_D}$ | $\left(\dfrac{\partial\Theta}{\partial T}\right)_p$ |
| 0 | 0 | 0 | 0 | | | | 0 | | | 0 | | |
| 20 | 2.652 | 0.001 | 2.651 | 5.865 | 1.099 | 6.44 | 1.01 | 130.9 | 6.544 | 2.645 | 1.00227 | −0.0148 |
| 60 | 10.21 | 1.13 | 9.08 | 9.78 | 0.511 | 5.00 | 7.45 | 137.3 | 2.289 | 9.30 | 0.978 | 0.050 |
| 100 | 14.34 | 3.86 | 10.48 | 10.86 | 0.336 | 3.65 | 12.45 | 139.8 | 1.398 | 10.82 | 0.970 | 0.042 |
| 140 | 18.44 | 7.20 | 11.24 | 11.595 | 0.247 | 2.86 | 16.10 | 141.1 | 1.008 | 11.33 | 0.995 | 0.005 |
| 180 | 22.97 | 11.02 | 11.95 | 12.495 | 0.199 | 2.47 | 18.96 | 141.3 | 0.785 | 11.55 | 1.035 | −0.027 |
| 220 | 28.12 | 15.22 | 12.90 | 13.525 | 0.166 | 2.24 | 21.43 | 139.4 | 0.634 | 11.67 | 1.105 | −0.066 |
| 260 | 33.81 | 19.66 | 14.15 | 15.18 | 0.144 | 2.19 | 23.67 | 135.8 | 0.522 | 11.75 | 1.203 | −0.106 |
| 300 | 40.43 | 24.22 | 16.21 | 17.21 | 0.124 | 2.14 | 25.86 | 130.5 | 0.435 | 11.81 | 1.374 | −0.162 |
| 340 | 46.90 | 28.69 | 18.21 | | | | 28.00 | 124.0 | 0.365 | 11.83 | 1.539 | −0.196 |

| 1 | 14 | 15 | 16 | 17 | 18 | 19 | 20 | 21 |
|---|---|---|---|---|---|---|---|---|
| $T(°K)$ | $V$ (cm³/mole) | $\alpha \cdot 10^6$ $(\text{deg}^{-1})$ | $\alpha V$ (cm³/mole deg) | $\beta^S \cdot 10^{10}$ (cm³/dyne) | $\dfrac{\alpha^2 VT}{C_p}$ | $\beta^T \cdot 10^{10}$ (cm²/dyne) | $\partial S/\partial V$ (cal/cm³ deg) | $\left(\dfrac{\partial(\ln\Theta)}{\partial V}\right)_T$ (mole/cm³) |
| 0 | 101.7 | 0 | 0 | 0.095 | 0 | 0.095 | | |
| 20 | 102.0 | 35 | 0.00356 | 0.097 | 0.000 | 0.097 | 0.0860 | −0.0322 |
| 60 | 102.6 | 133 | 0.01365 | 0.106 | 0.003 | 0.109 | 0.298 | −0.0321 |
| 100 | 103.2 | 166 | 0.0171 | 0.111 | 0.004 | 0.115 | 0.347 | −0.0320 |
| 140 | 103.8 | 181 | 0.0188 | 0.116 | 0.005 | 0.121 | 0.358 | −0.0316 |
| 180 | 104.7 | 204 | 0.0214 | 0.125 | 0.007 | 0.132 | 0.372 | −0.0322 |
| 220 | 105.9 | 240 | 0.0254 | 0.132 | 0.011 | 0.143 | 0.403 | −0.0346 |
| 260 | 107.1 | 308 | 0.0330 | 0.150 | 0.020 | 0.170 | 0.436 | −0.0371 |
| 300 | 108.7 | 404 | 0.0440 | 0.175 | 0.032 | 0.207 | 0.467 | −0.0394 |
| 340 | 110.2 | 525 | 0.0580 | 0.205 | 0.055 | 0.260 | 0.484 | −0.0408 |

# 4. Thermodynamic Functions of a Naphthalene Crystal

Table 1—continued

| $T(°K)$ | 22 $(\partial\Theta/\partial V)_p$ (deg mole/cm³) | 23 $\left(\frac{\partial\Theta}{\partial V}\right)_T$ | 24 $\left(\frac{\partial\ln\Theta}{\partial V}\right)_p$ | 25 $\gamma_T$ | 26 $\gamma_p$ | 27 $C_p - C_V$ | 28 $\overline{C_p - C_V}$ | 29 $\Delta(\Delta U)$ (cal/mole) | 30 $\Delta U$ (cal/mole) |
|---|---|---|---|---|---|---|---|---|---|
| 0 | | | | | | 0 | | | 0 |
| 20 | −4.2 | −4.2 | −0.0322 | 3.28 | 3.28 | 0.006 | 0.003 | 0.06 | 0.06 |
| 60 | 3.66 | −4.41 | 0.0267 | 3.30 | −2.74 | 0.240 | 0.123 | 4.94 | 4.98 |
| 100 | 2.46 | −4.48 | 0.0176 | 3.31 | −1.82 | 0.596 | 0.418 | 16.72 | 21.7 |
| 140 | 0.27 | −4.46 | 0.0019 | 3.28 | −0.20 | 0.948 | 0.772 | 30.88 | 52.6 |
| 180 | −1.26 | −4.55 | −0.0089 | 3.37 | 0.93 | 1.413 | 1.189 | 47.56 | 100.1 |
| 220 | −2.60 | −4.82 | −0.0186 | 3.66 | 1.97 | 2.25 | 1.84 | 73.6 | 173.5 |
| 260 | −3.22 | −5.04 | −0.0237 | 3.97 | 2.54 | 3.71 | 2.98 | 119.2 | 292.7 |
| 300 | −3.68 | −5.14 | −0.0282 | 4.27 | 3.06 | 6.15 | 4.93 | 197.2 | 489.9 |
| 340 | −3.38 | −5.06 | −0.0273 | 4.50 | 3.01 | 9.53 | 7.84 | 313.6 | 803.5 |

| $T(°K)$ | 31 $\frac{C_p - C_V}{C_D}$ | 32 $\left(\frac{\partial\Theta}{\partial V}\right)_T V\alpha$ | 33 $C_V$ | 34 $C_V - C_M$ | 35 $\overline{(C_V - C_M)\Delta T}$ (cal/mole) | 36 $H$ (cal/mole) | 37 $F^{vib}$ (cal/mole) | 38 $\frac{C_V - C_{mol}}{C_D}$ | 39 $\left(\frac{\partial\Theta}{\partial T}\right)_V$ |
|---|---|---|---|---|---|---|---|---|---|
| 0 | | 0 | 0 | 0 | | 0 | 0 | | 0 |
| 20 | 0.0023 | 0.015 | 2.646 | 2.645 | 26.5 | 26.5 | +6.3 | — | 0 |
| 60 | 0.026 | 0.060 | 9.97 | 8.84 | 229.7 | 261.1 | −186.0 | 0.950 | 0.114 |
| 100 | 0.055 | 0.077 | 13.74 | 9.88 | 374.4 | 652.2 | −592.8 | 0.912 | 0.123 |
| 140 | 0.083 | 0.084 | 17.49 | 10.29 | 403.4 | 1086.5 | −1167.5 | 0.906 | 0.095 |
| 180 | 0.124 | 0.097 | 21.54 | 10.52 | 416.2 | 1550.2 | −1862.6 | 0.913 | 0.068 |
| 220 | 0.193 | 0.122 | 25.87 | 10.65 | 423.4 | 2047.0 | −2667.6 | 0.913 | 0.055 |
| 260 | 0.316 | 0.166 | 30.10 | 10.44 | 421.8 | 2588.0 | −3566.2 | 0.887 | 0.059 |
| 300 | 0.520 | 0.226 | 34.28 | 10.06 | 410.0 | 3195.2 | −4562.8 | 0.850 | 0.065 |
| 340 | 0.803 | 0.294 | 37.37 | 8.68 | 374.8 | 3883.6 | −5636.4 | 0.732 | 0.098 |

Column 24 contains the values

$$\left(\frac{\partial \Theta}{\partial V}\right)_p = \frac{1}{\alpha V}\left(\frac{\partial \Theta}{\partial T}\right)_p$$

Equating for the lowest temperatures

$$\frac{1}{\alpha V}\left(\frac{\partial \Theta}{\partial T}\right)_p = -\frac{\alpha}{\beta_T C_D}$$

helps us to extrapolate the quantity $\alpha$ to zero.

The values of $\gamma_T$ and $\gamma_p$ in columns 25 and 26 are equal by definitions to

$$-\frac{V}{\Theta}\left(\frac{\partial \Theta}{\partial V}\right)_T \quad \text{and} \quad -\frac{V}{\Theta}\left(\frac{\partial \Theta}{\partial V}\right)_p$$

In column 27 are given the calculated and slightly rounded values of $C_p - C_V = \alpha^2 V T / \beta_T$.

Column 29 gives the values of $\Delta(\Delta U) = \overline{(C_p - C_V)}\,\Delta T$; $\Delta U$ is the work of expansion of a lattice.

In compiling column 32, one can check the correctness of the figures and the unambiguity of interpolation. Column 32 is calculated, on the one hand, as a product of columns 23 and 16, and on the other,

$$\left(\frac{\partial \Theta}{\partial V}\right)_T V\alpha = -\frac{\Theta}{T}\frac{C_p - C_V}{C_D}$$

i.e., it is equal to the product of columns 10 and 31.

The values of $C_V - C_{\text{mol}}$ must lie on a smooth curve, since the $C_p - C_{\text{mol}}$ and $C_p - C_V$ curves have been smoothed out. In column 35 are given the values of $\overline{(C_V - C_{\text{mol}})}\,\Delta T$. In our case the enthalpy is equal to the internal energy (since the table has been compiled for atmospheric pressure). Column 36 for $H = E$ can be computed either from column 5 or as

$$\sum_0^T (C_V - C_{\text{mol}})\,\Delta T + \Delta U$$

The free energy (the vibrational crystal part) $F = E - TS$ is included in column 37.

Finally, column 39 is calculated by the formula

$$\left(\frac{\partial \Theta}{\partial T}\right)_V = \frac{\Theta}{T}\left(1 - \frac{C_V - C_{\text{mol}}}{C_D}\right)$$

As we have said above, these results are of interest because they are typical. It is quite clear that molecular crystals have their own specific features. To these belong the high values of $C_p - C_V$. The smoothing of the crystal part of the heat capacity $C_V$ is characteristic of them. The value 12 cal/mole, which is ideal for a lattice composed of particles with six degrees of freedom, has not been reached because of the deviation from quasi harmonicity. Indeed,

$$C_V - C_{mol} = C_D\left[1 - \frac{T}{\Theta^{exp}}\left(\frac{\partial\Theta^{exp}}{\partial T}\right)_V\right] = C_D\eta$$

where $\eta$ is the degree of quasi harmonicity.

It is worthy of notice that the vibrational parts of the crystal energy constitute, at temperatures below 100°K, only a small part of the lattice energy

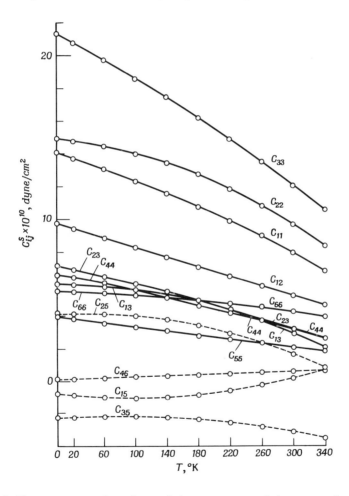

*Fig. 3.* The temperature dependence of the components of the tensor of adiabatic elasticity moduli for naphthalene.

(which is equal to 16,700 cal/mole for naphthalene); they do not play a leading role at room temperature either.

It is interesting that a small percentage of the departure from the quasi harmonicity at low temperatures can lead to the passage of $\Theta$ through a minimum and a maximum. The rise of $\Theta$ within a certain interval of low temperatures is not caused by the quasi-harmonic effect. In fact, as has already been pointed out, $C_p - C_V > 0$, but in the quasi-harmonic model

$$C_p - C_V = T \frac{\partial S}{\partial \Theta} \left( \frac{\partial \Theta^q}{\partial T} \right)_p = -\frac{T}{\Theta} C_D \left( \frac{\partial \Theta^q}{\partial T} \right)_p$$

i.e., $\Theta^q$ must fall over the whole temperature range.

### b. Tensor Quantities

On the basis of measurements of the adiabatic elasticity tensor (Teslenko and Afanassieva) and of the expansion tensor (Ryzhenkov), we can give tables and graphs of a number of tensor quantities characterizing the behavior of a naphthalene crystal. Figures 3 and 4 show the temperature dependences

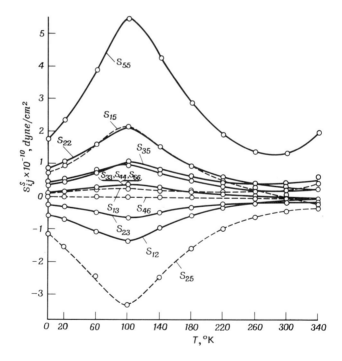

**Fig. 4.** The temperature dependence of the components of the tensor of adiabatic compliance moduli for naphthalene.

*Table 2*

DIFFERENCE BETWEEN ADIABATIC
AND ISOTHERMAL COMPLIANCES
(MODULI OF FLEXIBILITY) OF NAPHTHALENE

| | $\alpha_i \alpha_j \dfrac{TV}{C_p} \cdot 10^{-10}$ (cm²/dyne) | | |
|---|---|---|---|
| $ij$ | 100°K | 220°K | 300°K |
| 11 | 0.001 | 0.002 | 0.006 |
| 22 | 0.000 | 0.000 | 0.000 |
| 33 | 0.001 | 0.003 | 0.006 |
| 44 | 0.000 | 0.000 | 0.000 |
| 55 | 0.000 | 0.002 | 0.005 |
| 66 | 0.000 | 0.000 | 0.000 |
| 12 | 0.000 | 0.001 | 0.002 |
| 13 | 0.001 | 0.002 | 0.006 |
| 23 | 0.000 | 0.001 | 0.002 |
| 15 | 0.001 | 0.002 | 0.006 |
| 25 | 0.000 | 0.001 | 0.001 |
| 35 | 0.001 | 0.003 | 0.005 |
| 46 | 0.000 | 0.000 | 0.000 |

of the adiabatic tensors of elasticity and compliance. Table 2 gives the values of the differences between the elements of the isothermal and adiabatic tensors of compliances calculated with the aid of Eq. (1.15) for three temperatures.

The tensor $s_{ijkl}^T$ has been calculated from the values of these differences and the isothermal elasticity tensor obtained by inversion of the matrix.

In Table 3 are presented the values of all the four tensors for the following temperatures: 100°, 200°, and 300°K.

Formula (1.13) was used to compute the tensor $\partial S/\partial \varepsilon_{ij}$, and the tensors $(\partial \Theta/\partial \varepsilon_{ij})_T$ and $\gamma_T$ were obtained by the use of (3.7). The path of the four elements of these tensors versus temperature is plotted in Fig. 5.

The considerable difference in the diagonal components of the tensors $\gamma_c$ points to the large anisotropy of lattice vibrations.

And, finally, in Fig. 6 we compare the curves of the derivatives $(\partial \Theta/\partial T)_\varepsilon$ and $(\partial \Theta/\partial T)_V$. The derivative $(\partial \Theta/\partial T)_V$ is also given in Table 1, and the derivative $(\partial \Theta/\partial T)_\varepsilon$ is computed from formula (3.8).

We thus compare the effect, on the degree of quasi harmonicity, of the

## Table 3

COMPONENTS OF THE TENSORS OF ADIABATIC AND ISOTHERMAL MODULI OF ELASTICITY AND COMPLIANCE (FLEXIBILITY) OF NAPHTHALENE AT DIFFERENT TEMPERATURES

$C^s_{ijkl} \times 10^{10}$ (dyne/cm²)

**100°K**

| | | | | | |
|---|---|---|---|---|---|
| 12.45 | 8.32 | 5.78 | −0.95 | 0 | 0 |
| | 14.12 | 6.08 | 4.12 | 0 | 0 |
| | | 18.63 | −2.10 | 0 | 0 |
| | | | 5.67 | 0 | 0.34 |
| | | | | 3.44 | 0 |
| | | | | | 5.40 |

**220°K**

| | | | | | |
|---|---|---|---|---|---|
| 9.98 | 6.58 | 4.53 | −0.46 | 0 | 0 |
| | 11.96 | 4.55 | 3.09 | 0 | 0 |
| | | 14.94 | −2.44 | 0 | 0 |
| | | | 4.38 | 0 | 0.58 |
| | | | | 2.73 | 0 |
| | | | | | 4.89 |

**300°K**

| | | | | | |
|---|---|---|---|---|---|
| 8.03 | 5.42 | 3.10 | 0.31 | 0 | 0 |
| | 9.78 | 3.38 | 1.82 | 0 | 0 |
| | | 12.19 | −2.98 | 0 | 0 |
| | | | 3.37 | 0 | 0.74 |
| | | | | 2.26 | 0 |
| | | | | | 4.42 |

$S^s_{ijkl} \times 10^{-10}$ (cm²/dyne)

**100°K**

| | | | | | |
|---|---|---|---|---|---|
| 0.982 | −1.371 | 0.385 | 2.149 | 0 | 0 |
| | 2.119 | −0.638 | −3.306 | 0 | 0 |
| | | 0.258 | 1.029 | 0 | 0 |
| | | | 5.473 | 0 | −0.011 |
| | | | | 0.177 | 0 |
| | | | | | 0.186 |

**220°K**

| | | | | | |
|---|---|---|---|---|---|
| 0.302 | −0.335 | 0.094 | 0.514 | 0 | 0 |
| | 0.605 | −0.239 | −0.955 | 0 | 0 |
| | | 0.185 | 0.451 | 0 | 0 |
| | | | 1.938 | 0 | −0.027 |
| | | | | 0.232 | 0 |
| | | | | | 0.208 |

**300°K**

| | | | | | |
|---|---|---|---|---|---|
| 0.210 | −0.133 | 0.004 | 0.084 | 0 | 0 |
| | 0.319 | −0.166 | −0.458 | 0 | 0 |
| | | 0.236 | 0.445 | 0 | 0 |
| | | | 1.386 | 0 | −0.051 |
| | | | | 0.308 | 0 |
| | | | | | 0.235 |

## *Table 3—continued*

$C^T_{ijkl} \times 10^{10}$ (dyne/cm²)

**100°K**

$$
\begin{pmatrix}
12.024 & 7.815 & 5.337 & 0 & -1.009 & 0 \\
 & 13.542 & 5.525 & 0 & 4.070 & 0 \\
 & & 18.225 & 0 & -2.189 & 0 \\
 & & & 5.670 & 0 & 3.449 \\
 & & & & 0.335 & 0 \\
 & & & & & 5.397
\end{pmatrix}
$$

**220°K**

$$
\begin{pmatrix}
9.469 & 6.098 & 3.870 & 0 & -0.414 & 0 \\
 & 11.494 & 4.072 & 0 & 3.090 & 0 \\
 & & 13.744 & 0 & -2.201 & 0 \\
 & & & 4.376 & 0 & 2.653 \\
 & & & & 0.568 & 0 \\
 & & & & & 4.880
\end{pmatrix}
$$

**300°K**

$$
\begin{pmatrix}
7.116 & 4.479 & 2.159 & 0 & 0.313 & 0 \\
 & 8.836 & 2.366 & 0 & 1.847 & 0 \\
 & & 11.189 & 0 & -2.982 & 0 \\
 & & & 3.368 & 0 & 2.270 \\
 & & & & 0.731 & 0 \\
 & & & & & 4.413
\end{pmatrix}
$$

$S^T_{ijkl} \times 10^{-10}$ (cm²/dyne)

**100°K**

$$
\begin{pmatrix}
0.983 & -1.371 & 0.386 & 0 & 2.150 & 0 \\
 & 2.119 & -0.638 & 0 & -3.306 & 0 \\
 & & 0.259 & 0 & 1.030 & 0 \\
 & & & 0.177 & 0 & -0.011 \\
 & & & & 5.473 & 0 \\
 & & & & & 0.186
\end{pmatrix}
$$

**220°K**

$$
\begin{pmatrix}
0.304 & -0.334 & 0.096 & 0 & 0.516 & 0 \\
 & 0.605 & -0.238 & 0 & -0.954 & 0 \\
 & & 0.188 & 0 & 0.448 & 0 \\
 & & & 0.232 & 0 & -0.027 \\
 & & & & 1.940 & 0 \\
 & & & & & 0.208
\end{pmatrix}
$$

**300°K**

$$
\begin{pmatrix}
0.216 & -0.131 & 0.010 & 0 & 0.090 & 0 \\
 & 0.319 & -0.164 & 0 & -0.457 & 0 \\
 & & 0.242 & 0 & 0.450 & 0 \\
 & & & 0.308 & 0 & -0.051 \\
 & & & & 1.391 & 0 \\
 & & & & & 0.235
\end{pmatrix}
$$

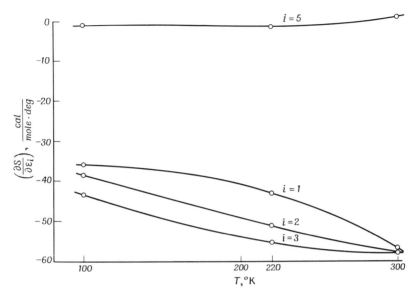

*Fig. 5.* The tensor of strain derivatives of entropy for naphthalene.

retention of the cell volume only with that of the retention of the cell volume and shape.

The behavior of these derivatives is very specific. One would expect that they would rise smoothly. The "jump" of anharmonicity at low temperatures cannot yet be explained within the framework of the quasi-harmonic model.

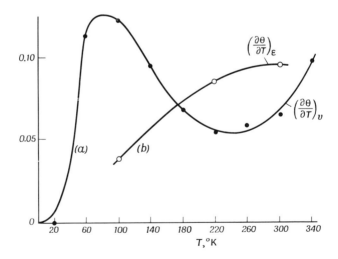

*Fig. 6.* The isochoric temperature derivative of the characteristic temperature of naphthalene: (a) in scalar variant; (b) in tensor variant.

## 5. CHOICE OF AN OPTIMAL QUASI-HARMONIC MODEL

The thermodynamic functions can be calculated in the quasi-harmonic approximation, without any special difficulty, by the use of atom–atom potentials. Further development of these investigations, however, will acquire meaning only when analysis of a thermodynamic experiment indicates that the quasi-harmonic model is sufficiently effective. The success of the quasi-harmonic model is by no means obvious *a priori* since it may be assumed that the anharmonicity in organic crystals is high (because of the small mass of molecules and weak intermolecular forces).

Anharmonicity in the atom–atom potential model can also in principle be taken into account by expanding the lattice energy according to low powers of the displacements of the molecules and by the use of higher (following the second) powers of these displacements. The complications involved in this approach however make it almost unsuitable for practical applications. It is for this reason that the quasi-harmonic approximation, which makes use of the harmonic equation of motion and takes into account the anharmonicity through the dependence of the force coefficients on the lattice parameters, is of particular interest. The characteristic temperature is calculated from the crystal part of the entropy by the Debye formulas

$$S_{cr} = -6R \ln(1 - e^{-\Theta/T}) + 8R D(\Theta/T)$$

The computed $\Theta$ is a function of two variables in the general case, e.g., $\Theta = \Theta(V, T)$.

If a real crystal were quasi harmonic, there would be no explicit dependence of $\Theta$ on temperature, i.e., explicit anharmonicity. In this case $\Theta = \Theta(V)$. It has been shown above that a real crystal is not described quite satisfactorily by the quasi-harmonic model.

In this section we wish to give an answer to the following question [7]: What function $\Theta(V)$ will be optimal, i.e., will provide the best agreement with experiment, if Debye's expressions are used for all the thermodynamic functions? Let us set the Grüneisen coefficient for the optimal model equal to

$$\gamma^{oq} = \gamma_T$$

where $\gamma_T = -(V/\Theta)(\partial\Theta/\partial V)_T$. The point is that $\gamma_T$ describes the variation of $\Theta$ in an isothermal process, which corresponds to the absence of explicit anharmonicity. The behavior of $\gamma_p$ or $\frac{1}{2}(\gamma_T + \gamma_p)$ proves, however, to be incompatible with the quasi-harmonic model.

The model becomes completely defined if the initial conditions are specified at $0°K$, but, from the practical point of view, it is convenient to specify them for $20°K$. After doing this, we can compute any thermodynamic function using the general formulas, for instance, the isothermal compressibility $\beta_T$, the difference of the heat capacities $(C_p - C_V)$, the work of expansion of a crystal by $\Delta V$ which is equal to the increment of the lattice energy on thermal expansion of a crystal, and also the thermal quantities—entropy, $S_{cr}$, heat capacities $C_p$ and $C_V$, and others.

For a model, which we have called the optimal quasi-harmonic model, a number of thermal and calorific quantities have been computed.

Tables 4 and 5 present the results of calculation of a number of quantities for the optimal model; the corresponding experimental values are also included for comparison. The thermal quantities practically coincide with experiment.

It is natural that, in spite of the excellent agreement of the value of $(C_p - C_V)^{oq}$ with experiment, the heat capacities themselves, both experimental and calculated, may differ greatly, since in the optimal quasi-harmonic model $C_V^{cr} = C_D$, which is not the case for a real crystal. The quantity $C_D$ is only slightly dependent on the form of the function $\Theta(V)$, particularly when $T > \Theta$, and it is therefore impossible to improve, to any extent, the agreement between the heat capacities themselves by the variation of that function. This is the argument which justifies our modifying the name of the proposed model by the word "optimal."

From the foregoing it follows that the calculated values of the enthalpy, entropy, and free energy may also differ strongly from the experimental data (up to a few percent). Thus, in calculations using atom–atom potentials within the framework of the quasi-harmonic model, excellent agreement can be

### Table 4

THERMAL CHARACTERISTICS OF NAPHTHALENE

| T (°K) | $C_p - C_V$ (cal/mole deg) | | $\Delta U$ (cal/mole) | | $\beta_T \times 10^{10}$ (cm³/erg) | |
|---|---|---|---|---|---|---|
| | exp.[a] | o.q. | exp. | o.q. | exp. | o.q. |
| 0 | 0 | 0 | 0 | 0 | 0.095 | ? |
| 20 | 0.006 | 0.006 | 0.06 | 0.06 | 0.097 | 0.097 |
| 60 | 0.240 | 0.248 | 4.98 | 5.14 | 0.109 | 0.104 |
| 100 | 0.596 | 0.604 | 21.7 | 22.2 | 0.115 | 0.113 |
| 140 | 0.948 | 0.950 | 52.6 | 53.3 | 0.121 | 0.120 |
| 180 | 1.413 | 1.445 | 100.1 | 101 | 0.132 | 0.130 |
| 220 | 2.25 | 2.28 | 173.5 | 176 | 0.143 | 0.141 |
| 260 | 3.71 | 3.76 | 292.7 | 296 | 0.170 | 0.168 |
| 300 | 6.15 | 6.15 | 489.9 | 495 | 0.207 | 0.208 |
| 340 | 9.53 | 9.53 | 803.5 | 808 | 0.260 | 0.261 |

[a] Here and in Table 5 the abbreviation exp. signifies "experimental values," and o.q. the values calculated for the optimal quasi-harmonic model.

## Table 5

### Thermal Characteristics of Naphthalene

| $T(°K)$ | $\Theta(°K)$ | | $S_{cr}$ | | $C_p - C_{mol}$ | | $C_p$ | | $C_v^{cr}$ | | $H_{cr}$ | | $F_{vib}^{cr}$ | |
| | exp. | o.q. | exp. | o.q. | exp. | o.q. | exp. | o.q. | exp. | o.q. | exp. | o.q. | exp. | o.q. |
| | | | | | | | | cal mole/deg | | | | | | |
| 1 | 2 | | 3 | | 4 | | 5 | | 6 | | 7 | | 8 | |
| 0 | ? | ? | 0 | 0 | 0 | 0 | 0 | 0 | 0 | 0 | 0 | 0 | 0 | 0 |
| 20 | 130.9 | 130.9 | 1.01 | 1.01 | 2.651 | 2.651 | 2.652 | 2.652 | 2.645 | 2.645 | 26.5 | 26.5 | +6.3? | +6.3? |
| 60 | 137.3 | 129.4 | 7.45 | 8.00 | 9.08 | 10.1 | 10.21 | 11.2 | 8.84 | 9.55 | 261 | 280 | −191 | −200 |
| 100 | 139.8 | 126.8 | 12.45 | 13.51 | 10.48 | 11.5 | 14.34 | 15.4 | 9.88 | 11.01 | 652 | 710 | −614 | −640 |
| 140 | 141.1 | 123.9 | 16.10 | 17.58 | 11.24 | 12.7 | 18.44 | 19.9 | 10.29 | 11.46 | 1086 | 1200 | −1220 | −1300 |
| 180 | 141.3 | 120.7 | 18.96 | 20.76 | 11.95 | 13.1 | 22.97 | 24.1 | 10.52 | 11.64 | 1550 | 1700 | −1963 | −2000 |
| 220 | 139.4 | 116.9 | 21.43 | 23.48 | 12.90 | 14.3 | 28.12 | 29.5 | 10.65 | 11.75 | 2047 | 2300 | −2841 | −2900 |
| 260 | 135.8 | 112.0 | 23.67 | 25.97 | 14.15 | 15.7 | 33.81 | 35.4 | 10.44 | 11.81 | 2088 | 2900 | −3859 | −3900 |
| 300 | 130.5 | 105.4 | 25.86 | 28.43 | 16.21 | 18.5 | 40.43 | 42.7 | 10.06 | 11.84 | 3195 | 3500 | −5053 | −5000 |
| 340 | 124.0 | 96.8 | 28.00 | 30.83 | 18.21 | 20.1 | 46.90 | 48.8 | 8.68 | 11.86 | 3884 | 4300 | −6440 | −6200 |

expected between the thermal quantities and experiment. At the same time it is clear that the computation of the heat capacity requires that anharmonicity should be taken into account explicitly.

It should be noted that the analysis just made refers only to a single isobaric cross section ($p = 0$). At pressures of the order of thousands of atmospheres the characteristic temperature and its dependence on the parameters of state may undergo a substantial change.

## 6. CALCULATION OF THE QUASI-HARMONIC MODEL BY THE ATOM–ATOM POTENTIAL METHOD

In the rigorous approach, such a calculation must be carried out through a full computation of the vibrational problem. The dispersion surfaces $\omega(\mathbf{k})$ permit strict ("strict" in the quasi-harmonic approximation) calculation of the vibrational part of the free energy with the aid of formula (2.2); the lattice energy is computed independently by using the methods outlined in Chapter II.

Despite the possibility of performing the most laborious calculations by means of computers, the wide use of such computations, however, is hardly justifiable. They are too cumbersome to be applied to a model that cannot claim (as is evident from the preceding section) to provide a precision higher than a few percent.

The quasi-harmonic model may be given a Debye form. This means that the variation of the spectrum with crystal cell dimensions is reduced to the change of the frequency of the cut-off of the Debye parabola $\omega_D$, i.e., the characteristic temperature $\Theta = \hbar\omega_D/k$.

As has already been shown (see Chapter III), we can compute, by the use of the atom–atom potential method, the mean square of the spectrum frequency $\overline{\omega^2}$ as a function of unit cell dimensions. There exists a simple relation between $\omega_D$ and $\overline{\omega^2}$. Since the density of the frequency distribution is given by the expression

$$(3\omega^2/\omega_D{}^3)\,d\omega$$

then

$$\overline{\omega^2} = (3/\omega_D{}^3) \int_0^{\omega_D} \omega^4\,d\omega = \tfrac{3}{5}\omega_D{}^2$$

or

$$(\overline{\omega^2})^{\frac{1}{2}} = 0.775\Theta$$

A comparison of the calculated values of $\overline{\omega^2}$ with the values of $\Theta$ found from thermodynamic experiments has been carried out by Kitaigorodsky and Mukhtarov [8].

The choice of the crystals, for which the characteristic temperature has been calculated, is explained, on the one hand, by the availaiblity of structural data for as wide a temperature range as possible, and, on the other hand, by the availability of data on $\Theta$ determined for these substances by means of calorimetric measurements. For these reasons, we selected crystals of naphthalene, anthracene, diphenyl mercury, benzene, and ethylene for our study. These compounds are very typical representatives of molecular crystals. The characteristic temperatures and constants $\gamma$ for these crystals have been computed with the aid of a program providing for the computation of $\Theta$ for crystals with molecules each containing up to 80 atoms of up to five different species.

A comparison of the calculated and experimental values of the characteristic temperature and constant $\gamma$ shows (Table 6; see also Fig. 1, where data on $\Theta$ are given for naphthalene only; these data are rather typical) that the maximum difference between theory and experiment does not exceed 20%. Such differences are observed in the region of low temperatures; at moderate and high temperatures they are considerably lower and are of the same order as the experimental errors. The qualitative nature of the experimental and theoretical temperature dependence curves is the same, the only exception being the region of temperatures below 100°K for a naphthalene crystal [9]: The calculated $\Theta$ increases in this region with rise of temperature, whereas the experimental $\Theta$ decreases. This circumstance, as well as the increase of the difference between the experimental and theoretical curves with fall of temperature observed for all the crystals studied, may probably be attributed to the deterioration of the accuracy of the Debye model in the region of low temperatures.

In the work cited the elements of the Grüneisen tensor have also been computed for crystals of naphthalene and anthracene [10].

Comparison of the calculated values of $\gamma_{ik}$ with the experimental values reveals rather large discrepancies, which amount to 30% in different regions

*Table 6*

| Naphthalene | | | Anthracene | | | Diphenyl mercury | | | Benzene | | |
|---|---|---|---|---|---|---|---|---|---|---|---|
| $V(\text{Å}^3)$ | $\gamma_{calc}$ | $\gamma_{exp}$ | $V(\text{Å}^3)$ | $\gamma_{calc}$ | $\gamma_{exp}$ | $V(\text{Å}^3)$ | $\gamma_{calc}$ | $\gamma_{exp}$ | $V(\text{Å}^3)$ | $\gamma_{calc}$ | $\gamma_{exp}$ |
| 345 | 2.41 | 2.05 | 461 | 4.10 | 3.51 | 445 | 5.04 | 4.52 | 492 | 4.45 | 4.32 |
| 350 | 2.48 | 2.15 | 463 | 4.18 | 3.83 | 450 | 5.45 | 5.03 | 495 | 4.30 | 4.25 |
| 355 | 2.53 | 2.25 | 472 | 4.25 | 4.02 | 455 | 5.89 | 5.57 | 501 | 4.25 | 4.05 |
| 360 | 2.60 | 2.35 | 487 | 4.30 | 4.25 | 460 | 6.37 | 6.09 | 508 | 4.20 | 3.85 |

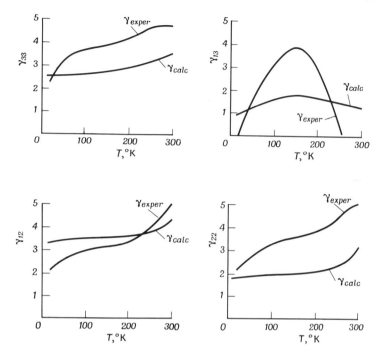

***Fig. 7.*** Experimental and calculated values of the components of the Grüneisen tensor for naphthalene.

of the temperature interval (see Fig. 7). However, taking into account the number of assumptions underlying the theoretical model, as well as the procedure for determining $\gamma_{ik}$ from the experimental data, we may regard the resulting agreement as quite satisfactory.

The problem of comparing the results of calculation of the lattice energy by the atom–atom potential method with the thermodynamic data has been considered in the literature [11]. We shall discuss the comparison with the experimental data in the scalar aspect. If, as considered above, the energy surface of a crystal has been constructed, we can imagine any thermodynamic process as the movement of a point on the surface from the bottom of the energy well (the absolute zero of temperature) along a certain curve $U$ $(a, b, c, \alpha, \beta, \gamma)$. Since we have in mind only the comparison with scalar thermo-dynamic quantities, it is sufficient to represent this curve as $U(v)$. It is con-venient to compare with isobaric data not only $U(v)$, i.e., the change in the lattice energy in the process of thermal expansion, but also the first and second derivatives of the lattice energy with respect to the unit cell volume, namely, $\partial U/\partial v$ and $\partial^2 U/\partial v^2$. The character of the agreement between the theoretical and experimental $U(v)$ curves and the two derivatives $\partial U/\partial v$ and $\partial^2 U/\partial v^2$

must clearly indicate the possibilities of the atom–atom potential method in predicting various thermodynamic quantities.

The values of the lattice energy and its derivatives cannot be found from the thermodynamic measurements alone. In order to compare the $U(v)$ curve calculated by the atom–atom potential method with thermodynamic data a model of the solid has to be used.

Let us assume, as above, that the crystal part of the internal energy can be represented as a sum

$$E(v, T) = U(v) + E_{\text{vib}}(v, T)$$

where $E_{\text{vib}}(v, T)$ is the energy of the intermolecular vibrations. Since the zero-point energy of the intermolecular vibrations is negligible for the overwhelming majority of molecular crystals, we may assume that $E_0 = U_0$. Furthermore, it is evident that the derivatives $(\partial E/\partial v)_T$ and $(\partial^2 E/\partial v^2)_T$ must not differ greatly from $\partial U/\partial v$ and $\partial^2 U/\partial v^2$, because the explicit dependence of the vibrational energy on the unit cell volume characterizes the deviation from the harmonic approximation which must be valid, at least roughly. From these general physical considerations it is also clear that the lower the temperature (or the smaller the volume) the better the agreement between the values of the corresponding derivatives of $E$ and $U$.

In the work cited above a comparison of the thermodynamic data obtained by integration and differentiation of the experimental curve $(\partial E/\partial v)_T = (C_p - C_V)/\alpha v$ with the results of calculation by the atom–atom potential method was carried out for naphthalene and anthracene crystals.

The results obtained for the naphthalene crystal are presented in Fig. 8. The values of $\Delta U$ and $\Delta_V E$ are measured (according to the available data) from 78°K. The main result of this comparison is rather encouraging: The experimental and theoretical curves are close to each other. Without performing additional computations we cannot say whether this discrepancy is due to a poor choice of the potentials or is a consequence of neglecting the vibrational part of the energy in the calculations. However, the fact that the $\Delta U(v)$ curve lies above the $\Delta_v E(v)$ curve can only be associated with an inexact choice of the potential curves, which must be corrected according to these results.

Let us now consider the procedures for calculating the vibrational component of the crystal part of the internal energy. Let us assume that we have succeeded in finding the exact atom–atom potentials and the values of $\Delta_v E_{\text{vib}}(\partial E_{\text{vib}}/\partial v)$ and $\partial^2 E_{\text{vib}}/\partial v^2$ from the difference of the ordinates of the curves of the type given above. An independent estimate of these quantities can be carried out in the following ways:

1.  The dynamic (and consequently thermodynamic) problem can be solved quite strictly on the basis of the atom–atom potentials.

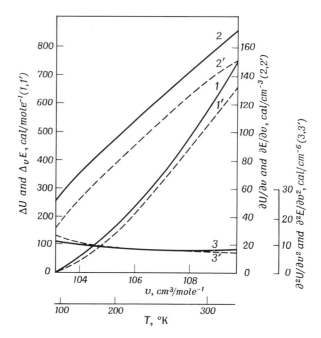

**Fig. 8.** Comparison of the lattice energy and its volume derivatives calculated by the atom–atom potential method with the internal energy and its derivatives obtained from experimental data for a naphthalene crystal. 1, $\Delta U$; 1′, $\Delta_V E$; 2, $(\partial U/\partial v)$; 2′, $(\partial E/\partial v)_T$; 3, $(\partial^2 U/\partial v^2)$; 3′, $(\partial^2 E/\partial v^2)_T$.

2.   The vibrational spectrum of a crystal and also the crystal parts of the thermodynamic functions can be calculated in the quasi-harmonic approximation.

3.   And, finally, use may also be made of the quasi-harmonic $[\Theta = \Theta(v)]$ Debye approximation for the crystal parts of the thermodynamic functions. In this case there are four ways of calculating the $\Theta(v)$ curve. First, it can be computed through the average square of the frequencies of the intermolecular vibrational spectrum $\overline{\omega^2}$, which is determined by the atom–atom potential method, followed by calculation of the tensor $(\partial \overline{\omega^2}/\partial \varepsilon_i)$, the expansion tensor, and the introduction of temperature with the aid of this tensor, (a). Secondly, the computation may be made through the same average square of frequencies $\overline{\omega^2}$ determined for different cell volumes and by the use of the experimental curve of $v(T)$, (b). Thirdly, the $\Theta(v)$ curve may be chosen in an "optimal" way directly from the thermodynamic data, (c). And, finally, (d), if the curve $(\partial U/\partial v)$ has been calculated by means of the atom–atom potential method and the value of $\Theta$ is known from the thermodynamic data at least for a single temperature point, the $\Theta(v)$ curve can be computed from the equation of state

for the crystal, which has the following form in the Debye approximation:

$$\frac{\partial U}{\partial v} = -6R \frac{T}{\Theta} D\left(\frac{\Theta}{T}\right) \frac{\partial \Theta}{\partial v}$$

(here we ignore the atmospheric pressure as compared with $\partial U/\partial v$).

In cases 1, 2, and 3(a) the thermodynamics of a crystal is calculated successively by the atom–atom potential method without using any data from experiment. Problem 1 is too complicated. The possibilities of the second variant are not yet evident. In cases 3(b), (c), and (d) the experimental data on the thermal expansion of a crystal are to be used for comparing the theoretical and experimental values of the energy. We have tried the last three methods of case (3) for the naphthalene crystal. The results obtained may be considered negative.

The vibrational components of the internal energy of a crystal and its derivatives are very sensitive to the choice of the $\Theta(v)$ curve, and since the curve itself is determined on the basis of a number of assumptions, the computations discussed above can only yield an estimate of the order of magnitude. The use of the atom–atom potential method for calculating a number of thermodynamic quantities, however, need not wait for precise methods of estimating the vibrational components, because, as the curves of Fig. 8 show, the vibrational corrections to the values of the lattice energy and its volume derivatives are small. The general validity of this calculation must, of course, be confirmed by calculations for other crystals.

The question of the role of the vibrational corrections to the values of the strain derivatives of the free energy arises in calculations of the elasticity moduli of molecular crystals made by the atom–atom potential method [12]. Expressing the free energy of a molecular crystal as a sum of the potential energy of the interaction of molecules (the lattice energy $U$) and the energy of vibrations ($E_{\text{vib}}$), we can rewrite formula (1.13) for the isothermal moduli of elasticity as follows:

$$C_{ij} = \frac{\partial}{\partial \varepsilon_i}\left(\frac{\partial U}{\partial \varepsilon_j}\right) + \frac{\partial}{\partial \varepsilon_i}\left(\frac{\partial E_{\text{vib}}}{\partial \varepsilon_j}\right) \tag{6.1}$$

where $U$ and $E_{\text{vib}}$ are energy per unit volume and $\varepsilon_i$ is the relative strain.

Since the intramolecular frequencies vary with volume approximately two orders of magnitude more slowly than the intermolecular frequencies [1], the energy of the intramolecular vibrations at small deformations may be considered invariant and hence the second term in expression (6.1) refers only to the change in the energy of the intermolecular vibrations during the process of deformation.

The effects of the zero point intermolecular vibrations are usually ignored [12] at the absolute zero of temperature and the elasticity tensor is determined by the second strain derivatives of the lattice energy density. The temperature dependence of the elastic properties of a crystal is associated both with the change of the lattice "statics," i.e., of the unit cell dimensions and the orientation of the molecules in the lattice, and with the change of the free energy of the intermolecular vibrations. Since the vibrational part cannot be found directly from the experimental data and, as indicated above, it is a very time-consuming procedure to find it theoretically (though in principle this can be done with the help of the atom–atom potential method), it is probably reasonable first: (1) to choose the interatomic potential curves, with the aid of which we can compute, with sufficient accuracy, the elasticity tensor for a molecular crystal at the absolute zero of temperature (this problem is quite strict); and (2) knowing the thermal expansion tensor for a crystal, and using the potential curves chosen for absolute zero, to calculate the variation of the elasticity moduli with temperature without taking into account the vibrational terms. From the discrepancy between the theoretical and experimental curves $c_{ij}(T)$ it will be possible to establish the effect of lattice vibrations on the values of the elasticity moduli.

The changes of the dimensions and angles of the unit cell of a molecular crystal in the process of homogeneous deformation—the so-called external deformations $\varepsilon_i$, coinciding with the parameters of the macroscopic theory of elasticity—are accompanied by a change in the position of the centers of gravity and orientations of the molecules in the deformed cell, i.e., the translational and orientational internal deformation $U_k$. Internal deformations are induced in such a manner as to reduce to a minimum the changes in the lattice energy caused by the external deformations, i.e.,

$$\partial U^{\text{lat}}/\partial U_k = 0 \qquad (6.2)$$

From the condition (6.2) it follows that the internal deformations are independent of each other and are functions of external deformations: $U_k = f(\varepsilon_i)$. Taking into account that on double differentiation of $v(\varepsilon_i, U_k)$ with respect to $\varepsilon_i$, we obtain

$$c_{ij} = \left(\frac{\partial^2 v}{\partial \varepsilon_i \partial \varepsilon_j}\right)_\circ + \sum_k \left(\frac{\partial^2 v}{\partial U_k^2}\right)_\circ \left(\frac{\partial U_k}{\partial \varepsilon_i}\right)_\circ \left(\frac{\partial U_k}{\partial \varepsilon_j}\right)_\circ \qquad (6.3)$$

The subscript $\circ$ signifies that the values of the derivatives refer to the point of equilibrium.

If the surface of the lattice energy $U(a_i, x_e, \varphi_k)$ has been calculated, where $a_i$ are the parameters of the unit cell, $a, b, c, \alpha, \beta, \gamma; x_e$ ($e = 1, 2, 3$) are the coordinates of the centers of gravity of the molecules, and $\varphi_k$ ($k = 1, 2, 3$)

are the Euler angles characterizing the orientation of the inertial axes of the molecules relative to the crystallographic axes, the elasticity moduli of the crystal can be found directly from this surface. We shall now show how this can be done for monoclinic crystals, the molecules of which are located at the center of inversion. In such crystals translational internal deformations do not arise (the center of inversion remains intact if the lattice undergoes a homogeneous external deformation [14]. The internal deformations are therefore reduced to changes of the three Euler angles $\varphi_k$. The lattice energy is a function of the seven variables $a$, $b$, $c$, $\beta$, $\varphi_1$, $\varphi_2$, and $\varphi_3$. Using these variables, we can describe four of the six elements of the relative external strain tensor $\varepsilon_1$, $\varepsilon_2$, $\varepsilon_3$, and $\varepsilon_5$ and the orientational internal strains $U_k$ ($k = 1, 2, 3$) as follows:

$$\varepsilon_1 = \frac{a - a_o}{a_o}, \qquad \varepsilon_2 = \frac{b - b_o}{b_o}, \qquad \varepsilon_3 = \frac{c' - c_o'}{c_o'}$$

$$\varepsilon_5 = \frac{s - s_o}{c_o'}, \qquad U_k = \frac{\varphi_k - T_{k_o}}{\varphi_{k_o}}$$

(6.4)

where $c' = c\cos(\beta - 90°)$, $s = c\sin(\beta - 90°)$, and the subscript o refers to the equilibrium point.

Differentiating in expression (6.3), the energy density $U$ with respect to the strain components as a complex function

$$(U = \varphi_1(\varepsilon_i, U_k), \quad \varepsilon_i = f_2(a_i), \quad U_k = f_3[f_k(a_i)]),$$

we obtain, with the help of (6.4), a formula for calculating the elasticity moduli of a crystal in the form

$$c_{ij} = a_{oi} a_{oj} \left( \frac{D^2 U}{Da_i\, Da_j} \right)_o$$

$$= a_{oi} a_{oj} \left[ \left( \frac{\partial^2 U}{\partial a_i\, \partial a_j} \right)_o + \sum_{k = 1,2,3} \left( \frac{\partial^2 U}{\partial \varphi_k^2} \right)_o \left( \frac{D\varphi_k}{Da_i} \right)_o \left( \frac{D\varphi_k}{Da_j} \right)_o \right]$$

(6.5)

The quantities $D^2 U/Da_i\, Da_j$ and $D\varphi_k/Da_i$ are derivatives of the lattice parameters along the surface of equilibrium Euler angles $\varphi_k$ [see condition (6.2)]. Formula (6.5) is suitable for calculating, from the energy surface, the elasticity moduli for crystals of monoclinic and any other higher system, except the moduli for $C_{44}$, $C_{46}$, and $C_{66}$.

Such a calculation has been carried out for the naphthalene crystal. The theoretical moduli for $C_{11}$, $C_{22}$, $C_{33}$ as functions of temperature are shown in Fig. 9, where they are compared with the experimental data reported in the literature [15]. The interatomic potential curves are selected for this calculation so that at absolute zero there is but a slight difference between theory

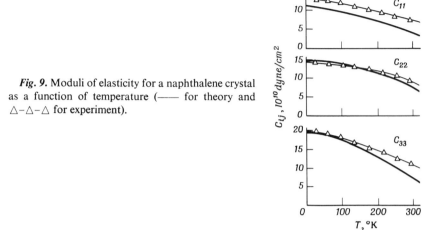

**Fig. 9.** Moduli of elasticity for a naphthalene crystal as a function of temperature (—— for theory and △-△-△ for experiment).

and experiment. This discrepancy increases as the temperature rises, which probably is an indication of the increasing role of the intermolecular lattice vibrations. At a temperature of 323°K, which is close to the melting point of naphthalene, the static term in expression (6.1) makes up 60–70% of the experimental values of $C_{ij}$.

REFERENCES

1. A. I. Kitaigorodsky, *Kristallografiya* **7**, 195 (1962); A. I. Kitaigorodsky and B. D. Koreshov, *Fiz Tverd. Tela* **8**, 62 (1966).
2. B. D. Koreshkov, Thesis, Univ. of Moscow, 1968.
3. J. Timmermans, "Physico-Chemical Constants of Pure Organic Compounds." Elsevier, Paris, 1950.
4. See the reference in M. Suzuki and M. Ito, *Spectrochim. Acta Part A* **24**, 1091 (1968).
5. A. P. Ryzhenkov and V. M. Kozhin, *Kristallografiya* **13**, 6, 1028 (1968).
6. G. Afanassieva, Thesis, Moscow State Univ., Moscow, 1970.
7. A. I. Kitaigorodsky and B. D. Koreshkov, *Fiz. Tverd. Tela* **11**, 3203 (1969).
8. E. I. Mukhtarov, Thesis, Univ. of Baku, 1970.
9. A. I. Kitaigorodsky and E. I. Mukhtarov, *Fiz. Tverd. Tela* **10**, 3474 (1968).
10. A. I. Kitaigorodsky and E. I. Mukhtarov, *Kristallografiya* **13**, 889 (1968).
11. A. I. Kitaigorodsky and K. V. Mirskaya, *Mol. Cryst.* **4**, (1969).
12. K. V. Mirskaya, *Fiz. Tverd. Tela* "Soviet Physics–Solid State".
13. H. Huntington, *Solid State Phys.* **7**, 213 (1958).
14. M. Born and K. Huang, "Dynamical Theory of Crystal Lattices." Oxford Univ. Press, London and New York, 1954.
15. G. Afanassieva, Synopsis of Thesis, Moscow State Univ., Moscow, 1970.

# Chapter VII
# Conformations of Organic Molecules

## 1. The Mechanical Model of a Molecule

In the preceding chapters the problems pertaining to the mutual positions of the molecules in a crystal have been discussed. This chapter deals with the mutual positions of the atoms in a molecule, i.e., the conformation of the molecule. Conformational concepts have become one of the most important fields of organic chemistry. The properties of the molecule, e.g., reactivity, reaction rate, bond strength, the heats of formation and hydrogenation of unsaturated compounds, etc., are ascribed by chemists to the molecular geometry. Thus they disguise the relationship between geometry and properties of the molecule by the somewhat indefinite term "steric effects." Nowadays the theory of conformational analysis makes possible quantitative evaluation of these steric effects.

If a molecule is looked upon as a system of electrons and nuclei, then, by solving the Schrödinger equation in every case, we may derive all the properties of the molecule, including its geometry. But, nonempirical calculations involve extremely complicated, sometimes insurmountable, mathematical difficulties; hence the attempts to find more expedient—empirical—solutions. The successful application of atom–atom potentials to the estimation of intermolecular forces gives us grounds to believe that this approach will be no less fruitful in the case of the interatomic potentials within molecules.

The concept of the theoretical conformational analysis of organic molecules was put forward as early as late forties and early fifties independently by Hill [1, 1a], Westheimer [2–6] and Kitaigorodsky [6–8]. These early contributions are analyzed in [9, 10]. The mechanical model underlying theoretical conformation analysis is based on the Born–Oppenheimer approximation. This means that the potential energy of a molecule may be represented with sufficient accuracy as a continuous function of the nuclear coordinates. The problem is to find empirically such potential functions as can adequately describe the geometry of a molecule. An essential condition for such potential functions is to employ the smallest possible number of empirical parameters; otherwise the method will lose its predictative ability.

The following assumptions are made in the method of atom–atom potentials to be applied within molecules:

1. A molecule may be viewed as a system of interacting atoms; for a large class of problems the atoms may be assumed to be connected by absolutely rigid bonds (rods).

2. The interatomic interactions in a molecule may be described by the potential curves (6-exp, 6-12 etc.) used to characterize intermolecular interactions. These interactions are generally of a central nature, although in some cases it is advisable to take into account deviations from centrality (see page 398).

3. For each atom, there exist "ideal" bond angles, i.e., from the center of each atom, valence lines may be drawn just as they would have been directed had there been no nonbonded interactions, or bonds between neighboring atoms.

4. Generally speaking, ideal bond angles should be found empirically. But it is only the simplest valency schemes of quantum chemistry that may serve as initial values, i.e., the $90°$ angle between the O, S, Se, N, P, and As bonds, the tetrahedral angle for the aliphatic carbon atom, the $120°$ angle for the trigonal and aromatic carbon atoms.

5. Any deviation of the bond angles from the ideal values requires some energy.

6. The sum of the angular deformation energy and that of nonbonded interactions will be called strain, or conformational, energy

$$U_{st} = U_{ang} + U_{nonbonded} \qquad (1.1)$$

By varying the atoms' mutual positions in the case of rigid bonds, we may build up the $U_{st}$ surface as a function of independent geometrical parameters. This surface will have one or several minima corresponding to one or several conformations. If one minimum is by far deeper than the rest of them, the molecule may have one stable conformation. If there are several minima separated by $RT$-commensurate barriers, there should exist some conformers capable of interconversion, and, in the gaseous or liquid phases, conformational equilibrium will be achieved.

This model is of no particular interest if one believes that there are no universal methods to calculate $U_{st}$ and also that $U_{ang}$ and $U_{non\,bonded}$ are different in each molecule. But the mechanical model, like any semiempirical theory, is of value if it can predict the conformational and strain energies of a great number of molecules using some theory in which the required parameters are to be determined experimentally. For this scheme to be feasible, the number of values to be determined should exceed the number of theoretical parameters by at least two or three orders.

The energy of the nonbonded interactions is calculated like that for the crystal, i.e., it is equal to the sum of the interactions of all pairs of nonbonded atoms:

$$U_{nonbonded} = \sum_{i>j}\sum f_m(r_{ij}) \tag{1.2}$$

where $f_m$ are the atom–atom potentials (the index $m$ indicates the type of interaction); $r_{ij}$ is the distance between atoms $i$ and $j$; as in the case of inter-molecular interactions, if the 6-exp potential is used, then

$$f(r) = -Ar^{-6} + B\exp(-Cr) \tag{1.3}$$

where $A$, $B$, and $C$ are empirical parameters.

As to the angular deformation energy, when the ideal bond angles are equal to $\alpha_0$ and the true ones are equal to $\alpha$, this energy will be

$$\sum_{all\,angles} f(\alpha-\alpha_0) \tag{1.4}$$

i.e., it is assumed to be additive (the sum of the energies of all the bond angles).

For small deviations $(\alpha-\alpha_0)$, it is possible to confine ourselves to the quadratic term of the expansion of the function $f(\alpha-\alpha_0)$, and, presented as quasi-elastic energy, the formula will look like

$$U_{ang} = \tfrac{1}{2}\sum_i C_i(\alpha_i-\alpha_{i0})^2 \tag{1.5}$$

where the $C_i$ are elastic constants.

Kitaigorodsky and Dashevsky [11, 12] made an attempt to use the smallest number of constants. Therefore, $C$ and $\alpha_0$ were considered to be universal constants for a wide range of compounds. It is believed that good agreement with experimental results may be attained by assuming that all carbon atoms could be divided into tetrahedral (sp$^3$-hydridization) and trigonal (sp$^2$). For the former, $\alpha_0 = 109°28'$, for the latter $\alpha_0 = 120°$.

The values of the constants $C$ lie in the 20–90 kcal mole$^{-1}$ rad$^{-2}$ range. The $C$ constants for O and N atoms are also within this range, the ideal angles being assumed to be 90°. (It may also be assumed that in the case of pyramidal nitrogen the ideal angle is 90°, while for planar nitrogen it is 120°).

The $C_\alpha$ constants are associated with $K_\alpha$ spectroscopic deformation force constants. By definition

$$K_\alpha = \frac{1}{N}\left(\frac{\partial^2 U}{\partial \alpha^2}\right)_{\alpha = \alpha_e} \tag{1.6}$$

where $N$ is the number of equivalent interactions and $\alpha_e$ is a bond angle of the equilibrium conformation (i.e., essentially not an ideal angle). Introducing into (1.6) the strain energy we have

$$K_\alpha = C_\alpha + \frac{\partial f(r)}{\partial \alpha}\left(\frac{\partial r}{\partial \alpha}\right)^2 + \frac{\partial^2 f(r)}{\partial \alpha^2}\frac{\partial r}{\partial \alpha} \tag{1.7}$$

In (1.7) summation signs are omitted for the sake of simplicity; equivalent interactions are not taken into account either.

Calculations have shown that the spectroscopic deformation constants are approximately twice as large as the "conformation" elastic constants, because the former, in accordance with Eq. (1.7), takes into account the nonbonded interactions.

Expressing the strain energy by means of independent geometrical parameters $\varepsilon_e$ one may find the minimum of this energy, either by solving the system of equations

$$\frac{\partial U}{\partial \varepsilon_e} = \sum_i C_i \Delta\alpha_i \frac{\partial \Delta\alpha_i}{\partial \varepsilon_e} + \sum_{i>j}\sum f_m'(r_{ij})\frac{\partial r_{ij}}{\partial \varepsilon_e} \tag{1.8}$$

(where $\Delta\alpha_i = \alpha_i - \alpha_{i0}$), or by minimizing a function of several variables with the help of a computer. The energy minimum will correspond to an equilibrium conformation that is established as a result of a compromise between the tendency of nonbonded atoms to move apart to equilibrium distances and that of bond angles to maintain their ideal values.

So far, we have considered two terms in the potential function of an organic molecule: the energy of the nonbonded interactions and the bond-angle deformation energy. In the following sections, dealing with applications of the mechanical model to various types of molecules, we shall discuss other components of the potential function. The term "nonbonded interactions" will be used tentatively, i.e., meaning only the terms involved because of dispersion attractions and the repulsions of atoms due to the overlap of their electron shells; these effects are usually described by potentials of the 6-exp or 6-12 type.

It is evident that electrostatic energy and torsion interactions, etc., should also be classified as nonbonded interactions. But, since they have analytical forms which make them different from the atom–atom potentials indicated above, they will be classed among other components of the potential function.

It should be stated a priori that the role of the electrostatic interactions in forming the optimal geometry (but not its relative stability) is, as a rule, very small. But in a simple molecule, as, e.g., $H_2O$, $NH_3$, $H_2S$, $Cl_2O$, $F_2O$, $PH_3$, $(CH_3)_2O$ etc., the contribution of the electrostatic forces to the equilibrium conformation may be quite significant.

Obviously, it is possible to elaborate a semiempirical analysis of the deformations of simple inorganic molecules, but this analysis is bound not to be too convincing, because many empirical parameters are indispensable. However, to analyze large classes of organic molecules, a comparatively small number of empirical parameters is required. For example, knowing only the potential curves for $C \cdots C$, $C \cdots H$, and $H \cdots H$, and the elastic constants for the bond angles, one may estimate the conformation of many thousands of hydrocarbons; hence, the great importance of semiempirical methods of conformational analysis for making predictions in organic chemistry. So, it is quite natural that the majority of the papers on theoretical conformational analysis deals with organic molecules.

As a rule, intermolecular forces are smaller by one order than those of non-valent repulsion in molecules. Even in macromolecules whose helical con-formation is determined by the 0.5 kcal mole$^{-1}$ per monomeric unit energy difference between the trans and gauche forms, the role of the intermolecular forces in attaining optimal geometry is almost always insignificant, as will be shown below. In so-called "overcrowded" molecules, whose ideal model contains pairs of too closely located atoms, intermolecular forces may be neglected also. The sublimation energy of most organic crystals is approxi-mately 10–25 kcal mole$^{-1}$; taking into consideration that each molecule is surrounded with at least 10–14 neighboring ones, the contribution of each molecule should be estimated as 2–3 kcal mole$^{-1}$. Since the strain energy exceeds that value by one–two orders, the conformational calculation may be restricted to estimating only intermolecular forces.

Exception should be made only for those molecules in which the order of the strain energy difference between any two conformations is comparable with the intermolecular interaction energy. For example, in crystalline di-phenyl, the planar conformation exists [13, 14], whereas in the vapor phase, the phenyl rings are turned to each other by 42°, as is shown by an electron-diffraction investigation [15].

Adrian [16] calculated the energy difference between the planar and non-planar diphenyl conformations to be 0.8 kcal mole$^{-1}$, hence the intermolecular forces play the decisive role in determining the optimal conformation of this molecule.

2. PARAMETERS FOR CONFORMATIONAL CALCULATIONS

Since the method described is empirical, special attention should be paid to the parameters to be used. First, it is sensible to assume that the atom–atom potential parameters are determined only by the type of interacting atoms. These parameters should describe both the intra- and intermolecular interactions (i.e., the potentials do not depend on the kind of atomic interactions analyzed—those between atoms of different molecules or between atoms belonging to one molecule); they should be transferable from one molecule to another. Thus, the energy of interaction of any pair of atoms not bound with a covalent bond depends only on the distance between them. Otherwise, the predicting power of the method would be very low.

The present physicochemical literature deals with a great number of parameters put forward by various authors. Unfortunately, no preference may yet be given to any of them, because the aim of every paper is to investigate one or two features of a definite class of substances. For example, the potentials of Scott and Scheraga [17] were verified only by calculations of the internal rotation barriers in simple molecules containing single bonds and polypeptide conformations [18, 19]; the potentials by Poltev and Sukhorukov [20], were checked only by calculations of the sublimation heats of several organic crystals; the potentials suggested by Williams [21], were used only in a comparative analysis of 70 estimated and experimental constants—the parameters of the unit cells of several hydrocarbons, and their sublimation energies (the number of such parameters may be increased by one order).

For the reasons described above, it is very important to determine the parameters of the atom–atom potentials based on all the available physicochemical data and using definite mathematical criteria. Universal potentials are used to estimate quite different properties, such as, the parameters of the unit cells of crystals, their sublimation heats, the thermodynamic functions of crystals, the thermochemical properties of gases, the conformations of molecules, the frequencies of the vibration spectra of molecules and crystals, the second virial coefficients and transport properties of polyatomic gases, and evidence on the dispersion of molecular beams. What should optimal potentials be like? What will be the error of optimal three-parameter potentials in determining the above properties? Will quite different parameters be useful to any degree to describe inter- and intramolecular interactions? Is there any sense in discriminating between aliphatic and aromatic carbon atoms? Unfortunately, none of these questions has received a clear answer. Ramachadran's attempt to calculate the conformations of polypeptides and polysaccharides using the potentials of different authors is stimulating [22], but is still not sufficiently decisive to answer any of the above questions.

Molecular conformations are chiefly determined by the repulsions between nonbonded atoms at small distances; but for making up a crystal both attraction and repulsion are essential, the principal role belonging to the position and depth of the potential well. Therefore, when calculating conformations, to use potentials derived from the analysis of intermolecular interactions will mean taking chances. Consequently, there arises the problem of obtaining potential curves from the data on the geometry of overcrowded molecules [23] or the data on their energies of formation [24]. One may expect that in the future attempts will be made to "combine" potential curves to make them operative over a wide range of internuclear distances. A brief survey will now be made of the methods for obtaining and refining the atom–atom potentials used by some authors.

The simplest atom–atom potentials are walls of infinite height corresponding to just one empirical parameter: $r = r_{min}$. If $r \geqslant r_{min}$, contacts between atoms are allowed, otherwise they are forbidden. Such potentials, corresponding to the method of rigid spheres in conformational analysis or to the principle of dense packing of molecules in crystal, furnish only information on allowed or forbidden regions in the space of the independent geometrical parameters describing the geometry of the molecule. These potentials provide neither the accurate position of the minima, nor the relative stability of different conformations; but the shape of the potential wells can be roughly estimated. It should be noted that the method of rigid spheres is not a very useful one in the case of small overcrowded molecules possessing no internal rotation. But, for molecules like peptides, in which conformational freedom is rather high, these methods may be used to interpret some interesting facts. For example, the points corresponding to real polypeptides and proteins should not get into forbidden regions of conformation maps of dipeptides; that this is very often the case is shown in the next chapter.

Ramarandran and Sasisekharan [25] analyzed extensive experimental evidence on the structures of amino acids, peptides, and similar substances; they established characteristic distances between the atoms corresponding to the shortest intra- and intermolecular contacts. These distances are shown in Table 1; in the first column the "normal" distances encountered in many compounds are given; in the second, "extreme," rare distances are given, but which do occur, especially in the case of H bonds. As a rule, extreme distances are shorter by 0.5–0.8 Å than equilibrium ones (corresponding to the minimum of the 6-exp or 6-12 potential curves). Of course, it should be taken into account that conformations in which H bonds arise may fall into forbidden regions. In fact, the average O⋯H distance for the C–O⋯H bond is as low as 1.8 Å which is less by 0.4–0.6 Å than the allowed distance.

Now, let us discuss potential curves which have no discontinuities in their derivatives, e.g., 6-exp or 6-12. The most common potentials of this kind now used are those of Kitaigorodsky [26], Hendrickson [27], Liquori and

## Table 1

VALUES OF MINIMUM CONTACTS BETWEEN NONBONDED
ATOMS[a]

| Type of contact | Normal limits (Å) | Extreme limits (Å) |
|---|---|---|
| H ··· H | 2.0 | 1.9 |
| H ··· O | 2.4 | 2.2 |
| H ··· N | 2.4 | 2.2 |
| H ··· C | 2.4 | 2.2 |
| O ··· O | 2.7 | 2.6 |
| O ··· N | 2.7 | 2.6 |
| O ··· C | 2.8 | 2.7 |
| N ··· N | 2.7 | 2.6 |
| N ··· C | 2.9 | 2.8 |
| C ··· C | 3.0 | 2.9 |

[a] G. N. Ramachandran and V. Sasisekharan [25].

co-workers [28], Scott and Scheraga [18], Flory [29], and Dashevsky [30].
There exist about ten other independent procedures for obtaining potential
curve parameters, but they are less reliable and are not sufficiently verified by
a priori geometric calculations.

The universal potential of Kitaigorodsky

$$f(r) = 3.5[8600 \exp(-13r/r_0) - 0.04(r_0/r)^6] \qquad (2.1)$$

contains just one parameter, the equilibrium distance $r_0$. Ramachandran et al.
[22] have shown that, even with such strict limitations on the 6-exp potential
parameters, conformational calculations of peptides and, especially, sugars
yield quite good results. Two sets of parameters were considered: A, B, and C
[see (1.3)], satisfying Eq. (2.1), i.e., $K_1$ and $K_2$, Table 2. The first set, $K_1$, was
found assuming that $f(r) = 0$ when $r$ is equal to the sum of the van der Waals
radii (i.e., the mean intermolecular distances found by Bondi [31] as a result
of an analysis of extensive crystallochemical data). The second set, $K_2$, is
found on the assumption that in that point the potential curve has its
minimum.

Liquori et al. [28] tried to find in different papers suitable potentials for
various types of interactions. But this attempt did not prove successful, and
the system of potentials proved inconsistent. In any case it was shown in [22]
that with dipeptides these potentials give an additional minimum in the region
which is intermediate between the conformation corresponding to the right-
hand α-helix and the one corresponding to the β-structure; this does not
agree with the experimental evidence, e.g., with the IR spectra. Hence, at least
one of the curves should be reconsidered.

*Table 2*

EQUILIBRIUM DISTANCES FOR THE UNIVERSAL POTENTIAL
OF KITAIGORODSKY[a]

| Interaction | $r_0(\text{Å})$ $K_1 K_2$ | | Interaction | $r_0(\text{Å})$ $K_1 K_2$ | |
|---|---|---|---|---|---|
| H $\cdots$ H | 2.66 | 2.40 | N $\cdots$ CH$_3$ | 3.78 | 3.40 |
| H $\cdots$ N | 3.06 | 2.75 | O $\cdots$ O | 3.33 | 3.00 |
| H $\cdots$ O | 3.00 | 2.70 | O $\cdots$ C | 3.56 | 3.20 |
| H $\cdots$ C | 3.22 | 2.90 | O $\cdots$ CH$_3$ | 3.72 | 3.35 |
| C $\cdots$ CH$_3$ | 3.39 | 3.05 | C $\cdots$ C | 3.78 | 3.40 |
| N $\cdots$ N | 3.44 | 3.10 | C $\cdots$ CH$_3$ | 3.94 | 3.55 |
| N $\cdots$ O | 3.39 | 3.05 | CH$_3$ $\cdots$ CH$_3$ | 4.11 | 3.77 |
| N $\cdots$ C | 3.61 | 3.25 | | | |

[a] C. M. Venkatachalam and G. N. Ramachandran [22].

As was shown in the case of the universal potential (2.1), it is most important to choose the $r_0$ equilibrium distance. If the choice is good it only remains to find two parameters for the 6-exp potential and one for the 6-12 potential. Taking account of the fact that in a crystal the closest atoms belonging to different molecules are, as a rule, located at distances shorter than equilibrium ones, Scott and Scheraga [18] and Flory *et al.* [29] obtained $r_0$ by adding 0.2 Å to the mean value of the intermolecular contacts found by Bondi (this difference is likely to be 0.3–0.4 Å).

These authors derived the second parameter indispensable to their potentials from some experimental data based on the theory of dispersion forces. They used the Slater–Kirkwood equation [32] for the energy of dispersion interactions

$$A = \frac{\frac{3}{2}e(\hbar/m)\alpha_i \alpha_j}{(\alpha_i/N_i)^{1/2}+(\alpha_j/N_j)^{1/2}} \tag{2.2}$$

where $A$ is the coefficient at $r^{-6}$ in the 6-exp or 6-12 potentials, $\alpha_i$ and $\alpha_j$ are the polarizabilities of atoms $i$ and $j$. $N_i$ and $N_j$ are the effective numbers of electrons capable of being polarized (sometimes this is said to be the number of electrons in the outer shell); $\hbar$, $e$, and $m$ are the fundamental constants: the Plank constant, the charge, and the mass of the electron. Polarizabilities of atoms were taken from the table in [33] containing the experimental results; the dependence of $N$ on the atomic number was found by Scott and Scheraga who used the data on inert gases, and interpolated it for other atoms.

In the case of the 6-12 potentials the constants found, i.e., $A$ and $r_0$ are sufficient, since the $B$ coefficient of $r^{-12}$ is determined by the condition that $f(r)$ is a minimum at $r = r_0$:

$$B = \tfrac{6}{12}Ar_0^6 \tag{2.3}$$

Flori *et al.* who used the 6-exp form, also dealt with only two empirical parameters, assuming $C$ in (1.3) to be equal to 4.60.

It is important to note that the optimal conformations and their relative stabilities depend not so much on the attractions between the nonbonded atoms as on their repulsions. Therefore, it is advisable to pay attention to the parameters describing repulsion, rather than attraction (whereas Scheraga and Flory derived their potentials based on the empirical values characterizing attraction). Equation (2.2) is very approximate and, which is the main thing, the values of this equation cannot be obtained directly from experiment. Hence the inevitability of a considerable error in the value of the repulsion forces obtained automatically by minimization of the potentials at $r = r_0$.

Dashevsky in some of his papers [11, 23] tried to refine the potential curves of nonbonded atoms using structural and thermochemical data. The 6-exp potentials were used; to simplify parametrization the $Cr_0$ product was assumed to be equal to 13, and the values of the equilibrium distances were borrowed from the crystallochemical data, using the same principle as [18]. Thus, every potential contained but one variable parameter. For interactions of atoms of different kinds, combination rules were found empirically, thus enabling one to obtain, e.g., the $C \cdots Cl$ curve by using the data on $C \cdots C$ and $Cl \cdots Cl$ interactions.

If

$$f_{11} = -A_{11}r^{-6} + B_{11}\exp(-C_{11}r) \tag{2.4a}$$

and

$$f_{22} = -A_{22}r^{-6} + B_{22}\exp(-C_{22}r) \tag{2.4b}$$

then

$$B_{12} = \left(\frac{B_{11}^{\frac{1}{2}} + B_{22}^{\frac{1}{2}}}{2}\right)^2, \qquad C_{12} = 2\frac{C_{11}C_{22}}{C_{11}+C_{22}}, \qquad A_{12} = 23.63B_{12}/C_{12}^6 \tag{2.4c}$$

These relationships were obtained as a result of several attempts, and the rule for $B$ given in (2.4c) gives more satisfying results in estimation of the energies of formation and isomerization, than the $B_{12} = 2B_{11}B_{22}/(B_{11}+B_{22})$ rule used previously in conformation calculations [12] and known to other authors [34]. The somewhat unusual expression for $A$ in (2.4c) is due to taking $Cr_0 = 13$.

Table 3 lists the parameters of the Dashevsky potentials and in Fig. 1 are

## Table 3

THE POTENTIALS OF DASHEVSKY

| Interaction | $A(\text{kcal mole}^{-1} \text{Å}^6)$ | $B(\text{kcal mole}^{-1})$ | $C(\text{Å}^{-1})$ | $r_0(\text{Å})$ |
|---|---|---|---|---|
| H $\cdots$ H | 40.1 | $2.86 \cdot 10^4$ | 5.200 | 2.43 |
| C $\cdots$ C | 476 | $3.77 \cdot 10^4$ | 3.513 | 3.70 |
| C $\cdots$ H[a] | 121 | $3.28 \cdot 10^4$ | 4.130 | 3.15 |
| O $\cdots$ O | 354 | $9.65 \cdot 10^4$ | 4.333 | 3.00 |
| N $\cdots$ N | 395 | $7.62 \cdot 10^4$ | 4.063 | 3.20 |
| F $\cdots$ F | 183 | $6.23 \cdot 10^4$ | 4.483 | 2.90 |
| Cl $\cdots$ Cl | 2900 | $2.29 \cdot 10^5$ | 3.513 | 3.70 |
| Br $\cdots$ Br | 3350 | $1.62 \cdot 10^5$ | 3.250 | 4.00 |
| I $\cdots$ I | 6020 | $2.42 \cdot 10^5$ | 3.025 | 4.30 |

[a] The parameters of the C $\cdots$ H curve were determined independently without using combination rules.

given calculations of the conformations of several simple hydrocarbon molecules which were used to refine the potential curve parameters.

The 6-exp potentials without limitations are not very convenient for computer calculations since $\lim_{r \to 0} f(r) = \infty$ (see Fig. 2). It is clear that the minimum $r_m$ values for each curve should be such that at $r < r_m$ it is prohibited to use the 6-exp potentials. The $r_m$ values should be so chosen that: (1) they correspond to the diminishing parts of the potential curves; and (2) they do not exceed the distances between the "pressed" atoms in the overcrowded molecules. Let us add the parabola

$$f(r) = ar^2 + br + c \tag{2.5}$$

to the 6-exp potential at point $r_m$, demanding that both the values of the function and of its derivatives should coincide at this point (which is sometimes essential for minimization). Then

$$a = f''(r_m)/2$$

$$b = f'(r_m) - f''(r_m)r^2 \tag{2.6}$$

$$c = f(r_m) - f'(r_m) + \tfrac{1}{2}f''(r_m)r^2$$

where $f(r)$ is the 6-exp potential considered. If, for example, in minimizing the potential function, one has some distance between the nonbonded atoms less than $r_m$, expressions (2.5) and (2.6) should be used instead of the usual 6-exp potentials.

For computer calculations, the 6-12 potentials are much more convenient

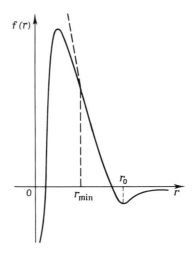

**Fig. 1.** Equilibrium conformations of some hydrocarbon molecules calculated with Dashevsky's potentials and elastic constants. In square brackets, accepted values of bond lengths; in round brackets, experimental data.

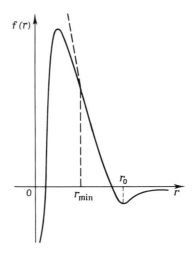

**Fig. 2.** Curve of interaction between nonbonded atoms (6-exp). Parabola is shown with broken line. Proportions are not observed: the peak of the $f(r)$ curve exceeds by three or four orders the depth of the minima at $r_0$.

because, first, $\lim_{r \to 0} f(r) = \infty$ and, secondly, $r^{-12}$ takes less time to calculate than the exponential. Therefore, the 6-exp potentials were transformed into 6-12 potentials [35] and were then used by some authors to calculate peptide conformations [36]. In search of the $A$ and $B$ parameters for the Lennard-Jones potential

$$f(r) = -Ar^{-6} + Br^{-12}$$

the

$$S = \int_a^\infty [-A'r^{-6} + B' \exp(-C'r) + Ar^{-6} - Br^{-12}]^2 \, dr \qquad (2.7)$$

integral was minimized, where $A'$, $B'$ and $C'$ are constants corresponding to the Dashevsky potentials, the $a$ values were chosen so as to ensure $\sim 5$ kcal mole$^{-1}$ energy at $r = a$. The results are listed in Table 4.

One should mention also the potentials of Hendrickson [27] which are used to calculate the conformations of cycloalkanes [37, 38]. These potentials were derived from intermolecular potentials obtained from the experimental data on the scattering of He and Ne molecular beams; the H$\cdots$H curves were compared with He$\cdots$He, C$\cdots$C with Ne$\cdots$Ne; the C$\cdots$H curve was obtained by means of the combination rules.

Some authors made their conformation calculations using spectroscopic deformation force constants. But, as is clear from Eq. (1.7), such a procedure is basically wrong. In [23] the author attempted to refine independently the elastic constants. As a result of a simultaneous variation of the elastic constants and parameters of the potential curve, and application of the experimental data on some "standard" hydrocarbon molecules, Dashevsky and Kitaigorodsky [11] obtained the following constants: $C_{C,aliph} = 30$, $C_{C,ethylene} = 70$, and $C_{C,arom} = 52$ kcal mole$^{-1}$ rad$^{-2}$. These values were used to calculate the conformations of the molecules shown in Fig. 1. It should be emphasized that these values of the elastic constants are valid only for the potentials listed in

### Table 4

THE 6–12 POTENTIALS

| Interaction | $A$ (kcal mole$^{-1}$ Å$^6$) | $B$ (kcal mole$^{-1}$ Å$^{12}$) | $r_0$ (Å) |
|---|---|---|---|
| H$\cdots$H | 31.4 | $2.97 \cdot 10^3$ | 2.40 |
| C$\cdots$C | 95 | $2.24 \cdot 10^5$ | 4.06 |
| O$\cdots$O | 164 | $6.25 \cdot 10^4$ | 3.03 |
| N$\cdots$N | 228 | $1.35 \cdot 10^5$ | 3.25 |
| CH$_3$ $\cdots$ CH$_3$ | 5069 | $1.267 \cdot 10^7$ | 4.16 |

Tables 2 and 3, i.e., they are reliable only when being used with this set of potentials. However, it is possible to elaborate such a scheme operating both with the atom–atom potentials and the constants derived from the vibration spectra frequences of the molecules by solving the reverse problem. (Of course the question of whether these constants are unambiguous and transferable from molecule to molecule should be discussed specially.) For example, Allinger *et al.* [39], using spectroscopic deformation constants, did not include the 1 ⋯ 3 interactions, i.e., the nearest nonbonded interactions, and thus met the requirements of Eq. (1.7).

3. INTERNAL ROTATION IN MOLECULES

It is well known that rotation around ordinary C–C bonds in ethane, butane, and other aliphatic molecules is not free; passing through the eclipsed conformations the molecules have to overcome a potential barrier. The question arises whether it is possible to describe the height and form of these barriers by means of the potential functions dealt with above?

It turned out that the potentials of nonbonded interactions obtained from various physicochemical data provide no rotation barrier values for small molecules which are comparable with the experimental ones. Mason and Kreevoy [40, 41] were the first to demonstrate this when they made an attempt to use the potentials they had obtained from experiments on the scattering of inert gases for estimating the internal rotation barriers in haloid derivatives of ethane, and also in molecules of the $Cl_2B–BCl_2$ and $H_3Si–SiH_3$ type. The calculated barriers proved to be almost twice smaller than the experimental ones, and the values for methane and methylsilane were equal to zero. Hence the conclusion that atom–atom potentials depending only on the distances between the nuclei are not sufficient for a correct description of these barriers.

The data by Mason and Kreevoy gave impetus to a number of suggestions and concepts as to the "nature" of the internal rotation barrier which were considered in detail in some monographs [42, 43]. *Ab initio* calculations for simple hydrocarbons in terms of the Hartree–Fock method allow one to obtain values of the barriers close to the experimental ones, in spite of the fact that the difference between the calculated total energy and the true energy exceeds by about two orders of magnitude the value of the barrier. Pitzer and Lipscomb [44] used the usual Slater functions and computed over 1000 integrals in order to calculate the barrier in ethane to be equal to 3.3 kcal mole$^{-1}$ (the experimental value is 2.9 kcal mole$^{-1}$). Later Goodisman [45] showed that the energy difference between the eclipsed and staggered forms exceeds the true one by 500 kcal mole$^{-1}$, which is 150-fold higher than the height of the barrier. Clementi and Davis [46] used a Gaussian basis to calculate a 3 kcal mole$^{-1}$ value for the barrier in ethane. These authors demonstrated that the barrier value does not depend on the basis set chosen, which added to the validity of their results.

Hoyland [47] obtained excellent results for the internal rotation barriers in butane. Using Gaussian functions, he estimated the barrier to be equal to 3.54 kcal mole$^{-1}$ and the energy difference between the *trans* and *gauche* form to be equal to 0.82 kcal mole$^{-1}$, which is in complete agreement with the available experimental data [48].

Unfortunately, the *ab initio* calculations in all these examples show how they can be made to work, rather than make a real contribution to empirical conformational analysis, since the question of what interactions give rise to a barrier remains obscure. By way of example, we should like to note that the rotational barrier in as simple a molecule as $H_2O_2$ is still a stumbling block for theoreticians dealing with *ab initio* calculations [49–51]. Some of them believe the unshared pairs of the oxygen atoms play a specific role in this molecule.†

The question arises: How, by adding a certain term to a potential function, or by modifying it, does one obtain values and forms of the barriers comparable with the experimental data? There seem to be three possibilities:

1. To introduce into the potential function not only a term which depends on the distances between the nuclei (the sum of atom–atom interactions) but also a term depending on the mutual positions of the vectors connecting the nuclei, to be more precise, on the angles between groups of certain vectors.

2. To change the interaction potentials of nonbonded atoms so as to make them automatically give internal rotation barriers without introducing torsion energy. The potentials will still remain central, i.e., depending only on the distances between the nuclei, but they will not be universal any longer. For example, it will not be possible to use these potentials to predict the conformations of aromatic molecules, energy of intermolecular interactions, etc.

3. To use only atom–atom interaction potentials, but to make them non-central, i.e., to assume that atoms have forms other than spherical, for example, that of rotational ellipsoids.

The Born–Oppenheimer theorem does not say anything about what the molecular potential function should be. It is essential that it should be a con-

---

† Recently, significant progress has been made both in calculation and revealing of the physical origin of the barriers to internal rotation; see, e.g., F. M. Lehn *in* "Conformational Analysis. Scope and Present Limitations" (G. Chiurdoglu, ed.). Academic Press, New York and London, 1971. Also, A. Veillard, *Chem. Phys. Lett.* **4**, 51 (1969); T. H. Dunning, Jr., N. W. Winter, *Chem. Phys. Lett.* **11**, 194 (1971).

tinuous function of the nuclear coordinates. Therefore all three possibilities above are correct from the point of view of quantum mechanics.

The majority of investigators chose the first possibility. It has been experimentally proved that a staggered conformation is more favorable than an eclipsed one, and the shapes of the barriers are roughly cosine-like [48]; hence for ethane, its haloid derivatives, and, in general, for molecules of the $A_1A_2A_3X$–$YA_4A_5A_6$ type the additional energy of interaction of the $A_1A_2A_3$–$A_4A_5A_6$ groups (i.e., bond orientation energy) is usually described as

$$U(\varphi) = (U_0/2)(1+\cos 3\varphi) \qquad (3.1)$$

where $U_0$ is an empirical constant and $\varphi$ is the rotation angle measured from the eclipsed conformation. This expression is usually called a torsion potential.

In fact, by expanding the rotation potential function into a Fourier series and having in mind that in the case of ethane and its analogues this is a periodic function with period $2\pi/3$ (the bond vectors $A_1$–X, $A_2$–X, etc. have threefold symmetry in relation to the X–Y rotation axis), we shall have

$$U(\varphi) = (U_0/2)(1+\cos 3\varphi) + (U_0'/6)(1+\cos 6\varphi) + \cdots \qquad (3.2)$$

The microwave spectra [48] of small molecules show that in most cases it is only the first harmonic of the series which is really essential; therefore all terms except the first one are usually neglected.

The above approximation makes the following formula useful in calculating the internal rotation barriers of ethane-like molecules:

$$U(\varphi) = (U_0/2)(1+\cos 3\varphi) + \sum_{i>j}\sum f(r_{ij}) \qquad (3.3)$$

It should be added that the bond angles are assumed not to change in the course of the rotation. In [17, 52] good agreement with the experimental data was obtained for the $H_3C$–$CH_3$, $H_3C$–$SiH_3$, and $H_3Si$–$SiH_3$ molecules and their haloid derivatives. The bond angles were assumed to be tetrahedral, the $U_0$ values were taken equal to 3.0 kcal mole$^{-1}$ for the C–C bond, 1.5 kcal mole$^{-1}$ for the C–Si bond, and 1.1 kcal mole$^{-1}$ for the Si–Si bond.

It should be noted that without taking into account variations of the bond angles it is hardly possible to obtain precise barrier values, especially in the case of large substituents. That was shown in paper [53] with haloid derivatives of ethane. For example, in hexachloroethane the barrier calculated for tetrahedral bond angles equals 26.7 kcal mole$^{-1}$, whereas on minimization of the potential function by varying the bond angles it decreased to 13.7 kcal mole$^{-1}$ (the experimental values are approximately 10.8 [54] and 17.5 kcal mole$^{-1}$ [55]).

It is of interest that in ethane the HCH bond angle practically does not

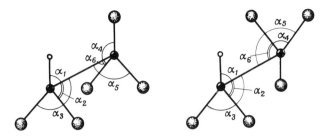

Fig. 3. Staggered and eclipsed conformation of hexachlorethane molecule.

change on rotation; both in the eclipsed and staggered forms it is 107.67° (the experimental value [56] is 107.8°). In molecules with large substituents the bond angle deformation amounts to several degrees. For example, the ClCC bond angle in hexachlorethane is 114° for the eclipsed and 110° for the staggered conformations. In pentachlorethane the equilibrium angles of the eclipsed form have the following calculated values (Fig. 3): $\alpha_6 = 106.8°$, $\alpha_4 = \alpha_5 = 112.7°$, $\alpha_3 = 109.3°$, $\alpha_1 = \alpha_2 = 111.4°$; and in the staggered form: $\alpha_6 = 108.5°$, $\alpha_4 = \alpha_5 = 111.8°$, $\alpha_3 = \alpha_2 = 110.3°$, $\alpha_1 = 111.4°$. Such seemingly insignificant deformations of bond angles result in a decrease in the barrier height from 19.5 to 12.4 kcal mole$^{-1}$ (the experimental values are 14.2 [56] and 12.8 [57] kcal mole$^{-1}$).

Let us return to potential functions in which, besides the interaction energy between pairs of nonbonded atoms and bond angles deformation, we intro- duce torsion energy as (3.1), i.e.,

$$U_{st} = U_{nonbonded} + U_{ang} + U_{tors} \qquad (3.4)$$

All further examples concerning conformations of aliphatic molecules, including vinyl polymers will be based on potential functions of this kind.

The second possibility was made use of by Magnasco [58]. Having intro- duced the Morse potentials, he selected for them such parameters that the calculated barriers of 30 molecules with one rotational degree of freedom were in almost complete agreement with the experimental values. Such pre- cision was unprecedented, but it should not be forgotten that the universality of atom–atom potentials was sacrificed.

Finally, the third possibility was recently put forward by Kitaigorodsky [59]. It is possible to obtain internal rotation barriers in molecules in terms of atom–atom potentials, but they will not be spherically symmetric potentials any more. It is assumed that the equilibrium radius of the univalent atom depends upon the angle formed by this radius and the valence bond. Assum- ing that $r_0$ is equal to $a$ and $a/\varepsilon$, respectively, along and across the bond, and assuming thereby the univalent atom to have the form of an ellipsoid of rotation, we can readily derive an equation for the equilibrium radius:

$$r_0 = (a/2) [(\varepsilon \sin^2 \psi_1 + \cos^2 \psi_1)^{-\frac{1}{2}} + (\varepsilon \sin^2 \psi_2 + \cos^2 \psi_2)^{-\frac{1}{2}}] \qquad (3.5)$$

where $\psi_1$ and $\psi_2$ are the angles between the intraatomic vector and the bonds (pp. 165, 166, Chapter 2).

In this formula there is one empirical parameter $\varepsilon$ just like $U_0$ in (3.1). Assuming $\varepsilon = 0.82$ and utilizing the universal Kitaigorodsky potential [26], we shall obtain a value of 3 kcal mole$^{-1}$ for the barrier to internal rotation in ethane. No calculations were carried out to verify the transferability of the empirical constant $\varepsilon$ from one molecule to another; therefore it is difficult to predict the validity of this approach.

### 4. CONFORMATIONS OF ALIPHATIC MOLECULES

Equation (3.4) is quite sufficient to calculate the conformations of aliphatic molecules; if the necessary parameters are known, the optimal geometry of the molecule can be obtained by minimization of the potential function. But in some cases the problem of numerous minima becomes a serious handicap.

One should start by deciding what result is sought. Keeping this in mind, one can divide all conformation problems into two types. The first type comprises the problems where the true geometry is to some extent known and may be assumed to be a zero approximation in the search for the local minimum of potential energy: the conformation so estimated may be compared with the results of structural investigations. The potential functions of the overcrowded aromatic molecules, which will be considered in the next section, have, as a rule, one deep minimum, and hence it is reasonable to find it by means of minimization of a function of several variables so as to obtain the geometry a priori.

In molecules with internal rotation there are always several or many minima in the potential surface. In this case it is interesting not only to find the precise coordinates of the atoms of the various conformers (i.e., to solve the problems of the first type) but also to predict the population of the various conformations and the energy differences of the rotation isomers, since this is very important for the configuration statistics of polymer chains. Finally, it is also essential to find the absolute minimum, so as to predict the most probable conformation of the molecule in the crystal.

In the potential energy surfaces of open hydrocarbon chains there is a great number of local minima corresponding to various conformations; it rapidly increases as the number of C atoms in the chain rises.

The absolute energy minimum of normal alkanes always corresponds to the trans conformation of neighboring units, i.e., the C atoms form a planar zigzag (since we count rotation angles from the eclipsed cis conformation, for the optimal conformation $\varphi_1 = \varphi_2 = \varphi_3 = \cdots = 180°$). But besides this conformation, there may exist ones in which at least some rotation angles differ from 180°, and, as follows from the torsion energy expression (3.1), conformations with $\varphi = 60°$ and $300°$ should make a considerable contribution. The latter should, evidently, correspond to local minima of the potential surface, their depths being comparable to that of the absolute minimum, and their percentage occupancy may be calculated by means of the Boltzman factor:

$$n_1/n_2 = \exp[-(U_1 - U_2)/RT] \qquad (4.1)$$

where $n_1$ and $n_2$ are relative populations of the two conformers, $U_1$ and $U_2$ their strain energies, $R$ is the gas constant, and $T$ is absolute temperature. Bartell and Kohl [60] used electron diffraction data to estimate the percentages of the various conformations of lower $n$-alkanes at room temperature:

|          |          |          | %    |              |     | %    |
|----------|----------|----------|------|--------------|-----|------|
| $n$-butane | T     | (trans)  | 59.1 | $n$-hexane   | TTT | 24.5 |
|          | G        | (gauch)  | 40.9 |              | TTG | 33.6 |
| $n$-pentane | TT    |          | 38.4 |              | TGT | 16.8 |
|          | TG       |          | 52.7 |              | TGG | 11.6 |
|          | GG       |          | 9.0  |              | GTG | 11.6 |
|          |          |          |      |              | GGG | 2.0  |

McCullough and McMahon [61] investigated theoretically the rotational states of $n$-hexane. For the sake of simplicity they assumed that $\varphi_1 = \varphi_2 = \varphi_4 = \varphi_5 = \varphi_b$ and $\varphi_3 = \varphi_d$ (i.e., a two-dimensional potential surface). All the $C \cdots H$ and $H \cdots H$ interactions were calculated using the curves of the same authors [62], ($C \cdots C$ interactions were not taken into account) the bond angles were assumed to be ideal. It was found that, besides the trans form $\varphi_b = \varphi_d = 180°$, corresponding to the absolute energy minimum, there are two more pairs of structures corresponding to the local minima: 168°–168°–60°–168°–168° and 185°–185°–80°–185°–185°, their energy exceeding by 1 kcal mole$^{-1}$ that of the planar zigzag. These conformations are similar to the Bartell and Kohl's TGT forms. Finally, one more conformation is possible: 52°–52°–160°–52°–52° (GTG) with a higher energy. It is interesting that the equilibrium rotation angles in these conformations differ from pure trans and gauche ones; this means that both torsional forces tending to realize pure conformations and nonbonded interaction potentials responsible for some small deviations play their role.

Scott and Scheraga [63] calculated with much higher precision the conformations of the lower alkanes—pentane, hexane, heptane, and also polyethylene. For bond lengths and bond angles the experimental values were used [64], the rotation angles around the C–C bonds were varied, and for pentane, hexane, and heptane four-, five-, and six-parameter problems, respectively, were solved (for all possible rotation angles).

It was proved that even lower alkanes have a great number of rotational states corresponding to local minima in the potential surface (there are 11 in pentane and 43 in hexane). In the case of heptane, the 28 most important states were analyzed; the energy differences for the majority of conformations are not great.

It is of interest that no energy minima were found in the $\varphi$ range indicated by McCullough and McMahon. Their suggestion that in hexane $\varphi_1 = \varphi_2 = \varphi_4 = \varphi_5$ does not seem to be correct.

Conformation calculations of saturated cycles have been very numerous; this is understandable, because the strain theory for them was elaborated as early as in the works of Bayer, and the thermochemical data made it possible to estimate the energies of angle strains.

In cycloalkanes, owing to the conditions of cyclization, there are not as many strain energy minima as exist in open chains. Besides, as is shown by calculations, in small ($n = 3\text{--}7$) and medium ($n = 8\text{--}12$) rings the population of the most favorable conformation is rather high.

For a long time there has been no unanimous opinion about the conformation of cyclobutane; but recently IR specroscopy data allowed some authors [65–67] to make an unambiguous conclusion about the nonplanar conformation of the ring. The potential function of this molecule depends on the dihedral angle, as is shown in Fig. 4. $\varphi \approx 30°$ corresponds to the energy minimum, and at $\varphi = 0$ the barrier is about 1 kcal/mole. It is easy to conclude that the puckering of the ring is due to torsion energy, whereas the nonbonded repulsions of the opposite C atoms hinder puckering. It is interesting to note that the potentials of many authors predict cyclobutane to have a planar equilibrium conformation and it was only Allinger et al. [68] who could offer the parametrization which gave an energy minimum for the nonplanar ring and the correct value of the barrier at $\varphi = 0$.

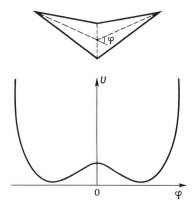

*Fig. 4.* Dependence of potential energy of cyclobutane upon dihedral angle.

Cyclobutane derivatives are of great interest. For example, the microwave spectra of chloro- [69] and bromocyclobutanes [70] showed both cyclobutane molecules to have a puckered ring, the angles being equal, respectively, to 20° and 29.3°. A nonplanar ring was also found in octachlorocyclobutane [71], anemonine [72], 1,2-dibromo-1,2-dicarbomethoxycyclobutane [73], octafluorocyclobutane [74], and 1,1-difluorocyclobutane [75]. But the four-member ring in tetraphenylcyclobutane [76], the photodimer of cyclopentanone [77], tetracyanocyclobutane [78], and *trans*-1,3-dicarboxicyclobutane [79] was proved to have a planar conformation. It is not easy to predict the conformation of the four-member heterocycles. For example, trimethylene oxide [80] has a planar molecule, whereas trimethylene sulphide [80] and silanecyclobutane [81] have nonplanar conformations. No predictions of equilibrium conformations for these compounds have been made.

Cyclopentane drew the attention of many investigators as early as at the end of the forties when its nonplanar conformation had been unambiguously established from entropy measurements [82]. Assuming that the equilibrium conformation of this molecule is the result of competitive angular and torsion tensions, Pitzer *et al.* [83, 84] found two equilibrium conformations for cyclopentane: $C_s$ and $C_2$ (Fig. 5). The interactions of the nonbonded atoms were not taken into account since for bond angles, spectroscopic deformation constants were used. $U_0$ was assumed to be equal to 2.8 kcal mole$^{-1}$ and the puckering of the ring was described by the formula

$$z_j = (2/5)^{\frac{1}{2}} q \cos [(2\pi/5) + \varphi] \tag{4.2}$$

where $z_j$ is the displacement of the $j$th atom normal to the plane of the ideal ring, $q$ is the amplitude of puckering, and $\varphi$ is the phase angle of puckering.

It was shown by calculations that the energy minimum has the corresponding value of $q = 0.50$ Å, the $C_s$ and $C_2$ conformations having approximately the same energy, 4 kcal/mole less than that of the planar conformation. Thus, in cyclopentane the planar conformation is obviously unfavorable due to torsion strains. It was also shown that the conformation transition $C_s \rightarrow C_2$ occurs without overcoming the potential barrier.

Hendrickson [27] calculated in detail the conformations of five-, six and seven-membered cycloalkanes. According to his estimation, the half-chair conformation, $C_s$, is by 0.53 kcal more favorable than the envelope $C_2$. It is

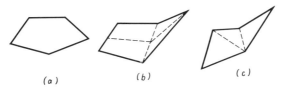

*(a)*                    *(b)*                    *(c)*

*Fig. 5.* Conformations of cyclopentane: a, planar; b, envelope; c, half-chair.

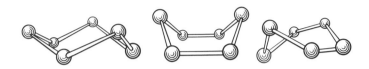

*Fig. 6.* Three possible conformations of cyclohexane: (from left to right) chair, boat, and twist-chair.

interesting that in cyclopentane derivatives, e.g., in furanose rings [85], a $C_2$-like conformation occurs more frequently. Ribose and deoxiribose in nucleotides and nucelosides have on an average 80% of $C_2$ and 20% of $C_s$ conformation (see Section VIII.3).

There is rich experimental and theoretical data on the conformations of six-membered rings. For example, the monograph [9] describes mostly cyclohexane and its derivatives. Therefore we shall not discuss this problem here. However, the fundamental result of Hendrickson which was later confirmed by other authors should be mentioned. He found that the energy difference between the chair and boat conformations is 6.93 kcal mole$^{-1}$ and the potential barrier of transition from one conformation to the other is 14 kcal mole$^{-1}$. However, this is not because the classical boat conformation corresponds to the energy minimum, as was believed by some authors, but it arises in the course of the transition from the chair to the twist form having $C_2$ symmetry (Fig. 6), the twist form being more favorable than the boat by 1.60 kcal mole$^{-1}$. Thus, the energy difference between the chair and twist conformations is 5.3 kcal mole$^{-1}$ (the experimental value is 5.9 kcal mole$^{-1}$ [86]). As to the optimal geometry of the cyclohexane molecule, Kitaigorodsky [87] predicted it to have CCC bond angles larger than tetrahedral ones (112°, according to the calculation). This was later confirmed experimentally [88, 89] and by calculations of other authors.

For cycloheptane, Hendrickson analyzed five conformations, two of them being traditional: boat and chair. The energy minimum corresponds to the chair conformation with the CCC bond angles increased to 112° (which agrees with the X-ray data [90]).

Later Hendrickson applied his method to medium cycles: cyclooctane (I), cyclononane (II) and cyclodecane (III). The most stable conformations of these molecules (Fig. 7) are not the only possible ones; there exist several other conformations slightly differing in strain energy.

Wiberg [37], using the same parameters as Hendrickson, found the equilibrium conformations and strain energies for cyclooctane, cyclononane, and cyclodecane. It should be noted that he also succeeded in finding a convenient way of introducing independent geometrical parameters: instead of the commonly used internal coordinates he employed the Cartesian atomic coordinates.

Fig. 7. Optimal conformations of cyclooctane, cyclononane, and cyclodecane.

The transition from the latter to the former coordinates was achieved by means of comparatively simple formulas. Minima were sought by the steepest descent method using cartesian displacements of the atoms.

Figure 8 shows six possible conformations of cyclooctane. Ideal conformations were assumed as initial approximations in the search for the minimum.

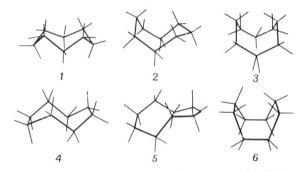

Fig. 8. Six possible conformations of cyclooctane after Wiberg.

The energies of these six conformations after minimization are listed in Table 5.

Thus, (5), (1), and (2) are the most favorable conformations. The bond angles of the equilibrium conformations are essentially different from the

### Table 5

STRAIN ENERGIES OF DIFFERENT CONFORMATIONS
OF CYCLOOCTANE

| Conformation | Nonbonded interactions | Torsion energy | Total energy (kcal mole$^{-1}$) |
|---|---|---|---|
| 1 | −9.85 | 10.99 | 3.62 |
| 2 | −9.79 | 5.00 | 4.03 |
| 3 | −8.89 | 1.91 | 5.06 |
| 4 | −9.50 | 13.78 | 9.40 |
| 5 | −9.57 | 9.25 | 2.65 |
| 6 | −9.61 | 13.06 | 12.15 |

ideal tetrahedral ones. For example, the calculated CCC angles for conformation (5) are 116.9° and 110.1° and for conformation (2), 117.7°, 109.6°, 114.0° and 114.4°.

Bixon and Lifson [38] made an attempt to find the optimal parametrization for conformation and enthalpy of formation (see Section 6) for $(6\text{-}12)n$-cycloalkanes. Varying $C_{HCH}$, $C_{HCC}$, and $C_{CCC}$ as well as $U_0$ in expression (3.1) they obtained geometry which agrees with the experimental data and also quite reasonable values for the enthalpies of formation. It is quite natural that in their calculation scheme these authors used various elastic constants for the bond angles with tetrahedral C atoms at their vertices: the angular constants had values similar to the spectroscopic deformation constants. However, interactions with adjacent nonbonded atoms were not taken into account, which is roughly in agreement with Eq. (1.7). These authors varied the torsion component and showed that better agreement with the experimental data is achieved if the value for $U_0$ is 3.4 kcal mole$^{-1}$ (instead of 2.8 or 3.0 kcal mole$^{-1}$).

As to the calculation procedure, it goes without saying that a computer is required to calculate conformations of complex molecules with sufficient accuracy; otherwise the calculations are too labor-consuming and their low accuracy renders them useless. The most important step in making up a program is to express the atomic coordinates via chosen independent geometrical parameters. (Sometimes it is possible to express the required distances between the nonbonded atoms without calculating the coordinates.) Then one calculates interatomic distances, bond angles, dihedral angles, and draws up the expression for strain energy. Calculation of the potential function makes up a special block, if the program is written in machine code, or it is a procedure if algorithmical language is used. Then there is the problem of minimization of a potential function with several variables.

Formulas required for calculating atom coordinates will be given in Section VIII.1. These formulas, or their modifications, have often been used to calculate cycloalkane conformations. However, an important requirement is that the cycle be closed, since in this case the number of independent parameters decreases by six. If, besides the three terms indicated in Eq. (3.4), one introduces into the potential functions the deformation energy of the valence bonds

$$U_b = \tfrac{1}{2} \sum_i K_i \Delta l_i^2 \tag{4.3}$$

where $\Delta l_i$ is the change in the bond lengths from their ideal values, and $K_i$ are elastic constants approximately equal to the spectroscopic constants for valence vibrations, then closing of the cycle will occur automatically in the course of the minimization.

Deformations of the lengths of valence bonds, as is shown by relevant calculations, do not exceed 0.01–0.02 Å, and are very rarely interesting from the structural point of view. Therefore, to decrease the number of independent parameters by which the function is minimized, it is usually feasible to assume the lengths of the valence bonds to be perfectly fixed. However, for cyclic molecules a "cyclization" potential, like

$$U_{\text{cycl}} = K(l - l_0)^2 \qquad (4.4)$$

should be introduced, where $l_0$ is the bond length, $l$ is the distance calculated for this bond in this set of independent geometric parameters, and $K$ is a constant ideally equal to $\infty$. The latter constant must be varied in the course of minimization, otherwise potential (4.4) prevails over the other components of the potential function.†

Below, methods of minimization will be briefly described (they are discussed in detail in reviews [91–93]). If minimization begins far from the minimum, it is advisable to use linear methods, e.g., that of the steepest descent. But in the vicinity of the minimum these methods give slow convergence and quadratic methods prove to be more effective. The majority of authors engaged in conformational calculations have employed the steepest-descent method. Scheraga *et al.* [94] succeeded in using the method of Davidon [95] for minimizing the potential functions of oligopeptides; this method is in principle a quadratic one, though it does not require the calculation of second derivatives. Generally speaking, it is desirable to have a complex of programs using both linear and square methods; in this case it is possible to solve conformational problems of various kinds. Such programs have been written in ALGOL-60 [96].

Returning to the conformations of cycloalkanes, it should be mentioned that not only the precise position of the potential function minimum, but also the isomerization paths, e.g., the paths of conversion from one conformer to another, should be found by an appropriate technique.

The most favorable isomerization paths correspond to movement along the bottoms of potential-surface valleys. They may be found in the following way.‡ Let $A_0$ be a point belonging to a bottom of a valley (it may be found

---

† Another procedure was put forward by A. A. Lugovskoy and V. G. Dashevsky [*Zh. Strukt. Khim.* **13**, 122 (1972)]; the authors found expressions for dependent geometrical parameters through independent parameters provided the bond lengths are fixed and bond angles are soft (this algorithm is especially useful in searching for conformational isomerization paths). On the other hand, N. Go and H. A. Scheraga [*Macromolecules* **3**, 178 (1970)] elaborated the algorithm for cyclic systems with fixed bond lengths and angles.

‡ Transition states of cyclohexane and its analogues have been investigated theoretically in the papers [97, 98]. V. G. Dashevsky and A. A. Lugovskoy [99] employed an automatic procedure for finding the isomerization paths and the corresponding barrier heights.

going from a local minimum). Choosing the most important parameter $x_i$ (for which $\Delta U/\Delta x_i$ is the least) and giving it an increment $h$, execute a local descent over $n-1$ variables, excluding this one, and find a point $A_1$ which belongs to the bottom if $h$ is sufficiently small. The next step is made in the direction of the vector $A_0 A_1$, after which local descent proceeds with the exclusion of the most important variable. The most effective minimization technique in this case is quadratic; the increment $h$ may be selected in such a way that the process is of a relaxation character. The saddle points thus found correspond to the barriers to interconversion. Application of the above procedure to $\alpha$-D-glucose arrived at the lowest barrier to chair-to-chair isomerization ($1C \rightleftharpoons C1$) of 10.8 kcal mole$^{-1}$ (with Dashevsky's functions) which agrees with the experimental values.

In the simple cycles considered above (except cyclobutane) the bond angle deformations are not very great, and even with ideal values of bond angles one can always choose conformations without inadmissibly short

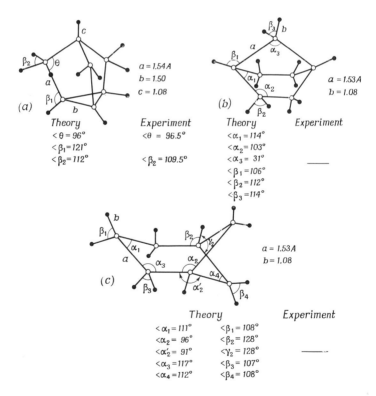

**Fig. 9.** Equilibrium conformations: a, nortricyclene; b, norcamphane; c, norpinane, calculated by Kitaigorodsky.

contacts between nonbonded atoms. The situation is different in bi- and polycyclic systems: as a rule, the ideal models of these molecules have very high values of strain energies and repulsions of nonbonded atoms result in serious deformations.

Kitaigorodsky [87] was the first to calculate such "tied-up" molecules. Figure 9 shows the results of calculation of equilibrium conformations for nortricyclene, norcamphane, and norpinane. In spite of the fact that the calculations were simplified (no torsion strains were taken into account, since they are not very important in determining the conformations of molecules of these types), rather good agreement with the experimental data was obtained. Attention should be paid to the great deformations of the bond angles as compared to ideal tetrahedral models.

Zaripov, *et al.* [100] calculated conformations of many "tied-up" molecules using the expression for the potential functions given above. In Table 6 there are calculated and experimental data for four molecules of this type. (The table includes one molecule with double bonds: bicyclo-[2,2,1]-heptadiene-1,2 to be discussed in Section 6 in connection with thermochemical data.) Instead of $U_{st}$, $U_{st}^*$ are given; this indicates that only interactions of neighboring nonbonded atoms and $C \cdots C$ interactions of type $1 \cdots 3$ were taken into account in the calculations; the values of the bond and dihedral angles obtained thereby are rather accurate and $U_{st}^* > U_{st}$ by approximately 8% for cyclic systems (only the attraction of atoms the distance between which exceeds $r_0$ is neglected).

One more remark can be made about the calculation of the conformations of polycyclic compounds: If expression (3.1) is used for the torsion energy, there could be no unambiguity in choosing the rotation angles (e.g., for the C1–C2 bond in bicyclo [2,1,1]-hexane the dihedral angle can be determined either as C3C2C1C5 or as C3C2C1C6). Therefore the following expression was suggested.

$$U_{\text{tors}} = \sum_{\text{CC bonds}} \sum_{i=1}^{9} (0.333/2)(1 + \cos 3\varphi_i) \qquad (4.5)$$

where $\varphi_i$ is one of the nine dihedral angles of the CCCC, HCCC, or HCCH type. In the case of acyclic hydrocarbons, in which the bond angles are almost tetrahedral, the latter formula automatically becomes (3.1).

It is interesting that in polycyclic systems the deviations of the bond angles from the ideal values may amount to 30°; nevertheless, the simple expression (1.5) for the angular deformation energy works well. But if agreement between the structural and thermochemical data is desired, some more complex expressions may be required, as is shown by certain authors [68, 101]. For example, in the case of great deformations deviations from Hooke's law should be taken into consideration.

### Table 6

CONFORMATIONS OF SOME POLYCYCLIC SYSTEMS

| Molecule | $U_{st}$ (kcal mole$^{-1}$) | Angle | Value of the angle | | Reference |
|---|---|---|---|---|---|
| | | | Calculated | Experimental | |
| | | Bicyclo-[2, 1, 1] hexane | | | |
| | | $\theta$ | 132.3° | 129.5° | |
| | | $C_1C_2C_3$ | 99.3° | 100.5° | |
| | 151.1 | $HC_5H$ | 108.6° | 98.8° | a |
| | | $HC_2H$ | 107.8° | 117.3° | |
| | | $HC_4C_5$ | 119.2° | 119.6° | |
| | | Bicyclo-[2, 2, 1] heptane | | | |
| | | $\theta$ | 113.2° | 113.0° | |
| | | $C_1C_7C_4$ | 93.5° | 93.2° | |
| | 146.3 | $HC_7H$ | 108.8° | 110° | b |
| | | $HC_2H$ | 107.3° | 110° | |
| | | $HC_1C_7$ | 119.2° | | |
| | | Bicyclo-]2, 2, 1] heptadiene-2, 5 | | | |
| | | $\theta$ | 114.6° | 115.0° | |
| | | $C_1C_7C_4$ | 94.4° | 92.0° | |
| | 141.5 | $HC_7H$ | 108.6° | 112° | b |
| | | | | (assumed) | |
| | | $HC_2C_3$ | 127.4° | 123° | |
| | | $HC_1C_7$ | 119.2° | | |
| | | Tricyclo-[1, 1, 1, 0$^{4, 5}$] pentane | | | |
| | | $C_2C_1C_4$ | 90.3° | 91.3° | |
| | | $C_1C_2C_3$ | 81.5° | 81.5° | |
| | 163.9 | $HC_2H$ | 110.8° | | c, d |
| | | $HC_1H_2$ | 119.2° | | |
| | | $HC_2C_4$ | 119.2° | | |

<sup></sup>

[a] G. Dallinga and L. H. Toneman. *Rec. Trav. Chim. Pays-Bas* **86,** 171 (1967)
[b] Y. Morino, K. Kuchitsu, A. Yokozeki. *Bull. Chem. Soc. Jap.* **40,** 1552 (1967).
[c] C. S. Gibbons, J. Trotter, *J. Chem. Soc. A* p. 2027 (1967).
[d] J. Trotter and C. S. Gibbons. *J. Amer. Chem. Soc.* **89,** 2792 (1967).

### 5. ETHYLENIC, CONJUGATED, AND AROMATIC SYSTEMS

Hill [1] calculated the equilibrium conformation and strain energy of *cis*- and *trans*-2-butene as early as 1948. The bond angle deformations were estimated by using the force constants obtained from an analysis of the

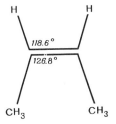

**Fig. 10.** Equilibrium geometry of cis-2-butene, calculated by Hill.

propylene spectrum [102]. The methyl groups were assumed to be two inter-acting atoms for which a semiempirical potential was calculated; in addition, the $H \cdots H$ and $CH_3 \cdots H$ nonbonded interactions were taken into account. (As was mentioned above, it is not consistent to employ both spectroscopic constants and nonbonded interaction potentials.) The geometry of cis-2-butene calculated by Hill is shown in Fig. 10. The calculated energy difference between the cis and trans configurations is 1.852 kcal mole$^{-1}$; this is com-parable with the difference in their heats of formation, which is 1.288 kcal mole$^{-1}$ [103].

Cycloalkenes are undoubtedly very interesting, since considerable bond angle deformations are very likely to occur in them due to the conditions of cyclization and the tendency of the ethylene group to maintain planar geometry structure. High strain energies are responsible for the specific thermo-chemical properties of these systems. Some of the simplest cycloalk-enes are described in the paper [104]; let us consider their conformations.

The potential functions required for the conformational analysis of cyclo-alkenes include expression (3.4) and an additional term to take account of the rigidity of the ethylene system:

$$U_{eth} = (U_0^{eth}/2)(1 - \cos 2\theta) \qquad (5.1)$$

where $\theta$ are the dihedral C–C=C–C angles; the $U_0^{eth}$ constant may be assumed to be equal to 60 kcal mole$^{-1}$ (the activation barrier for the cis-trans con-version of ethylene-like compounds). The introduction of (5.1) decreases the number of possible cycloalkene conformations. For example, for cyclo-octatetraene only conformation (III) out of the five possible nonplanar ones has the angles $\theta$ equal to zero; this corresponds to the minimum of (5.1); the four other conformations have very high strain energies:

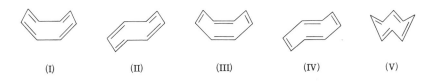

(I)         (II)         (III)         (IV)         (V)

It is difficult to take into account the torsion strains arising from the mutual orientations of the single and double bonds. In [105] this effect, in polymers of the diene type, was taken into account by the expression

$$1.0(1-\cos 3\psi) \tag{5.2}$$

where $\psi$ is the C=C–C–C dihedral angle measured from the cis orientation of the single and double bonds, i.e., the one which corresponds to the energy minimum. (Expression (5.2) describes the rotation potential in propylene.) However, it was shown by relevant calculations that, in practice, the use of (5.2) does not change the optimal geometry; neither does this term affect the strain energy, with the sole exception of cyclohexene for which the agreement between the calculated and experimental isomerization energies for the half-chair–boat conformations is a little worse. Therefore in the examples given below expressions of the type (5.2) are not considered; further investigation is required to discover whether they should be taken into account.

Thus, the following expression was used when the potential functions of cycloalkenes were minimized:

$$U_{st} = U_{nonbonded} + U_{ang} + U_{tors} + U_{eth} \tag{5.3}$$

with the Dashevsky parameters; for bond lengths the standard values were used (see the classification of bonds in Section 6). Results of some of the calculations are given below.

*a. Cyclopentene and Cyclopentadiene*

Figure 11 shows the calculated conformations of these molecules. The molecule of cyclopentene is nonplanar (the dihedral angle is 165°) due to the torsion strain (in the equilibrium conformation it is 4.86 kcal mole$^{-1}$). The energy difference between the planar and nonplanar forms (for the planar conformations the equilibrium bond angles have the following values:

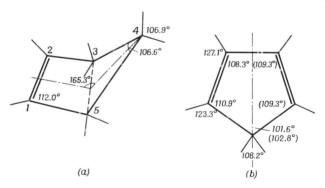

$(a)$                                    $(b)$

*Fig. 11.* Calculated conformations of: a, cyclopentene; b, cyclopentadiene. In parentheses, experimental data.

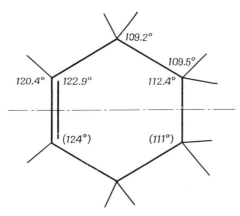

**Fig. 12.** Conformation of cyclohexene. Electron diffraction results in parentheses.

$C1C2C3 = 112.2°$, $C2C3C4 = 104.1°$, and $C3C4C5 = 107.4°$) is 0.55 kcal mole$^{-1}$; this agrees with the measured value of the inversion barrier, 0.66 kcal mole$^{-1}$ [106, 107]. Thus, inversion of cyclopentene occurs rather readily; at room temperature the C4 atom vibrates in relation to the C1C2C3C5 plane and since no other conformations are possible, pseudorotation does not occur, unlike the case of cyclopentane.

In the cyclopentadiene molecule the equilibrium conformation is a planar one, and the bond angle deformations inside the cycle are rather high—108° and 111° instead of 120°, also 101.6° instead of 109.5° (the angular deformation energy is 6.3 kcal mole$^{-1}$).

*b. Cyclohexene*

In [108] the equilibrium conformations of the two possible forms of this molecule, $C_2$ and $C_S$ (Fig. 12), were calculated. The half-chair ($C_S$) proved to be more stable by 5.4 kcal mole$^{-1}$ than the boat ($C_2$)†; similar results were obtained earlier by French authors [109]. An electron-diffraction investigation of this molecule carried out simultaneously [108] showed that the calculated bond angles of the half-chair conformation agree with the experimental radial distribution curve within the limits of error of the experiment; the boat conformation was found to be entirely unsatisfactory.

*c. 1,4 and 1,3-Cyclohexadienes*

The equilibrium conformations of these molecules calculated in [104] are shown in Fig. 13; for the former molecule both the nonplanar form corres-

---

† As shown in [99], the boat form is the transition state on the interconversion path from chair to chair.

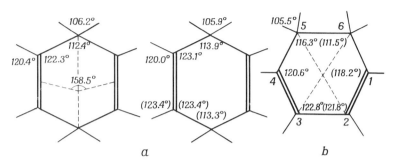

**Fig. 13.** Calculated conformations of (a) 1,4- and (b) 1,3-cyclohexadienes. Bond angles for planar and nonplanar forms of the former are obtained by minimization of the potential function.

ponding to the energy minimum and the planar one are shown. The energy difference between them is 0.1 kcal mole$^{-1}$ and, due to the small barrier, 1,4-cyclohexadiene exists in a continuous spectrum of conformations with the planar form most prevalent. From this point of view it is easy to interpret the results of the electron diffraction study [110], whose authors came to the conclusion that the planar conformation is the most probable but nonplanar forms are also likely to exist and cannot be unambiguously found from the electron intensity curve. The electron-diffraction data do not agree with Herbstein's [111] estimation of the equilibrium dihedral angle, 140°; the change of this angle requires several kcal per mole.

The $C_2$ symmetry of 1,3-cyclohexadiene was established from the microwave spectra [112, 113] and the electron diffraction study [113a]. In this paper the dihedral angle between the C2C3C4C5 and C3C2C1C6 planes was found to be 17°. Calculation confirmed the $C_2$ symmetry (the dihedral angle was 18°) and proved it to be 0.25 kcal/mole more stable than the planar form. However, large out-of-plane displacements require but small amounts of energy, and therefore the atomic coordinates of this molecule should be rather "diffuse." This may account for the incomplete agreement between the calculated and experimental values of the bond angles.

### d. 1,3,5-Cycloheptatriene

The following conformations of this molecule were studied by electron diffraction [114]. It was shown that the experimental scattering intensity curve corresponds best of all to the nonplanar conformation (VIII) (boat) with $\theta_1 = 139.5 \pm 2°$, $\theta_2 = 143.5 \pm 2°$ (Fig. 14). Another investigation [115] has lead to rather different angles, $150.5 \pm 4°$ and $130.0 \pm 5°$. Potential-function minimization gives the two equilibrium conformations (VI) and (VIII), the latter being more stable by 6.85 kcal mole. Its parameters are: $\theta_1 = 148.4°$,

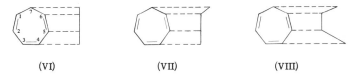

(VI)         (VII)         (VIII)

$\theta_2 = 131.2°$, $\angle C_2 C_3 C_4 = 125.0°$, $\angle C_1 C_1 C_1 = 110.8°$, $\varphi = 1.0°$; These are very close to Butcher's data and differ substantially from Traetteberg's. Note that Traetteberg's data correspond to $\varphi = 22°$ which is quite unfavorable.

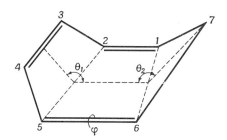

**Fig. 14.** Planar and nonplanar equilibrium conformations of 1,3,5-cycloheptriene.

*e. Cyclooctatetraene*

Figure 15 shows the optimal conformation of this molecule calculated in [104]. The results of this calculation are in complete agreement with the experimental data obtained by the electron diffraction method [116].

The examples given above show that the mechanical model of the molecule predicts fairly well the structures of such systems, which could not have been described earlier without the concepts of resonance and conjugation. Moreover, as will be shown in the next section, the thermochemical properties of these systems can also be derived from strain energies corresponding to the potential function minimum, without any additional assumptions.

Now let us consider the conformations of the aromatic molecules. These systems are interesting in that steric hindrance in them (close interatomic contacts in ideal models) may result in deformations of various types: changed bond angles, out-of-plane distortions, rotations of polyatomic groups around the bonds connecting them with the benzene nuclei. Structural evidence on overcrowded aromatic molecules is discussed in detail in review [117].

It is obvious that, besides nonbonded interactions and bond angles deformations,† the potential function should account for nonplanar deformations

---

† Planar deformations of aromatic molecules were calculated for the first time by Kitaigorodsky on the example of acenaphthene [87].

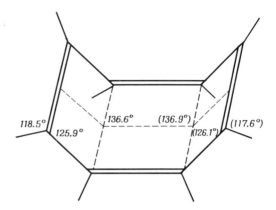

**Fig. 15.** Conformation of cyclooctatetraene. In parentheses, experimental data.

of the benzene nuclei, and also the energy of departure of the substituents from the mean plane of the benzene ring. In terms of quantum chemistry this means a decreased overlapping of $p$-$\pi$-electrons of neighboring bonds, or a decrease in the delocalization energy, which is obviously unfavorable for aromatic molecules. This is also the reason rotation around bonds that are conductors of conjugation is unfavorable.

Coulson *et al.* [118, 119] suggested a calculation scheme for the nonplanar deformations of aromatic and some other conjugated systems. Let us analyze the fragments of the molecules presented in Fig. 16a,b,c. For example, the ethylene molecule may correspond to a; to b, a fragment of the benzene molecule, its derivatives, naphthalene etc.; to c, certain fragments of the naphthalene, acenaphthene, or phenanthrene molecules etc. If $Z_1, ..., Z_6$ indicate the departure of the atoms from the plane; $K_1$, the spectroscopic force constant for nonplanar vibrations; $K_2$, the torsional force constant for rotation around the C–C bond (an assumption is made at this point that all types of atomic displacements can be described by these two constants alone); then the strain energy for the cases a, b, and c, without accounting for the repulsion

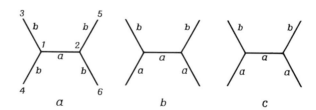

**Fig. 16.** Fragments of ethylene, conjugated, and aromatic molecules after Coulson and Senent.

between the nonbonded atoms, may be presented as

(a)
$$U = \frac{1}{2}K_1\left[\frac{a}{b}Z_3 + \frac{a}{b}Z_4 + Z_2 - \left(1 + \frac{2a}{b}\right)Z_1\right]^2$$

$$+ \frac{1}{2}K_1\left[\frac{a}{b}Z_5 + \frac{a}{b}Z_6 + Z_1 - \left(1 + \frac{2a}{b}\right)Z_2\right]^2$$

$$+ \frac{1}{2}K_2[Z_3 - Z_4 - Z_5 + Z_6]^2 \tag{5.4}$$

(b)
$$U = \frac{1}{2}K_1\left[\frac{a}{b}Z_3 + Z_4 + Z_2 - \left(2 + \frac{a}{b}\right)Z_1\right]^2$$

$$+ \frac{1}{2}K_1\left[\frac{a}{b}Z_5 + Z_6 + Z_1 - \left(2 + \frac{a}{b}\right)Z_2\right]^2$$

$$+ \frac{1}{2}K_2\left[\frac{a}{b}Z_3 - Z_4 - \frac{a}{b}Z_5 + Z_6 + \left(1 - \frac{a}{b}\right)(Z_1 - Z_2)\right]^2 \tag{5.5}$$

(c)
$$U = \frac{1}{2}K_1\left[\frac{a}{b}Z_3 + Z_4 + Z_2 - \left(2 + \frac{a}{b}\right)Z_1\right]^2$$

$$+ \frac{1}{2}K_1[Z_5 + Z_6 + Z_1 - Z_2]^2$$

$$+ \frac{1}{2}K_2\left[\frac{a}{b}Z_3 - Z_4 - Z_5 + Z_6 + \left(1 - \frac{a}{b}\right)Z_1\right]^2 \tag{5.6}$$

($a$ and $b$ are bond lengths).

It is easy to derive Eq. (5.4)–(5.6) if the molecule is considered to be distorted at random (i.e., all $Z_1, \ldots, Z_6$ are not equal to zero) and all the atoms are transferred into ideal planar position by means of several consecutive operations. The constants $K_1$ and $K_2$ were estimated from the spectra of ethylene and benzene: for ethylene, $K_1 = 0.131 \times 10^5$ dynes cm$^{-1}$, $K_2 = 0.157 \times 10^5$ dynes cm$^{-1}$; for benzene, $K_1 = 0.1474 \times 10^5$ dynes cm$^{-1}$, $K_2 = 0.00553 \times 10_5$ dynes cm$^{-1}$.

The determination of an equilibrium conformation consists of a search for the minimum of the potential energy surface which is expressed by Eqs. (5.4)–(5.6) for the molecular fragments. To obtain unambiguous results, additional assumptions must be introduced, because Eqs. (5.4)–(5.6) do not contain independent geometrical parameters of the molecule, and the number of equations determining the position of the minimum is always less than the number of unknown quantities.

This difficulty is overcome by application of the rigid-sphere method to

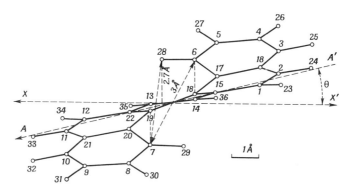

definite fragments of the molecule. If the molecule is overcrowded due, chiefly, to the interaction of some two atoms localized in its periphery, these atoms, in accordance with the above approximation, should deviate from their ideal positions enough to make the distance between them equal to the sum of their van der Waals radii. Then the positions of the other atoms can be found on the assumption that all potential function derivatives of displacements of atoms are equal to zero. The rigid-sphere method has an essential limitation, *viz*, even the interaction of the above peripheral atoms is, in fact, less rigid than that described by the rigid-sphere method. All other interactions are assumed to be soft, and that is why the method is inconsistent.

Figure 17 shows the conformation of the 3:4–5:6 dibenzophenanthrene molecule as calculated by Coulson and Senent. The ideal model of this molecule contains impermissibly short H28···H29, C6···C7, and C7···H28 contacts. An X-ray study [120, 121] has proved the C6···C7 distance to be equal to 3.0 Å. If this is the case, the van der Waals radius of the "compressed" carbon atom system may be assumed to be equal to 1.5 Å, and, if the hydrogen radius is 1.2 Å, then $r(C7···H28) = 2.7$Å. Now it is easy to obtain the values of the displacement of the C6, C7, H28, and H29 atoms from the mean plane of the molecule: $Z_6 = -Z_7 = 1.32$Å; $Z_{28} = -Z_{29} = 1.36$Å. Having made these assumptions, Coulson and Senent solved 32 linear equations of the $\partial U/\partial Z = 0$ type to find the $z_i$, the departures of the remaining 32 atoms from the plane; these authors obtained in this way carbon atom coordinates which agree well with the experimental data.

The calculations of the following five polynuclear condensed hydrocarbons made by Ali and Coulson [122, 123] by the same method are worth mentioning: 5:6-7:8 dibenzoperylene (IX), 1:12-5:6-7:8 tribenzoperylene (X), tetrabenzoperopyrene (XI); chrysene (XII) and 20-methylcholanthrene (XIII).

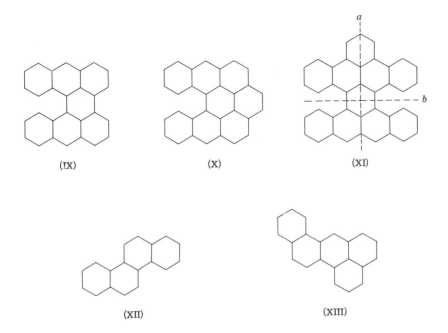

(IX)  (X)  (XI)

(XII)  (XIII)

For tetrabenzoperopyrene two conformations are possible: (1) on one side of the $b$ axis the atoms have positive displacements, on the other side, negative ones; this conformation is symmetrical with respect to the $a$ axis; (2) the upper half of the molecule is turned relative to the lower half around the $a$ axis; this could be called a "propeller" conformation. Calculations showed that conformation (1) has a corresponding strain energy of 41.5 kcal mole$^{-1}$ while the "propeller" conformation has 36.3 kcal mole$^{-1}$. Thus the latter conformation is more favorable.

The strain energies for the following molecules were calculated: dibenzophenanthrene, 17.9; dibenzoperylene, 18.54; tribenzoperylene, 18.60 kcal mole$^{-1}$. Naturally, the strain energy of tetrabenzoperopyrene proved to be almost twice as high as that of the above molecules.

In a later paper Coulson and Haigh [124] developed a general analysis using potential curves of nonbonded interactions.

Let us assume that in an overcrowded molecule there are $t$ equivalent interactions of nonbonded atoms, $U_x$ is the deformation energy of the bond angles and $U_z$ is the energy change for nonplanar deformations. Then the strain energy may be expressed as

$$U = tf(R) + U_x + U_z \qquad (5.7)$$

where $f(R)$ is the interaction potential of the nonbonded atoms localized at the distance $R$ from each other;

$$U_x = \tfrac{1}{2}\mathbf{x}'\mathbf{K}\mathbf{x}, \qquad U_z = \tfrac{1}{2}\mathbf{z}'\mathbf{A}\mathbf{z} \qquad (5.8)$$

where $\mathbf{x}$ is the vector corresponding to the bond angle's deviation from $120°$ and the change in bond lengths as compared to their standard values in the absence of strain; $\mathbf{z}$ is the vector of displacement normal to the mean plane; $K$ and $A$ are symmetrical square matrices of force constants.

The deformation effect is expressed by the change in the distance between the nonbonded atoms as compared to the ideal distances

$$R = R_0 + R_x + R_z + R_{xz} \qquad (5.9)$$

where $R_0$ is the ideal distance; $R_x$ and $R_z$ are, respectively, the planar and nonplanar displacements of the atoms; $R_{xz}$ is a term accounting for the mutual influence of these displacements (when the atoms depart from the mean plane the projection of the molecule onto this plane is somewhat distorted). If we assume that the departures of atoms from the plane do not depend on the in-the-plane deformations (which is valid for small nonplanar deformations), then the $R_{xz}$ term may be neglected. Then the following equations may be derived from (5.8) and (5.9)

$$H(R, \mathbf{x}, \mathbf{z}) = \alpha'\mathbf{x} + \tfrac{1}{2}\mathbf{z}'\mathbf{B}\mathbf{z} - (R - R_0) = 0 \qquad (5.10)$$

where $\mathbf{B}$ is a symmetrical square matrix, and $\alpha'$ is some constant.

In addition, a relationship can be found between the angles and bond lengths. Using the rectangular $l \times m$ matrix $\mathbf{C}$ ($l$—the number of bond conditions, $m$—the order of the $\mathbf{x}$ vector), and the vector $\mathbf{d}$ this equation may be expressed in a general form as

$$\mathbf{g}(\mathbf{x}) = \mathbf{C}\mathbf{x} + \mathbf{d} \qquad (5.11)$$

For further analysis, the Lagrangian method of indefinite coefficients will be employed with $\rho$ and $\sigma'$ as the multipliers

$$W \equiv U(R, \mathbf{x}, \mathbf{z}) + \rho H(R, \mathbf{x}, \mathbf{z}) + \sigma'\mathbf{g}(\mathbf{x}) \qquad (5.12)$$

Now, if $U(R, \mathbf{x}, \mathbf{z})$ is substituted by its value from (5.7), the problem will be confined to a search for the minimum of the function

$$W = tf(R + \tfrac{1}{2}\mathbf{x}'\mathbf{K}\mathbf{x} + \tfrac{1}{2}\mathbf{z}'\mathbf{A}\mathbf{z} + \rho\alpha'\mathbf{x} + \tfrac{1}{2}\rho\mathbf{z}'\mathbf{B}\mathbf{z} - \rho(R - R_0) + \sigma'(\mathbf{x} + \sigma'\mathbf{d})) \qquad (5.13)$$

with respect to $\mathbf{x}$, $\mathbf{z}$, and $R$.

Differentiating (5.13) with respect to $R$, all $\mathbf{x}$ and all $\mathbf{z}$, we shall derive an equation in matrix form, from which, by means of nontrivial operations, the values of the geometric parameters may be estimated and the meaning of the indefinite multipliers may be clarified. Coulson and Haigh [124] suggested an inequality which is a criterion for the atom's departure from the plane.

This inequality means that if the value of $df/dR$, i.e., the repulsion forces between the nonbonded atoms are great enough, departure of the atoms from the plane occurs; if $df/dR$ does not exceed some critical value, the molecule maintains its planar conformation.

This analysis and, especially, Eq. (5.7) should be extended to the non-equivalent interactions, i.e., repulsions between the members of all the pairs of nonbonded atoms should be accounted for. In this case the equations become extremely complicated, without being new in principle.

The $K$ matrix is considered diagonal (bond angle interactions are left out of account) and spectroscopic force constants are its elements; according to Syvin [125] they have the following values: C–C, 6.40, C–H, 5.8; CCC, 0.740, CCH, 0.849 mdyne $\mathring{A}^{-1}$.

The elements of the $A$ matrix are constants analogous to $K_1$ for out-of-plane atom displacements (see equations (5.4)–(5.6)) and $K_2$ for torsion movements. These constants are refined in paper [126].

(XIV)                    (XV)                    (XVI)

Using essentially the above method Coulson and Haigh calculated the equilibrium conformations for phenanthrene (XIV), triphenylene (XV), and chrysene (XVI). In all the three molecules the ideal model is distorted due to repulsion of hydrogen atoms localized close to each other. (It is also probable that the C$\cdots$C and C$\cdots$H interactions left out of account also play some role.) The conformation of each molecule was calculated using the eleven H$\cdots$H curves described in the literature. The phenanthrene conformation was shown to be always planar, independent of the potential curve chosen. In the case of triphenylene and chrysene with ten of the H$\cdots$H curves (excluding the most rigid one) the planar conformation is obtained. X-ray studies carried out after the calculation had been published proved the phenanthrene molecule to be in fact planar [127]. In triphenylene [128] the peripheral benzene nuclei are slightly displaced with respect to the central nucleus plane. It is interesting to note that the changes in the bond lengths are on an average several thousandths of an angstrom, and it is only for the most rigid curves of nonbonded H$\cdots$H interactions that the changes are 0.01–0.02 $\mathring{A}$, i.e., beyond the accuracy of the X-ray experiment. This proves the practical possibility of using absolutely rigid valence bonds.

In the studies of Kitaigorodsky and Dashevsky [12, 129] a somewhat different approach was employed. The potential functions of aromatic systems are essentially similar to those discussed above and it is only for the non-planar deformations that additional terms were introduced.

For the nonplanar deformation energy, the following expression was used:

$$U_{nonpl} = \tfrac{1}{2}\sum_j C_j^* \sin^2 \beta_j \qquad (5.14)$$

where $\beta_j$ are bond deflection angles and $C_j^*$ are the respective elastic constants.

The parameters $\beta_j$ are introduced in the following way. For the peripheral atoms, the $\beta_j$ are the angles formed by the peripheral bond and the plane of the two adjacent bonds of the benzene nucleus. (If we deal with naphthalene or polynuclear aromatic systems, every benzene ring is a peripheral one with respect to the adjacent nuclei.) For the displacements of the benzene nuclei, $\beta_j$ are the angles formed by the planes of the neighboring bond pairs, e.g., the angles between the planes C1C2C3 and C2C3C4.

Parameters determined in this way, unlike those of Coulson, are more convenient, because they can be applied to complicated molecules of low symmetry. Coulson's parameters require the mean plane of the molecule, and it is not always possible to draw it. With small out-of-plane deformations and high symmetry $\beta_j$ may be easily compared with Coulson's parameters. (Coulson uses the atoms' departures from the mean plane in angstroms as independent parameters.)

It should be noted that, for small out-of-plane bond distortions, (5.14) becomes Hooke's law:

$$U_{nonpl} = \tfrac{1}{2}\sum_j C_j^* \beta_j^2 \qquad (5.15)$$

Now consider the potential functions for rotations of the polyatomic groups in relation to the benzene nuclei planes. It is known only that this function must be periodic, because, when the group makes a complete turn (360°), the energy remains unchanged. Besides, it is an even function as rotations in two opposite directions are energetically equivalent. An even periodic function expanded in a Fourier series will be

$$U(\varphi) = U_0/2 + \sum_{n=1}^{m} U_n \cos n\omega\varphi \qquad (5.16)$$

where $U_0, U_1, \ldots, U_m$ are the coefficients of the expansion of the energy $U(\varphi)$ in terms of the angle of rotation, $\omega = 2\pi/T$; $T$ is the period of rotation. (For even functions, all the expansion coefficients of sinusoidal harmonics are equal to zero.) It is obvious that for nitro, amino, and phenyl groups with their twofold symmetry $T = \pi$, and that for the asymmetric COOH groups $T = 2\pi$.

To simplify parametrization, only one harmonic of the Fourier series was used, i.e., expressions like

$$U_{rot} = (U_0/2)(1 - \cos\varphi) \qquad (5.17)$$

or

$$U_{rot} = (U_0/2)(1 - \cos 2\varphi) \qquad (5.18)$$

which are derived from (5.16).

The $U_0$, obviously, should be different for different polyatomic groups; in terms of quantum chemistry this means that these groups have different conjugations with the benzene nucleus. There are many examples in the literature where parameters for the rotations of polyatomic groups have been determined by the method of molecular orbitals [16, 130–132].

With the above assumptions, the strain energy of an aromatic molecule is defined as

$$U_{st} = U_{nonbonded} + U_{ang} + U_{nonpl} + U_{rot} \qquad (5.19)$$

All the components of the potential function should be expressed via independent geometric parameters, and the problem of determining the optimal conformation is confined to the search for the minimum in (5.19) by varying these parameters.

The use of Dashevsky's potentials for calculating the conformations of overcrowded aromatic molecules requires the knowledge of several elastic constants. In papers [11, 12, 30] it was shown that if $C_1^* = 128$ kcal mole$^{-1}$ rad$^{-2}$ for nonplanar deformations of the benzene rings, $C_2^* = 64$ kcal mole$^{-1}$ rad$^{-2}$ for the dihedral angles formed by the peripheral bonds with the planes of the adjacent bonds of the benzene ring, and $U_0$ is of the order of 2–10 kcal/mole ($U_0 = 3.6$ for phenyl rings, see (5.18), 6.7 for nitrogroups, see (5.18), and 2.4 for the carboxyl groups, see (5.17)); the calculated conformations have fairly good agreement with the experimental data. It should be added that $C_2^*$ does not depend upon the nature of the atom associated with the benzene ring (the elastic constants of the bond angles were indicated above). Let us consider some examples from the paper [11].

*f. Octachloronaphthalene*

The calculated conformation of the molecule is shown in Fig. 18. Within the limits of experimental error, the bond angles and those of the departure of the bonds from the adjacent bond planes agree with the experimental values [133], the $C_9C_1Cl$ and $ClC_1C_2$ angles being the only exception.

It should be noted that parameters $\beta_1$, $\beta_2$, $\beta_3$, and $\beta_4$ are introduced for the deviations of the $C_9$–Cl, $C_1$–Cl, $C_1$–$C_2$, and $C_2$–Cl bonds from the adjacent

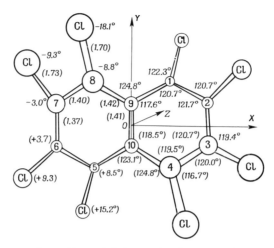

**Fig. 18.** Geometry of octachlornaphthalene molecule. In parentheses, experimental data. For nonplanar deformations, the values of independent parameters are given.

planes. Atomic deviations from the mean plane determined in the X-ray experiment can be easily calculated by means of these parameters:

$$Z(C_1) = 1.42 \sin \beta_1, \qquad Z(C_2) = 1.42 \sin \beta_1 + 1.75 \sin \beta_2, \qquad \text{etc.}$$

Naphthalene derivatives obtained by substitution of all H atoms by halogens and also 1,4,5,8-substituted derivatives can have the two conformations shown in Fig. 19. Regardless of the zero-order approximation chosen, the search for the minimum always leads to the conformation obtained experimentally. The depth of the minimum is very great: the strain energy of the ideal planar model of octachloronaphthalene is 188 kcal mole$^{-1}$ whereas for the equilibrium conformation it is 136 kcal mole$^{-1}$.

It is of interest that if one fixes the $C_9C_1Cl$ bond angle assuming it to be

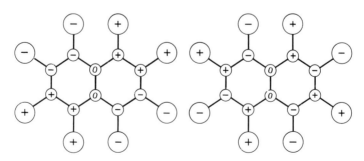

**Fig. 19.** Two possible conformations of octa- and 1,4,5,8-haloidnaphthalenes. The sign of + indicates displacement of the atom out of drawing plane.

124.8° (experimental value) and minimize the remaining parameters, the resulting conformation turns out to be very close to the experimental one, with an energy exceeding that of the strain energy of the conformation in Fig. 18 by only 80–100 kcal mole$^{-1}$. Hence the conclusion that the optimum conformation is not very sensitive to the value of the $C_9C_1Cl$ angle, or, to be more exact, the minimum in the cross section of the potential surface with respect to the deformations of this angle is rather flat and the small effect of molecular packing can affect the value of the $C_9C_1Cl$ angle.

*g. Dichloracenaphthene*

According to calculations, this is a planar molecule (Fig. 20). Mutual repulsion of the Cl atoms results in pronounced planar deformations, the $C_3C_4C_5$, $C_4C_5Cl$, and $ClC_5C_6$ bond angles being distorted most of all. There is very good agreement with the experimental data [134], but for one detail: The calculation gives an entirely planar molecule, while experiments have shown the Cl atom to deviate from the mean plane by 0.05 Å. If the $C_2{}^*$ is decreased to 60 kcal mole$^{-1}$ rad$^{-2}$, i.e., by 6% only, then the calculated deviation of Cl atoms will be identical to the experimental one. This exemplifies the fact that the out-of-plane deformations of aromatic systems are very sensitive to the values of the constants $C_1{}^*$ and $C_2{}^*$.

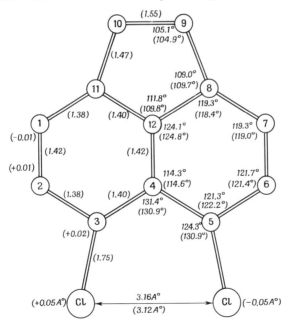

*Fig. 20.* Conformation of dichloracenaphthene. In parentheses experimental data.

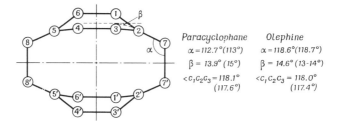

Paracyclophane        Olephine
$\alpha = 112.7°(113°)$   $\alpha = 118.6°(118.7°)$
$\beta = 13.9°\ (15°)$    $\beta = 14.6°\ (13\text{-}14°)$
$<c_1c_2c_3 = 118.1°$     $<c_1c_2c_3 = 118.0°$
$\qquad (117.6°)$         $\qquad (117.4°)$

**Fig. 21.** Conformation of 2,2′-paracyclophane and its unsaturated analog.

## h. Paracyclophanes

The molecules of paracyclophanes and their unsaturated analogs with $C_7$–$C_{7'}$ and $C_8$–$C_{8'}$ double bonds (olefine) are interesting in that no ideal models can be devised for them: If one assumes their benzene rings to be planar with the exocyclic $C_2$–$C_7$ bond localized in that plane, then the bond angles in the bridges are bound to be distorted; and vice versa; with ideal bond angles no planar benzene nuclei can exist. The mutual repulsion of the benzene nuclei in these molecules causes both deformations of the nuclei themselves (they acquire the boat conformation) and deformations of the bond angles in the bridges (Fig. 21).

The deformations of these molecules may be interpreted in terms of the mechanical model. Three independent geometric parameters will be enough for the calculations: $\alpha$, the bond angle in the bridge; $\beta$, the distortion angle of the benzene ring; and $\varepsilon$, the change of the bond angles in the nucleus ($\angle C_1 C_2 C_3 = 120° - 2\varepsilon$). Then the angle of departure of the $C_2$–$C_7$ bond ($\gamma$) from the $C_1 C_2 C_3$ plane is equal to $\alpha - \beta - 90°$. Figure 21 shows the calculated and experimental [135, 136] data.

Though the positions of the H atoms were not calculated, the deformations of the HCC and HCH bond angles were taken into consideration. (The constants for the bond angles' deformations were increased approximately twofold.)

The equilibrium conformation corresponds to a rather deep energy minimum. For the quasi-ideal model of paracyclophane (the benzene nuclei are planar and the bond angles in the bridges are tetrahedral), the strain energy is 178.8 kcal mole$^{-1}$ whereas for the equilibrium conformation it is 155.3 kcal mole$^{-1}$.

## i. Dichlorodiphenylnaphthacene

In the molecule of 5,6-dichloro-11,12-diphenylnaphthacene the steric hindrances are due to the proximity of phenyl rings and chlorine atoms in the ideal models. The carbon atoms of the phenyl groups are superimposed upon each other. Though rotation of the phenyl groups around the single C–C bonds attaching them to the naphthalene nucleus sharply decreases the strains, it is quite clear that one type of deformation will not be sufficient. Even if the phenyl rings turn by 90°, the nonbonded distances between the carbon atoms of the two phenyl rings will still be impermissively small, 2.42 Å. Therefore several types of deformations should be expected.

The conformational problem in the case of dichlorodiphenylnaphthacene was to search for the minimum in a function of ten variables. In fact, complete elucidation of the geometry requires a still greater number of variables. But as this number increases, the computer time required to search for the minimum will grow rapidly. (The square method was used which is a modification of the well-known procedure of Newton–Rafson.) Ten independent geometric parameters are sufficient to reveal the main features of the geometry of this molecule. These geometric parameters were introduced in the following way: four parameters for the deformations of the bond angles, $C_2 C_1 C_2$, $C_1 C_2 Cl$, $C_9 C_{10} C_9$, and $C_{10} C_9 C_{11}$, respectively (the atoms are numbered "structurally") [137]; five parameters for the out-of-plane deformations, i.e., the departures of the $C_1$–$C_2$, C–Cl, $C_{10}$–$C_9$, $C_9$–$C_{11}$, and $C_8$–$C_7$ bonds from the adjacent planes; and, finally, the angle of rotation of the phenyl nuclei with respect to the main plane of the adjacent benzene ring. Some insignificant assumptions had to be made to fix some of the structural details. For example, it was assumed that the $C_7$–H bond lies in the plane of the extreme benzene nucleus, and that the nucleus itself was planar. For bond lengths, the experimental values were used [138].

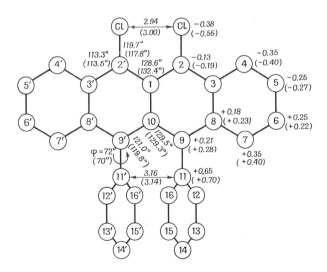

*Fig. 22.* Conformations of 5, 6-dichloro 11, 12-diphenylnaphthacene. In parentheses, experimental data.

Figure 22 shows the calculated and experimental data. In some cases the discrepancy between the calculated and experimental results exceeds the experimental error (e.g., for the $C_2C_1C_2$, and $ClC_2C_1$ bond angles); however the calculation may be considered to be satisfactory. The calculations describe not only the main features of the conformation of this molecule, but also some of its finer details: The central benzene rings in the naphthacene nucleus are almost planar; the phenyl rings are mutually turned, and the rotation angle coincides with the experimental value; the $C_9$–$C_{11}$ bond is considerably out of the plane of the adjacent benzene ring, etc.

Kitaigorodsky and Dashevsky [12] studied a number of interesting polynucleic aromatic molecules. In papers [30, 139] the conformations of aromatic molecules containing nitro groups were predicted; the conformations of halide derivatives of naphthoic acids are analyzed in [140]; in [141] the deformations of these molecules are correlated with their electronic spectra.

The examples described above, and all the available data, lead to the conclusion that the mechanical model makes it possible to interpret and predict the conformation of overcrowded aromatic molecules with various types of deformations.

## 6. Geometry of Molecules and Thermochemical Properties of Substances

Heats of combustion of a great number of organic compounds have been measured in calorimetric bombs with 0.02% accuracy. Using these results, one can easily calculate [142] the energies of formation of molecules from atoms (atomization energies) which can furnish rather useful information on the properties and structure of the molecules. To a first approximation, the energy of formation from atoms may be expressed as the sum of the bond energies; to this end each bond (C–C, C=C, C–H, etc.) should be assumed to have a definite energy and this value should be considered universal. But it has long been known that the additivity rule is not accurate enough [143–145], the discrepancy being several kilocalories per mole, i.e., large enough, so that they cannot be neglected in view of the high accuracy of the measurements.

The most important factor producing the deviations from additivity of the thermochemical properties are the strain energies; it therefore seems natural that these properties can be calculated with the help of the mechanical model.

The methods of calculations of the thermochemical properties of molecules, e.g., the formation energies of open hydrocarbon chains, have been developed for twenty years now. According to the method of Tatevsky et al. [146, 147], four types of C–H and ten types of C–C bonds should be distinguished, depending on their immediate environment. In this case one can obtain equations containing nine constants which can be determined from the heats of formation of lower alkanes. The heats of formations of higher alkanes are calculated with the help of these nine constants, the accuracy of calculation being almost equal to that of measurement. This is an excellent way to explain the different heats of formation of structural isomers.

Another approach, developed by Allen [148], consists in assuming that the C–C and C–H bond energies are equal in all paraffins, and that the differences in heats of formation are accounted for by interactions between nonbonded atoms. When interactions between the adjacent atoms were taken care of, a $0.2\ \text{kcal mole}^{-1}$ per bond average difference was established between the calculated and experimental heats of formation. A more refined theory, accounting for the interactions between more distant atoms also, suggests an average difference of $0.14\ \text{kcal mole}^{-1}$ per bond, a figure that practically coincides with the average accuracy limits of the experiment.

The latter approach has more in common with theoretical conformational analysis, because it enables one to estimate the nonbonded interactions directly from the thermochemical data.

The schemes of Bernstein [149] and Skinner [150, 151] are essentially similar to that of Allen. It should be noted that all the authors used constants determined from heats of formation to describe the nonbonded interactions. So it is only logical (as was proved by Tatevsky and Papulov [152]) that both approaches, being based on equivalent classifications of bonds and interactions, yield equivalent equations for the energies of atomization of alkanes.

In other papers dealing with this problem one or another approach is developed. For example, Laidler [153] distinguished four types of C–H bonds and one type of C–C bond, hence there is less accuracy in his predictions as compared to the data of Tatevsky; group additivity [154, 155] is essentially equivalent to a rough estimate of the total energy, that of the bonds and interactions; the interactions of the valence bonds [156] are mathematically equivalent to the interactions of nonbonded atoms.

It is quite probable that the successful application of the two above approaches to prediction of the atomization energies of alkanes is due primarily to the similar geometry of molecules of homologous series. In the case of molecules in which steric hindrances cause strong deformations as compared to the ideal models, neither Tatevsky's nor Allen's scheme will be expedient. Therefore an attempt to devise a more general scheme, applicable to any molecule, is welcome.

Deviations of additivity are due not only to nonbonded interactions, but also to bond angle deformations. It is on the basis of the angular strains that the thermochemical properties of the cycloalkanes [157] were previously calculated; the nonbonded interactions were entirely neglected or estimated too crudely. Pedley [158] ascribed the deviations from additivity in the atomization energies of alkanes to bond-angle deformations, but failed to obtain adequate parametrization. It is quite obvious that a general calculation scheme should take into account both bond-angle deformations and nonbonded interactions.

Based on this conclusion, Dashevsky [24] calculated the atomization energies of alkanes on the assumption that the C–C and C–H bond energies are transferable from one molecule to another, and that strain energies are responsible for the deviations from additivity. The principal equation describing the thermochemical properties will be

$$-\Delta H_{at} = \sum E_{bonds} - U_{st} \qquad (6.1)$$

where $-\Delta H_{at}$ is the atomization energy, $\sum E_{bonds}$ is the sum of the bond energies, and $U_{st}$ is the strain energy corresponding to the optimal conformation. (If the molecule has several conformers or a continuous conformation spectrum, the averaging over all possible conformations should be carried out using the Boltzman statistics.)

The C–H bond energy was estimated from the heat of formation of methane†

$$-\Delta H_{at,CH_4}^{298.15°K} = 397.17 = 4E(C-H) - 6f_{H\cdots H}(1.496\text{Å}) \qquad (6.2)$$

and

$$E(C-H) = 101.27 \text{ kcal mole}^{-1}$$

Then, calculating the equilibrium conformation of propane (Fig. 1) and estimating its strain energy (40.18 kcal mole$^{-1}$) from the equation

$$954.30 = 2E(C-C) + 8E(C-H) - U_{st} \qquad (6.3)$$

we shall have

$$E(C-C) = 92.16 \text{ kcal mole}^{-1}‡$$

Using these values of the bond energies, one can calculate the atomization energies of any saturated hydrocarbon. For example, for $n$-butane, whose equilibrium conformation is shown in Fig. 23, the nonbonded interaction energy will be 53.81 kcal mole$^{-1}$ and the angular deformation energy, 1.60 kcal mole$^{-1}$. Then, it is possible to show that the torsion contributions from the nonbranched chains are additive with quite good accuracy, and are equal on the average to 0.29 kcal mole$^{-1}$ per C–C bond. (This contribution is due to the fact that the Boltzman statistics at 298.15°K correspond to a continuous conformation spectrum; of course, the probability density is greatest for the planar zigzag, but the "average" conformer corresponds to a rotation angle

---

† For the energies of formation of H and C atoms the values 170.92 and 52.03 kcal mole$^{-1}$, respectively, were assumed; values for energies of formation of molecules were borrowed from "Physico-Chemical Properties of Individual Hydrocarbons," [159].

‡ Less satisfactory results are obtained if the C–C bond energy is found from the atomization energy of ethane.

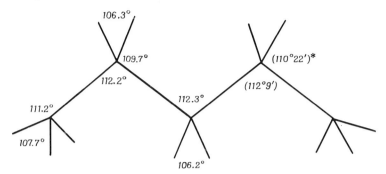

*Fig. 23.* Equilibrium conformation of n-butane. In parentheses, experimental data. The asterisk shows the mean value of the HCC angle. (R. A. Bonham, L. S. Bartell, D. A. Kohl. *J. Amer. Chem. Soc.* **81**, 4765 (1959)).

of 168°.) Using the figures given above and Eq. (6.1) we find a value of 1234.64 kcal mole$^{-1}$ for the atomization energy of $n$-butane, which is close to the experimental value of 1234.73 kcal mole$^{-1}$.

Such calculations are laborious. Consequently, the author of the above-cited paper obtained approximate expressions allowing the prediction of the atomization energies of $n$-alkanes, and of their structural isomers, with good accuracy (the method becomes thereby one of the versions of Allen's scheme). For the atomization energies of 15 isomers with three to six C–C bonds, a 0.21 kcal mole$^{-1}$ average difference between the calculated values and the experimental ones was obtained.

Furthermore, in paper [160] it was shown that even in the case of small saturated cycles, as well as bi- or polycyclic systems (not to mention the medium cycles for which other authors [38, 68, 161] have obtained good results), the value of 92.16 kcal mole$^{-1}$ for the C–C bond and 101.24 kcal mole$^{-1}$ for the C–H bond will give satisfactory agreement with the experimental data. For example, for cubane (XVII), whose strain energy is 275.50 kcal mole$^{-1}$, the calculated atomization energy is 1640.1 kcal mole$^{-1}$, which is only 5 kcal mole$^{-1}$ more than the experimental value of 1635.4 kcal mole$^{-1}$ [162].

Similarly, for hexamethylprismane (XVIII) the calculated strain energy is

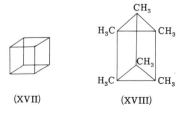

(XVII)                    (XVIII)

284 kcal mole$^{-1}$ and the atomization energy is 2921 kcal mole$^{-1}$, which is again very close to the experimental value [163] of 2923.4 kcal mole$^{-1}$. Thus, even for systems with great angular deformations, there is quite satisfactory agreement, though not as good as for the alkanes, between the calculated and experimental atomization energies; the difference does not exceed several kilocalories per mole.

Let us now consider systems with multiple bonds. Generally speaking, the atomization energies of hydrocarbons with multiple bonds also obey additive schemes. But if one assumes transferability of individual bond energies from molecule to molecule, the deviations from additivity will be greater than for saturated hydrocarbons. This is why such notions as additional stabilization, conjugation, or resonance energies are also usually invoked when dealing with multiple-bond systems.

In paper [164] the above principles were generalized to be applied to systems with multiple bonds. The calculations are again based on Eq. (6.1), and the following two assumptions are made: (1) the length and energy of an X–Y bond is determined only by the sort and hybridization of the X and Y atoms; (2) there is mutual and unambiguous correspondence between the energies and lengths of bonds.

The first assumption was made by Coulson [165] and Walsh [166]; Dewar and Schmeising [167] as well as Bernstein [168] based their bond energy systems on these assumptions.

The validity of the second assumption may be verified only by calculation. Dewar and Schmeising's system of bond energies does not cover all hydro-carbons (e.g., the atomization energy of allene calculated by means of their scheme exceeds the experimental value by 8.5 kcal mole$^{-1}$). The lengths and energies of bonds were mutually and unambiguously correlated, but strain energies were never included in calculations by such schemes. On the other hand, Bernstein [168], having calculated the atomization energies of a great number of hydrocarbons, and having compared them with the relevant experimental data, arrived at the conclusion that there is no unambiguous dependence between the bond length and its energy; his $E(l)$ curve branches into two at $l > 1.30$ Å.

It is clear that if only Eq. (6.1) and the two assumptions above are used in calculations, the resonance energy of any molecule will be equal to zero, i.e., the conjugation concept can be ignored when dealing with the molecule's geometry and energy of formation. This is not a new point of view. Dewar [169], for example, distinguished between the properties of molecules conditioned by the behavior of individual electrons (ionization potentials, electron spectra, etc.) and properties arising due to "collective" interactions of electrons and nuclei (geometry, energies of formation, vibration spectra frequencies, etc.). In calculating properties of the latter type we probably may consider a molecule as a system of interacting atoms without involving electronic concepts.

Table 7 is a classification of the C–C and C–H bonds based on structural data (the procedure for obtaining the energies of individual bonds indicated in the fourth column will be described later).

To find the dependence of the bond energies on their lengths, let us choose such "standard" molecules as contain almost all the bonds indicated in Table 7. The strain energies and equilibrium conformations of 14 standard molecules calculated by Dashevsky are given in Table 8.

### Table 7

| Bond structure | Designation[a] | Bond length (Å) | Bond energy (kcal mole$^{-1}$) |
|---|---|---|---|
| C–C bonds | | | |
| $>$C–C$<$ | $S_{11}$ | 1.535 | 92.16 |
| $>$C–C$<$ | $S_{12}$ | 1.507 | 98.83 |
| $>$C–C$\equiv$ | $S_{13}$ | 1.459 | 111.83 |
| $>$C–C$<$ | $S_{22}$ | 1.476 | 107.22 |
| $>$C–C$\equiv$ | $S_{23}$ | 1.426 | 120.01 |
| $\equiv$C–C$\equiv$ | $S_{33}$ | 1.376 | 129.18 |
| $>$C=C$<$ | $D_{11}$ | 1.333 | 133.78 |
| $>$C=C= | $D_{12}$ | 1.305 | 135.95 |
| =C=C= | $D_{22}$ | 1.284 | 137.85 |
| $-$C$\equiv$C$-$ | T | 1.205 | 160.82 |
| C$-$C$_{arom}$ | A | 1.396 | 126.04 |
| C–H bonds | | | |
| $>$C–H | $h_1$ | 1.100 | 101.27 |
| $>$C–H | $h_2$ | 1.085 | 105.10 |
| $\equiv$C–H | $h_3$ | 1.059 | 115.51 |

[a] S stands for the single bond, D for the double bond, T for the triple bond; A means aromatic; the two indices indicate the maximum bond multiplicities on the left and right.

## Table 8

GEOMETRY AND ATOMIZATION ENERGY OF MULTIPLY BONDED HYDROCARBONS

| Molecule | Equilibrium conformation (calculated) | Equilibrium conformation (experimental date) | Strain energy (kcal mole$^{-1}$) | Sum of bond energy | Atomization energy (calculated) (kcal mole$^{-1}$) | Atomization energy (experimental data) (kcal mole$^{-1}$) |
|---|---|---|---|---|---|---|
| Ethylene H$_2$C=CH$_2$ | HCH 116.8° | HCH 151.6°[a] HCH 117.2°[b] HCH 117.3°[c] | 16.08 | D$_{11}$ + 4h$_2$ | 538.11 | 537.70 |
| Propylene H$_2$C=CH-CH$_3$ | CCC 122.4° HC=C 120.6° HC-C 117.0° HCH$_{ethylene}$116.9° HCH$_{methyl}$107.6° | CCC 124.8° HC-C, HC=C$_{ethylene}$120° HCH$_{methyl}$107.2°[d] | 31.12 | D$_{11}$ + S$_{12}$ + 3h$_1$ + 3h$_2$ | 820.60 | 820.42 |
| 1,3-Butadiene H$_2$C=CH-CH=CH$_2$ | CCC 121.9° HC=C 120.6° HC-C 117.3° HCH 116.6° | CCC 122.9°, assumed HCH, | 35.76 | 2D$_{11}$ + S$_{22}$ + 6h$_1$ | 869.62 | 969.89 |
| Cis-2-butene H$_3$C-CH=CH-CH$_3$ | CCC 123.4° HC=C 120.7° HC-C$_{ethylene}$115.9° HCH 107.4° | HCC 120°[e] | 47.25 | D$_{11}$ + 2S$_{11}$ + 6h$_1$ + 2h$_2$ | 1102.01 | 1102.07 |
| Isobutylene H$_2$C=C(CH$_3$)$_2$ | C-C-C 118.1° C-C-C 121.0° HCH$_{ethylene}$116.0° HCH$_{methyl}$107.4° HCC$_{methyl}$112.0° | C-C-C 112° HCC$_{methyl}$110.4°[f] C-C-C 115.3° HCC$_{methyl}$112.9° HCH$_{ethylene}$118.5°[g] | 45.47 | D$_{11}$ + 2S$_{11}$ + 6h$_1$ + 2h$_2$ | 1103.79 | 1104.44 |
| Allene H$_2$C=C=CH$_2$ | HCH 116.3° HCC 121.9° HCH$_{ethylene}$118.8° | HCH 118.9°[h] 116.0°[i] 119.0°[j] | 18.12 | 2D$_{12}$ + 4h$_2$ | 674.18 | 675.20 |
| 1,2-Butadiene |  |  | 30.21 | 2D$_{12}$ + S$_{12}$ + 3h$_1$ + 3h$_2$ | 959.63 | 957.45 |

| Molecule | Angles | | | | |
|---|---|---|---|---|---|
| $H,CH_3 / H(1)\;C{=}C{=}C\; H(2)$ | H(1)CC 121.9°<br>H(2)CC 120.6°<br>HCC$_{methyl}$ 107.5° | | | | |
| Acetylene HC≡CH | HCH 108.8°[k] | 0 | $T+2h_3$ | 391.85 | 391.83 |
| Propyne HC≡C-CH$_3$ | HCH 106.8° | 15.07 | $T+S_{13}+3h_1+h_3$ | 676.90 | 676.80 |
| Butyne HC≡C-CH$_2$-CH$_3$ | HCH$_{methylene}$ 105.9°<br>HCH$_{methyl}$ 107.8°<br>CCC 112.7° | 30.27 | $T+S_{13}+S_{11}+Sh_1+h_3$ | 956.40 | 956.74 |
| Benzene φ-H | all angles 120° | 69.04 | $6A+6h_2$ | 1317.80 | 1318.24 |
| Toluene φ-CH$_3$ | HCH$_{methyl}$ 107.5°<br>angles of trigonal<br>C atoms~120° | 82.75 | $S_{11}+3h_1+6A+5h_2$ | 1601.21 | |
| 1,4-Dimethylbenzene H$_3$C-φ-CH$_3$ | Same | 97.66 | $2S_{12}+6h_1+6A+4h_2$ | 1884.27 | 1883.97 |
| Diphenyl φ-φ | C$_2$C$_1$C$_6$ 117.6°<br>HC$_2$C$_1$ 120.8°<br>HC$_3$C$_4$ 119.7°<br>$φ^m=26°$<br>$φ=42°$[l] | 140.91 | $S_{22}+12A+10h_2$ | 2529.81 | 2529.34 |

[a] L. S. Bartell and R. A. Bonham, *J. Chem. Phys.* **31**, 400 (1959).
[b] L. S. Bartell, E. A. Roth, C. D. Hollowell, K. Kuchitsu, and J. E. Young, *J. Chem. Phys.* **42**, 2603 (1965).
[c] N. C. Allen and E. K. Plyle, *J. Amer. Chem. Soc.* **80**, 2673 (1958).
[d] D. R. Lide and D. E. Munn, *J. Chem. Phys.* **27**, 868 (1957).
[e] A. Almenningen and O. Bastiansen, *Acta Chem. Scand.* **12**, 1221 (1958).
[f] L. S. Bartell and R. A. Bonham, *J. Chem. Phys.* **32**, 824 (1960).
[g] H. Scharpen, L. V. Le Roy, and V. W. Laurie, *J. Chem. Phys.* **39**, 1732 (1963).
[h] A. Almenningen, O. Bastiansen, and M. Traetteberg, *Acta Chem. Scand.* **13**, 1699 (1959).
[i] K. N. Rao, A. H. Nielsen, and W. H. Fletcher, *J. Chem. Phys.* **26**, 1572 (1957).
[j] J. Overend and B. Crawford, *J. Chem. Phys.* **29**, 1002 (1958).
[k] D. R. Lide, *J. Chem. Phys.* **29**, 864 (1958).
[l] A. Almenningen and O. Bastiansen, *Kgl. Nor. Vidensk. Selsk. Skr.* **4**, (1958).
[m] The angle between the planes of phenyl nuclei.

Now the problem is to find the $E(l)$ function for C–C and C–H bonds. Let us assume the $\Delta H^i_{calc}$ to be the atomization energy of the molecule calculated from Eq. (6.1) using the set of bond energies and strain energies given in Table 8.

The $E(l)$ functions are to be found from the expression

$$\sum_i g^i \left[(\Delta H^i_{calc} - \Delta H^i_{exper})/\Delta H^i_{exper}\right]^2 \tag{6.4}$$

where the $g^i$ are weight factors. Ideally, one should have searched for the minimum of this expression not for the 14 molecules chosen, but for all the molecules whose heats of formation are known. But one may believe that the introduction of new molecules will not affect the results too much. In the paper cited all the $g^i$ were assumed to be equal to unity, with the exception of benzene for which $g = 3$.† The following dependences were obtained when expression (6.4) was verified and minimized:

$$E(C\text{–}C) = 857.634\,l_1{}^5 + 10264.689\,l_1^{-4} - 6606.481\,l_1{}^3$$
$$+\ 13212.531\,l_1{}^2 + 20194.833\,l_1 - 10562.740 \tag{6.5}$$

$$E(C\text{–}H) = 1615.109\,l_2^{-2} + 2222.195\,l_2 - 3678.000 \tag{6.6}$$

(the values of 92.16 and 101.27 kcal mole$^{-1}$ for ordinary C–C and C–H bonds, respectively, were used). The corresponding curves are given in Fig. 24a, b, and the energies of the individual bonds appear in the fourth column of Table 7. The $E(l)$ curves are different from those given in the literature [167, 168, 170–172]; it is obvious that simple functions like $l^{-1}$ or $l^{-3}$ do not agree with the experimental data. At all points of the curve given in Fig. 24a $dE/dl < 0$, i.e., the energy of the C–C bond decreases monotonically as the bond length increases.

The atomization energies of the 14 standard molecules agree very well with the experimental values, the average difference is less than 0.5 kcal mole$^{-1}$, and the only exceptions are the 1–2 kcal mole$^{-1}$ discrepancies for allene and 1,2-butadiene. This system of bond energies allows one to calculate the atomization energies for any hydrocarbon with sufficient accuracy. Let us consider two examples.

*a. Cis-1,3-butadiene*

The calculated optimal conformation is shown in Fig. 25a. With the strain energy of the equilibrium conformation of 37.58 kcal mole$^{-1}$, the atomization energy will be 967.80 kcal mole$^{-1}$, more than the experimental value by only 0.2 kcal mole$^{-1}$. The cis–trans isomerization energy for 1,3-butadiene is,

---

† With $g = 1$ the difference between the calculated and experimental atomization energy of benzene was 1.5 kcal mole.

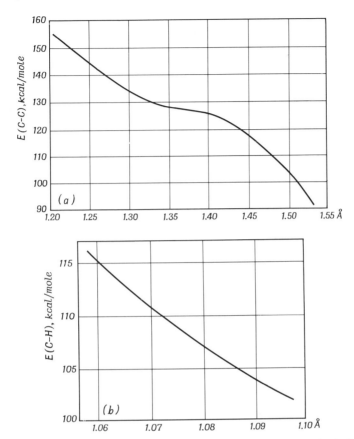

*Fig. 24.* Dependence of bond energy upon bond length: a, for C–C bonds; b, for C–H bonds.

according to calculations, 1.82 kcal mole$^{-1}$; the experimental value is 2.3 kcal mole$^{-1}$.

### b. 1,3-pentadiene

The bond angles of the equilibrium conformation of the cis and trans forms are given in Fig. 25b,c. The strain energies of these forms are equal to 52.69 and 51.08 kcal mole$^{-1}$, respectively; the atomization energies are 1251.84 and 1253.45 kcal mole$^{-1}$; the experimental values are 1251.55 and 1253.21 kcal mole$^{-1}$. The calculated value for the cis–trans isomerization energy (1.61 kcal mole$^{-1}$) almost coincides with the difference in the formation energies of these molecules (1.66 kcal mole$^{-1}$).

The results obtained show that the geometry and atomization energies of

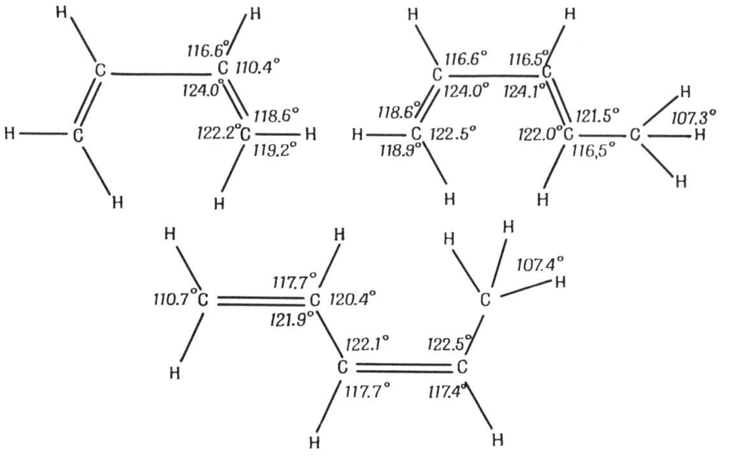

a

b

H         H
 \       /
  C——————C 110.4°
 116.6°
 124.0°
         118.6°
H——C   122.2°C——H
   /       119.2°
  \
   H        H

H          H
 \        /
  C  116.6°  116.5°
 // 124.0°  124.1° \\
118.6°            C
H——C 122.5°   121.5°   H
   118.9°   122.0°C——C   107.3°
  \              116,5°  ——H
   H    H              \
                        H

H          H
 \        /
110.7°C======C 120.4°
    117.7°
    121.9°
       \ 122.1°   122.5°/
        C==========C
       / 117.7°   117.4° \
      H                   H

H      H
 \    / 107.4°
  C    ——H
  \
   C
  /
 H

c

*Fig. 25.* Calculated equilibrium conformations of the molecules of: a, cis-1,3-butadiene; b, cis-1,3-pentadiene; and c, trans-1,3-pentadiene.

hydrocarbon molecules may be interpreted in terms of the scheme adopted without any additional concepts. One may believe that these properties of molecules are determined by the so-called "steric" effects (i.e., nonbonded interactions, angular deformations, etc.) and by hybridization (i.e., types of bonds). But it is hardly sensible to state what factors are "decisive" since the scheme is empirical.

Although the resonance energies in the scheme are considered to be equal to zero, it is easy to answer the question about the relative stabilities of the real and hypothetical molecules. The molecules of benzene and 1,3,5-cyclohexatriene will be good examples. In the latter molecule the strain energy is 67.44 kcal mole$^{-1}$ (it is a planar molecule, the C–C bond lengths are 1.476 and 1.333 Å, all CCC angles are equal to 120°) and the atomization energy is 1286.18 kcal mole$^{-1}$. Hence benzene (Table 8) is more stable by 31.8 kcal mole$^{-1}$ than cyclohexatriene which agrees with well-known values of the resonance energy of benzene.

Besides atomization energies, there are other thermochemical properties of compounds connected with strain energies which can be of interest to theoreticians.

1. To estimate isomerization energy it is not necessary to determine atomization energies by summing up bond energies; it is sufficient to compare the strain energies obtained by potential function minimization.

Kitaigorodsky and Dashevsky [12] have given calculations of this kind.

The molecule of 2,7-dimethylphenanthrene (XIX) is planar, according to calculations; the strain decreases owing to the insignificant deformations in the plane. The spatial proximity of the methyl groups in the molecule of 4,5-dimethylphenanthrene (XX) causes serious out-of-plane deformations. The C4 and C5 atoms depart from the plane by 0.25 Å in opposite directions, and methyl groups are shifted even more. The difference in the strain energy between 2,7- and 4,5-dimethylphenanthrene is 14.2 kcal mole$^{-1}$, which agrees rather well with the difference in the combustion heats (i.e., isomerization heat) of these compounds, 12.6 kcal mole$^{-1}$ [173].

Similar calculations were made for 1,12- and 5,8-dimethylbenzophenanthrene, (XXI) and (XXII). Both molecules are nonplanar, but the latter one is distorted more, due to steric hindrances. The isomerization energy is 11.5 kcal mole$^{-1}$, according to calculations; the experimental value is 11.0 kcal mole$^{-1}$ [173].

Finally, the maximum isomerization heat of dimethylbenzophenanthrene (XXIII) is as low as 4.3 kcal mole$^{-1}$, while that of dimethylhexahelicene (XXIV)

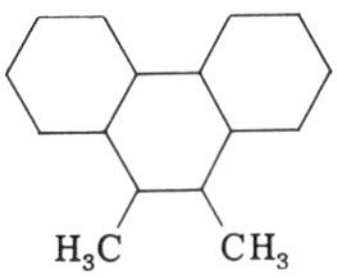

$H_3C$    $CH_3$

(XIX)

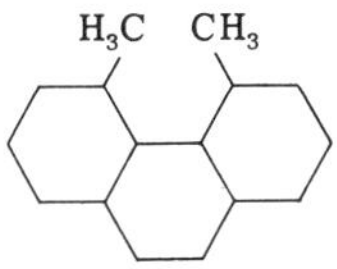

$H_3C$    $CH_3$

(XX)

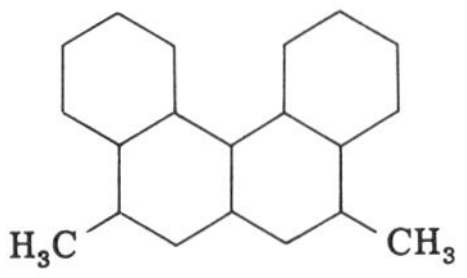

$H_3C$    $CH_3$

(XXI)

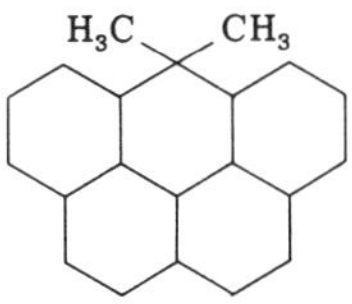

$H_3C$    $CH_3$

(XXII)

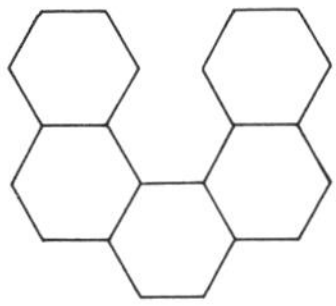

(XXIII)

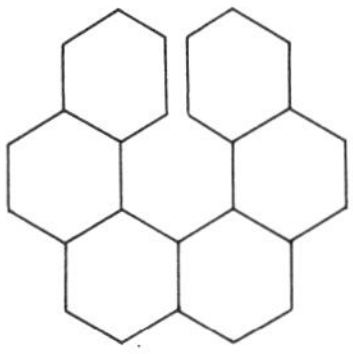

(XXIV)

is about 2 kcal mole$^{-1}$. (The $CH_3$ groups, which are not indicated, are in positions corresponding to the maximum strain.)

Such low values of isomerization heats are due to the specific geometry of these molecules: the existing deformations are so strong that new substitutions do not involve high energy losses on deformations. (It should be noted that the combustion heats of these compounds have not been determined.)

2. If exo- and endo-isomerization energies are to be calculated, it is indispensable to determine the atomization energies since exo- and endo-isomers contain bonds of very different hybridization.

It is known, for example, that the isomerization heat of 1,4-pentadiene (XXV) into 1,3-pentadiene (XXVI) is $-6.7$ kcal mole$^{-1}$.

(XXV)                    (XXVI)

Rough calculation of the atomization energies of these molecules gives figures of 1245.12 kcal mole$^{-1}$ (experimental value, 1245.91 kcal mole$^{-1}$) and 1250.07 kcal mole$^{-1}$; therefrom one derives the value of the isomerization energy $-4.2$ kcal mole$^{-1}$.

Of great interest are calculations of the exo–endo isomerization heats of various cyclic hydrocarbons, since these systems are associated with two competitive effects: the change in hybridization and the change in strain energy. For example, the measured heat of isomerization of 1,3-dimethylenecyclobutane (XXVIIa) 1-methyl-3-methylcyclobutene (XXVIIb) is $-5.0$ kcal mole [157]:

(a)                    (b)

(XXVII)

The sums of the bond energies of these two compounds are equal to 1461.68 and 1477.86 kcal/mole, respectively, i.e., by bond energies (XXVIIb) is more stable by 16 kcal mole$^{-1}$. On the other hand, the strain energy of molecule (XXVIIb) is less by several kilocalories per mole, since in it there is a more favorable situation for trigonal atoms of the cycle. (Instead of 90°, due to the valence bonds being unequal, the angles in the cycle are 82° and 98°; the latter are situated at the double bonds.) A simple calculation for planar models, without minimization of the potential functions, yields strain energies of 132 and 122 kcal mole$^{-1}$ for (XXVIIa) and (XXVIIb) respectively, and the isomerization heat will be $-6$ kcal mole$^{-1}$; this agrees well with the experimental figure.

3. The method also makes it possible to calculate hydrogenation heats of unsaturated compounds. If $C_n H_m$ is the initial compound and $C_n H_p$ is the reaction product, then the equation for the hydrogenation heat will be

$$-\Delta H_{\text{hydr}} = -\Delta H_{\text{at}}(C_n H_p) - [-\Delta H_{\text{at}}(C_n H_m)] - (p-n)52.09 \quad (6.7)$$

in which (6.1) is valid for the atomization energy of each substance. For example, when benzene is hydrogenated into cyclohexane:

$$-(1680.22-317.80-52.09 \times 6) = 49.88 \text{ kcal mole}^{-1},$$

which coincides with the experimental value of $-49.8$ kcal mole$^{-1}$ [174]. Similarly, for the hydrogenation of $CH_3C\equiv CCH_3$ into $n$-butane the value will be $-64.59$ kcal mole$^{-1}$. (The experimental value is $-65.12$ kcal mole$^{-1}$ [175].

Several examples of hydrogenation heat calculations for bi- and polycyclic systems are given in paper [101]. For hydrogenation of bicyclo-[2,2,1]-heptadiene-2,5 into bicyclo-[2,2,1]-heptane, using the data of Table 4, the value of $-66.2$ kcal mole$^{-1}$ was obtained; the experimental result was $-68.1$ kcal mole$^{-1}$ [176].

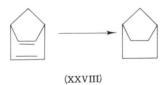

(XXVIII)

Cycloalkenes whose conformations were considered in the preceding section will be good illustrations of this method of calculation of thermochemical properties. Using the values of $U_{\text{st}}$ obtained when calculating the equilibrium conformations, one can easily estimate the thermochemical properties of these molecules (Table 9).

The data listed in Table 9 display the satisfactory agreement between the calculated and experimental results, the average difference being 2 kcal mole$^{-1}$. But it should be noted that the calculated atomization energy for the majority of molecules is less than the experimental; this means that either the values of the bond energies are too low, or the strain energies are too high (the latter depend on the potential function parameters). However, it should be borne in mind that both the bond energies and the parameters of the potential functions were obtained from the analysis of acyclic compounds. Therefore,

## Table 9

COMPARISON OF CALCULATED AND EXPERIMENTAL VALUES OF ATOMIZATION ENERGIES
AND HYDROGENATION HEATS OF CYCLOALKENES

| Molecule | $\sum E_{bond}$ (kcal mole$^{-1}$) | $U_{st}$ (kcal mole$^{-1}$) | $-\Delta H_{at}$ (kcal mole$^{-1}$) | | $-\Delta H_{hydr}$ (kcal mole$^{-1}$) | |
|---|---|---|---|---|---|---|
| | | | Calculated | Experimental | Calculated | Experimental |
| Cyclopentene | 1333.58 | 73.63 | 1259.95 | 1263.45[a] | 29.89 | 26.92[e] |
| Cyclopentadiene | 1195.38 | 62.70 | 1132.68 | 1134.43[b] | 52.92 | 50.87[e] |
| Cyclohexene | 1628.28 | 81.80 | 1546.48 | 1548.12[a] | 29.37 | 28.59[e] |
| 1,4-Cyclohexadiene | 1488.36 | 72.91 | 1419.45 | — | 52.26 | 27.10[f] |
| 1,3-Cyclohexadiene | 1490.08 | 73.43 | 1417.65 | 1416[c] | 54.06 | 55.37[e] |
| 1,3,5-Cycloheptatriene | 1646.58 | 78.91 | 1567.97 | 1569.69[a] | 73.78 | 72.85[e] |
| Cyclooctatetraene | 1804.80 | 91.63 | 1713.57 | 1712.96[d] | 100.57 | — |

[a] "Physico-Chemical Properties of Individual Hydrocarbons" (V. M. Tatevsky, ed.). Gostoptechizdat, Moscow, 1962.
[b] K. Mortimer, "Reaction Heats and Bond Strengths." Pergamon, Oxford, 1962.
[c] D. Cox, *Tetrahedron* **19**, 1175 (1963).
[d] N. E. Duncan and G. J. Janz, *J. Chem. Phys.* **23**, 434 (1955).
[e] H. A. Skinner, "Modern Aspects of Thermochemistry." Roy. Inst. of Chem. Lectures, monographs and reports, No. 3. London, 1958.
[f] R. B. Turner, W. R. Meador, and E. A. Smith, *J. Amer. Chem. Soc.* **79**, 4116 (1957).

in the case of the cycloalkanes, the fairly good agreement between the calculations and the experimental results testifies to the fact that the valence bonds in these molecules have almost the same properties as those in cyclic compounds; the specific physicochemical (including thermochemical) properties of these compounds are due to their higher strain energies and, primarily, to angular deformation energies.

Hitherto we have been discussing the application of the mechanical model to calculation of the thermochemical properties of hydrocarbons. In the paper [177] the atomization energies of the halogen derivatives of methane are given. It is shown that minimization of the potential functions using independent geometrical parameters gives much better agreement between the calculated values and the experimental results. (Earlier, in paper [178], Allen's scheme had been applied to the same compounds. The increments for the nonbonded intreactions had been taken into account, but this had practically no effect on the results. Other additive schemes for this homologous series are presented in [179–181].)

Thus, the scheme of atom–atom potentials has proved to be quite acceptable for calculating the thermochemical properties of such different systems as alkanes, olefines, aromatic molecules, cycloalkanes and cycloalkenes, bi- and polycyclic systems and, finally, halogen derivatives. Will this approach still be as fruitful in calculations of the atomization energies of more complex molecules, e.g., oxygen-containing compounds, heterocycles, nitro compounds? Certainly this question cannot be answered unless detailed calculations have been carried out, but, since rough additive schemes have proved to be fairly good with complex systems, there is a hope that the atom–atom potential scheme will also be successful. The advantage of this scheme over others is self-evident: The relationships between the geometry of the molecule and the thermochemical properties of the compound can be understood.

## 7. CONSISTENT FORCE FIELD

Hitherto we have been discussing in detail the atom–atom potential scheme as applied to calculation of the optimum geometries of molecules and the atomization enthalpies. Vibration spectra frequencies can also be discussed from this point of view. Research along this line has only just been started, and we have only a few remarks to offer.

The molecule's potential energy $U(\mathbf{x})$, where $\mathbf{x}$ consists of independent coordinates, may be expanded into a Taylor series near the equilibrium position

$$U(\mathbf{x}) = U(\mathbf{x}_0) + \sum_i (\partial U/\partial x_i)_{\mathbf{x}=\mathbf{x}_0} \delta x_i$$
$$+ \tfrac{1}{2}\sum_i \sum_j (\partial^2 U/\partial x_i \, \partial x_j)_{\mathbf{x}=\mathbf{x}_0} \delta x_i \delta x_j + \cdots \qquad (7.1)$$

where $\mathbf{x}_0$ is a vector of independent geometric parameters corresponding to the equilibrium conformation, and $\partial x_i$ are small displacements of the atoms from their equilibrium positions. The first term of the expansion gives the atomization enthalpy, which is the sum of the bond and strain energies. The second term is zero in the case of the equilibrium conformation, since the condition of the minimum of the potential function corresponds to the vanishing of the first derivatives

$$(\partial U/\partial x_i)_{\mathbf{x}=\mathbf{x}_0} = 0 \qquad (7.2)$$

Finally, the third term, which includes the second derivatives of the potential function, is necessary in order to calculate the harmonic frequencies of the molecules' vibration spectra; the following expansion terms, neglected in our case, carry information about the anharmonicity.

As is known [182], the solution of the Lagrangian equations for the molecular vibrations is finally given by the secular equation

$$|\mathbf{GF} - \lambda \mathbf{I}| = 0 \qquad (7.3)$$

where $\mathbf{G}$ is the matrix of kinematic coefficients accounting for the molecule's geometry and the mass of the atoms; $\mathbf{F}$ is the matrix of the dynamic coefficients, with components

$$F_{ij} = (\partial^2 U/\partial x_i \partial x_j)_{\mathbf{x}=\mathbf{x}_0} \qquad (7.4)$$

The vibration spectra frequencies are $2\pi v_i = (\lambda_i)^{1/2}$ where $v_i$ is the frequency of the $i$th vibration ($\lambda_i$ are the eigenvalues), $\mathbf{I}$ is the unit matrix.

The matrix $\mathbf{G}$ is determined unequivocally from the geometry of the molecule and all the necessary expressions are known for its elements, whereas the matrix $\mathbf{F}$ causes much controversy. Spectroscopists see their chief task in finding all the elements of the matrix $\mathbf{F}$ from the experimental data on the vibration frequencies of the molecule, i.e., by solving the inverse vibration problem, spectroscopists obtain information about the force field of the molecule. But it is not possible, in principle, to obtain all the elements of the matrix $\mathbf{F}$ from the spectrum of any given molecule, because the number of equations is always less than that of the unknown quantities. The way out is then the following: Besides the spectrum of the given molecule, the spectra of its isotopic derivatives are investigated, because they have the same force field (this assumption is true to a great degree of accuracy). Thus, additional equations are obtained allowing one to find all the elements of the matrix $\mathbf{F}$. Unfortunately, such a procedure is only practical for the simplest molecules. In almost all interesting cases the number of equations is very insufficient, and one is compelled to resort to criteria for the transferability of force constants from one molecule to another.

The problem of the transferability of the **F** matrix elements for force fields of various kinds is discussed in detail in [182–184] and will not be dealt with here. The only relevant remark to make is that force constants for stretching vibrations of certain bonds usually change very little from one to another similar molecule, on the other hand, the constants corresponding to angular deformations change to a greater degree, and the greatest changes occur in the force constants describing skeletal and other low frequency vibrations.

The potential functions used for calculating conformations and thermo-chemical properties allow one, as is seen from (7.4), to estimate unequivocally all the elements of the matrix **F**. For example, in the case of the aliphatic systems we may, using the above designations, write the following expression:

$$U = U_{\text{nonbonded}} + U_{\text{ang}} + U_{\text{tors}} + U_{\text{bonds}}$$

$$= \sum_{i>j} f(r_{ij}) + \tfrac{1}{2}\sum_i C_i(\Delta\alpha_i)^2 + \tfrac{1}{2}U_0\sum_j (1+\cos 3\varphi_j) + \tfrac{1}{2}\sum_s K_s(\Delta l_s)^2 \quad (7.5)$$

adding the deformation energy of the valence bonds as $\tfrac{1}{2}\sum_s K_s(\Delta l_s)^2$ ($K_s$ are the elastic constants; $\Delta l_s$ are the changes in bond lengths from their ideal values). Then, differentiating (7.5) twice with respect to the independent geometrical parameters, we derive all the elements of the matrix **F**.

The analytic expressions for the second derivatives are known to be ex-tremely complicated, but Galaktionov [36] suggested a handy method for their calculation by means of a computer. Instead of analytical differentiation one may use the finite difference approximations:

$$\partial^2 U/\partial x_i^2 = [U(x_0) + U(x_i+2h) - 2U(x_i+h)]/h^2 \quad (7.6)$$

$$\partial^2 U/\partial x_i\,\partial x_j = [U(x_0) + U(x_i+h, x_j+h)$$

$$- U(x_i+h) - U(x_j+h)]/h^2 \quad (7.7)$$

where $h$ is the small increment in the argument. If $h$ is of the order of 0.001–0.0001, energies are expressed in kilocalories per mole, angles in radians, and distances in ångstroms, then such an approximation will give quite satisfactory results, as was shown by Dashevsky and Dakhis [185].

It has been emphasized more than once that it is allowable to neglect changes in the bond lengths when calculating conformations. This term may also be excluded from calculations of the vibration spectra frequencies if one is interested in noncharacteristic low-frequency vibrations depending on the conformation of the molecule. And it is only when a complete description of the vibration spectra frequencies is required that changes in bond lengths should be included since otherwise the high frequencies are impossible to obtain.

The method of calculation described above is not new. As early as in 1931 Urey and Bradley [186] suggested expressions similar to (7.5) for tetrahedral molecules of the $CX_4$ type; the repulsion of the nonbonded atoms was described by potentials of the $K/r^n$ type where $K$ and $n$ are constants, and $r$ is the interatomic distance. For $n$, a range of values was tried and $n \approx 9$ was found to give satisfactory agreement between the calculated and experimental frequencies. Later Heath and Linnett [187, 188] introduced potentials for the interactions of nonbonded atoms when analyzing tetrahedral and nonlinear three-atom molecules.

The modified force field of Urey–Bradley [189] which has gained wide recognition, also contains terms corresponding to interactions of nonbonded atoms. For example, for tetrahedral molecules the usual expression will be

$$U = \sum_{i=1}^{4} [K_j' l_i \Delta l_i + \tfrac{1}{2} K_i (\Delta l_i)^2]$$

$$+ \sum_{i>j} \sum [H_{ij}' l_{ij}^2 \Delta \alpha_{ij} + \tfrac{1}{2} H_{ij} (l_{ij} \Delta \alpha_{ij})^2] \qquad (7.8)$$

$$+ \sum_{i>j} \sum [F_{ij}' q_{ij} \Delta q_{ij} + \tfrac{1}{2} F_{ij} (\Delta q_{ij})^2]$$

where $\Delta l_i$ are bond length changes, $\Delta \alpha_{ij}$ are bond angle changes, $\Delta q_{ij}$ are changes in the distances between nonbonded atoms; $K_i$, $K_i'$, $H_{ij}$, $H_{ij}'$, $F_{ij}$ and $F_{ij}'$ are the corresponding force constants.

When using this formula in calculations, one usually derives force constants from an analysis of the experimental data; it is at this stage that the question of the transferability of atom–atom potentials becomes that of changes in the constants $F_{ij}'$ and $F_{ij}$ from one molecule to another. Not everything is clear about the terms which are linear with respect to changes in bond lengths, bond angles, and distances between the nonbonded atoms. In the equilibrium conformation these terms should have turned out to be zero, if there had been no dependence between the parameters of (7.8). Therefore it seems more logical to calculate the vibration frequencies of molecules in a general harmonic force field, including no linear terms. To this end any dependent geometrical parameters employed in the potential function expression should be expressed via independent ones; then the matrix of force coefficients could be found from Eqs. (7.4), (7.6), and (7.7) unequivocally for any chosen set of independent parameters.

Vibration spectra frequencies are very sensitive to the parameters of potential functions, much more sensitive than the optimal conformations or the enthalpies. This is why it seems reasonable to refine the atom–atom potentials and the other parameters of the potential function using the experimental data, i.e., the vibration frequencies. On the other hand, spectroscopists may benefit from the potential functions, for example, in the case of assignment of frequencies, especially in the low-frequency range.

Lifson and Warshel [161] made an attempt to find the potential function parameters giving the best agreement with the experimental data on the geometries, enthalpies, and vibration frequencies of cycloalkanes and $n$-alkanes. As is known, the strain energy in such systems may be expressed as

$$U_{st} = \sum_{i>j}\sum f(r_{ij}) + \tfrac{1}{2}C_{CCC}\sum_i (\alpha_i^{CCC} - \alpha_0^{CCC})^2$$
$$+ \tfrac{1}{2}C_{HCC}\sum_i (\alpha_i^{HCC} - \alpha_0^{HCC})^2$$
$$+ \tfrac{1}{2}C_{HCH}\sum_i (\alpha_i^{HCH} - \alpha_0^{HCH})^2$$
$$+ \tfrac{1}{2}U_0 \sum_i (1 + \cos 3\varphi_i)$$
$$+ \tfrac{1}{2}K_{C-C}\sum_i (l_i^{C-C} - l_0^{C-C})^2$$
$$+ \tfrac{1}{2}K_{C-H}\sum_i (l_i^{C-H} - l_0^{C-H})^2 \qquad (7.9)$$

In this equation the first term corresponds to the energy of interaction of nonbonded atoms for which use was made of the 6-12 potentials

$$f(r) = \varepsilon(r_0/r)^{12} - 2\varepsilon(r_0/r)^6 \qquad (7.10)$$

($\varepsilon$ is the depth of the potential well, $r_0$ is the equilibrium distance); the second, third, and fourth terms give the angular deformation energy, the elastic constants for the CCC, HCC, and HCH angles being different; the fifth term describes the torsion energy, and the last two terms are the deformation energies of the C–C and C–H bonds, with $K_{C-C}$ and $K_{C-H}$ the corresponding elastic constants.

Besides, to achieve better agreement between the calculated and experimentally observed enthalpies of formation, the authors included electrostatic interactions; they believed that on the C and H atoms there are equal and opposite charges, $e_{eff}$, interacting in accordance with the Coulomb law.

In Table 10 are the parameters of the consistent force field which gives the best description of the enthalpies of formation, the geometries and the vibration frequencies of saturated hydrocarbon molecules.

In another paper Lifson *et al.* [190] obtained a consistent force field for

### Table 10

PARAMETERS OF CONSISTENT FORCE FIELD

| kcal mole$^{-1}$ | Å | kcal mole$^{-1}$ rad$^{-2}$ | kcal mole$^{-1}$ Å$^{-2}$ | |
|---|---|---|---|---|
| $\varepsilon_{C\cdots C}$ 0.0196 | $(r_0)_{C\cdots C}$ 4.228 | $C_{CCC}$ 44.0 | $K_{CC}$ 222.0 | $U_0$ 2.836 kcal/mole |
| $\varepsilon_{C\cdots H}$ 0.00939 | $(r_0)_{C\cdots H}$ 3.582 | $C_{HCC}$ 53.58 | $K_{C-H}$ 573.8 | $e_{eff}$ = 0.144e |
| $\varepsilon_{H\cdots H}$ 0.0045 | $(r_0)_{H\cdots H}$ 2.936 | $C_{HCH}$ 76.28 | | |

446

some systems containing amide groups. Their parameters will be useful for calculating the conformations and vibration spectra frequencies of dipeptides, polypeptides, and proteins (see Section VIII.2).

The examples described above are but the first steps along these lines; it should be expected that the atom–atom approach will enrich the theory of molecular vibration spectra. Besides, investigations of this kind will elucidate the question of whether the atom–atom potentials are universal and transferable when used to describe different properties of molecules.

REFERENCES

1. T. L. Hill, *J. Chem. Phys.* **14**, 465 (1946).
1a. T. L. Hill, *J. Chem. Phys.* **16**, 938 (1948).
2. F. H. Westheimer and J. E. Mayer, *J. Chem. Phys.* **14**, 733 (1946).
3. F. H. Westheimer, *J. Chem. Phys.* **15**, 252 (1947).
4. M. Riger and F. H. Westheimer, *J. Amer. Chem. Soc.* **72**, 19 (1950).
5. F. H. Westheimer, *in* "Steric Effects in Organic Chemistry" (M. S. Newman, ed.), Chapter 12. Wiley, New York and Chapman & Hall, New York, 1956.
6. A. I. Kitaigorodsky, *Izv. Akad. Nauk SSSR Ser. Fiz.* **15**, 157 (1951).
7. A. I. Kitaigorodsky, "Chemical Organic Crystallography." Consultants Bureau, New York, 1959.
8. A. I. Kitaigorodsky, *Dokl. Akad. Nauk SSSR* **124**, 1967 (1959).
9. E. L. Eliel, N. L. Allinger, S. L. Angyal, and G. A. Morrison, "Conformational Analysis." Wiley (Interscience), New York, 1965.
10. V. G. Dashevsky, Yu. T. Struchkov, and R. L. Avoyan, Crystallochemistry 1966. *Itogi Nauki Khim. Nauki* **9**, 117 (1968).
11. V. G. Dashevsky and A. I. Kitaigorodsky, *Teor. Eksp. Khim.* **3**, 43 (1967).
12. A. I. Kitaigorodsky and V. G. Dashevsky, Tetrahedron **24**, 5917 (1968).
13. G. B. Robertson, *Nature (London)* **191**, 593 (1961).
14. A. Hargreaves and S. H. Rizvi, *Acta Crystallogr.* **15**, 365 (1962).
15. A. Almenningen and O. Bastiansen, *Kgl. Nor. Vidensk. Selsk. Skr.* **4, 1958**, 4.
16. E. J. Adrian, *J. Chem. Phys.* **28**, 608 (1958).
17. R. A. Scott and H. A. Scheraga, *J. Chem. Phys.* **42**, 2209 (1965).
18. R. A. Scott and H. A. Scheraga, *J. Chem. Phys.* **44**, 8 (1966); **45**, 2091 (1966).
19. K. B. Gibson and H. A. Scheraga, *Biopolymers* **4**, 709 (1966).
20. V. I. Poltev and B. I. Sukhorukov, *Zh. Strukt. Khim.* **9**, 298 (1968).
21. D. E. Williams, *J. Chem. Phys.* **45**, 377 (1966).
22. C. M. Venkatachalam and G. N. Ramachandran, *in* "Conformation of Biopolymers" (G. N. Ramachandran, ed.), p. 83. Academic Press, New York, 1967; V. S. R, Rao, P. Sundararajan, C. Ramakrishnan, and G. N. Ramachandran, *Ibid*, p. 721.
23. V. G. Dashevsky, *Zh. Strukt. Khim.* **6**, 888 (1965); **7**, 93 (1966).
24. V. G. Dashevsky, *Zh. Strukt. Khim.* **9**, 289 (1968).
25. G. N. Ramachandran and V. Sasisekharan, *Advan. Protein Chem.* **23**, 283 (1968).
26. A. I. Kitaigorodsky, *Tetrahedron* **14**, 230 (1961).
27. J. B. Hendrickson, *J Amer Chem. Soc.* **83**, 4537 (1961).
28. P. De Santis, E. Giglio, A. M. Liquori, and A. Ripamonti, *Nature (London)* **206**, 456 (1965).
29. P. J. Flory, D. A. Brant, and W. J. Miller, *J. Mol. Biol.* **23**, 47 (1967).

30. V. G. Dashevsky, Yu. T. Struchkov, and Z. A. Akopyan, *Zh. Strukt. Khim.* **7**, 594 (1966).
31. A. Bondi, *J. Phys. Chem.* **68**, 44 (1964).
32. J. C. Slater and J. G. Kirkwood, *Phys. Rev.* **37**, 682 (1931).
33. J. Ketelaar, "Chemical Constitution," p. 91. Elsevier, Amsterdam, 1958.
34. E. A. Mason, R. J. Munn, and J. Smith, *Discuss. Faraday Soc.* **40**, 27 (1965).
35. V. G. Dashevsky, High molecular compounds. Theoretical aspects of conformations of macromolecules. *Itogi Nauki Khim. Nauki* **21**, 210 (1970).
36. "Conformational Calculations of Complex Molecules." Publ. House of Acad. Sci. of Belorussian SSR, Minsk, 1970.
37. K. Wiberg, *J. Amer. Chem. Soc.* **87**, 1070 (1966).
38. M. Bixon and S. Lifson, *Tetrahedron* **23**, 769 (1967).
39. N. L. Allinger, M. A. Miller, F. A. Van-Catledge, and J. A. Hirsh, *J. Amer. Chem. Soc.* **89**, 4345 (1967).
40. E. A. Mason and M. M. Kreevoy, *J. Amer. Chem. Soc.* **77**, 5808 (1955).
41. M. M. Kreevoy and E. A. Mason, *J. Amer. Chem. Soc.* **79**, 4851 (1957).
42. M. V. Volkenstein, "Configurational Statistics of Polymer Chains." Wiley (Interscience), New York, 1963.
43 T. M. Birshtein and O. B. Ptitsyn, "Conformations of Macromolecules." Wiley (Interscience), New York, 1966.
44. R. M. Pitzer and W. N Lipscomb, *J Chem. Phys.* **39**, 1995 (1963).
45. J. Goodisman, *J. Chem. Phys.* **44**, 2085 (1966).
46. E. Clementi and D. R. Davis, *J. Chem. Phys.* **45**, 2593 (1966).
47. J. R. Hoyland, *J. Chem. Phys.* **49**, 2563 (1968).
48. E. B. Wilson, *Advan. Chem. Phys.* **2**, 367 (1959).
49. W. H. Fink and L. C. Allen, *J. Chem. Phys.* **46**, 2276 (1967).
50. D. Pedersen and R. Morokuma, *J. Chem. Phys.* **46**, 3941 (1967).
51. W. E. Palke and R. M. Pitzer, *J. Chem. Phys.* **46**, 3948 (1967).
52. J. L. De Coen, G. Elefante, A. M. Liquori, and A. Damiani, *Nature (London)* **216**, 910 (1967).
53. V. G. Dashevsky, *in* "Conformational Calculations of Complex Molecules." Publ. House of Acad. Sci. of Belorussian SSR, Minsk, 1970, p. 33.
54. Y. Morino and E. Hirota, *J. Chem. Phys.* **28**, 185 (1958).
55. D. E. Shaw, D. W. Lepard, and H. L. Welsh, *J. Chem. Phys.* **42**, 3736 (1965).
56. G. Allen, P. N. Brier, and G. Zane, *Trans. Faraday Soc.* **63**, 827 (1967).
57. J. Carle, *J. Chem. Phys.* **45**, 4149 (1966).
58. V. Magnasco, *Nuovo Cimento* **24**, 425 (1962).
59. A. I. Kitaigorodsky, *Vysokomol. Soedin.* **10A**, 2669 (1968).
60. L. S. Bartell and D. A. Kohl, *J. Chem. Phys.* **39**, 3097 (1963).
61. R. L. McCullough and P. E. McMahon, *J. Phys. Chem.* **69**, 1747 (1965).
62. R. L. McCullough and P. E. McMahon, *Trans. Faraday Soc.* **60**, 2089 (1964).
63. R. A. Scott and H A Scheraga, *J Chem. Phys.* **44**, 8 (1966).
64. R. A. Bonham, C. L. Bartell, and D. A. Kohl, *J. Amer. Chem. Soc.* **81**, 4765 (1959).
65. S. Meiboom and L. C. Snyder, *J. Amer. Chem. Soc.* **89**, 1038 (1967).
66. V. T. Aleksanyan, G. M. Kuziyanz, M. Yu. Lukina, S. V. Zotova, and E. I. Vostokova, *Zh. Strukt. Khim.* **9**, 141 (1968).
67. V. Toyotoshi and S. Takeniko, *J. Chem. Phys.* **49**, 1470 (1968).
68. N. L. Allinger, J. A. Hirsch, M. A. Miller, I. J. Tyminski, and F. A. Van-Catledge, *J. Amer. Chem. Soc.* **90**, 1199 (1968).
69. H. Kim and W. D. Gwinn, *J. Chem. Phys.* **44**, 865 (1966).

70. W. G. Rotshild and B. P. Dailey, *J. Chem. Phys.* **36**, 2931 (1962).
71. T. B. Owen and J. L. Hoard, *Acta Crystallogr.* **4**, 172 (1961).
72. I. L. Karle and J. Karle, *Acta Crystallogr.* **20**, 55 (1966).
73. I. L. Karle, J. Karle, and K. Britts, *J. Amer. Chem. Soc.* **88**, 2918 (1966).
74. H. P. Lemaire and R. L. Livingston, *J. Amer. Chem. Soc.* **74**, 5732 (1952).
75. A. C. Luntz, *J. Chem. Phys.* **50**, 1109 (1969).
76. T. N. Margulis, *Acta Crystallogr.* **19**, 857 (1965).
77. T. N. Margulis, *Acta Crystallogr.* **18**, 742 (1965).
78. B. Greenberg and B. Post, Abstr. *Annu. Meet. Amer. Crystallogr. Ass., Austin, Texas, 1966.*
79. T. N. Margulis and M. S. Fisher, *J. Amer. Chem. Soc.* **89**, 223 (1967).
80. W. D. Gwinn and A. C. Luntz, *Trans. Amer. Crystallogr. Ass.* **2**, 90 (1966).
81. J. Loanem and R. C. Lord, *J. Chem. Phys.* **48**, 1508 (1968)
82. J. G. Aston, S. L. Schumann, H. L. Fink, and P. M. Doty. *J. Amer. Chem. Soc.* **63**, 2029 (1941).
83. J. E. Kilpatrick, K. S. Pitzer, and R. Spitzer, *J. Amer. Chem. Soc.* **69**, 2483 (1947).
84. K. S. Pitzer and W. E. Donath, *J. Amer. Chem. Soc.* **81**, 3213 (1959).
85. S. Furberg, *Acta Crystallogr.* **3**, 325 (1950).
86. N. L. Allinger and L. A. Freiberg, *J. Amer. Chem. Soc.* **82**, 2393 (1960).
87. A. I. Kitaigorodsky, *Tetrahedron* **9**, 183 (1960).
88. N. V. Alekseev and A. I. Kitaigorodsky, *Zh. Strukt. Khim.* **4**, 163 (1963).
89. M. Davis and O. Hassel, *Acta Chem. Scand.* **17**, 1181 (1963).
90. J. B. Hendrickson, *Tetrahedron* **19**, 1387 (1963).
91. H. A. Spang, *SIAM Ind. Appl. Math. Rev.* **4**, 343 (1962).
92. R. Fletcher, *Comput. J.* **8**, 33 (1965).
93. B. T. Polyak, *Ekon. Mat. Metodi* **3**, 881 (1967).
94. K. D. Gibson and H. A. Scheraga, *Proc. Nat. Acad. Sci. U.S.* **58**, 13 (1967).
95. R. Fletcher and M. J. D. Powell, *Comput. J.* **6**, 163 (1963).
96. V. G. Dashevsky, "Comformations of Organic Molecules." Chimia Publishing House, Moscow, to appear.
97. J. B. Hendrickson, *J. Amer. Chem. Soc.* **89**, 7036, 7043, 7047 (1967).
98. H. M. Pickett and H. L. Strauss, *J. Amer. Chem. Soc.* **92**, 7281 (1970).
99. V. G. Dashevsky and A. A. Lugovskoy, *J. Mol. Struct.* **12**, 39 (1972).
100. N. M. Zaripov, V. G. Dashevsky, and V. A. Naumov, *Izv. Akad. Nauk SSSR Ser. Khim.* p. 1963 (1970).
101. V. G. Dashevsky, *Voprosi Stereokhimii (Kiev)* **2**, 17 (1972).
102. E. B. Wilson, Jr. and A. J. Wells, *J. Chem. Phys.* **9**, 319 (1941).
103. F. Rossini, "Selected Values of Physical and Thermochemical Properties of Hydrocarbons and Related Compounds." Pittsburgh, Pennsylvania, 1953.
104. V. G. Dashevsky, V. A. Naumov, and N. M. Zaripov, *Zh. Strukt. Khim,* **11**, 796 (1970); **13**, 171 (1972).
105. N. P. Borisova, "Carbon Chain High Molecular Compounds" (in Russian), p. 84. 1964.
106. L. Laane and C. Lord, *J. Chem. Phys.* **47**, 4941 (1967).
107. T. Ueda and T. Shimanouchi, *J. Chem. Phys.* **47**, 5018 (1967).
108. V. A. Naumov, V. G. Dashevsky and N. M. Zaripov, *Dokl. Akad. Nauk SSSR* **185**, 604 (1969).
109. R. Bucourt and D. Haunaut, *Bull. Soc. Chim. Fr.* p. 1366 (1965); p. 501 (1966).
110 G. Dallinga and L. H. Toneman. *J. Mol. Struct.* **1**, 117 (1967–1968).
111 F. H. Herbstein, *J. Chem. Soc.* p. 2292 (1959).
112. S. S. Butcher, *J. Chem. Phys.* **42**, 1830 (1965).

113. G. Lass and M. D. Harmony, *J. Chem. Phys.* **43**, 3768 (1965).
113a. G Dallinga and L. H. Toneman, *J. Mol. Struct.* **1**, 11 (1967–1968).
114. M. Traetteberg, *J. Amer. Chem. Soc.* **86**, 4265 (1964).
115. S. S. Butcher, *J. Chem. Phys.* **42**, 1833 (1965).
116. O. Bastiansen, L. Hedberg, and K. Hedberg, *J. Chem. Phys.* **27**, 1311 (1957).
117. R. L. Avoyan, Yu. T. Struchkov, and V. G. Dashevsky, *Zh. Strukt. Khim.* **7**, 289 (1966).
118. C. A. Coulson and S. Senent, *J. Chem. Soc.* pp. 1813, 1819 (1955).
119. C. A. Coulson, S. Senent, and M. A. Herraez, *An. Real Soc. Espan. Fis. Quim. Ser. A* **52**, 515 (1956).
120. J. M. Robertson, *Nature (London)* **169**, 322 (1952).
121 C. Dean, M. Pollak, B. M. Craven, and G. A. Geffrey, *Acta Crystallogr.* **11**, 10 (1958).
122. A. A. Ali and C. A. Coulson, *J. Chem. Soc.* p. 1558 (1959).
123. C. A. Coulson, *Ind. Chim. Belge* **28**, 149 (1963).
124. C. A. Coulson and C. W. Haigh, *Tetrahedron* **19**, 527 (1963).
125. S. Syvin, *Acta Chem. Scand.* **11**, 1499 (1957).
126. C. A. Coulson and A. Golebiewski, *J. Chem. Soc.* p. 4948 (1960).
127 J. Trotter, *Acta Crystallogr.* **18**, 605 (1965).
128. F. R. Ahmed and J. Trotter, *Acta Crystallogr.* **16**, 503 (1963).
129. A. I. Kitaigorodsky and V. G Dashevsky, *Teor Eksp. Khim.* **3**, 35 (1967).
130. T. H. Goodman and D. A. Morton-Blake, *Teor. Chim. Acta* **1**, 458 (1962).
131. G. Favini and M. Simonetta, *Teor. Chim. Acta* **1**, 294 (1962).
132. G. Favini and A. Gamba, *Gaz. Chim. Ital.* **95**, 236 (1965).
133. G. Gafher and F. H. Herbstein, *Nature (London)* **200**, 130 (1968).
134. R. L. Avoyan and Yu T. Struchkov, *Zh. Strukt. Khim.* **2**, 719 (1961).
135. D. A. Bekoe and K. N. Trueblood, *Meeting Amer. Crystallogr. Ass., Boseman, Montana, 1964.*
136. C. L. Coulter and K. N. Trueblood, *Acta Crystallogr.* **16**, 667 (1963).
137. R. L. Avoyan, A. I. Kitaigorodsky, and Yu. T. Struchkov, *Zh. Strukt. Khim.* **4**, 633 (1963).
138. R. L. Avoyan, A. I. Kitaigorodsky, and Yu. T. Struchkov, *Zh. Strukt. Khim.* **5**, 420 (1964).
139. Z. A. Akopyan, Yu. T. Struchkov, and V. G. Dashevsky, *Zh. Strukt, Khim,* **7**, 408 (1966)
140 V. G. Dashevsky, R. L. Avoyan, L. A. Didenko, and V. N. Lisitsin, *Zh Org Khim.* **4**, 891 (1968).
141. V. N. Lisitsin, L. A. Didenko, and V. G. Dashevsky, *Zh. Org. Khim* **4**, 1086 (1968).
142. T. L. Cottrell, "The Strengths of Chemical Bonds." Butterworth, London, 1954.
143. K. Fajans, *Ber. Deut. Chem. Ges. B* **53**, 643 (1920); **55**, 2826 (1926); *Z. Phys. Chem.* **99**,395 (1921).
144. L. Pauling, *J. Amer. Chem. Soc.* **54**, 3570 (1932).
145. F. D. Rossini, *J. Chem. Phys.* **2**, 145 (1934); *Chem. Rev.* **27**, 1 (1940).
146 V M. Tatevsky, *Dokl. Akad Nauk SSSR* **75**, 819 (1950); **78**, 67 (1951); *Zh Strukt,* **25**, 241 (1951).
147. V. M. Tatevsky, N. F. Stepanov, and S. M. Yarovoy, "Methods of Calculations of Thermochemical Properties of Paraffin Hydrocarbons." Gostoptekhizdat, Moscow, 1960.
148. T. L. Allen, *J. Chem. Phys.* **31**, 1039 (1959).
149. H. J. Bernstein, *J. Chem. Phys.* **20**, 263 (1952).
150. H. A. Skinner, "Modern Aspects of Thermochemistry." Roy. Inst. of Chem. Lectures, monographs and reports, No. 3. London, 1958.

151. H. A. Skinner, *J. Chem. Soc.* **11**, 4396 (1962).
152. V. M. Tatevsky and Yu. T. Papulov, *Zh. Strukt. Khim.* **34**, 241 (1960).
153. K. J. Laidler, *Can. J. Chem.* **34**, 626 (1956).
154. J. L. Franklin, *Ind. Eng. Chem.* **41**, 1070 (1949).
155. S. W. Benson and J. H. Buss, *J. Chem. Phys.* **29**, 546 (1958).
156. K. Wiener, *J. Amer. Chem. Soc.* **69**, 17, 2636 (1947); *J. Chem. Phys.* **15**, 766 (1947).
157. C. T. Mortimer, "Reaction Heats and Bond Strengths." Pergamon, Oxford, 1962.
158. J. B. Pedley, *Trans. Faraday Soc.* **57**, 1492 (1961).
159. "Physico-Chemical Properties of Individual Hydrocarbons." Gostoptekhizdat, Moscow, 1961.
160 N. M. Zaripov, V. G. Dashevsky, and V. A. Naumov, *Izv. Akad. Nauk SSSR Ser. Khim*, p. 1642 (1971).
161. S. Lifson and A. Warshel, *J. Chem. Phys.* **49**, 5116 (1968).
162. B. D. Kybett, S. Carrol, P. Natalis, D. W. Bonnell, J. L. Margrave, and J. L. Franklin, *J. Amer. Chem. Soc.* **88**, 626 (1966).
163. J. F. M. Oth, *Rec. Trav. Chim. Pays-Bas* **87**, 1185 (1968).
164. V. G. Dashevsky, *Zh. Strukt. Khim.* **11**, 489 (1970).
165. C. A. Coulson, *J. Phys. Chem.* **56**, 311 (1952).
166. A. D. Walsh, *Trans. Faraday Soc.* **53**, 60 (1957).
167. M. J. S. Dewar and H. N. Schmeising, *Tetrahedron* **5**, 166 (1959); **11**, 96 (1960).
168. H. J. Bernstein, *Trans. Faraday Soc.* **57**, 466 (1961).
169. M. J. S. Dewar, "Hyperconjugation." Ronald Press, New York, 1962.
170. G. Glockler, *J. Chem. Phys.* **21**, 1249 (1953); *J. Phys. Chem.* **61**, 31 (1957).
171. H. Felchenfeld, *J. Phys. Chem.* **61**, 1133 (1957).
172. E. M. Popov and G. A. Kogan, *Usp. Khim.* **37**, 256 (1968).
173. M. A. Frish, C. Barker, J. L. Margrave and M. S. Newman, *J. Amer. Chem. Soc.* **85**, 2356 (1965).
174. G. B. Kistiakowsky, J. R. Ruchoff, H. A. Smith, and W. E. Vaughan, *J. Amer. Chem. Soc.* **58**, 137, 146 (1936).
175. T. Flitcroft, H. A. Skinner, and M. C. Whiting, *Trans. Faraday Soc.* **53**, 783 (1957).
176. R. B. Turner, W. R. Meador, and E. A. Smith, *J. Amer. Chem. Soc.* **79**, 4116 (1957).
177. V. G. Dashevsky, *Zh. Strukt. Khim.* **11**, 746 (1970).
178. G. P. Somayaijulu, A. P. Kudchadker and B. J. Zwolinski, *Ann. Rev. Phys. Chem.* **16**, 213 (1965).
179. H. J. Bernstein, *J. Chem. Phys.* **19**, 140 (1951); **20**, 263 (1952).
180. Yu. G. Papulov and V. M. Tatevsky, *Vest. MGU, Ser. Chim.* No. 2, 5 (1963); *Zh. Fiz. Khim.* **35**, 2695 (1961).
181. A. I. Vitvitsky, *Zh. Org. Khim.* **3**, 1354 (1967).
182. E. B. Wilson, J. C. Descius, and P. C. Cross, "Molecular Vibrations." McGraw-Hill, New York, 1955.
183. M. V. Volkenstein, M. A. Elyashevich, and B. I. Stepanov, "Vibrations of Molecules," Vol. 1, 2. Gos. Izd. Tekn.—Teor. Lit., Moscow, 1949.
184. T. Shimanouchi, *Pure Appl. Chem.* **7**, 131 (1963).
185. M. I. Dakhis and V. G. Dashevsky, *Dokl. Akad. Nauk. SSSR* **203**, 369 (1972); M. I. Dakhis, V. G. Dashevsky, and V. G. Avakyan, *J. Mol. Struct.* **13**, 339 (1972).
186. H. C. Urey and C. A. Bradley, *Phys. Rev.* **38**, 1969 (1931).
187. D. F. Heath and J. W. Linnett, *Trans. Faraday Soc.* **44**, 561 (1948).
188. J. W. Linnett and D. F. Heath, *Trans. Faraday Soc.* **48**, 592 (1952).
189. T. Shimanouchi, *J. Chem. Phys.* **17**, 245 (1949).
190. A. Warshel, M. Levitt, and S. Lifson, *J. Mol. Spectrosc.* **33**, 84 (1970).

# Chapter VIII
## Conformations of Macromolecules and Biopolymers

1. The Structure of Stereoregular Macromolecules in Crystals

This chapter describes the application of the atom–atom potential scheme to macromolecules—natural and synthetic. It is appropriate to warn the reader at once that the language of the interactions between atoms is not always sufficient to describe all the properties of polymeric molecules; sometimes less sophisticated models are more effective for such complex systems. For example, it is hardly possible that the structure of t-RNA or m-RNA in solution could be comprehensively described by means of atom–atom potentials only; as to protein, prediction of its complete three-dimensional structure is also doubtful, though some theoreticians have different opinions about that question.

We are going to confine ourselves to the aspects of the conformations of macromolecules which can be interpreted in terms of the mechanical model of the molecule discussed in Chapter VII. This refers primarily to the helical conformations of stereoregular macromolecules that can be crystallized.

Stereoregularity implies the equivalence of the monomeric units, hence the geometric parameters describing the conformation are repeated in every monomeric unit. Therefore, the theoretical conformational analysis of a macromolecule is the investigation of a potential surface in a small number of variables, even though the number of monomeric units may be great.

If the molecule is not stereoregular, e.g., a globular protein, conformational analysis allows one to find useful regularities in the interactions of adjacent monomeric units. Again, molecules of nucleic acids—DNA and RNA and those of fibrillar proteins—are not, strictly speaking, stereoregular but they are sequences of geometrically similar monomeric units and in the first approximation may be looked upon as stereoregular macromolecules.

Synthetic regular macromolecules, whose conformations will be discussed in detail here, are very handy models to verify the effectiveness of the scheme of atom–atom potentials. We shall limit ourselves to crystalline polymers alone, since in solution the potential energy of a macromolecule depends upon a great number of variables. For the theoretical investigation of the conformations of macromolecules in solution, approximate methods [1, 2] have been elaborated allowing one to obtain average characteristics of the molecules, and among others, $\langle h^2 \rangle$, the mean square of chain length. Most effective for this purpose is rotation isomer theory, which is based on the energy differences of rotation isomers. The rotation isomer model is, of course, a rougher approximation than that of a continuous conformation spectrum, which includes the interactions of all the atoms in terms of atom–atom potentials; nevertheless, it is quite good enough to describe almost all the properties of macromolecules in solution. (If one uses the rotation isomer model in calculations, the atom–atom potentials may give the differences in the energies of the rotamers.)

Stereoregular synthetic polymers have become widely known during the past 15 years. Ziegler et al. [3] did the pioneering research and reported the polymerization of ethylene at low pressure. These authors used a new catalyst, a mixture of solutions of $Al(CH_3)_3$ and $TiCl_4$. In the same year (1955) Natta et al. [4, 5] used Ziegler's polymerization method to synthesize some poly-α-olefines, including polypropylene and polystyrene. Thereafter innumerable stereoregular crystalline polymers have been synthesized. In only one paper, (Natta et al. [6]) the structures of about 30 vinyl isotactic polymers with aromatic side groups are described. It should be emphasized that Natta was the first not only to synthesize numerous stereoregular polymers but also to determine their three-dimensional structures by means of X-ray studies. Besides, Natta [7]† was one of the first investigators who employed the methods of theoretical conformational analysis.

The nomenclature of stereoregular polymers is dealt with in several papers and summed up in the paper [8] and review of Corradini [9]. Special rules have been suggested for indicating the order and position of atoms in an isolated polymer chain, and also for the mode of packing of neighboring chains. Let us take as an example the vinyl polymers, i.e., $(-CR_2-)_n$,

———————————

† The papers by Natta and the members of his school are collected in Ref. [7].

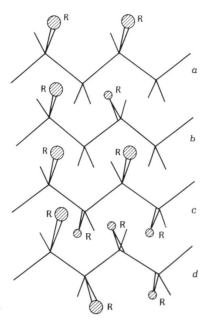

*Fig. 1.* Structures of vinyl polymers. a, isotactic; b, syndiotactic; c, threo diisotactic; d, erythro diisotactic. The trans chain of the C atoms is in the plane of the diagram, the side chains are in the normal plane; the larger circles indicating side radicals protrude forward from the diagram plane; smaller circles, backwards.

$(-CH_2-CHR)_n$ and $(-CHR-CHR-)_n$. The first type consists of carbon-chain polymers with identical side groups, such as polyethelene, R = H, or poly-tetrafluoroethylene, R = F. Another widely known group is $(-CH_2-CHR-)_n$. These may have two different configurations, isotactic and syndiotactic shown in Fig. 1a and 1b. If the polymer is tentatively represented as a planar trans chain, then isotactic polymers will exhibit parallel transfer of adjacent monomeric units, while the syndiotactic ones will have a mirror plane combined with the parallel transfer. Polymers of the $(-CHR-CHR-)_n$ type are called diisotactic, being threo- or erythrotactic depending on the positions of the substituents (Fig. 1c,d).

It is not difficult to conclude that regular polymers should crystallize into helical structures. Indeed, their adjacent units are equivalent and the mutual position of each pair of the adjacent monomeric units is identical. This seems to be a necessary and sufficient condition for the formation of a helical structure. In some cases the helix may degenerate into a planar zigzag. For example, the polyethylene molecule is a chain in which all the carbon atoms lie in one plane, owing to the fact that all the hydrogen atoms are located at equilibrium distances from each other (2.5 Å). Such macromolecules as polyvinyl chloride, polystyrene, polyvinylidene chloride, etc. cannot have their carbon backbone planar, due to the mutual repulsions of atoms in the side chains. Thus, if the carbon backbone of polyvinylidene chloride had been planar the shortest $H \cdots Cl$ distance would have been less by 0.6 Å than the equilibrium one. Obviously, if adjacent monomer units rotate with respect to each other forming thereby a helix, this distance increases.

Natta and Corradini [10, 10a] have put forward the basic principles of the geometric organization of stereoregular macromolecules. These they formulated as three postulates:

1. The axis of a macromolecule (i.e., that of the helix) is parallel to a crystallographic axis, and all the monomer units occupy geometrically equivalent positions in relation to this axis. This experimental fact is confirmed by a great deal of X-ray evidence, and has been called the equivalence postulate.

2. The conformation of the polymer chain in a crystal approximately corresponds to one of the minima of potential energy of the isolated chain. This means that the mode of packing of the molecules in the crystal is of secondary importance in determining the chain conformation, but some small deviations from the conformation of the isolated chain are possible.

3. The chains of the macromolecules are parallel and are separated from each other by distances characteristic of low molecular weight compounds. This postulate means essentially that the principle of close packing applies to macromolecules.

X-ray analysis of crystalline polymers may, in general, furnish evidence about the atomic coordinates in the unit cell; but, since the order is not absolutely perfect, the number of reflections is small, and no direct solutions of the structural problem are usually possible. X-ray diffraction patterns of a stretched sample give information about the identity period $c$ along the fiber axis. The trial and error method is usually employed to obtain the parameters of the helix, i.e., the translation of the monomer unit along the axis ($d$) and the angle of rotation in the plane normal to the helix axis ($\theta = 2\pi m/n$, where $m$ is the number of turns and $n$ is the number of monomer units per period); i.e., certain suggestions are made as to the symmetry of the helix, or (which is the same thing) as to the number of monomer units per turn. For example, it is suggested that the helix has the symmetry $3_1$ (i.e., three monomeric units per turn, $n/m = 3$), the symmetry $4_1$, or $7_2$, etc. Some types of symmetry of helices are shown in Fig. 2. A theoretical calculation of the intensity is then made for the chosen type of symmetry, and this is compared to the experimental results. The theory of X-ray scattering from helices has been developed by Cochran et al. [11] for application to the X-ray diffraction patterns of helical polypeptides; it has been used later to predict the structures of DNA, regular polymers, etc.

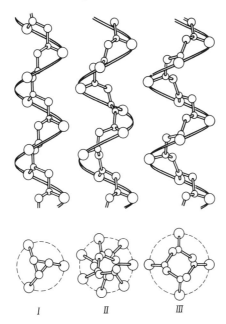

*Fig. 2.* Some types of helical structures occurring in crystalline polymers: I, $3_1$ helix; II, $7_2$; III, $4_1$.


To obtain information about the mutual positions of the atoms of the chain from X-ray fiber patterns, one should determine the relationships between the parameters of the helix and those pertaining to the mutual positions of the atoms. As was mentioned above, the conformation of the molecule in the crystal is determined, first and foremost, by the intramolecular interactions. Consequently, choosing potentials for the atoms' interactions and minimizing the potential function by varying independent geometric parameters, one can obtain the optimal conformation; the parameters of the helix are thus obtained automatically.

Let us consider a regular polymer as a polyatomic chain and find out what parameters describe the conformation of such a molecule. Figure 3 shows a "diatomic"† chain of the $(-M_1-M_2-)_n$ type. Such a chain is a model for all the vinyl polymers with single-atom side groups, and also for polyaldehydes. It is clear that the geometry of this chain without side groups will be fully described by only six independent parameters. They can be chosen in various ways, but the natural coordinates accepted in structural chemistry, and in the theory of molecular vibration spectra prove to be the most suitable. These

---

† This term is hardly a good one, but it is widely used in the literature.

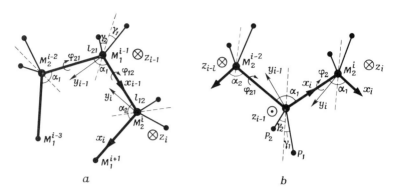

*Fig. 3.* Necessary designations and local coordinate systems in a double-atom chain:
a, cisoid conformation; b, transoid conformation.

parameters are the bond lengths $l_{12}$ (or $l_1$) or $l_{21}$ (or $l_2$), the bond angles,
$\alpha_1$ and $\alpha_2$, and also the dihedral angles $\varphi_{12}$ (or $\varphi_1$) and $\varphi_{21}$ (or $\varphi_2$).

The measuring of dihedral angles (rotation angles) should be clarified.
Authors of the Italian school stick to the conformation nomenclature of
Natta *et al.* [12]. According to this nomenclature the rotation angles are
measured from the cisoid conformation (i.e., $\varphi = 0$ in Fig. 3a) clockwise
regardless from which side the polymer chain is viewed, from either the right
or the left; the range of variation of $\varphi$ is the semisegment from 0 to $2\pi$.

In the configurational statistics of polymer chains (authors belonging to
Volkenstein's [13, 14] and Flory's schools [2, 15]) different designations are
employed. Dihedral angles are measured from the transoid conformation
($\varphi = 0$ in Fig. 3b and $\varphi = 180°$ in Fig. 3a), and both positive and negative
signs of angles are possible. These designations seem to be more natural,
since trans conformations of adjacent monomer units frequently occur in
macromolecules, whereas cis conformations are forbidden. However, for conformational maps this nomenclature is not convenient, because in this case
the energy minima occur on the boundaries of the maps. Therefore, we shall
use the nomenclature of the Italian school.

One can assume that the bond lengths in macromolecules are constant and
transferable from one molecule to another, but deformations and bond angles
should be dealt with more carefully. Of course, to make a rough prediction
of the form of the helix one can assume all CCC angles to be tetrahedral or
equal to 114°, as is done by many authors, but if higher accuracy is required
for the helix parameters and, especially, for the atomic coordinates for calculating intensity distributions in X-ray pictures, the assumption of unchanging
bond angles is too rough. This question will be discussed in detail below.

Undoubtedly, the rotation angles are the most important parameters. Thus, the conformation of a macromolecule may be presented, to a good approximation, as a sequence of internal rotation angles: $\varphi$ for monoatomic chains of the $(-M-)_n$ type (polyethylene, polytetrafluoroethylene); $\varphi_1$ and $\varphi_2$ for diatomic chains $(-M_1-M_2-)_n$ (vinyl polymers, polyaldehydes); $\varphi_1$, $\varphi_2$, $\varphi_3$ for triatomic chains $(-M_1-M_2-M_3-)_n$ (polypeptides); $\varphi_1$, $\varphi_2$, $\varphi_3$, and $\varphi_4$ for tetraatomic chains $(-M_1-M_2-M_3-M_4-)_n$ (diene polymers), etc. It should be noted that owing to the planar structure of the amide groups (I) in polypeptides

(I)

$\varphi_2 = 0$ or $\pi$; therefore, roughly, the potential function can be two-dimensional. In diene polymers, due to the planar structure of the group (II) there are three

(II)

rotation angles instead of four. In polypeptides, and, possibly, in diene polymers, these groups may be slightly nonplanar, but the rotation angles within them change approximately within the same range as do the bond angles.

For a priori calculations of the geometries of macromolecules and also to calculate the parameters of the helices, one should know the coordinates of all atoms within one coordinate system. It is quite easy to find coordinates of atoms in their "proper" systems. For example, for the monoatomic substituents one can write, using the designations of Fig. 3,

$$M_1{}^i(0,0,0),$$

$$P_1(-p_1 \cos \gamma_1 \cos \tfrac{1}{2}\alpha_1, \ -p_1 \cos \gamma_1 \sin \tfrac{1}{2}\alpha_1, \ p_1 \sin \gamma_1)$$
$$P_2(-p_2 \cos \gamma_2 \cos \tfrac{1}{2}\alpha_1, \ -p_2 \cos \gamma_2 \sin \tfrac{1}{2}\alpha_1, \ -p_2 \sin \gamma_2) \tag{1.1}$$

For polyatomic substituents the expressions will be more complicated (one should transfer consecutively the coordinates of side group atoms into the system of coordinates of the main chain); but this difficulty is not one of principle.

The procedure for calculating the atomic coordinates in one system is the following. After Eyring [16] one chooses the $i$th coordinate system so that the $x$ axis is directed along the main axis; the $y$ axis lies in the plane of the three atoms of the main chain; and the $z$ axis, so that the coordinate system is right-handed (Fig. 1); the system is centered on the $i$th atom of the main chain. Then by means of the transformation

$$\mathbf{X}_{i-1} = \mathbf{A}_1 \mathbf{X}_i + \mathbf{B}_1 \tag{1.2}$$

where $\mathbf{X}$ are the coordinate vectors, $\mathbf{A}_1$ is a matrix $(3 \times 3)$ and $\mathbf{B}_1$ is a vector,

$$
\mathbf{A}_1 = \begin{vmatrix}
-\cos\alpha_1 & -\sin\alpha_1 & 0 \\
\sin\alpha_1 \cos\varphi_{12} & -\cos\alpha_1 \cos\varphi_{12} & -\sin\varphi_{12} \\
\sin\alpha_1 \sin\varphi_{12} & -\cos\alpha_1 \sin\varphi_{12} & \cos\varphi_{12}
\end{vmatrix}, \quad
\mathbf{B}_1 = \begin{vmatrix}
l_{12} \\
0 \\
0
\end{vmatrix}
\tag{1.3}
$$

the coordinates are transferred into the $i$th system. In fact, this transformation corresponds to a rotation of the $M_1$–$M_2$ bond by the angle $\varphi$ which is given by the rotation matrix

$$
\mathbf{A}^\varphi = \begin{vmatrix}
1 & 0 & 0 \\
0 & \cos\varphi & -\sin\varphi \\
0 & \sin\varphi & \cos\varphi
\end{vmatrix}
\tag{1.4}
$$

by a further twist through an angle complementery to the bond angle given by the rotation matrix

$$
\mathbf{A}^\alpha = \begin{vmatrix}
-\cos\alpha & -\sin\alpha & 0 \\
\sin\alpha & -\cos\alpha & 0 \\
0 & 0 & 1
\end{vmatrix}
\tag{1.5}
$$

and, finally, by a translation along bond $l$ by its length. Thus,

$$\mathbf{A}_1 = \mathbf{A}_1^\alpha \mathbf{A}_{12}^\varphi \tag{1.6}$$

Similarly the atomic coordinates are transferred from the $(i-1)$th system to the $(i-2)$th system, and then to the next monomeric unit.

Analogous expressions can be easily found for a $p$-atomic chain (i.e., a chain in which every monomeric unit has $p$ rotation angles) (see formulas (1.17)–(1.21) below).

Having obtained the coordinates of the monomer unit, let us recall the transformation for the monomer unit,

$$\mathbf{Y}_{K-1} = \mathbf{A}\mathbf{Y}_K + \mathbf{B}_K \tag{1.7}$$

where $\mathbf{Y}$ are the coordinates of the monomer unit's atoms, $\mathbf{A}$ and $\mathbf{B}$ are transformation matrices related to the $\mathbf{A}_1$, $\mathbf{B}_1$, $\mathbf{A}_2$ matrices in the following way:

$$\mathbf{A} = \mathbf{A}_1 \mathbf{A}_2 \cdots \mathbf{A}_p \tag{1.8}$$

$$\mathbf{B} = \mathbf{B}_1 + \mathbf{A}_1 \mathbf{B}_2 + \mathbf{A}_1 \mathbf{A}_2 \mathbf{B}_3 + \cdots + \mathbf{A}_1 \mathbf{A}_2 \cdots \mathbf{A}_{p-1} \mathbf{B}_p \tag{1.9}$$

and, determining the coordinates of the helix atoms, we use transformation (1.7). Other modifications of this method have also been suggested. For example, Galaktionov [17] developed the method of dual quaternions; Eddy [18] suggested another coordinate system. In some cases the method of dual quaternions somewhat reduces the machine time, but it is essentially a more concise representation of the above transformation.

Now let us consider the parameters of the helices of stereoregular polymers. The following values were mentioned above: $c$, the identity period; $m$, the number of turns; and $n$, the number of monomer units per period. These parameters are obtained from analysis of X-ray diffraction patterns. One more convenient parameter $K = n/m$, the number of monomer units per turn, is introduced. Now, for the general case, the $(-M_1-M_2-\cdots-M_p)_n$ chain, we shall have the following set of helical parameters

$$\rho_1, \rho_2, ..., \rho_{p-1}, \rho_p$$

$$d_{12}, d_{23}, ..., d_{(p-1)p}, d_{p1} \qquad (1.10)$$

$$\theta_{12}, \theta_{23}, ..., \theta_{(p-1)p}, \theta_{p1}$$

and then the translation $d$ along the axis of the helix and the rotation angle $\theta$ on transition from one monomer unit to another will be expressed as

$$d = d_{12} + d_{23} + \cdots + d_{(p-1)p} + d_{p1}$$

$$\theta = \theta_{12} + \theta_{23} + \cdots + \theta_{(p-1)p} + \theta_{p1} \qquad (1.11)$$

For example, for the diatomic chain (Fig. 4) we have six parameters of the helix $\rho_1, \rho_2, d_{12}, d_{21}, \theta_{12}, \theta_{21}$; it is seen in the Fig. 4 that $d = d_{12} + d_{21}$, $\theta = \theta_{12} + \theta_{21}$ (as mentioned above, the conformation of a diatomic chain is characterized by six independent parameters—two bond lengths, two bond angles, and two rotation angles).

The parameters $d$ and $\theta$ are associated with $c$, $M$, and $n$ in a simple relationship:

$$d = c/n, \qquad \theta = 2\pi/K \quad (K = m/n) \qquad (1.12)$$

So, on the one hand we have the set of helical parameters of the macromolecule; on the other hand, the set of "internal" parameters—bond lengths, bond angles, rotation angles, etc.; i.e.,

$$l_{12}, l_{23}, ..., l_{(p-1)p}, l_{p1}$$

$$\alpha_1, \alpha_2, ..., \alpha_{p-1}, \alpha_p \qquad (1.13)$$

$$\varphi_{12}, \varphi_{23}, ..., \varphi_{(p-1)p}, \varphi_{p1}$$

Shimanouchi and Mizushima [19] have suggested a general method for calculation of the helix parameters via the internal parameters, and have

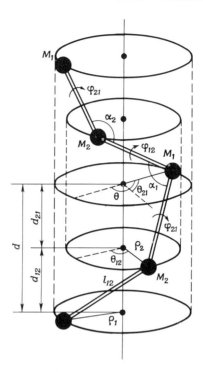

**Fig. 4.** Helical and internal coordinates of a diatomic chain.

derived general formulas for monoatomic chains. Later these formulas were generalized so as to apply to chains of any structure [20–24]. General expressions for the helical parameters are

$$\cos(\theta/2) = (1 + a_{11} + a_{22} + a_{33})^{\frac{1}{2}}/2 \qquad (1.14)$$

$$d\sin(\theta/2) = \frac{b_1(a_{13} + a_{31}) + b_2(a_{23} + a_{32}) + b_3(1 - a_{11} - a_{22} + a_{33})}{2(1 - a_{11} - a_{22} + a_{33})^{\frac{1}{2}}}$$

$$\qquad (1.15)$$

$$2\rho_1{}^2(1 - \cos\theta) + d^2 = b_1{}^2 + b_2{}^2 + b_3{}^2 = R^2 \qquad (1.16)$$

where $R$ is the distance between equivalent atoms $M_1$ of two adjacent monomeric units, and $a$ and $b$ are elements of the $\mathbf{A}$ and $\mathbf{B}$ matrices:

$$\mathbf{A} = \begin{vmatrix} a_{11} & a_{12} & a_{13} \\ a_{21} & a_{22} & a_{23} \\ a_{31} & a_{32} & a_{33} \end{vmatrix} = \mathbf{A}^\varphi_{12}\mathbf{A}_1{}^\alpha\mathbf{A}^\varphi_{23}\mathbf{A}_2{}^\alpha \cdots \mathbf{A}^\alpha_{p-1}\mathbf{A}^\varphi_{p1}\mathbf{A}_p{}^\alpha \qquad (1.17)$$

$$\mathbf{B} = \begin{vmatrix} b_1 \\ b_2 \\ b_3 \end{vmatrix} = \mathbf{B}_{12} + \mathbf{A}_{12}^{\varphi}\mathbf{A}_1{}^{\alpha}\mathbf{B}_{23} + \cdots + \mathbf{A}_{12}^{\varphi}\mathbf{A}_1{}^{\alpha}\mathbf{A}_{23}^{\varphi}\mathbf{A}_2{}^{\alpha}$$

$$+ \mathbf{A}_{(p-2)(p-1)}^{\varphi}\mathbf{A}_{p-2}^{\alpha}\mathbf{B}_{(p-1)p} + \mathbf{A}_{12}^{\varphi}\mathbf{A}_1{}^{\alpha}\mathbf{A}_{23}^{\varphi}\mathbf{A}_2{}^{\alpha}\cdots\mathbf{A}_{(p-1)p}^{\varphi}\mathbf{A}_{p-1}^{\alpha}\mathbf{B}_{p1}$$

$$(1.18)$$

$$\mathbf{A}_i{}^{\alpha} = \begin{vmatrix} -\cos\alpha_i & -\sin\alpha_i & 0 \\ \sin\alpha_i & -\cos\alpha_i & 0 \\ 0 & 0 & 1 \end{vmatrix} \qquad (1.19)$$

$$\mathbf{A}_{ij}^{\varphi} = \begin{vmatrix} 1 & 0 & 0 \\ 0 & \cos\varphi_{ij} & -\sin\varphi_{ij} \\ 0 & \sin\varphi_{ij} & \cos\varphi_{ij} \end{vmatrix} \qquad (1.20)$$

$$\mathbf{B}_{ij} = \begin{vmatrix} l_{ij} \\ 0 \\ 0 \end{vmatrix} \qquad (1.21)$$

Introducing corresponding elements of the matrices into Eqs. (1.14)–(1.16) and solving the system, one can obtain expressions for helical parameters. Given below are the final expressions for the monoatomic and diatomic chains.

Monoatomic chain $(-M-)_n$

$$\cos(\theta/2) = \cos(\varphi/2)\cos(\alpha/2) \qquad (1.22)$$

$$d = l\sin(\varphi/2)\sin(\alpha/2)/\sin(\theta/2) \qquad (1.23)$$

$$\rho = [0.5(l^2 - d^2)/(1 - \cos\theta)]^{1/2} \qquad (1.24)$$

Diatomic chain $(-M_1-M_2)_n$:

$$\cos(\theta/2) = \cos(\varphi_{12}/2 + \varphi_{21}/2)\sin(\alpha_1/2)\sin(\alpha_2/2)$$

$$-\cos(\varphi_{12}/2 - \varphi_{21}/2)\cos(\alpha_1/2)\cos(\alpha_2/2) \qquad (1.25)$$

$$d_{12} = [\sin(\varphi_{12}/2 + \varphi_{21}/2)\sin(\alpha_1/2)\sin(\alpha_2/2)$$

$$-\sin(\varphi_{12}/2 - \varphi_{21}/2)\cos(\alpha_1/2)\cos(\alpha_2/2)]\cdot l_{12}/\sin(\theta/2) \qquad (1.26)$$

$$d_{21} = [\sin(\varphi_{12}/2 + \varphi_{21}/2)\sin(\alpha_1/2)\sin(\alpha_2/2)$$

$$+\sin(\varphi_{12}/2 - \varphi_{21}/2)\cos(\alpha_1/2)\cos(\alpha_2/2)]\cdot l_{12}/\sin(\theta/2), \qquad (1.27)$$

$$d = d_{12} + d_{21} \tag{1.28}$$

$$\rho_1 = [0.5(l_{12}^2 - 2l_{12}l_{21}\cos\alpha_2 + l_{21}^2 - d^2)/(1 - \cos\theta)]^{1/2} \tag{1.29}$$

$$\rho_2 = [0.5(l_{12}^2 - 2l_{12}l_{21}\cos\alpha_1 + l_{21}^2 - d^2)/(1 - \cos\theta)]^{1/2} \tag{1.30}$$

$$\cos\theta_{12} = (\rho_1{}^2 + \rho_2{}^2 + d_{12}^2 - l_{12}^2)/(2\rho_1\rho_2) \tag{1.31}$$

$$\cos\theta_{21} = (\rho_1{}^2 + \rho_2{}^2 + d_{21}^2 - l_{21}^2)/(2\rho_1\rho_2) \tag{1.32}$$

Thus, knowing the internal parameters, and solving Eqs. (1.22)–(1.24) consecutively we obtain the parameters of a monoatomic chain; or solving Eq.
(1.26)–(1.37) we get the parameters of a diatomic chain. Sugeta and Miyazawa
[24] have developed the algorithm for computing the helical parameters of
any chain (including hex-atomic chains, for nucleic acids).

Let us determine now how the parameters of a helix depend upon bond
angles and rotation angles. This will be exemplified by the monoatomic chain.
Figure 5 shows a plot of $K$ versus $\varphi$. The curves for the different bond angles
of the main chain merge in the transoid conformation range. This means that
the main parameters of the helix, e.g., $K = 2\pi/\theta$ do not depend greatly upon
bond angle variations, but, for chains of the "cisoid type" (helices $4_1$, $5_1$) the
deviation of $K$ in some cases exceeds the experimental error.

For diatomic chains, diagrams are not as convenient for determining the
helical parameters as for the monoatomic chains. It is not difficult to present
on a plane the lines of equal $\theta$ (Fig. 6) or $d$ values (Fig. 7) with fixed bond

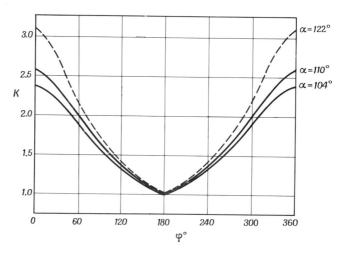

Fig. 5. Relationship between the number of monomeric units per turn of the helix and
the rotation angle of the monoatomic chain for three values of the CCC bond angles;
$l = 1.54$ Å.

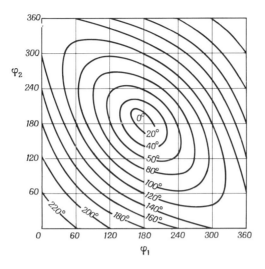

**Fig. 6.** The lines of equal $\theta$ values ($\theta$ is the angle of helical rotation) at different $\varphi_1,\varphi_2$ for diatomic chain with 114° bond angles ($l = 1.54$ Å).

angles, in our case $\alpha_1 = \alpha_2 = 114°$. But, in general, $\alpha_1$ is not equal to $\alpha_2$; and if this is taken into account, the corrections to the helical parameters may be quite significant (they are different in different regions of the maps of $\theta(\varphi_1, \varphi_2)$

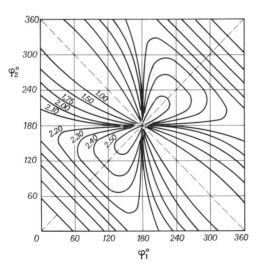

**Fig. 7.** The lines of equal $d$ values (projection of a monomeric unit on the axis of the helix, in angstroms) at different $\varphi_1,\varphi_2$ for diatomic chain with 113° bond angles (P. De Santis, E. Giglio, A. M. Liquori, and A. Ripamonti. *J. Polymer Sci. A1*, 1383 (1963)).

or $K(\varphi_1, \varphi_2)$. Hence, to determine the parameters of a diatomic chain from the bond and rotation angles, it is desirable to solve the system of equations (1.26)–(1.32) numerically.

Let us now consider the conformational analysis of stereoregular macromolecules. The potential functions for conformation calculations were discussed in the preceding chapter. For vinyl polymers, the expression for strain energy is the same as that of alyphatic molecules, viz.

$$U_{st} = U_{nonbonded} + U_{ang} + U_{tors} \tag{1.33}$$

For the sake of completeness, it should be noted that qualitative characteristics—the principles of staggered bonds and of maximum number of van der Waals contacts—may also account rather well for certain peculiarities of the geometry of helical macromolecules. Bunn and Holmes [25] and Natta and Corradini [10a] using these qualitative principles, have explained the main features of the structures of vinyl polymers. But, as will be shown below, the potential functions are finer instruments for the analysis of the conformational states of macromolecules than are the mere geometrical characteristics.

Let us consider the conformations of polymers of different types.

### a. Vinyl Polymers of $(-CX_2-)_n$ Type

Polyethylene (X = H) and polytetrafluoroethylene (X = F) are well-known representatives of this type which have found wide commercial application. Bunn [26] was the first to interpret the X-ray diffraction patterns of polyethelene fiber, and concluded that the macromolecule has the form of a planar zigzag. Later Huggins [27] suggested that the polyethelene molecule is not planar but slightly helical, since between hydrogen atoms there exists some repulsion, though very weak. But an analysis of the IR spectra of crystalline polyethelene [28] did not confirm this suggestion. A planar molecule has higher symmetry than a helical one; therefore in it some absorption bands are forbidden by symmetry, and this is confirmed by the experimental data. (It should be noted that the X-ray diffraction patterns furnish no unequivocal proof of the planar configuration of the chain.)

Figure 8a shows the results of three calculations of the polyethylene potential function with respect to the rotation angle $\varphi$ around the C–C bond, those of Liquori et al. [29], Scott and Scheraga [30], Dashevsky and Murtasina [31]. In spite of the fact that in these three calculations different potentials of nonbonded interactions were used, all these authors predicted the optimal conformations to be a planar zigzag. It should be noted that Liquory et al. made their calculations without using a torsion potential. Nevertheless, their results for polyethylene, as well as for other vinyl polymers, agree with the experimental data. Thus, the energy minimum of the nonbonded interactions (if

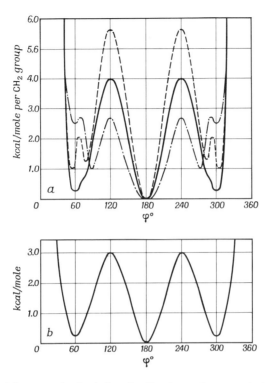

**Fig. 8.** a, Potential energy of polyethylene (kcal/mole per $CH_2$ group): (———) Dashevsky potentials, $U_0 = 3.0$ kcal/mole; (———) Scott and Scheraga potentials, $U_0 = 2.8$ kcal/mole; (—·—·—·) Liquori potentials without torsion term. b, Potential function of n-butane (V. G. Dashevsky. *Z. Struct. Khimii* **9**, 289 (1968)).

only central atom–atom potentials are meant) predetermines the conformation of the molecule, whereas the torsion term only slightly changes the position of the minimum.

It is interesting to compare the results of the calculations of the potential function for polyethelene with that for *n*-butane (Fig. 8b) which is the simplest analogue of the latter, because it has $C_1 \cdots C_4$ interactions. The *n*-butane function has as few as three minima (in accordance with the staggered bonds principle), whereas the polyethylene curve has either five minima or inflection points in the range $\varphi \approx 60°–80°$. Additional minima, or features, are due to more distant interactions than $1 \cdots 4$.

Crystalline polytetrafluoroethylene has the $13_6$ helix ($K = 2.17$) [32], its rotation angle is 163.5°, i.e., owing to the repulsion of the fluorine atoms, the deviation from planarity (180°) is significant. At temperatures exceeding 19°C the helix is likely to acquire the symmetry $15_7$ ($K = 2.14$). Potential functions for polytetrafluoroethylene are calculated in [29, 33–36].

Iwasaki [33] at first calculated the conformation of polytetrafluoroethylene only with 6-exp atom–atom potentials to obtain a rotation angle of 162.5°; then he introduced dipole–dipole interactions,† and the optimal conformation slightly shifted—the rotation angle was then 161°. Similar values were obtained by De Santis *et al.* [29], according to whose calculation $\varphi = 165°$. It is noteworthy that the hump at 180° is very small; this fact is used by the authors to account for the ready transition of the right handed helix into left handed above 20°C [35]. Iwasaki's value for the barrier was 1 kcal/mole. It should be noted that neither De Santis, nor Iwasaki, nor the authors of papers [35] included torsion potentials.

Bates [36] carried out an interesting investigation of the potential function of polytetrafluoroethylene. He suggested that one and the same force field should be valid for describing the internal rotation in hexafluoroethane (the barrier, 4 kcal/mole), a number of the properties of perfluoroalkanes, and the helical conformation of polytetrafluoroethylene. These properties are determined primarily by the relative effects of the torsion term and the F ··· F interaction. Using the value of the equilibrium radius of fluorine borrowed from [37] and having estimated the parameter $A$ (see formula (1.3), Chapter VII) from the Slater–Kirkwood equation (formula (2.2), Chapter VII), Bates varied only the parameter $B$ of the 6-exp potential. The dipole–dipole interactions of the C–F bonds [38] were also taken into account. It turned out that, if the $U_0$ constant is about 3 kcal mole$^{-1}$ (i.e., that in ethane), and $B$ is chosen so that the value of the internal rotation barrier is equal to the measured one, then a planar zigzag will be the equilibrium form of polytetrafluoroethylene, which does not agree with the experimental data. Having analyzed possible combinations of the F ··· F potentials and $U_0$ constants, the author concluded that the potential $f_{\text{F}\cdots\text{F}} = -118/r^6 + 1.042 \times 10^5 \exp(-4.60r)$ and a constant $U_0$ equal to 1.5 kcal mole$^{-1}$ are the best to describe the above properties. Furthermore, the equilibrium angle $\varphi$ in polytetrafluoroethylene lies between 163–169°.

The calculations made by Bates showed that the constant $U_0$ may prove to be untransferable. If this is the case, the predictability of a potential function containing a torsion term would not be so great, since every time one needs to choose a new constant $U_0$. But it is too early to make a final conclusion, as the restriction Bates imposed on the F ··· F potential is too rigid: In the first place, one could vary the F ··· F equilibrium distance within reasonable limits and, secondly, a deviation from the semiempirical rule of Slater–Kirkwood is also quite allowable. Also, the parametrization of the electrostatic component is not quite convincing (which affects the value of the internal rotation

---

† The role of electrostatic interactions will be discussed in greater detail in connection with conformations of peptides.

barrier of hexafluoroethane); finally, all calculations were made with tetrahedral bond angles though variations in bond angles can affect the results. It could also not be ruled out that the helical conformation of polytetrafluoroethylene is maintained by intermolecular interactions.

### b. Vinylidene Polymers

Let us consider the structure of three well-known vinylidene polymers $(-CH_2-CX_2-)_n$: $X = F$, Cl, and $CH_3$. These polymers are diatomic chains; their potential energy vs. $(\varphi_1, \varphi_2)$† maps at fixed bond angles give a good idea of the positions of minima. But if precise atomic coordinates are required, a minimum for at least four parameters must be sought, i.e., the two rotation angles, and the two nonequivalent bond angles of the main chain. (Still more precise atomic coordinates may be achieved by also varying the bond angles in the side groups.)

Polyvinylidene fluoride is only slightly more overcrowded than polyethylene, and it should be expected to have a slightly less tendency to twist than that of polytetrafluoroethylene. Since conformational freedom is great in this molecule, the structure of the polymer chain is affected to a greater extent by intermolecular interactions. Indeed, depending on the kind of treatment, polyvinylidene fluoride may have two crystalline modifications [39–41], in one of which ($\beta$) the molecule has the form of a planar zigzag and in the other ($\alpha$) it may be a slightly wound helix. According to the authors of [40] the deviation angles from the planar zigzag are close to 10°, but the authors of [39] have objected that such angles are incompatible with the space group established. Thus, the $\beta$ form of polyvinylidene fluoride requires further verification.

Figure 9 shows the potential map of polyvinylidene fluoride [31] (a similar map was given in [42]) with three symmetrically nonequivalent minima. The first one corresponds to a planar zigzag, the second to a helix with $K \approx 3$, and the third to a helix with $K \approx 2.15$. Structural realization of the third minimum is less probable, because its depth is less by 0.5 kcal mole$^{-1}$ than that of the first two minima and it has steeper walls, i.e., in addition, it yields to them in entropy. As to the first two minima, they are practically equally probable. It is possible that the $\beta$ form of polyvinylidene fluoride corresponds to the $K \approx 3$ helix.

It is interesting that the position of the minima for this molecule are in almost complete agreement with the principle of staggered bonds; the optimum rotation angles (180°, 180°), (170°, 60°), (55°, 55°), and other angles determined from the symmetry conditions, correspond, respectively, to the

---

† In what follows the $CH_2-CR'R$ ‡ $CH_2-CR'R''$ dihedral angle in all vinyl and vinyledene polymers will be designated as $\varphi_1$ and the $CR'R''-CH_2$ ‡ $CR'R''-CH_2$ as $\varphi_2$.

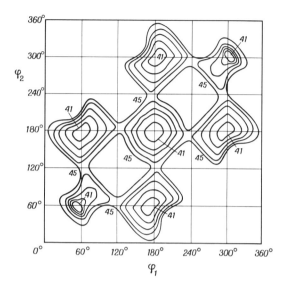

*Fig. 9.* Potential map of polyvinylidene fluoride. Bond angles of the main chain are assumed to be 114.6°; energy contours are given with 1 kcal/mole intervals per monomeric unit (two CC bonds).

trans–trans, trans–gauche, and gauche–gauche conformations of the monomer units. It is not so much that the torsion term predetermines these conformations, but that these conformations are also favorable for central nonbonded interactions.

Let us consider now a more overcrowded macromolecule, polyvinylidene chloride. The analysis of X-ray diffraction patterns of polyvinylidene chloride fibers [43] yielded rotation angles close to 0 or $\pi$, but for conformational calculations the structure of this polymer should, undoubtedly, be studied more thoroughly.

Liquori *et al.* [29] draw $(\varphi_1, \varphi_2)$ potential maps for polyvinylidene chloride with CCC bond angles of 114° and 120° and showed that the main conformational conclusions do not depend upon the bond angle chosen. But the two independent bond angles in this molecule should not be equal, since repulsion between the Cl atoms is stronger than that between the H atoms; hence it should be expected that the CCC angle will be close to tetrahedral at the $CX_2$ group and will be much larger at the $CH_2$ group (in this case the $X \cdots X$ distance increases). The search for the absolute minimum of the potential function [31] (for minimization, the valley method was used [44]) yielded the following values of the independent parameters: $\angle CCC(CH_2) = 125°$, $\angle CCC(CX_2) = 113°$, $\varphi_1 = 44°$, $\varphi_2 = 113°$. Figure 10 shows potential maps for this macromolecule for the bond angles: (a) $\alpha_1 = \alpha_2 = 114.6°$; (b) $\alpha_1 =$

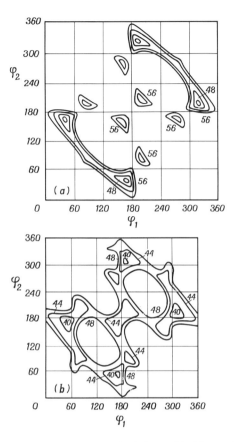

*Fig. 10.* Conformational map of polyvinyledene chloride. a, in the main chain $\angle CCC(CH_2) = \angle CCC(CCl_2) = 114.6°$; b, for $\angle CCC(CH_2) = 125°$, $\angle CCl(CCl_2) = 113°$. Equipotentials are drawn with 4 kcal/mole intervals per monomeric unit.

$125°$, $\alpha_2 = 113°$. The first map is (owing to the equality of the bond angles of the main chain) symmetrical across the diagonal, $\varphi_1 = 2\pi - \varphi_2$ (the right- and left-handed helices are equipotential) and the absolute minimum gives a helix with $K \approx 3.0$. The equipotentials of the second map are close to the level lines of the potential surface. But it should be born in mind that the true level lines can be obtained only when every point of the equipotentials corresponds to the energy minimum of parameters other than $\varphi_1$ and $\varphi_2$, i.e., the $\alpha_1$ and $\alpha_2$ minima. To build up true level lines is a very time-consuming problem, even with high-speed computers, and the information such lines furnish is not always important. But it is significant that near the minima isoenergetic contours are close to the true level lines.

Polyisobutylene is the most comprehensively studied vinylidene polymer, but its true conformation in the crystal has not been elucidated. After the identity period had been determined ($c = 18.6$ Å) from the analysis of a stretched sample [45], attempts were made to clarify its structure. Bunn [46] suggested that crystalline polyisobutylene has the helix of $8_5$ symmetry. Liquori [47] made the same suggestion and, by assuming that both rotation angles are the same and that the CCC bond angles of the backbone are equal to 114°, found that $\varphi_1 = \varphi_2 = 97°$. But the Japanese authors [48] pointed out that in such a conformation the distance between the $CH_3$ groups of the adjacent monomer units would be too short, 2.3 Å. (The calculations of numerous overcrowded molecules decsribed in Chapter VII, and the X-ray evidence shows that the minimum distance between the methyl groups must be 2.9–3.0 Å). Hence, the structure suggested by Liquori is unrealistic.

It is interesting to discuss the conformational calculations of polyisobutylene. The $(\varphi_1, \varphi_2)$ potential map calculated by De Santis *et al.* [29] is shown in Fig. 11. Without taking into account enantiomorphous helices, the potential map has six minima (there could have been more minima, should C and H atoms have been considered instead of effective potentials of the methyl groups). The deepest minima correspond to the rotation angles (45°, 155°), (45°, 85°), and (85°, 85°), all of which have approximately the

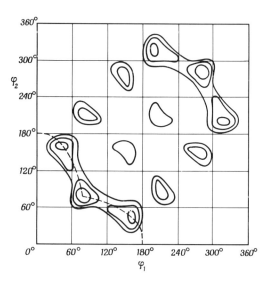

**Fig. 11.** Conformational map of polyisobutylene. Equipotentials are drawn with 10 kcal/mole intervals per monomeric unit. Bond angles in the main chain, 114°.

same energy. The authors of this paper chose a conformation with two equal rotation angles $(85°, 85°)$ and ascribed to it symmetry $8_5$.

The Japanese authors [48] proposed a different conformation. Assuming the helical symmetry $8_3$ and $120°$ angles in the backbone, they calculated the rotation angles $\varphi_1 = 48°$ and $\varphi_2 = 180°$. Moreover, they believe that this conformation has lower energy than that corresponding to the $8_5$ helix. In the map of Fig. 11 this conformation is near to one of the local minima. These authors also calculated the intensity distribution of the X-ray diffraction patterns for the $8_5$ and $8_3$ helices. They came to the conclusion that for both helices the observed layer intensities agree with the theoretical values.

It is noteworthy that, as was established in paper [25], better agreement between the theoretical and experimental intensity distributions is achieved if the bond angles are assumed to be different in the $8_5$ helix: In the main chain of polyisobutylene the CCC angle at the $CH_2$ group is taken to be $126°$ and that at the $C(CH_3)_2$ group to be $107°$. In this case the rotation angles will be $\varphi_1 = 51°$ and $\varphi_2 = 102.5°$ which is a rather unfavorable region on the map in Fig. 11.

Polyisobutylene exemplifies the difficulties encountered if the bond angles are not varied. It would be interesting, first, to calculate the polyisobutylene conformations by minimization of a potential function depending on several variables, e.g., two bond angles and two rotation angles; second, to take into account the C and H atoms of the methyl groups; third, to calculate the X-ray intensity distribution from the atomic coordinates obtained from the conformational calculation.†

### c. Isotactic Vinyl Polymers

The isotactic poly-$\alpha$-olefines $(-CH_2-CHR-CH_2-CHR-)_{n/2}$ have been the subject of extremely intensive study. Their most characteristic conformation is the $3_1$ helix, in which half the C–C bonds (every other one) are parallel to the helix axis (Fig. 2). The rotation angles are about $180°$ and $60°$, i.e., the conformations of the monomer units are close to trans or gauche conformations (the principle of staggered bonds is at work). If the side groups are not bulky or branched, the $3_1$ type of helix is always realized; otherwise the conformation is somewhat nearer to cisoid, because in this case the conformation is determined to a great extent by the repulsion of the side group atoms. With very large substituents, the symmetry of the helix is $4_1$ or even $5_1$; sometimes an intermediate, e.g., the $7_2$ type of helix is encountered, as is shown in Fig. 2.

---

† Recently Italian authors (Allegra *et al.* [48a]) made calculations clearing up all the above problems. They found the following coordinates of the potential function minimum: $\angle CCC(CH_2) = 124°$, $\angle CCC[(CH_3)_2] = 110°$, $\varphi_1 = 47.5°$, $\varphi_2 = 156.7°$. This conformation is not far from the $8_3$ helix and gives an excellent fit to the X-ray intensity distribution.

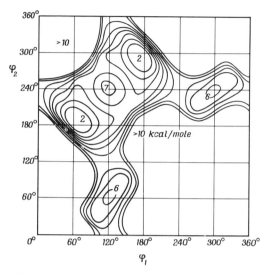

*Fig. 12.* Conformational map of isotactic polypropylene.

Isotactic polypropylene is the simplest polymer of this series. Natta *et al.* [49] introduced a potential function for isotactic polypropylene, the one earlier suggested to describe the conformational states of 1,1′-dimethylpropane:

$$U(\varphi) = \tfrac{3}{2}(1 + \cos 3\varphi) + \tfrac{2}{3}[2 + \cos \varphi + \cos(\varphi + \tfrac{2}{3}\pi)] \text{ kcal mole}^{-1}$$

with effective $CH_3$–$CH_3$ potentials by Mason and Kreevoy [50]. In the map of Fig. 12 the minima are close to $(\varphi_1, \varphi_2)$ equal 60° and 180°, which corresponds to the $3_1$ helix. Similar optimum conformations, slightly different in detail, was obtained by De Santis *et al.* [29]. Excluding enantiomorphous conformations, the map of De Santis contains three minima: (183°, 62°), (80°, 100°), and (60°, 160°); the first minimum is the deepest and corresponds to the helix with $K = 3.1$ (the bond angles were all assumed to be the same and equal to 114°).

A comparison of the polypropylene maps with those of polyvinylidene chloride and polyisobutylene shows that in the former the minimum occupies a wider region. (In the maps of the vinylidene polymers mentioned the 1 kcal mole$^{-1}$ contour would have been practically a point.) Consequently, in isotactic vinyl polymers the rotation angles and helix parameters may vary considerably. Rough calculations and an analysis of the known structures shows $\varphi_1$ to vary from 55° to 95° and $\varphi_2$ from 180 to 205°–210°.

The bond angles in isotactic polymers, like those in vinylidene polymers, should be different: Due to the repulsion of substituents the CCC angle at the CHR group should be less than that at the $CH_2$ group. In this connection it is interesting to note that the identity period of isotactic vinyl polymers, which have been unequivocally proved to have the $3_1$ helix, varies over a very wide range, from 6.3 Å, (e.g., poly-N,N'-di-$n$-butylacrylamide [51], R = C(O)N(CH$_2$CH$_2$CH$_2$CH$_3$)$_2$) to 6.55 Å (e.g., poly-5-trimethyl-silyl-pentene-1 [52], R = CH$_2$CH$_2$CH$_2$Si (CH$_3$)$_3$).

Considering the structures of isotactic vinyl polymers, one may divide them into three groups, i.e., those having aromatic side groups, alicyclic side groups, and branched aliphatic side groups. (Polyesters may be left out of consideration, since all of them form the $3_1$ helix.)

Benzene is the smallest aromatic substituent. Polystyrene (R = C$_6$H$_5$) crystallizes in the $3_1$ helix. If the benzene ring has substituents, the type of helix usually changes, but the type of helix for a given substituent is usually rather difficult to predict only on the basis of the intramolecular interactions. For example, it is natural for poly-o-methylstyrene [54, 55] to crystallize in the $4_1$ helix, because its CH$_3$ group is close to the main chain. (However, one should first prove that the most favorable position of this group in relation to the chain will not be as favorable for the interactions of the benzene rings of adjacent monomer units). It is also clear why poly-o-methyl-p-fluorostyrene [55], poly-2,5-methyl-styrene [12] have the $4_1$ helix and poly-m-methyl-styrene [56, 57] or poly-p-trimethyl-silylstyrene [56, 58] have the intermediate $11_3$ helix ($K = 3.67$) or $29_8$ ($K = 3.62$) for the former polymer and $29_9$ ($K = 3.22$) for the latter. But there are exceptions; e.g., poly-m-fluorostyrene [51, 55] has the $4_1$ and poly-o-fluorostyrene [51, 53, 55] the $3_1$ helix. Polymers having large aromatic substituents, such as polyvinylnaphthalene [59], have $4_1$ helices.

Intermolecular interactions play an important role in the structures of isotactic vinyl polymers. They are responsible for the type of helix in poly-m- and poly-o-fluorostyrenes. It is not very rare that two or three polymorphous modifications of a crystalline polymer have different helix parameters. For example, polyvinylcyclohexane [60] (cyclo-C$_6$H$_{12}$) has, depending on the temperature and conditions of synthesis, two crystalline forms, one with the $4_1$ ($K = 4$) helix; the other, $24_7$ ($K = 3.43$). In this case, again, intermolecular interactions are at work.

Vinyl isotactic polymers with large cycloalkyl side chains crystallize, as a rule, into a $4_1$, or $4_1$-like helix [60]. At usual temperatures and reaction conditions polyvinylcyclopentane, -hexane and -heptane also have the $4_1$ helix. Polyvinylcyclopentane has the smallest volume of them all, therefore at low temperatures there arise in it crystalline modifications in which the $11_3$, $10_3$ [60], or even $3_1$ [61] helix is realized. This is one more example of inter-molecular interactions.

Finally, polymers with branched side chains have been comprehensively studied. The $3_1$ helices are the most typical for the nonbranched chains, or those with branched side chains. If the branching begins at the second atom, intermediate helices occur at first, the $4_1$ helices. In the chains of such structures polymorphous modifications with various types of helix are quite frequent. Such cases are considered in detail in paper [60].

Intramolecular interactions play by far the greatest role in regular hydrocarbon chains with branched side chains. Borisova and Birstein [14] succeeded in proving by relevant calculations that branching is bound to bring about changes in the type of helix. According to their data on polypropylene, the rotation angles should be (60°, 180°), if the bond angles in the chain are equal to 114°. Furthermore, if, in the hydrocarbon side chains, the branching begins at the first atom, the rotation angles in the main chain will be (84°, 204°), which is almost the $4_1$ helical conformation; if the branching begins at the second atom of the side chain, the calculated rotation angles are (70°, 193°), which corresponds to an intermediate helix. Thus, the main features of the structures of regular crystalline hydrocarbons may be elucidated, without taking into account intermolecular interactions; however, a more thorough investigation is required in some cases (e.g., to explain conformational polymorphism).

### d. Syndiotactic Vinyl Polymers

The planar zigzag is the most typical conformation of syndiotactic poly-α-olefines $(-CH_2-CHR-CH_2-CHR-)_{n/2}$ In fact, in this conformation the side groups are at the greatest distance from each other, and, at the same time, the torsion energy has a minimum value. But it should be born in mind that an alteration to gauche conformations also corresponds to a minimum of the torsion energy, since the substituents are fairly far from each other. A gauche conformation of the all C–C bonds of the main chain is excluded, since then all the far-off atoms would bump into each other. Hence, there are two possibilities: (1) the helix is a planar or planarlike diatomic chain and (2) the molecule has a helical conformation; but the latter cannot be a diatomic one as the sequence $(\varphi_1, \varphi_2, \varphi_1, \varphi_2)$ would result in considerable overlap of the atoms. One of the crystalline modifications of syndiotactic polypropylene [62], is likely to have a tetraatomic helix with the sequence of rotation angles $(\varphi_1, \varphi_2, \varphi_2, \varphi_1)$. Helices with a greater number of nonequivalent rotation angles have not been found in syndiotactic polymers; however, their existence is not excluded in principle. Perhaps higher symmetry (fewer independent rotation angles) is favorable for molecular packing.

Polychlorvinyl (R = Cl) [63, 64], polyacrylonitryl (R = C≡N) [65], and polybutadiene-1,2 (R = CH=CH$_2$) are examples of syndiotactic vinyl polymers with planar trans chains.

Natta et al. [49] drew the $(\varphi_1, \varphi_2)$ potential map for polypropylene. As was to be expected, the $(180°, 180°)$, $(180°, 60°)$ and $(60°, 180°)$ bond pairs correspond to potential wells. Miyazawa [22], using the experimental data ($c = 7.3$ Å, $K = 2$), assuming the angle sequence $(\varphi_1, \varphi_2, \varphi_2, \varphi_1)$, and 114° bond angles, calculated the rotation angles to be $\varphi_1 = 57°$ and $\varphi_2 = 189°$. Thus, roughly speaking, the conformation of syndiotactic polypropylene corresponds to the gauche and trans forms of the monomer units. Borisova and Birstein [14] made detailed calculations of the conformations of syndiotactic vinyl polymers taking into account the internal rotations in the side groups. They showed, in complete agreement with the experimental data, that the planar trans form of polybutadiene-1,2 is somewhat more favorable as compared to the helical one, and that for polypropylene the helical conformation should be more stable.

Thus, the syndiotactic vinyl polymers of the $(-CH_2-CHR-)_n$ type are not "overcrowded." Their potential functions have no deep wells; the freedom of movement about the minima is rather great, which accounts for the higher flexibility of these polymers in solution.

### e. Diisotatic Vinyl Polymers

More than ten years ago Natta [62] was the first to synthesize diisotactic polymers (see Fig. 1). Using catalysts of the $TiCl_3-AlR_3$ type he polymerized *cis*- and *trans*-deuteropropylene ($CHD-CHCH_3$); the first isomer yielded erythro- the second, threo-diisotactic polydeuteropropylene. The existence of these two polymers was confirmed by IR spectroscopy [66]. As is shown in [67] Natta's method should, in principle, allow one to prepare diisotactic polymers of 1,2-substituted ethylene derivatives with the same side chains; but this has not been done. However, similar polymers with different side chains have been obtained. For example, diisotactic polymers have been

<br>

Cl⟍  ⟋OR
  C=C
H⟋  ⟍H

(III)

R′⟍  ⟋H
  C=C
H⟋  ⟍OR

R′ = $CH_3$ or $CH(CH_3)_2$

(IV)

<br>

synthesized from *cis*- and *trans*-chlorovinyl esters (III) [68] and also from *trans*-methyl and *trans*-isobutyl-propenyl esters (IV) [69].

Nothing is yet known about the structure of these polymers in crystals. Though diisotactic polychloroethylenes, probably, have not been obtained, their potential maps [31] should give some idea about their possible helical structures. As is shown in Fig. 13a,b, the conformational maps of threo diisotactic polymers with the same side chains have a plane and a symmetry

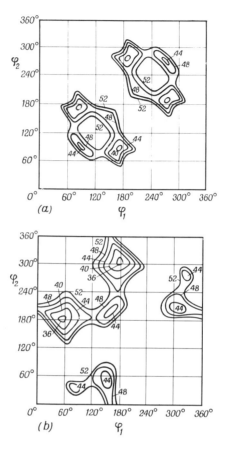

**Fig. 13.** Conformational maps: 1, threo diisotactic polychlorethylene; b, erythro diiso-tactic polychlorethylene. Equipotentials are drawn with 4 kcal/mole intervals per two C–C bonds. Bond angles in the main chain, 114.6°.

center ($\varphi_1 = 2\pi - \varphi_2$), whereas the erythro, has only a mirror plane. The deepest minimum of the threo diisotactic polymers corresponds to rotation angles of (170°, 70°) and the K = 3.0 helix. For the erythro diisotactic polymer, the rotation angles are (57°, 180°) and a helix with $K = 2.8$. Thus, both poly-mers are likely to be able to crystallize into the $3_1$ helix in spite of their dif-ferent chemical structures and dissimilar conformation maps.

*f. Polyoxides*

Polymethyleneoxide (O–CH$_2$–)$_n$ is the simplest example of the polyoxide polymer series (O–(CH$_2$)$_m$–)$_n$. Two crystalline modifications of this polymer

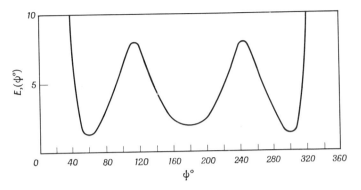

*Fig. 14.* Potential energy of polymethylene oxide versus rotation angle.

have been found to exist. One of them contains nine monomeric units in five turns [70, 71] and, if one assumes the bond length in the chain to be equal to 1.43 Å and both bond angles to be the same and equal to 111°, the rotation angle for such a monoatomic chain will be 77°. In another modification [72], by assuming the bond angles to be equal to 112°, one calculates the rotation angle to be equal to 63°. It may be expected that the freedom of movement in this molecule is rather high, and the rotation angle depends upon the molecular packing.

Calculations [29] made with rather rough assumptions about the $O \cdots O$ interactions actually revealed great conformational freedom (Fig. 14). The energy minimum corresponds to $\varphi = 67°$ and a helix with $K = 1.92$, a value just between two experimental ones for different crystalline modifications. The planar zigzag has a higher energy of intramolecular interactions, and, evidently for this reason, is not realized in the crystal. Actually, the distance between the H atoms of the adjacent monomer units in the planar zigzag of polymethyleneoxide would be much shorter than in polyethylene, therefore the formation of a helix is justified.

Although polyethyleneoxide [73, 74] has the helical conformation, the higher homologues are frequently planar [25]:—oxygen is, in fact, incorporated into the planar chain of polyethylene. It does not require much energy as the close $H \cdots H$ contacts in the $CH_2-O-CH_2$ fragments are rarer than in polyethyleneoxide.

Tadokoro [75] made a detailed analysis of polyoxide conformations based on the experimental data and qualitative speculations. Polymethyleneoxide is a chain consisting of monomer units having gauche conformations slightly shifted due to nonbonded interactions (77° compared with 60° in "pure" gauche conformations). It is interesting that variations of the bond angles resulted in better agreement between the calculated and measured intensity distributions of X-ray diffraction patterns. The following chain parameters gave the best agreement: $l(C–O) = 1.43$ Å, $\angle COC = 112°24'$, $\angle OCO = 110°49'$, $\varphi = 78°13'$.

In the next polymer of this series, polyethyleneoxide ($m = 2$), the most favorable conformation is the combination of one gauche (G) and two trans (T) conformations:

$$-CH_2-(-O-CH_2-CH_2-)-_n$$

$$T \quad T \quad G$$

It is not difficult to estimate that such a conformation will give a helix, similar to the experimentally found $7_2$ [74].

In polyoxacyclobutane the following conformations are possible:

$$-CH_2-(O-CH_2-CH_2-CH_2)-_n$$

| T T | G G | (I) |

$$T \quad T \quad G \quad G \qquad (I)$$

$$T \quad T \quad T \quad G \qquad (II)$$

$$T \quad T \quad T \quad T \qquad (III)$$

The two crystalline modifications of this polymer evidently correspond to conformations (I) and (II). Indeed, in one of them the identity period is 7.21 Å; whereas the calculated period of trans–trans–gauche–gauche combinations of the "nonshifted" conformation is near that value (6.86 Å). For the other modification the calculated and experimental periods are the same (8.40 and 8.41 Å respectively).

Finally, in polytetrahydrofurane ($m = 4$) only the T–T–T [76, 77] conformation was found, which seems to be the most favorable for close packing.

Summing up the analysis of the conformations of molecules of crystalline polymers, it is necessary to emphasize that in the majority of cases it is the intramolecular interactions that are the most important in determining the optimal structure. Valuable information about optimum conformations may be derived even from the nonbonded interactions alone. Some of the details of the structures of polymer chains may be understood, if the possibility of bond angle variations is taken into account. The role of the torsion term, and, consequently, of the staggered bond principle, seems to be but secondary: Almost always the energy minima of the central nonbonded interactions are close to the minima relevant to this principle, and, if these two factors ever compete, the first one usually wins. Finally, when the conformational freedom of the macromolecule is high, the effect of intermolecular interactions should not be overlooked. Therefore, calculations whereby the energy minimum is sought not only by varying the parameters of the polymer chain, but also those of the elementary unit are of special interest.

## 2. CONFORMATIONS OF PEPTIDES AND PROTEINS

During the last five or ten years theoretical calculations of conformations of peptides have been numerous and intensive; the number of papers published and ideas put forward in this field seems to be not less than that on the conformations of all small molecules taken together. The calculations pertaining to peptide conformations allow one to understand the nature of the intimate interactions in proteins, hence the great interest in peptides. Also, there is a hope, though not very strong, that the theoretical analysis developed for small peptide fragments will make it possible to predict the complete three-dimensional structures of globular proteins.

Four groups of investigators are engaged in the elucidation of peptide conformations: the groups of Ramachandran, Liquori, Flory, and Sheraga. Ramachandran and co-workers were the first to begin conformational calculations of peptides [78, 79], they found allowed and forbidden regions in the conformational maps [80, 81]. Liquori [82] was the first to apply atom–atom potentials to draw conformation maps. Flory [83] proved that electrostatic interactions should be taken into account in predicting the relative stability of the various conformations, and elaborated a method of calculating the flexibility of model peptides and copolymers. Scheraga has extensively studied the conformations of dipeptides and polypeptides using the method of hard spheres [84–88] and potential functions [89–91]; he has also made an attempt to find a method of calculating the optimal structures of cyclic peptides and irregular peptide fragments [92–94].

Recently two excellent reviews have been published analyzing the investigations of these four groups of theoreticians. The first, larger, review is by Ramachandran and Sasisekharan [95] and deals with the conformations of peptides and proteins. The second review, by Scheraga [96], analyzes only the calculations of oligo and polypeptides. Therefore we shall discuss in somewhat greater detail a number of papers dealing with the interactions of amino acid residues with each other and with the peptide chain. In addition, we shall consider the results obtained in the laboratory of Popov in Moscow which were not mentioned in the reviews [95, 96].

A polypeptide chain is a sequence of monomer units of the $-NH-C^\alpha HR-C'O-$ type called amino acid residues, which, when connected, form a regular polymer if $R^1 = R^2 = \cdots = R^n$, or an irregular polymer if the residues are not equivalent. As is clear from the general formula

$$(NH_2-C^\alpha HR^1-C'O)-(NH-C^\alpha HR^2-C'O)- \cdots -(NH-C^\alpha HR^n-C'OOH)$$

*Table 1*

STANDARD PARAMETERS OF A PEPTIDE UNIT

| | Bond angle (degrees) | Bond | (Å) |
|---|---|---|---|
| CC'N | 114 | C'–N | 1.32 |
| OC'N | 125 | N–C | 1.47 |
| CC'O | 121 | C–C' | 1.53 |
| C'NC | 123 | C'–O | 1.24 |
| C'NH | 123 | N–H | 1.00 |
| CNC | 114 | | |
| XC Y | 109.5 | | |

a polypeptide has an N-terminal (on the left) and a C-terminal (on the right). Proteins are irregular polypeptides. In natural polypeptides no branching is possible, since when a protein is being synthesized on a ribosome, peptidyl-*t*RNA adds every residue directly to the previous one.

As early as in 1953 Corey and Pauling [97] made a detailed analysis of all data available at that time on polypeptides and amino acids; they found the mean geometric parameters for a monomeric unit (Table 1). These parameters were confirmed by later structural investigations [98, 99]. Of course, the data of Table 1 are by no means absolute—in real molecules the bond lengths and bond angles may vary, but not significantly. The rotation angles around the bonds of the polypeptide chain are the most important conformational parameters, just as in the case of synthetic polymers.

According to the accepted nomenclature of polypeptide conformation [100], the rotation angles around the N–$C^\alpha$, $C^\alpha$–C', and C'–N bonds are designated $\varphi$, $\psi$, and $\omega$, respectively; these angles are measured from a fully extended trans chain ($\varphi = \psi = \omega = 0$); $\varphi = 0$ if the $C^\alpha$–C' bond is in the cis position with respect to the N–H bond and $\psi = 0$ if the $C^\alpha$–N and C'–O bonds are also in the cis position. The angles are counted clockwise, in the direction from the N-terminal to the C-terminal along the corresponding bond of the main chain.†

As the properties of the C'–N bond are intermediate between those of a single and double bond, the rotation around it ($\omega$-deformation) requires considerable energy. Hence, the two essential parameters are $\varphi$ and $\psi$, i.e., the

---

† Recently another way of conformational assignment was offered [100a]. Now only cisoid conformations (as in the case of synthetic macromolecules) are considered as corresponding to zero rotation angles, i.e., $\varphi = \psi = 180°$ for the extended form of Fig. 15. To take these rules into account, it is not necessary to redraw conformational maps; correcting the figures at the coordinate axis is sufficient.

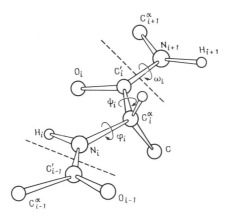

***Fig. 15.*** A fragment of peptide chain with accepted designations; broken lines cut off one residue.

angles of rotation around the $N-C^\alpha$ and $C^\alpha-C'$ bonds describe the conformation of dipeptide fragments with sufficient accuracy (Fig. 15). It should be also noted that the trans conformation of the $C^\alpha-C'O-NH-C$ group is more favorable by at least 2 kcal mole$^{-1}$ than the cis conformation; therefore, it exists in the open peptide systems and proteins.

A few words about the nomenclature of the conformations accepted for the R chain groups. The carbon atoms of the side chains, attached to $C^\alpha$, are designated as $\beta$, $\gamma$ etc.; the $C^\alpha-C^\beta$, $C^\beta-C^\gamma$, ... bonds as 1, 2, ..., the angles of rotation around these bonds as $\chi_1, \chi_2, ...$, these angles are again measured clockwise from $C^\alpha$ to $C^\beta$, from $C^\beta$ to $C^\gamma$, etc. But here the measuring is done from the eclipsed cis conformation. For example, for glutamine

$$C^\alpha \underline{\ \ 1\ \ } C^\beta \underline{\ \ 2\ \ } C^\gamma \underline{\ \ 3\ \ } C^\delta \Big\langle \begin{matrix} N \\ O \end{matrix}$$

the $\chi_1$, $\chi_2$, and $\chi_3$ angles correspond to rotations around the bonds 1, 2, 3.

The chemical formulas of amino acid residues are given in many textbooks on biochemistry and molecular biology, e.g., [101–103]. In [101] the authors give abbreviated designations for the residues and their space-filling models; designations of atoms, the numeration of bonds, and the rotation angles of side chains are given in the review of Scheraga [96]; Ramachandran and Sasisekharan [95] suggested extending the nomenclature with the aim of determining the coordinates of the hydrogen atoms also.

Knowing the geometric parameters of a monomer unit and the rotation angles, one may, by means of the matrix method described in the preceding section, express the coordinates of all the angles via the geometric parameters:

bond lengths, bond angles, and rotation angles. Ramakrishnan [104] was the first to do this for peptide chains. The transformation of atomic coordinates has become a standard procedure of computer calculations. Some details of such transformations are described in [83, 84, 91, 105].

The potential energy of a polypeptide may, with good accuracy, be presented as a sum of several terms:

$$U = U_{nonbonded} + U_{tors} + U_{el} + U_{Hb} + U_b + U_{ang} + U_{hydr} \qquad (2.1)$$

where $U_{nonbonded}$ is the energy of interaction of the nonbonded atoms; $U_{tors}$ is the torsion energy (i.e., an additional term to the interactions of the nonbonded atoms, which ensures better agreement between the calculated and experimental results); $U_{el}$ is the energy of the electrostatic interactions; $U_{Hb}$ is the energy of formation of the hydrogen bonds; $U_b$ and $U_{ang}$ are, respectively, the energy of deformation of bonds and bond angles; and, finally, $U_{hydr}$ is the hydration energy, i.e., the interactions between the molecules of dissolved compounds and those of the surrounding water or other solvent.

As is seen, there are many terms and, before discussing each of them, it should be mentioned which properties of the molecule depend on each term and to what extent.

The energy of interaction between the nonbonded atoms furnishes important information about the possible conformations of the polypeptide chain, namely, about "allowed" and "forbidden" regions in the space of the geometric parameters. Moreover, the precise positions of the minima in the space of these parameters is also determined by this component which, therefore, can be called the "number one" component. However, with regard to the relative stabilities of the different conformations, serious errors may be expected if one uses only the nonbonded interaction potentials.

The electrostatic interactions cannot, as a rule, change the positions of the minima, since they are functions that change slowly as the distances between the atoms change; but they may greatly increase or decrease the energy that corresponds to a certain local minimum. The nonbonded and electrostatic interactions play an approximately similar role in determining the relative stabilities of the different conformations.

Hydrogen bonds are stronger interactions than the attractions and repulsions between nonbonded atoms. Their energies depend to a greater degree on the nature of the solvent, because hydrogen bonds may arise between the molecules of the dissolved compound and the solvent.

The torsion energy is not great, but it has a definite (though small) effect on both the positions of the minima and the relative stabilities of the conformations.

Angular deformations in polypeptides do not play such an important role as in overcrowded polymers (e.g., as in polyvinylidene chloride). But, it should be born in mind that if the angular deformations are not taken into account, the energies of the nonbonded interactions corresponding to the local minima of the potential surface may be estimated incorrectly, and, in the regions between the local minima, higher energy values are then inevitable.

Changes in bond lengths may be safely neglected, because they require too much energy.

Finally, interactions with the solvent are not essential for small oligopeptides, but in polypeptides and proteins they can prove to be more important than all the other interactions taken together (if one speaks about the complete three-dimensional structure).

Let us now consider each term.

### a. Nonbonded Interactions

To calculate the peptide conformations both rigid spheres and the 6-exp and 6-12 potentials, as well as the potentials of Liquori, of Scott and Scheraga, of Flory, and of Dashevsky have been used. For a detailed analysis of atom–atom potentials see Section VII.1.

### b. Electrostatic Interactions

Any attempt to include the electrostatic energy into a potential function gives rise to a number of difficulties, and it is this component that is responsible for the greatest indeterminacy in the total energies of the different conformations. The first difficulty consists in the fact that the spatial distribution of electronic density in real molecules is continuous, and that since we do not know what it is, we are compelled to resort to rather rough approximations: dipole–dipole or monopole. The second difficulty is associated with the role of the solvent. The molecules of the solvent that penetrate into the space between the atoms of the peptide, or even those localized far from the peptide, change the electrostatic force field, and this change depends largely on the nature of the solvent. All these effects are difficult to take into account with high accuracy, and the approximations that will be described below may result in considerable error in determining the relative stabilities of different conformations.

It is known that the dipole moment of the amide group is 3.7 D [83]; those of the ester groups in aspartate and glutamate are 1.7 D, and that of the phenol group of tyrosine is 1.7 D. Knowing the directions of the dipole moment vectors, one could calculate the electrostatic energy in the dipole–dipole approximation. But, because these directions are not well known, and also because the dipole–dipole approximation gives wrong results when the interatomic distances are small, it is the monopole approximation that has found wider application recently. The electrostatic energy in this approximation is the sum of the Coulomb interactions of partial charges centered on the atomic nuclei

$$U_{el} = \sum_{i>j} \sum \frac{q_i q_j}{\varepsilon r_{ij}} \qquad (2.2)$$

where $q_i$, $q_j$ are the charges; $r_{ij}$ is the distance between atoms $i$ and $j$; $\varepsilon$ is effective dielectric permeability of the medium.

The partial charges are determined from the dipole moments of the bonds [38], which, in their turn, are determined from the experimental dipole moments of various compounds by means of an additive vector scheme. As is shown in Table 2, the values of the partial charges estimated by a number of authors do not differ very much. In the first three columns are the values (in $e$ units) for the atoms of the amide group obtained from the bond dipole moments; in the fourth column are the values calculated by the Hückel method with the parametrization suggested by Del Re [106] and also by Berthod and Pullman [107]; in the fifth column are the values recommended by Ramachandran [95].

*Table 2*

PARTIAL CHARGES ON PEPTIDE ATOMS

| Atom | I | II | III | IV | V |
|------|------|-------|-------|-------|------|
| C | +0.43 | +0.450 | +0.394 | +0.318 | +0.4 |
| O | −0.39 | −0.417 | −0.394 | −0.422 | −0.4 |
| N | −0.30 | −0.304 | −0.281 | −0.202 | −0.3 |
| H | +0.26 | +0.271 | +0.281 | +0.204 | +0.3 |

But, although one can somehow estimate the partial charges, it is almost impossible to estimate the effective dielectric permeability. Generally, $\varepsilon$ is determined by the atomic polarizabilities of the interacting atoms, the effect of the surrounding field, and the solvent. At small distances, $\varepsilon$ should be close to, but always more than unity. In reality, force lines pass mostly from atom to atom (in vacuum), but some of them will cross other atoms in the peptide and in the molecules of the solvent; as the distance increases, $\varepsilon$ should increase as well, but it is not clear to what value. For example, if water is the solvent, one may believe that even with great distances $\varepsilon$ should not reach the macroscopic value of 81.

It goes without saying that one should not ascribe to the factor $\varepsilon$ in formula (2.2) a specific physical meaning; it has nothing to do with the macroscopic dielectric permeability. As a matter of fact, $\varepsilon$ is a "fudge factor"; by varying it one can sometimes improve the agreement between the calculated and experimental results. The only correct way to present the problem is the following: In terms of the monopole approximation one should consider the interactions of point charges centered on the atoms of the dissolved compound and of the solvent. This requires averaging over all possible orientations and positions of the solvent molecules. Then, for interactions described by expressions of the type (2.2), $\varepsilon$ should be, naturally, equal to unity. Since this problem is practically impossible to solve, the only way out for theoreticians is to introduce some arbitrary value of $\varepsilon$ in (2.2) and to vary it depending on a solvent.

Brant and Flory [83], like other authors, considered $\varepsilon$ to be a constant independent on the distance between a pair of atoms and to have a value of 3.5 i.e., close to that of the high-frequency dielectric permeability of peptide. Scheraga et al. [85, 91] used values from 1 to 4 for $\varepsilon$. Ramachandran [95] suggested that $\varepsilon$ should be assumed to be equal to unity based on the successful calculations of the lattice energies of ionic crystals with the vacuum value of the dielectric constant.

The question of the value of $\varepsilon$ will be considered in detail for the case of alanine dipeptide.

So, from the above it is clear that the results of calculations are sensitive to changes in the parameters of the electrostatic term of the potential function; hence any changes should be made very carefully; in some cases one may vary the parameters within reasonable limits to find out how much the results depend upon the assumptions made.

## c. Hydrogen Bonds

The nature of the hydrogen bond has been widely discussed and it was not until recently that it was assumed to be a donor-acceptor, i.e., a weak chemical bond. Lippincott and Schroeder [108, 109] were the first to suggest a hydrogen bond potential that had a complex analytical form. Ooi et al. [91] slightly changed this potential and estimated the parameters for the N–H $\cdots$ O–C bond which often arises in peptides,

$$U_{\mathrm{Hb}} = A \exp(-bR) - (A/2)(R_0/R)^6 \exp(-bR_0)$$
$$- D^* \left\{ \left[ \frac{1+6^{1/2}\cos\theta_1}{1+6^{1/2}} \right]^2 + \left[ \frac{1+6^{1/2}\cos\theta_2}{1+6^{1/2}} \right]^2 \right\} \exp\left[ \frac{-h^*(R-r-r_0^*)^2}{2(R-r)} \right]$$
$$(2.3)$$

Here $A = 4{,}941 \cdot 10^6$, $b = 4.8$, $D^* = 82.4$, $h^* = 13.15$, $r = 1.01$, $r_0^* = 0.97$, $R_0 = 2.85$, $R$ is the N–O distance, $\theta_1$ and $\theta_2$ are the angles formed by the H $\cdots$ O direction and those of the unshared pairs of the oxygen atoms.

The number of parameters included in this potential is unsatisfying; another objection is the behavior of this function when $R$ increases, i.e., when $R \to \infty$, $U \to -2$ kcal mole$^{-1}$. Therefore, Lipkind et al. [110] suggested for the hydrogen bond potential the well-known Morse curve

$$U_{Hb} = D[1 - \exp(-n\Delta r)]^2 - D \qquad (2.4)$$

where $D$ is the dissociation energy of the hydrogen bond, $\Delta r = r - r_0$, $r_0$ is the equilibrium $H \cdots O$ distance, $n$ is an empirical parameter.

Formula (2.4) seemingly does not include an angular dependence of the hydrogen bond energy, which is usually believed to be very important. But, as was shown by Kitaigorodsky [111], by taking into account nonbonded interactions we automatically take into account the angular dependence of the hydrogen bond energy (see also Chapter II). Indeed, when the four atoms N–H$\cdots$O–C deviate from a straight-line, strong N–O interaction arises, to say nothing about the interactions with other atoms.

For the equilibrium distance $r(H \cdots O)$ the value of 1.80 Å has been accepted based on X-ray data [112]; for $D$, the value of 4 kcal mole$^{-1}$, from thermochemical estimates (if hydrogen bonds are formed in organic solvents, like $CCl_4$); finally, $n = 3$ Å$^{-1}$, from spectroscopic data. The curve is shown in Fig. 16. As $r$ increases, the curve asymptotically tends to zero; in the 2.7–

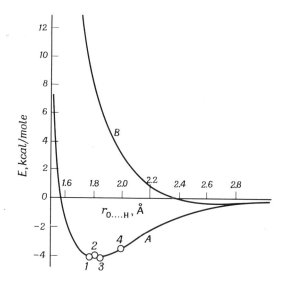

**Fig. 16.** O$\cdots$H potentials of the hydrogen bond (A) and nonbonded interactions (B). Points correspond to experimental data: 1-N-methylacetyl amide, 2-N-methyl-formamide, 3-acetoamide (E. M. Popov, V. G. Dashevsky, G. M. Lipkind and S. F. Arkhipova. *Molek. Biol.* **2**, 612 (1968)).

2.9 Å range it practically coincides with the curve of nonbonded interaction (B). This fact allows one to use the (A) curve as a universal $O \cdots H$ potential in peptides. The available experimental data on the lengths of hydrogen bonds in amides [113] are in the region of the minimum of this curve. In addition, some experimental results indicate the absence of an activation barrier in the hydrogen bond if the $O \cdots H$ distances are between 2.2–2.45 Å [114].

However, one should bear in mind that hydrogen bonds, like electrostatic interactions, are responsible for great indeterminacies in energy, because their contributions to the total energy depends greatly upon the nature of the solvent. Also, it is not yet clear if at least a portion of the electrostatic interactions should be included in the effective potential of the hydrogen bond.

### d. Torsion Potentials

In peptides there are three bonds for which potentials of this type are necessary: $N-C^\alpha$, $C^\alpha-C'$, and $C'-N$ (the respective rotation angles: $\varphi$, $\psi$, and $\omega$). In addition, for the side chains one should introduce potentials for rotations around the $C-C$ and certain other bonds (the rotation angles $\chi$).

Unfortunately, it is impossible to measure the internal rotation barriers in peptides with sufficient accuracy, and one has to resort to analogy with small molecules. Scott and Scheraga [89] suggested the following parametrization

$$U(\varphi) = (U_\varphi/2)(1+\cos 3\varphi) \qquad (2.5)$$

$$U(\psi) = (U_\psi/2)(1-\cos 3\varphi) \qquad (2.6)$$

$U_\varphi = 0.6$, $U_\psi = 0.2$ kcal/mole. These authors also used $U_\varphi = 1.5$ kcal/mole; it was shown that even if the maxima and minima are exchanged (the plus and minus signs in expressions (2.5, (2.6)), the main conclusions will not be affected. Hence, the $U$ values of Brant and Flory [83] ($U_\varphi = 1$ kcal mole$^{-1}$, $U_\psi = 1.5$ kcal/mole) and the minus in (2.5) can be justified; it is also understandable why Popov *et al.* [110], after having analyzed the role of the torsion terms in the optimal conformations, did not include them at all.

As to side groups, it is obvious that for the aliphatic ones the following expression is valid:

$$U(\chi) = (U_\chi/2)(1+\cos 3\chi) \qquad (2.7)$$

and $U_\chi = 3$ kcal/mole. A more detailed analysis of the torsion terms can be found in the review of Sheraga [96].

Special attention should be paid to rotation around the $C'-N$ bond, or, to be more exact, the energy required for the $C^\alpha-C'-N-C^\alpha$ peptide group to lose its planarity. As was shown in a number of papers published recently by Ramachandran [115–117] $\omega$-deformations are not only possible, but actually inevitable in small cyclic peptides; no cyclization can occur without them.

For the torsion energy associated with changes in $\omega$, it is reasonable to assume the expression:

$$U(\omega) = (U_\omega/2)(1 - \cos 2\omega) \tag{2.8}$$

where $U_\omega = 20$ kcal mole$^{-1}$ (cf. formula (5.18), Chapter VII). It is clear that nonplanar deformations of up to $10°$ correspond to an increase in energy of only 0.6 kcal/mole.

### e. Bond Angles' Deformations

Using expression (1.7) from Chapter VII and spectroscopic data [118, 119] one can obtain a value of 80 kcal mole$^{-1}$ rad$^{-2}$ for the average $C$ value for peptides. This same value was also recommended by Ramachandran [95]. But, as has been indicated above, one should not use spectroscopic deformation constants together with nonbonded interaction potential. The authors of [110] give values of the elastic constants sufficient for calculations of peptide conformations, i.e., for tetrahedral carbon, 30; for trigonal carbon, 70; and for nitrogen, 50 kcal mole$^{-1}$ rad$^{-2}$. Calculations taking into account the deformation energies of bond angles allowed Gibson and Scheraga [120] and Lipkind et al. [121] to reveal some interesting facts.

### f. Hydrophobic Interaction Energy

It is well known that it is these interactions that determine the three-dimensional structures of proteins [123]. Native protein exists in aqueous solution; therefore the polar amino acids are localized at the surface of the globule and interact with the water, whereas the nonpolar ones are hidden inside the globule, and come into contact with each other. A similar situation may occur in polypeptides; if the solvent is polar (water), the polar amino acids tend to be outside; in nonpolar solvents the opposite picture should be expected: the nonpolar amino acids will tend to get to the surface.

What is, then, the physical nature of these hydrophobic interactions? Let us consider a cube which contains 1000 molecules of water and two molecules of methane (the volume of the cube is such that its average density is close to 1 gm cm$^{-3}$). Probably, one can demonstrate, operating only with atom–atom potentials, that the minimum of free energy of such a system will correspond to the configurations in which two methane molecules are not in different regions of the cube, but are in close contact with each other. As a matter of fact, thermodynamic measurements show that dissolution of one molecule of methane in water requires several kilocalories per mole. It is obvious that for two molecules the loss of free energy of solution will be less if these molecules are in contact and, consequently, occupy less volume. Hydrophobic residues in proteins, such as alanine [R=CH$_3$], valine [R=CH(CH$_3$)$_2$], etc. are like methane in our example; consequently, the above is also true for them.

It goes without saying that the energy of the hydrophobic interactions should not be considered as one of the equal terms of the potential function because it is a contribution to the free energy of the system polypeptide–solvent. But, if one is interested in the conformation of a polypeptide, one may make an attempt to find appropriate expressions to add to the potential function of a peptide to take into account, roughly, the tendency of nonpolar residues to come into contact in aqueous solution.

In [123, 124] it was shown that it is interactions with the nearest molecules of the solvent that make the main contribution to the free energy of the system polypeptide–solvent. Roughly speaking, if $d$ is the diameter of a water molecule, and the distance between a given pair of atoms is $r < d + r_0$ ($r_0$ is the sum of the van der Waals atomic radii), the water molecules are forced out, and the contribution to the free energy becomes equal to zero. On the other hand, if one atom is brought close to another, the former will force out a definite number of solvent molecules proportional to the volume of this atom, $V$; but, when the distance becomes less than $d + r_0$, the quantity of the solvent driven out essentially does not increase. Reasoning of this kind made Gibson and Scheraga [94] seek analytical expressions for the hydration energy. Out of the abundant data reported in this paper we would like to mention here that calculations of the conformations of the cyclic peptides, oxitocin, vasopressin, and the cyclic octapeptide of ribonuclease [94] showed that the contribution of the hydration energy to the total energy is not small; but the search for the minimum without taking this energy into account results in the same optimal structure. This fact, obviously, shows that the molecules in question do not belong to systems containing external and internal sites; hence, unlike globular peptides, in these structures the hydrophobic interactions do not play a decisive role.

Now let us discuss the conformations of peptides. Fragments consisting of two peptide units, or the corresponding molecules with methyl groups at the ends, will be called dipeptides (see Fig. 15).

Almost all the most interesting and important features of the structures of polypeptides and proteins that depend on the mutual positions of the separate atoms (but not those of polyatomic structures, such as the $\alpha$-helices or $\beta$-structures) may be revealed by analyzing, first, the conformations of dipeptides, and, secondly, the interactions of adjacent residues with each other, and with the peptide chain. This is why we are going to describe in detail some conformational calculations of dipeptides and certain fragments that include some of the above interactions. Unfortunately, space does not permit discussion of many interesting features of the spatial arrangements of the various helical structures of polypeptides, the $\beta$-folded form, some fragments of lysozyme, myoglobin, etc. All these data may be found in the review by Ramachandran and Sasisekharan [95], which is little less than encyclopædic.

We are going to dwell on the problems pertaining to the allowed and forbidden regions in the space of the parameters describing the geometry of molecules, the forms of the potential wells, and the relative stabilities of various conformations. Of course, if some regions are forbidden in dipeptides, they will be forbidden in polypeptides and proteins. Furthermore, there is no difference in the forms of the wells and the relative stabilities of conformations of dipeptides and of the corresponding conformations of fragments of large molecules; hence, what is true for dipeptides will be true for other molecules.

The conformations of dipeptides are determined by the nature of the side radical R. It is reasonable to analyze in detail only the four radicals: glycine $(R = H)$, alanine $(R = CH_3)$,† valine $[R = CH(CH_3)_2]$ and proline; the latter incorporates into the peptide chain to form the group shown in (V).

$$\begin{array}{c} \underset{\text{H}}{\overset{\gamma\text{C}^{\text{H}_2}}{}} \\ \text{H}_2\text{C}^{\delta} \qquad {}^{\beta}\text{CH}_2 \\ -\text{CO}-\underset{\text{H}}{\text{N}}-\text{C}^{\alpha}\text{CO}- \end{array}$$

(V)

On the addition of glycine the chain retains its symmetry and, as in the case of polymers with symmetrical side chains (see the preceding section), the conformational map of glycine dipeptide (or a molecule of N-acetylglycine methylamide) should be centrosymmetric (the center of symmetry is at $\varphi = \psi = 180°$). Alanine is the residue with a $C^{\beta}$ atom having the least volume. Its map should have no symmetry elements and, due to the $CH_3$ group, the conformational freedom of the corresponding dipeptide is restricted as compared with that with glycine. The greater the volume of the side chain, the less the conformational freedom; it is least for the large valine radical.

Other amino acids are intermediate in this sense, and are close to alanine; only isoleucine is an analogue of valine. Finally, proline and hydroxyproline, another imino acid, have a specific feature: Their conformation has only one rotation angle, $\psi$.

Dipeptides may have both stretched (without H bonds) and bent forms. One of the bent forms, the form of Midzusima (M), corresponds to formation of the hydrogen bond $N-H\cdots O2-C2$ (the indices 2 mean that the atoms belong to the second peptide unit from the N-terminal) and has the angles $(\varphi, \psi) \approx 120°$ and $240°$; the other, Huggins form (H), has the same hydrogen bond but the $(\varphi, \psi)$ angles are $\approx 240°, 120°$.

---

† Unless otherwise stated, L-amino acids, i.e., those occurring in nature, are meant.

Bent forms are usually more favorable from the point of view of enthalpy (especially in organic solvents), but less favorable as far as entropy is concerned (the region where formation of H bonds is allowed is rather small). As is proved by experiments, in nonpolar solvents bent forms are usually in equilibrium with stretched ones (the percentage of the former in polar solvents is negligible due to weak H bonds). Conformations corresponding to bent forms of dipeptides do not occur in polypeptides and proteins; hence, it is the stretched forms that are of interest.

### g. Glycine

Figure 17 shows the conformational map of glycine dipeptide [96]; the authors use allowable limits for the interatomic contacts (Ramachandran, Table 1, Chapter 7) and Corey and Pauling's parameters (Table 1). Completely allowed regions amount to 45% of the whole space, 61% are within the extreme limits. Almost all the points corresponding to glycine residues in peptides fit into these regions.

Figure 18 shows the conformational map of N-acetylglycine methylamide [121]. This is the same dipeptide, but with methylated terminals: Instead of

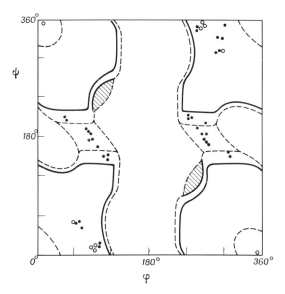

*Fig. 17.* Conformational map of glycine dipeptide. Completely allowed conformations are in the regions shown by the solid line, partially allowed by the broken line; circles correspond to conformations found in small unclosed peptides containing Gly; black dots show conformations of Gly in some cyclic peptides (G. N. Ramachandran and V. Sasisekharan. *Adv. Protein Chem.* **23**, 283 (1968)).

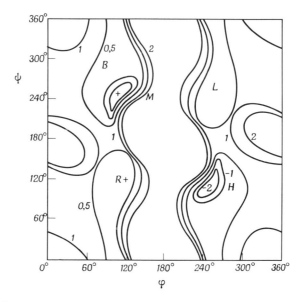

**Fig. 18.** Conformational map of N-acetylglycine methylamide. Nonbonded interactions and hydrogen bonds are taken into account (E. M. Popov *et al. Molek. Biol.* **2**, 622 (1968)).

$C^\alpha$ atoms there are $CH_3$ groups. This map was drawn not with the Corey–Pauling parameters, but with bond angles estimated by minimization of the potential function by varying independent geometric parameters. Comparing Figs. 17 and 18, one can see that the level lines in the latter, especially the 1 kcal/mole line, are quite similar to the allowable limits in Ramachandran's map. For this and for the other molecules described in [121] the conformational energy corresponding to the righthanded $\alpha$-helix (R) is assumed to be zero; in our case it is $\varphi = 115.9°$, $\psi = 122.1°$. The region of $\varphi \approx 9°$ and $\psi \approx 180°$ corresponds to an energy slightly above 1 kcal mole$^{-1}$ and, hence, is quite permissible. (In Ramachandran's map this region is forbidden, but some experimental points get into it.) The space within the 1 kcal mole$^{-1}$ contour occupies 60% of the whole map; so it is not surprising that this contour almost coincides with the extreme limits encircling 61% of Ramachandran's map.

Bent forms, which are symmetrical in the case of glycine, possess the lowest energy. Of course, if the H bond had not been taken into account in the calculation, there would have been no minima corresponding to such bent forms.

*h. Alanine*

Let us consider the conformational map of alanine dipeptide. In Fig. 19 one can see the normal and extreme limits of Ramachandran and the 1 kcal mole$^{-1}$ potential energy contours [96]. It is clear that the incorporation of the $C^{\beta}$ atom greatly hinders the conformational freedom of this dipeptide. The allowed region now covers only 8% of the whole area of the map, and the extreme limits include only 23%. The extreme limits are close to the 3 kcal mole$^{-1}$ contour.

The map of alanine dipeptide has three allowed regions which are designated as R, B and L. R corresponds to the righthanded α-helix which is apparently inherent in polypeptides and proteins, although R sometimes contains other types of helices. The large B region contains the β-structure, which is typical of many proteins and of some other structures, e.g., β-keratin, and silk. Finally, the region of the lefthanded α-helix, L, is partially allowed. It is clear from Fig. 19 that it is due to the asymmetry of the amino acid residues that the righthanded α-helix seems to be more favorable than the lefthanded one; this is confirmed by extensive experimental evidence. Of course, for right amino acid residues (at least the nonpolar ones, like alanine) the lefthanded helices would be more favorable.

Figure 20 shows the conformational map of N-acetylalanine methyl-amide

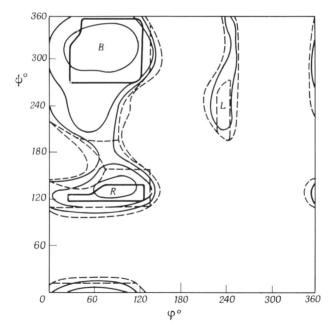

*Fig. 19.* Conformational map of alanine dipeptide. Allowed (solid line) and partially allowed (broken line) regions are shown and also energy contours calculated with atom-atom potentials (G. N. Ramachandran and C. M. Venkatachalam, and S. Krimm. *Biophys. J.* 6, 849 (1966)).

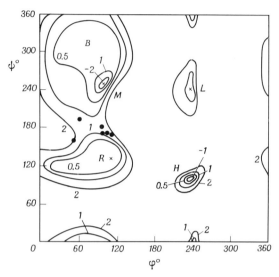

**Fig. 20.** Conformational map of N-acetyl-L-alanine methylamide (electrostatic energy is not taken into account).

made with Dashevsky's potentials and those of Morse for H bonds ($D =$ 4 kcal mole$^{-1}$). Besides the above three regions, there are two bent forms, one of which, M, is on the boundary of the B region, the other, H, is in the absolutely forbidden region. Yet, such a conformation is possible in non-polar solvents due to deformations of the bond angles and the formation of H bonds; this conformation has been proved experimentally [125].

It is seen in the map in Fig. 20 that R has somewhat less energy than B (the absolute minimum is shown by a cross). Since B has a larger area, these two forms may be considered equally probable. Now it is understandable why the frequency of occurrence of these two forms in polypeptides and proteins is about the same.

Similar maps have been drawn by Flory *et al.* [126, 129] using his own potentials; in their maps the absolute minimum also corresponds to the R region. But if one takes into account the electrostatic energy (see charges in Table 2 and $\varepsilon = 3.5$), the absolute minimum will be shifted to the B region. The reason for this is that the dipoles of the two peptide groups in conformation B are antiparallel and they attract, whereas in R they repel one another.

Lipkind *et al.* [128] using general considerations and comparing the experimental and calculated results, made an attempt to find a parametrization which would adequately describe the conformations of alanine dipeptide in various solvents. They varied only the values of the constant $\varepsilon$ and the depth of the well, $D$, in the H-bond potential; the other parameters remained

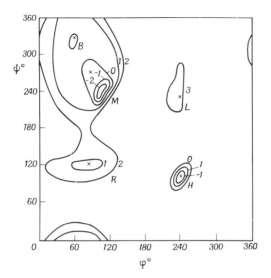

*Fig. 21.* Conformational map of N-acetyl-L-alanine methylamide for nonpolar solvents.

constant: Dashevsky's nonbonded interaction potentials and elastic constants, the torsion potentials, and the partial charges of Scheraga.

Three solvents were considered, $CCl_4$, $CHCl_3$, and water. The first solvent is nonpolar and forms no hydrogen bonds with peptide atoms; the second solvent is a polar one; it forms weak hydrogen bonds; water is an extremely polar solvent and the bonds it forms with a peptide are almost as strong as those of the peptide itself.

It is obvious that the effect of the solvent is not directly accounted for. Actually, one uses only the empirical parameters $\varepsilon$ and $D$ to improve the agreement between the calculated and experimental results. This procedure does not have a clear physical meaning; but it is useful for explaining and predicting some experimental data.

In [128] the dissociation energy $D$ of the hydrogen bond was assumed to be equal to the energy of dimerization of N-methylacetamide, which, according to the authors of [129] is 4 kcal/mole in $CCl_4$, 2 kcal mole$^{-1}$ in $CHCl_3$ and about zero in water (in the calculations it was 0.5 kcal mole$^{-1}$ for water). Furthermore, $D$ was assumed to be 4 in $CCl_4$, 10 in water,† and 6–7 in $CHCl_4$.

Figure 21 shows the map of alanine dipeptide in $CCl_4$. One can see a wide B region, with a lower energy than the R region, and an especially deep minimum in the M region. Hence, the conclusion that a nonpolar solvent should contain mostly a bent form, a few stretched forms of the B type, and there

---

† The same value was used in the paper of Krimm and Mark [130].

should be no R conformations at all. Calculation of the thermodynamic functions in accordance with the conformation map resulted in 60–70% of bent forms, a figure in good agreement with the experimental one.

The conformation of alanine dipeptide in water is readily visualized from an analysis of the map in Fig. 20. (In fact, in water the electrostatic interactions do not greatly influence the general picture, and this map does not take them into account.) The minima corresponding to bent forms should not be taken into consideration in this case. First, in water, according to the calculated and experimental results, the quantity of bent forms is practically equal to zero; secondly, the energies of the R and B forms are equal and the areas of the corresponding regions are comparable.

$CHCl_3$ is in an intermediate position; the proportion of bent forms is not great, but is not equal to zero. IR spectroscopy gives a figure of 25%, whereas the calculated value is 20%.

The examples considered above show that the relative stability of the various conformations largely depends on the choice of parameters and, primarily, on the parametrization of the electrostatic component of the potential function. However, many authors do not specify to what solvent their parametrization applies.

The map of alanine dipeptide is the most suitable for an analysis of the dihedral angles $\varphi$ and $\psi$ in known proteins. (Tables of $\varphi$ and $\psi$ values are given in review [96].) This is because all their residues, except glycine and proline, which is rather special, contain $C^\beta$ atoms which hinder the conformational freedom of each dipeptide fragment. In [96, 131] the following analysis was made: The points corresponding to some fragments of lysozyme and myoglobin were plotted on the maps; it turned out that the majority of these points really are in allowed or partially allowed regions. In the map of Fig. 19 all the points are within the 3 kcal mole$^{-1}$ contour and in the map of Fig. 20, the 2 kcal mole$^{-1}$ contour.

To sum up all this discussion of the conformation of alanine dipeptide it should be said that conformational maps are very important for the elucidation of the structure of proteins. However, they assist in interpreting rather than predicting, since first, their prediction of relative stability is rather questionable, and, secondly, the three-dimensional structures of proteins *in toto* depend also on distant interactions, although the interactions of adjacent peptide units, as has been shown above, diminish the number of possibilities.

### i. Valine

Valine and isoleucine limit the freedom of rotation of a peptide chain to a greater degree than any other residues, because the side radical branches off as early as at the $C^\beta$ atom. (In this sense one may compare them with the isotactic polymers which branch off at the first atom of the side chain, so that,

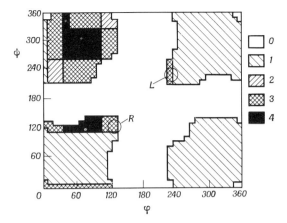

**Fig. 22.** Allowed regions of the conformational map for various dipeptides. Conformations for Gly are allowed in regions 1–4; for Ala, in regions 2–4; for other $C^\beta$-containing residues, in regions 3–4; for Val and Ileu, in region 4. Circles indicate R and L regions, corresponding to standard $\alpha$-helix.

owing to steric hindrances, in the crystal the macromolecule acquires a $4_1$ helix conformation instead of $3_1$.)

In [88] allowed and forbidden regions in the dipeptide conformational map were found for some residues and, among them, valine (Fig. 22). The area occupied in this map by valine (or isoleucine) is very small. As was mentioned above, other residues are intermediate between alanine and valine, but, as one can see in Fig. 22 and will see in the case of phenyl alanine, they resemble alanine more; the reason for this is evident—no branching at the $C^\beta$ atom.

Figure 23 shows the conformational map of N-acetyl-L-valine methylamide [121] calculated with the Dashevsky potentials without the electrostatic component. There are, as in the previous cases, five conformations: R, B, L, M, and H. It is interesting that here, as in alanine dipeptide, the B form has no minimum of its own; from any point of the B region a search for a minimum leads to a point corresponding to the M conformation.

The conformational map in Fig. 23 shows that the energies of the R and B forms are approximately equal, and that the electrostatic component would have decreased the energy of the B form. It should be noted that the relative stability of the different conformations depends not only on the electrostatic energy but also on the freedom of movement of the side radicals. Greater freedom of the side group corresponds to greater entropy and, consequently, to a gain in free energy.

Galaktionov [132] had calculated the free energy of the valine side radical $(-RT\ln z)$ at all $(\varphi,\psi)$ values of the conformational map (Fig. 24). This unusual map shows that the R conformation has limited freedom whereas the

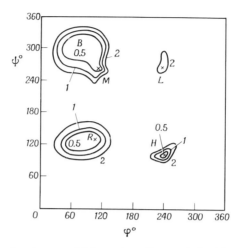

*Fig. 23.* Conformational map of N-acetyl-L-valine methylamide; electrostatic interactions are not taken into account.

B conformation has great freedom; there turns out to be a 0.5 kcal mole$^{-1}$ free energy gain for the B form. The same conclusion was made by Ooi *et al.* [91], who studied poly-L-valine. The greater stability of the B form accounts for the stretched conformation of poly-L-valine. However, in proteins, due to the small energy difference between the R and B forms, a valine residue may readily be incorporated into the helix.

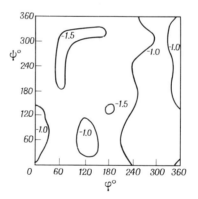

*Fig. 24.* Free energy levels of valine side radical in the conformational map $(\phi, \psi)$.

Thus, most amino acid residues are similar to alanine in that the peptide chain has limited freedom (but not in the sense of the relative stability of conformations). This is also demonstrated by the detailed study [133] of N-acetyl-L-phenylalanine conformations. The map shown in Fig. 25 is drawn

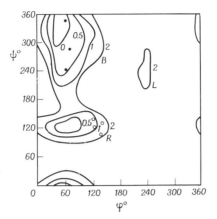

**Fig. 25.** Conformational map of N-acetyl-L-phenylalanine methylamide with optimal conformations of side radical. Points in the map correspond to conformations of phenyl alanine residues in lysozyme.

so that every point in it has a corresponding optimum conformation of the side radical. Out of nine possible conformations responsible for a local minimum, the optimum conformation was chosen in all cases. The authors used parameters meant for aqueous solutions (see above). It is obvious that this map is similar to that in Fig. 20, if the bent forms are left out of consideration in the latter. The maps of N-acetyl-L-tyrosine and probably, tryptophan methylamides, must be also similar to the above maps. Unfortunately, it is difficult to predict the relative stability of the R and B forms. As we have just seen, not only the parametrization, but also the freedom of movement of the side radical is of importance here.

*j. Proline*

In proline rotation around the $N-C^{\alpha}$ bond is impossible, and the $C^{\alpha}-C'$ angle is the only conformational parameter. The energy versus $\psi$ curve for L-acetyl-L-proline methylamide (Fig. 26) reveals two wells, one of which corresponds to the stretched form ($\psi = 132°$) and the other to the bent form ($\psi = 266°$). The $\varphi$ dihedral angle is approximately 120°. In this molecule the bent form is more favorable (of course, in nonpolar solvents), but in polypeptides and proteins, certainly, stretched forms are realized.

Theoretical calculations make it possible to predict not only the conformations of small fragments, like dipeptides, but also those of rather big molecules: cyclic oligopeptides, irregular oligopeptides possessing biological activity, and also polypeptides. Besides, one can predict the structures of

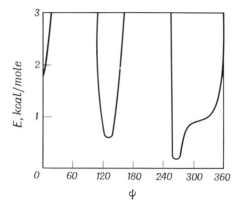

**Fig. 26.** Dependence of conformational energy of N-acetyl-L-proline methylamide on the $\psi$ angle (E. M. Popov *et al.*, *Molek. Biol.* **2**, 622 (1968)).

fibrillar proteins whose amino acid sequences are close to regular. These questions are discussed in detail in reviews [95, 96, 134].

Let us consider, by way of example, a rather complex depsipeptide cycle†: enniatine B (Fig. 27) which is a transmembrane ion-carrier. In [136] it was experimentally proved that the molecule of this six-member ring may have, depending on the polarity of the medium, two conformations, "nonpolar", N; and "polar," P. Conformational calculations [136] unequivocally predict these two structures and help to analyze extensive experimental evidence on this compound (including its NMR spectra).

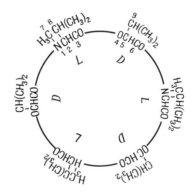

**Fig. 27.** Structural formula of enniatine B.

---

† Depsipeptides are intensively studied in the Institute for Chemistry of Natural Compounds of the USSR Academy of Sciences. See, e.g., [135]; their conformations are theoretically calculated in the laboratory of Popov.

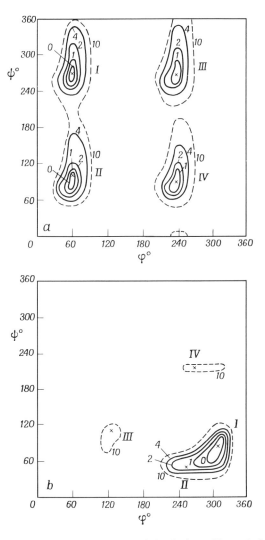

*Fig. 28.* Conformational maps of fragments of the depispeptide enniatine B: a, map of N acetyl-N-methyl-L-valine; b, map of O-acetyl-D-L-oxyisovaleric acid dimethylamide.

Conformational maps of N-acetyl-N-methyl-L-valine (I) and O-acetyl-D-L-oxyisovaleric acid dimethylamide (II) which are models of two peptide fragments of enniatine B, are shown in Fig. 28. Fragment (I) may have four conformations of approximately equal energy, whereas fragment (II) has only one conformation. To search for the optimal conformation of the whole enniatine B molecule it is not necessary to minimize the potential function over many variables: It is sufficient to build skeletal models with allowed conformations [combinations of the four conformations of fragment (I) and the one conformation of fragment (II)]. Analysis of these models shows that the molecule may really have two conformations if the six-member ring is closed. One of them is the "polar" form shown in Fig. 29 together with a $K^+$ ion incorporated into the molecule and forming coordination bonds with oxygen atoms. If the ions are absent, the molecule of enniatine B acquires a different conformation, a "nonpolar" one. Thus, both the calculations and the experimental data (not presented here) show that it is the conformational rearrangements that are responsible for the ion transport.

Nevertheless, the successful prediction of the enniatine B structure is an exception. The problem of numerous minima is the main difficulty encountered by theoreticians engaged in conformational calculations of irregular oligopeptides, to say nothing of proteins. If for every peptide unit the three conformations, R, B, and L, are possible, in a decapeptide there can be $3^{10}$ conformations. In cyclic molecules the number of conformations is somewhat smaller, since it is limited by the cyclization conditions; it is much smaller in symmetrical cyclic peptides (gramicidine S or enniatine B described above), but is still sufficient to make the problem of the search for the optimal structure extremely difficult.

This is exemplified by gramicidine S (L-Val-L-Orn-L-Leu-D-Phe-L-Pro)$_2$. An X-ray study [137] of this molecule revealed it to have a twofold symmetry axis. Unfortunately, this structure was not fully elucidated; however, the geometric model was suggested on the basis of some indirect data. Ten years later Liquori [138] and Scheraga [92, 96] used the stereochemical data (allowed contacts) to obtain their structures. Scheraga had to choose from 282 different conformations by minimizing potential functions.

All the three structures suggested are quite different. Balasubramanian [139]

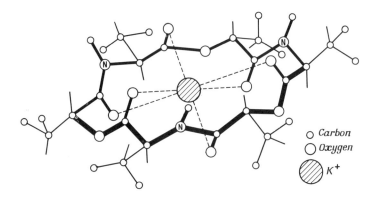

*Fig. 29.* Three-dimensional model of enniatine B-potassium ion complex.

analyzed the extensive experimental evidence on gramicidin S and came to the conclusion that neither Liquori's, nor Scheraga's structure can account for all of what is known about this compound.

In spite of the fact that nonlocal methods of search for minima have found wide application during recent years [140–142], it is scarcely possible to solve the problem of determining the absolute minimum of functions with 15–20 essential variables. The trouble is that as the number of variables increases the computer time required for seeking the absolute minimum grows exponentially, as does also the time required for calculating the potential function. Therefore it is more expedient to confine oneself to the most favorable conformations of very small fragments (dipeptides or tripeptides) according to the given primary structure. Then the optimal structure can be chosen from several versions by searching local minima.

Furthermore, one should bear in mind that not only the depth, but also the shape, of the well is important when choosing the optimal structure, since this choice depends not on the enthalpy, but on the free energy; the shape of the well gives the entropy portion of the free energy.

But even if the best versions have been found, the problem of the relative stability of the various conformations remains a stumbling block. As we have seen, the error in relative stability calculations, caused by all kinds of approximations and parametrizations may amount to 0.5 kcal mole$^{-1}$ per peptide unit. That will mean 5 kcal mole$^{-1}$ per molecule of a decapeptide, which is enough to make the choice of the optimal structure difficult. It is quite probable that it will be more difficult to solve the problem of the conformations of medium-size oligopeptides that to predict the structures of globular proteins, since, in the latter case, some general principles may suggest themselves.

Let us now discuss the conformations of the globular proteins, large irregular peptides. As has been shown above, the conformational maps of dipeptides furnish a good explanation of the distributions of the rotation angles in myoglobin and lysozyme. Now we shall discuss the considerations and conformational calculations which help predict the structures of macromolecules.

It was suggested long ago that the primary structure of protein, its amino acid sequence, determines its three-dimensional structure [143, 144]. In recent years the structures of three proteins: myoglobin, lysozyme and α-chymotrypsin have been unraveled. X-ray study has made it possible to reveal the secondary and tertiary structures of four or five other proteins; therefore, many investigators try to establish a correlation between the amino acid code (i.e., the composition and sequence of the amino acid residues) and the conformational code (i.e., the three-dimensional structure).†

---

† By the middle of 1972 the three-dimensional structures of more than 25 proteins had been elucidated.

Attempts to find this correlation are limited, due to the difficulties, at present insuperable, which make prediction of a complete protein structure from its amino acid sequence scarcely possible. Therefore it seems useful to elucidate some of the details of a protein structure associated with its amino acid code. Of great interest is the theoretical estimation of the percentage of helical structure, and the establishment of the helical and nonhelical sites. There are grounds to believe that in the future such an approach will allow one to predict structures of proteins with a large fraction of α-helices or β-structures. In this case the problem of prediction of protein structure may be divided into two parts: first, determination of the helical and nonhelical sites, then one chooses such mutual positions for the helical and nonhelical sites as give a minimum energy for a protein with the secondary structure so formed.

For prediction of the secondary structure it is natural to make use of the known protein structures. Knowing which residues are encountered in the helical and nonhelical sites, and having analyzed the available structural evidence, one can obtain some quantitative characteristics indicating the ability of some residues or their combinations to make or disintegrate regular sites of the secondary structure. The figures obtained may be used to predict the secondary structures of other proteins.

### k. Statistical Approach

In some papers [145–149] the experimental data or a statistical analysis of protein structures are used to distinguish between "helix-making" and "helix-breaking" residues. For example, in [145] it was shown that proline randomly incorporated into a helical polypeptide completely despiralizes it as soon as its content reaches 8 %. In [146] the residues, Ser, Thr, Val, Ileu, Cys were assigned as antihelical, based on the experimental data. Cook [148] made statistical calculations for whale myoglobin, the α- and β-chains of horse hemoglobin, and the albumen lysozyme, and pointed out that Ala, Leu, Val are helical residues, while Arg, Asn, Pro are antihelical ones. Recently Ptitsin [150] made statistical calculations for six proteins, i.e., about 1000 residues. Besides the above three proteins analyzed by Cook, he also included bovine ribonuclease A and α-chymotrypsin; as a result, the helical residues were identified to be: Leu, Ala, Glu, and the antihelical to be: Thr and Asn; it is interesting that the fact that two homologous proteins—(globins)—were left out of account did not influence the results.

There are some papers in which more complex criteria are suggested for the helical and nonhelical sites in proteins. The Italian authors [157, 152] introduce for every residue (based on its environment) the so-called helical potentials. Statistics pertaining to the ability of some residues, or their combinations, to form helices is developed in [153–156]. For example, Prothero

[153] suggested a rule that any site of a polypeptide chain consisting of five amino acids will be helical if three residues in it are Ala, Val, Leu or Glu.

Unfortunately, the results of statistical analysis of the data on several proteins are not satisfactory. For example, all theories predict a great percentage of helical structure for chymotrypsin, however, it contains only 4% of helical structure. Things are still worse in the case of $\beta$-structures; no generalisations can be made in this case. The reason for this is, probably that the available experimental data are insufficient for reliable statistics; secondly, that all the calculations resulting in this or that criterion apply to homologous proteins—(globins) whose secondary structures are, as yet, unrevealed, and last but not least, that secondary structure partly depends on long-range interactions.

*l. Interactions of Adjacent Structural Units and Their Role in the Formation of the Three-Dimensional Structures of Proteins*

The statistical approach is not the only one for determining the helical and nonhelical sites; a method based on atom–atom potentials may be applied as well. To this end one should consider the interactions of adjacent structural units; these interactions are operative in the dipeptides, or other fragments making up the protein.

Kotelchuck and Scheraga [157] analyzed "nonstandard" fragments of polypeptide chain (VI) in which the rotation angles $\psi_i$ and $\varphi_{i+1}$ (instead of

(VI)

$\varphi_i$ and $\psi_i$) are the essential parameters. Calculations for such fragments give important information about interaction between the side radicals, R′ and R″, and their interaction with the peptide chain. The main conclusion that these authors drew from calculations with various R′ and R″ is that the conformation of this fragment depends only on the nature of R′ and does not depend on R″.

The calculations were carried out in the following way: For various combinations of R′ and R″ the potential function was calculated which accounted for the nonbonded and electrostatic interactions, the torsion potentials ($C_i^\alpha \cdots C_{i+1}^\alpha$ were included in the torsion potential to reduce computer time), and the hydration energy. Then the function was minimized over $\psi$, $\varphi$ and $\chi$ ($\chi$ are the rotation angles in the side groups) from three zero points corresponding to the R, B, and L conformations in the main chain. It turned out that if R″ is replaced or even neglected, the character of the calculated optimal conformation of a given fragment of the main chain will not change (i.e., it will be still R, B, or L.)

All amino acid residues may be divided into two groups: helix-making (h) and helix-breaking (c) judging from the conformation R, B, or L that corresponds to the deepest minimum. The h group includes: Ala, Val, Leu, Ileu, Met, Gln, Phen, Cys, His, Arg; the c group: Ser, Asp, Asn, Try, Tyr, Lys. For all residues, except Arg and Lys, the energy difference between the absolute and the second-deepest local minimum exceeds 0.5 kcal mole$^{-1}$. The division given in the paper cited corresponds to that obtained as a result of the statistical analysis of proteins, but the latter could not be comprehensive because of insufficient statistical material.

Gly (certainly, if Gly is the R′ residue) and Pro are rather peculiar. As was to be expected, Gly is insensitive to spiralization, i.e., it is as much h as c (R, L, and B conformations have approximately equal energies). Pro, unlike other residues, interacts more with the atoms localized closer to the N terminal than to the C terminal. If R′ = Pro, Pro belongs to the h group (in fact, in proteins helices often begin with Pro), if R″ = Pro, Pro belongs to the c group. (It has been noticed that Pro is never encountered in the middle of a helix; besides, helices break at Pro.)

The results obtained by Kotelchuck and Scheraga are in good agreement with the concepts underlying the configurational statistics of polypeptide chains, and also with the experimental values of $(h^2/nl^2)$. This ratio characterizes the flexibility of a macromolecule; $h^2$ is the mean square of the chain length, $n$ is the number of peptide units, and $l$ is the virtual bond length, or the length of a vector drawn from a $C^\alpha$ atom to an adjacent one (3.80 Å); this ratio has been shown theoretically to have a limit as $n \to \infty$. According to Brant and Flory [83] who were basing their thinking on the analysis of spatial models, the R′–R″ interaction does not play any role, should both radicals contain $C^\beta$ atoms. The measured value of $(h^2/nl^2)$ for all polypeptides containing such residues is close to 9, regardless of the nature of the side groups; it is much lower only for polyglycine and for Gly copolymers. In the case of the proline copolymers which were studied by Schimmel and Flory [158] the configurational statistics are different. It should be noted that these authors came to a conclusion about the role of Pro in the helix before Kotelchuck and Sheraga.

Can one infer from an analysis of fragments beginning with R′ and ending with R″ anything new for understanding the interactions in a peptide chain, as compared with those in dipeptide fragments?

Figure 30 shows the conformational map of the NR′C$^\alpha$H–CO–NH–C$^\alpha$–HR″C′ fragment for R′ = R″ = Ala drawn with Dashevsky's potentials and

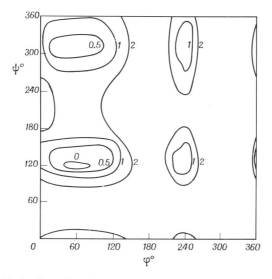

*Fig. 30.* Conformational map of the R'···R" fragment of Ala–Ala.

without taking into account the electrostatic interactions and the hydrogen bonds [134]. Comparing this map with that shown in Fig. 20 one can see that the energy contours are similar but, instead of three minima (R, B, and L typical of dipeptide maps), there are four minima, the fourth being in the H region (Huggins' conformation).

The R ··· R" fragment differs from the dipeptide fragments described above in that the latter includes the interactions of two amide groups; this renders the H conformation unfavorable for the stretched forms and also distorts the energy contours and changes the positions of the minima.

It is probable that a conclusion similar to that of Kotelchuck and Scheraga about the stability of the helical and nonhelical conformations may be obtained with dipeptide fragments by minimization of their potential functions. In fact, it is the dihedral angle $\psi$ which determines the R or B conformation; this angle is contained in both fragments. The L conformation, which is rarer than R or B, depends upon both angles. However, the difference between the absolute energy values of the R and B forms is higher in the case of a dipeptide, because in the latter the strong Coulomb interactions between the amide groups are taken into account. Therefore, if only dipeptide fragments are to be considered, much greater numbers of residues, including Ala, belong among the anti-helical ones. Nevertheless, there is no doubt that dipeptide fragments show more correctly the relative stabilities of different conformations (due to including the interactions of amide groups which exist in proteins) than the R' ··· R" fragments. That is why one is apt to think that the parametrization used by

Scheraga somewhat overestimates the electrostatic interactions. Maybe, $\varepsilon$ should be taken close to 10, as was done in the papers of Lipkind et al. [128].

Thus, the main conclusion to be made from an analysis of the two types of fragments is that the conformations of adjacent monomer units are practically independent and an irregular peptide may be assumed (with some accuracy) to consist of fragments having independent conformations. If that were the case, the problem of numerous minima for proteins or, at least, for oligopeptides would be solved. Then, for molecules like gramicidin S, a zero approximation could be found unequivocally, and minimization of the potential function by varying several parameters would not cause serious difficulties.

The simplified model, which takes account only of the local conformations of the $R' \cdots R''$ fragments gave about 75% correct predictions of the conformations of the peptide units in four proteins [157]. In fact, however, one cannot deny the presence of weak long-range interactions along the chain, and these make any predictions much less accurate.

*m. General Principles of Formation of Protein Structure*

As has been shown above, conformational calculations of small fragments may furnish much information useful for predictions of protein structure, but still it is not exhaustive. Ideally, having information about the optimal geometry of each peptide fragment, or their tripeptide combinations, one could predict the conformation of an irregular polypeptide from its amino acid sequence. But it is not so simple for several reasons: (a) for some residues two conformations of the $R' \cdots R''$ fragments are almost equally favorable (accordings to calculations [153] these residues are Arg and Lys); (b) Gly, owing to its conformational freedom, causes much indeterminacy; (c) even insignificant scatter in the rotation angles (3–5°) and bond angles (1–2°) result in a decreasing role of the unit-to-unit interactions as the length of the chain grows; (d) at some optimal combinations of rotation angles far-off residues may "bump" into closer ones—in configurational statistics this phenomenon is called the excluded volume effect. All this is, obviously, sufficient for the three-dimensional structure of protein to be formed as a result of interactions of distant sites along the chain, at the expense of hydrophobic effects.

Does the native structure of protein correspond to the absolute minimum of the potential function or to one of the local minima separated from the former by a high potential barrier? This question is very important for predictions of the spatial structures of proteins. Scheraga's [159] working hypothesis is that the native structure corresponds to some narrow distribution of conformations close to the absolute minimum, this distribution having the greatest statistical weight. Otherwise, according to Scheraga, a priori calculations would have been hopeless.

In reality most proteins do not recover all their properties after renaturation; the fact that their functional activity is preserved does not mean that their conformation is unchanged. For example, in [160] it is shown that in many proteins (arginase, hexaginase, deoxyribonuclease, chymotrypsinogen, urease, pepsin, lysozyme, etc.) denaturation proceeds via two stages, the first corresponding to increased freedom of movement of the side radicals, the second to an irreversible rearrangement of the protein's structure. So far, only one of the proteins, myoglobin, the most helical one, is known to recover almost entirely its physicochemical characteristics [161]. Evidently, most proteins, particularly those in which irregular sites are large, lose some of their properties after renaturation; therefore it may be suggested that their native structure corresponds to a metastable conformation developed during the course of protein synthesis in the ribosome.

Does that mean that any prediction of the structures of such proteins is hopeless? No, probably not, because the criteria derived from an analysis of the interactions of adjacent peptide units not only facilitate this task, but even make it simpler than the search for the absolute extremum. The idea that the structure of a protein is formed as it comes out of a ribosome has been put forward more than once [162, 163]. Phillips [164] has suggested a speculative scheme for the formation of the spatial structure of lysozyme from the N to the C terminal.

All the above considerations are hypothetical and, certainly, not sufficient to solve this problem in toto. As was mentioned above, the indeterminacy due to the specificities of the conformations of some residues, and the decreasing role of unit-to-unit interactions as the length of the peptide chain grows, makes hydrophobic interactions the major ones in determining the formation of the tertiary structures of proteins. If one suggests that these interactions are the only ones active in forming the three-dimensional structure, quite different criteria will be required for the search of the optimal conformation. For this purpose, a protein molecule should be roughly visualized as a thread of beads with a more or less random sequence of white and black ones corresponding to the hydrophilic and hydrophobic residues. The optimal conformation in this case will be a three-dimensional structure in which the greatest possible number of white beads are localized on the surface of the globule, and the greatest possible number of black beads inside the globule and in contact with each other. Of course, the concept that the structure is formed from the N terminal will require a search for the corresponding metastable state. This is not an easy problem, since it necessitates mathematical expressions for the "inside" and "outside" criteria; in addition, the effects of the volumes of the residues should be taken into account, and suitable algorithms should be found for calculation of the function and its optimization (for this purpose, the combinatorial method and discrete programming might be expedient).

In reality, both mechanisms are likely to be operative in the formation of the spatial structures of proteins; on the one hand, the tendency of a peptide unit to maintain its specific conformation (the secondary structure consisting of helices, $\beta$-structures and irregular sites is formed thereby), and on the other hand, the interactions between certain sites of the secondary structure creating optimal conditions for hydrophobic forces. That is why in this field of knowledge there exists a rare situation in which the scientists have good experimental evidence (the known spatial structures of some proteins) but no constructive theory to interpret the known and to predict the unknown.

3. Conformations of Polynucleotides and Nucleic Acids

The structure of the DNA molecule, which has been established by Watson and Crick as a result of an analysis of three-dimensional models and the X-ray evidence, consists of two chains; it is only in the course of replication that sites not connected with H bonds are formed in these chains. Double-stranded nucleic acids make up the genetic material of most organisms; their structure is known well enough, since they have been crystallized and subjected to X-ray study.

However, some phages have been shown to contain single-stranded DNA in which the content of adenine and thymine, and of guanine and cytosine are not equal [165]; as to RNA, its single-stranded form is well known [166].

Hence, the question of the conformations of single-stranded polynucleotides is important, not only from the point of view of understanding the structures and functions of "common" nucleic acids, but it also has a special interest.

Three questions pertaining to the conformations of polynucleotides will be considered below: (1) unravelling the structures of the smallest structural elements determining the conformations of macromolecules; (2) unravelling the conformations of single-stranded polynucleotide chains and the conditions affecting them; (3) the structural evidence on double-stranded nucleic acids and the possibility of its theoretical interpretation.

It should be noted that in both the single- and double-stranded polynucleotides, beginning from dinucleosidephosphates, the usual van-der-Waals and electrostatic forces (i.e., the nonbonded interactions), result in an "unusual" strong base stacking responsible for the three dimensional structure of polynucleotides consisting of several, or even of two, monomer units. Since the bases are stacked together, there is no empty space between them for molecules of the solvent.

Naturally, as long as we are dealing with mononucleotides and nucleosides containing a single base, this effect does not reveal itself, and the conformation is determined by the mutual attraction and repulsion of the nonbonded atoms of the molecule itself; in this case the effect of the solvent may be neglected.

What is the difference between polynucleotides and the usual polymers and polypeptides? First, the hydrophobic interactions, which in the latter emerge only when the chain is rather long and "inner" and "outer" regions can form, make a considerable contribution to the stacking interactions even in di-nucleotides, which, as will be shown below, practically does not increase as the chain becomes longer. Furthermore, it is the interactions of the nearest monomer units that play the decisive roles in the conformation of poly-nucleotides, whereas the interactions of more distant units may be considered negligible. This is why the dinucleosidephosphates, for example, may furnish important evidence about the conformations of single stranded polynucleo-tides.

In this sense mononucleotides and dinucleotides are widely different: the former possess very high conformational freedom, due to rotations around the five single bonds of the ribose phosphate backbone and the glycosidic bond between the base and the furanose ring; in dinucleotides the conforma-tional freedom is limited, because the bases tend to be located one under another. For this reason, an analysis of the conformations of the smallest structural units of nucleic acids (i.e., fragments whose size does not exceed that of a mononucleotide) does not imply a complete understanding of the conformations of macromolecules.

*a. Structural Units of Nucleic Acids*

A nucleoside is a five member sugar ring connected to a base with a glyco-sidic bond. Any single stranded polynucleotide consists of a number of mononucleotides, i.e., a nucleoside with a phosphate group, which is its monomer unit. A nucleotide is called for example, a nucleoside-2'-phosphate, a nucleoside-3'-phosphate, or a nucleoside-5'-phosphate, depending on the atom of ribose or deoxyribose 2', 3' or 5'—to which the phosphate group is bound (Fig. 31).

The conformation of a monomer unit is determined by the geometry of the furanose ring, the structure of the base, of the phosphate group, and, finally, by the relative positions of the base and the five-membered ring.

Let us begin with the conformation of the five-member ring. The structure of 2-deoxyribose was first roughly determined as a result of an X-ray study of 5'-bromo-deoxythymidine [167]. This was the only structure Spencer [168] could use in 1959 for building a model which revealed some of the struc-tural peculiarities of the furanose ring. For example, he found that the most

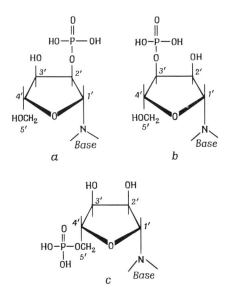

**Fig. 31.** Structural formulas of nucleotides. a, nucleoside-2′-phosphate; b, nucleoside-3′-phosphate; c, nucleoside-5′-phosphate.

favorable conformations correspond to a displacement of the $C_{2'}$ and $C_{3'}$ atoms from the plane of the other four atoms.

The conformation of the furanose ring is chiefly determined by repulsion between the nonbonded atoms of the ring itself, plus torsion strains, and, to some extent, by interactions with more distant atoms. A furanose ring resembles cyclopentane in which the nonplanar conformation is more favorable owing to the torsion energy.

The potential function for ribose and deoxyribose may be presented, to a rather good approximation, as

$$U = \sum_{i>j}\sum f_m(r_{ij}) + \tfrac{1}{2}C_0\,\Delta\alpha_0 + \tfrac{1}{2}C_C\sum \Delta\alpha_C^2$$

$$+ \tfrac{1}{2}U'_{C-C}\sum_{i=1}^{3}(1+\cos 3\varphi) + \tfrac{1}{2}U'_{C-O}\sum_{i=4}^{5}(1+\cos 3\varphi)$$

where $f_m(r_{ij})$ are the potentials of the nonbonded interactions, $\Delta\alpha_0$ is the deformation of the $C_{1'}O_{1'}C_{4'}$ angle, $\Delta\alpha_C$ is the deformation of other bond angles whose vertices are carbon atoms; the last two are torsion terms depending on the dihedral angles. Also, an electrostatic term in the monopole approximation may be included in the expression for the potential function.

In Kitaigorodsky's laboratory the optimal conformations of ribose [169] were calculated using the following potential function parameters: Dashevsky's potentials, $C_C = 30$ and $C_O = 65$ kcal mole$^{-1}$ rad$^{-2}$ (the ideal value of the COC angle is 90°), $U'_{C-C} = 3$ kcal mole$^{-1}$, $U'_{C-O} = 1.1$ kcal mole$^{-1}$. It was shown that the energy minimum corresponds to a conformation in which either the $C_{2'}$ or the $C_{3'}$ atom is displaced by 0.3–0.6 Å from the mean plane of the four other atoms. (If the differences in substituents are not taken into account, departures from the plane of one or the other atom are equally probable.) The four other atoms do not lie exactly in one plane: for example, if the $C_{2'}$ atom departs from the $C_1O_5C_4$ plane by 0.5 Å, the $C_{3'}$ atom will be located at a distance of 0.1 Å from this plane, but on the opposite side (Fig. 32). So, a furanose ring acquires an intermediate conformation between the envelope and half-chair ($C_2$ and $C_s$) typical of cyclopentane, being close to the $C_2$ form.

If the $C_{2'}$ (or $C_{3'}$) atoms are localized together with $C_{5'}$ on one side of the $C_1O_5C_4$ plane, one will have an endo conformation; and if the two atoms are localized on different sides, an exo conformation (Fig. 32). The latter conformation has been found only in two structures: deoxyadenosine [170] and thymidine [171]; in all the other nucleosides and nucleotides the endo conformation has been established.

The detailed structural data on nucleotides and nucleosides accumulated in recent years has allowed Sundaralingam and Jensen [172, 173] to suggest a classification of furanose conformations and to reveal some regularities. The following compounds were analyzed: cytidylic acid [174], 5-fluoro-2'-deoxy-β-uridine [175], adenylic acid [176], calcium thymidylate [177], deoxyadenosine [170].

Nonplanar deformations of the furanose ring seem to affect to some (though not a very great) extent the bond lengths and bond angles; for example, the

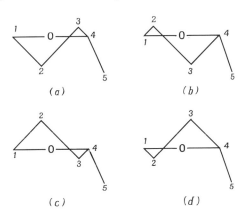

**Fig. 32.** Four conformations of the sugar ring: a, $C_{2'}$-endo; b, $C_{3'}$-endo; c, $C_{2'}$-exo; d, $C_{3'}$-exo; the $C_1O_5C_4$ plane is normal to the plane of drawing.

internal CCC angle for a carbon atom deviating from the ring plane is, on an average, 100.9°, i.e., 1.6° less than if the furanose ring were plane. The mean value of the CCC angles is 105.8° and the COC angle is the only one close to tetrahedral (109.3°). It should be noted that the same figures are derived from conformational calculations.

The bases are additional very important elements of nucleic acids. From all the X-ray studies of the purine and pyrimidine bases of nucleic acids it is clear that the atoms of which the rings are composed lie approximately in one plane (deviations from the mean plane are within the limits of experimental error). But the exocyclic bonds form small angles with the plane. For example, in the molecule of adenosine-5′-phosphate [176] the plane of the amino group forms a 25° angle with the base, the $C_{5'}$ atom lies 0.043 Å lower and the $N_1$ atoms, 0.051 Å higher than the base plane. In all nucleoside and nucleotide structures the deviation of the $C_{1'}$ sugar atom from the base plane is rather significant, of the order of 0.2 Å; due to this fact the angle between the glycosidic bond and the base plane is 5–8°.

Figure 33 shows the estimated values of the mean bond lengths and bond angles in bases [178]. The great differences in the values obtained by different authors for one and the same base depend in many cases upon whether the base is protonated or nonprotonated [173]. Other important factors are the presence of intermolecular hydrogen bonds and the mode of packing of the

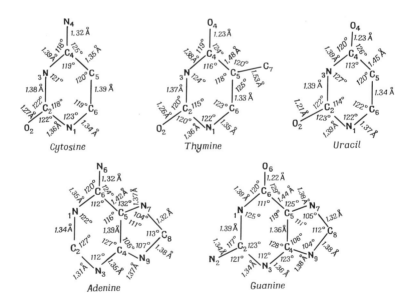

**Fig. 33.** Mean values of bond lengths and bond angles of heterocyclic residues of nucleic acids.

molecules in the crystal. In spite of all this it is advisable to use averaged values to calculate the conformations of mono- and polynucleotides.

Some authors [178, 179] have made a detailed analysis of the structure of the phosphate group. Thus, rather accurate mean values of the bond lengths and bond angles in nucleotides and the respective polymers are already known. Figure 34 shows the structural parameters for nucleotides used by Arnott [179] in the X-ray studies of nucleic acids. One can expect no more than a 0.02 Å difference in bond lengths and a 1–2° difference in bond angles between the values given in Fig. 34 and those in real nucleotides and nucleic acids.

It is obvious that bond lengths and bond angles are indispensable for an understanding of polynucleotide conformations. But the dihedral angles: The rotation angles around the single bonds of the riboso-phosphate backbone, as well as the angle of rotation around the glycosidic bond, are also essential parameters of the potential functions of these macromolecules, both in the case of stereoregular polymers and polypeptides.

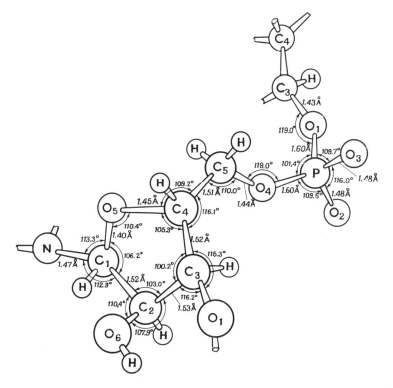

*Fig. 34.* Bond lengths and bond angles of a fragment of polynucleotide chain for $C_{3'}$-endo conformation of sugar (for $C_{2'}$-endo conformation bond angles at $C_{2'}$ should be substituted by the angles at $C_{3'}$ and vice versa). Primes in atom designations are omitted.

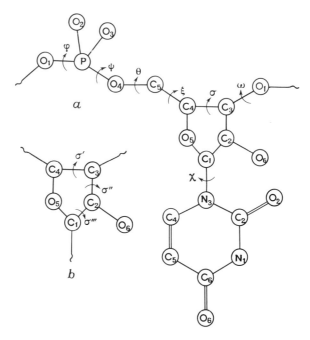

**Fig. 35.** a, Numeration of atoms and nomenclature of conformational angles of poly-nucleotide monomeric units (no primes in atom designations); b, three conformational angles responsible for the shapes of ribose and deoxyribose rings.

Arnott and Hukins [180] have suggested a handy nomenclature for poly-nucleotide conformations. Figure 35 shows a monomer unit of a polynucleo-tide (of course, it is not a simple mononucleotide) with designations of the rotation angles.

According to this nomenclature the movement along the polynucleotide chain is from $C_{5'}$ to $C_{3'}$. The conformational angles for the $i$th residue are designated as $\varphi_i$, $\psi_i$, $\theta_i$, $\xi_i$, $\sigma_i$, $\omega_i$. Rotation angles are counted from the cis-conformations (similarly to the nomenclature for the stereoregular macro-molecular conformations described in Section 1); the positive direction is a clockwise rotation; the $\sigma$ angle is counted from the cis-conformation of the $N-C_2$ (or the $N-C_4$) bond of the pyrimidine or purine bond with respect to the $C_{1'}-C_{2'}$ sugar bond. Some ambiguity arises when the conformation of the sugar ring is described: if the $C_{5'}-C_{4'}-C_{3'}-O_{1'}$ dihedral angle is designated as $\sigma$, and the $C_{5'}-C_{4'}-C_{3'}-C_{2'}$ as $\sigma'$, then $\sigma' \approx 125° + \sigma$.

Let us consider first the dihedral $\chi$ angle characterizing the mutual positions of the furanose ring and the base. Before the above nomenclature of con-formations had been suggested, the $\varphi_{CN}$ angle was used to designate the rota-tion around the glycosidic bond; Donohue and Trueblood [181] had suggested that it be defined as the angle formed by the trace of the base plane and the projection of the ribose (or deoxyribose) $C_1$–$O_1$ bond. Positive angles are counted clockwise in the direction of the $C_2$–N bond from $C_1$ to N. $\varphi_{CN}$ is considered to be equal to zero when the angle between the projections of the $C_1$–O and $N_9$–$C_4$ bonds in purine (or the $N_1$–$C_2$ bond in pyrimidine) is equal to 180°.

These authors inferred from an analysis of molecular models that there are two favorable torsion angle regions: the first, from $\varphi_{CN} = -30 \pm 45°$ (the anti region); and the second, from $\varphi_{CN} = 150 \pm 45°$ (the syn region) (Fig. 36).

The data will be discussed below in terms of $\varphi_{CN}$, and one should bear in mind that $\chi$ and $\varphi_{CN}$ are related in the following simple way: $\varphi_{CN} \approx \chi + 63°$; to be more exact, for the $C_2$-endo-sugar conformation $\varphi_{CN} = \chi + 64.8°$ and for the $C_3$-endo-conformation $\varphi_{CN} = \chi + 61.6°$.

It follows from the X-ray data that, in most nucleotides and nucleosides, the bases are anti-oriented, the only exception being deoxyguanosine [182] in which $\varphi_{CN} = 138°$. The conformations of nucleotides and nucleosides in solution may be established by using optical rotatory dispersion techniques. In [183] it was confirmed that the pyrimidine mononucleotides are in the anti-conformation.

Haschemeyer and Rich [184] have studied in detail the problem of the dihedral angles in nucleotides. It was supposed that freedom of rotation and allowed conformations are determined by distances between the atoms of the bases and the sugar ring. The regions of allowed values were found using Ramachandran's conditions. It should be noted that in some cases the atomic radii were somewhat diminished to improve the agreement between the calculated and experimental data.

It has been shown by calculations that in pyrimidine nucleosides steric

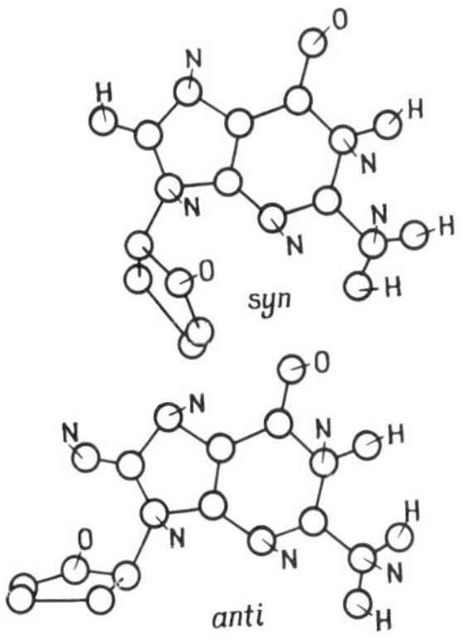

*Fig. 36.* Guanine in syn- and anticonformation in relation to furanose ring.

hindrances to rotation around the glycosidic bonds arise chiefly from the interactions of side substituents at the sugar carbons with the $C_2-O_2$ and $C_6-H$ base bonds. In purine nucleosides rotation is limited by shortened contacts with the $N_3$ atom of the six-membered ring. In this case $\varphi_{CN}$ may acquire any values in the range between the syn and anti conformations if diminished van der Waals radii are used. The results of calculations are presented in Table 3.

### *Table 3*

ALLOWED REGIONS OF TORSION ANGLES IN PYRIMIDINE AND PURINE
NUCLEOSIDES OBTAINED BY THE RIGID-SPHERES METHOD USING
VAN-DER-WALLS RADII AND NONBONDED INTERACTION POTENTIALS

| Conformation of sugar | Regions determined with normal van-der-Walls radii | Regions determined with nonbonded interaction potentials | Values found in the crystal |
|---|---|---|---|
| Purine nucleosides | | | |
| $C_2'$ endo | from $-85°$ to $-60°$<br>from $+110°$ to $-155°$ | from $-150°$ to $-36°$<br>from $+100°$ to $+163°$ | $+138°$ |
| $C_3'$ endo | from $-140°$ to $-85°$<br>from $-60°$ to $+10°$ | —— | $-20°$ |
| $C_3'$ exo | from $-85°$ to $+25°$ | from $-20°$ to $+30°$<br>from $+115°$ to $+171°$ | $-3°$ |
| Pyrimidine nucleosides | | | |
| $C_3'$ endo | from $-80°$ to $-50°$ | from $-35°$ to $-65°$<br>from $+143°$ to $+166°$ | $-24°$ |
| $C_2'$ endo | from $-80°$ to $-25°$<br>from $+125°$ to $+145°$<br>from $-140°$ to $-120°$ | from $-60°$ to $+23°$<br>from $+128°$ to $+142°$<br>from $-177°$ to $-165°$ | $-42°$ |

In [185–187] the energy of rotation around the glycosidic bond was calculated using nonbonded interaction potentials. It was shown that for uridine, adenosine, and cytidine the anti conformation is the most favorable one; in guanosine the syn conformation is the most stable. According to the authors of [185], the energy difference between the syn and anti conformations for adenosine and guanosine is 1–2 kcal mole$^{-1}$ and for uridine and cytidine, 5–7 kcal mole$^{-1}$. Thus, pyrimidine nucleosides and nucleotides always possess the anti conformation, whereas the purine nucleosides, both the syn and anti. Figure 37 shows the potential functions for rotation around the glycosidic bond for the ribose conformations which have been established in the various nucleosides (the calculations were made with Dashevsky's potentials). Table 3 contains the optimal values of the torsion angle $\varphi_{CN}$ obtained by the rigid

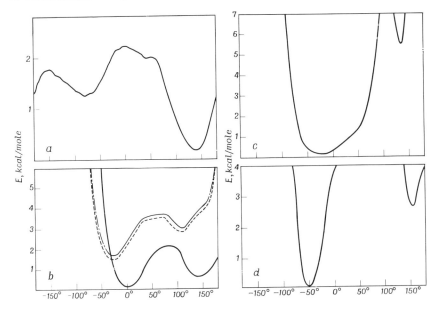

*Fig. 37.* Potential curves for rotation around the glycosidic bond in nucleosides: a, guanosine; b, adenosine; (---) $N_3$ atom is at 0.2 Å distance from sugar plane; (-·-) planar position of $N_3$ atom valence bonds; (—) electrostatic interactions are taken into account. c, Uridine; d, cytidine.

spheres method [182] and by calculating the potential energy of the base-sugar interaction [187]. The region of allowed rotation angles may be believed to correspond to the $RT$ range near the minimum of the energy function.

It is interesting that the energy of the electrostatic interactions does not appreciably affect the character of the curves; hence the indeterminacy of the parametrization (due to the monopole approximation and the charges calculated after Del Re) is not very significant. A more detailed analysis of the rotation potentials for various conformations of the sugar ring is to be found in [186].

Assuming now that the rotation around the glycosidic bond and the dihedral angles in the riboso-phosphate backbone are independent, let us consider the possible conformations of polynucleotide chain fragments. Figure 38 shows a fragment which has been thoroughly investigated by Ramachandran *et al.* [188] with the help of the rigid spheres method. One can see that it is somewhat different from the monomer unit suggested by Arnott and Hukins, Fig. 35. It should be noted that a complete description of the conformation of a nucleoside-3'-phosphate will require information about the $\varphi$ angle and of a nucleoside-5'-phosphate about the $\psi$ angle. But, since the rotations are not correlated, at least within the accuracy limits of the rigid spheres method,

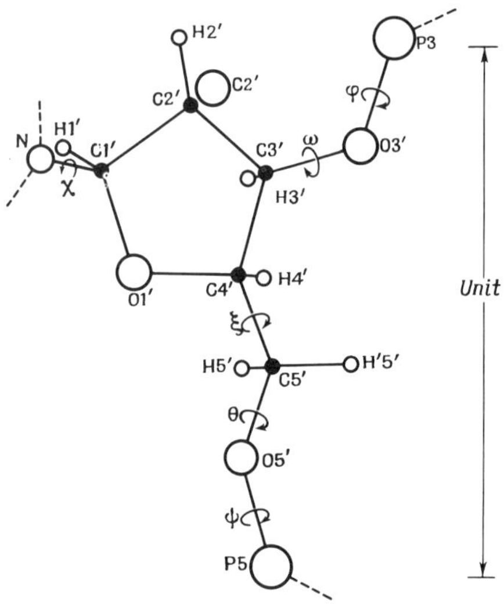

***Fig. 38.*** Ribose phosphate unit and conformational parameters: $\theta, \xi, \omega$.

and $\varphi$ and $\psi$ are not very important in the case of a monomer, one may believe that this fragment furnishes information about the allowed and forbidden regions of mononucleotides.

To find the allowed conformations every rotation is presented as a matrix, and the coordinates of the rotamer atoms are calculated as described in Section 1 of this chapter. Hydrogen atoms markedly affect the allowed regions. Furthermore, it is essential that the addition of a purine or pyrimidine base does not change the conformational rotation freedom of the fragments, hence the independent rotation angles in the riboso-phosphate backbone and the $\varphi_{CN}$ or $\chi$ angles. The possible conformations have the following set of torsion angles: $\theta$, about 180°; $\varepsilon$, about 60°, 180°, and 300°; $\omega$ in the range of 210°–260°.

Lakshminarayanan and Sasisekharan [189] made a conformational analysis of the above fragment using atom–atom potentials. They plotted energy against each rotation angle ($\theta$, $\xi$, and $\omega$) with fixed values of the other angles; in addition, two-dimensional conformational maps ($\psi, \theta$) and ($\theta, \xi$) were drawn. It is noteworthy that the fragment in question has many minima: There are nine minima in the ($\psi, \theta$) map with energy differences not exceeding 1 kcal/mole, and there are four minima in the ($\theta, \xi$) map. Thus, the search for the absolute potential energy extremum of the fragment presented in Fig. 38 is no trivial problem, to say nothing of the fact that its choice is determined by the parametrization.

In the regular single-stranded polymer (e.g., poly A) the riboso-phosphate backbone contains as many as five independent parameters, i.e., in addition to the angles described above $\varphi$ and $\psi$ should be added. These angles are operative as early as in dinucleoside phosphates; therefore the allowed conformations of the macromolecule may be found from an analysis of dinucleoside phosphates. (A similar situation exists for oligo- and polypeptides; all the main regularities can be derived from an analysis of the conformational maps of dipeptides).

Sasisekharan and Lakshminazayanan [190] applied the atom–atom potentials method to the conformations of a dimethyl phosphate which consists of two sugar rings connected with phosphoesteric bonds. The $(\varphi,\psi)$ potential map, according to their calculations has seven minima comparable in depth, the choice of the optimal minimum depending almost entirely upon the parameters selected, i.e., the partial charges, the $\varepsilon$ constant and the "immanent" potentials of the P–O bonds. One may only state that it is preferable to use gauche conformations with the following $(\varphi,\psi)$ rotation angles: $(60°,60°)$ or $(300°,300°)$; other gauche conformations are less favorable. This type of conformation was proved to exist by experimental investigations of such molecules.

To sum up the above, it should be noted that, unlike the case with peptides, for which the two-dimensional conformational maps make it possible to find the positions and forms of the monomer unit minima, when dealing with nucleotides one has to face the problem of numerous minima, and this is not very easy to solve. However, polynucleotides may be looked upon as stereoregular polymers, as stereochemically the two purine, as well as the two pyrimidine bases, are very much alike. Hence, one can expect that a conformational analysis of small polynucleotide fragments will result in the prediction of the three-dimensional structure of the macromolecule. (It is appropriate to recall that the conformational analysis of dipeptides and similar fragments is no key to disclosing the three-dimensional structure of a protein.)

### b. Conformations of Single-Stranded Oligo- and Polynucleotides

When monomer units, or fragments containing not more than one base were being considered, we could confine ourselves to atom–atom potentials only. When dealing with oligomers and polymers one is compelled to take into account hydrophobic, or, to be more exact, solvophobic interactions. In fact, native nucleic acids with their stacked bases can exist only in aqueous solutions over a certain temperature range; in organic solvents or at higher temperatures they undergo denaturation.

Sinanoğlu and Abdulnur [191] were the first to emphasize and explain the role of hydrophobic interactions in the formation of the spatial structures of oligo- and polynucleotides. This role is due to the fact that in the polynucleotide–solvent system there arise forces which prevent contacts between the nonpolar bases and the polar liquid (water); these forces are responsible for bringing the bases together and stacking them up (similarly to the tendency of nonpolar residues in proteins to come into contact).

It should be also noted that for estimating the free energy of a polynucleotide, its configurational entropy should be taken into account. In principle this could be calculated with the help of the potential functions, as was done by Flory *et al.* for polypeptides; the only difference will be that the integration should be done not for two-dimensional but for six-dimensional space. The great role of the configurational entropy follows from the fact that in a rigid-stacking conformation rotation around the single bonds is severely limited. In any case, it is the configurational entropy which is chiefly responsible for the free energy difference between the double-stranded nucleic acids (in which internal rotation around ten bonds is practically totally blocked) and the single-stranded polynucleotides formed in the course of denaturation.

The change in the free energy of a polynucleotide as a result of any conformational rearrangement (e.g., violation of the ordered base-stacking) can be formulated thus

$$\Delta F = \Delta E_{\text{nonbonded}} + \Delta E_{\text{el}} - T\Delta S,$$

where $\Delta E_{\text{nonbonded}}$ is the energy of the nonbonded interactions provided by the atom–atom potentials, $\Delta E_{\text{el}}$ is the electrostatic energy, and $\Delta S$ in the change in entropy including the solvation entropy (i.e., the hydrophobic interactions) and the configurational entropy.

De Voe and Tinoko [192] were the first to raise the question about the quantitative contribution of the various conformation-building forces to the total energy of a polynucleotide. The nonbonded interactions were estimated in a rather complex and questionable way; nevertheless, the values so obtained were of the same order of magnitude as those presented in later investigations. The electrostatic energies were calculated in the dipole–dipole approximation; the entire bases were taken as point dipoles. The value of the entropy change ($\Delta S = +28$ entropy units per mole) was borrowed from the paper of Kauzmann [193] who studied the solvation of benzene in water. The configurational entropy change was estimated from the formula $\Delta S = -RT \ln Z$ where $Z$ is the "sum over states" which amounts to $-4.6$ or $-13.1$ entropy units, according to different authors, depending on the number of stable conformational states occurring in the polynucleotide on rotation around the single bonds.

Later quantitative estimates of De Voe and Tinoko have been criticized by many authors [194–197], one of the items being the point-dipole approximations; instead, these authors took into account the Coulomb interactions between charges centered on the base atoms. It is almost impossible to prove at present whether any of the parametrizations suggested for the components contributing to the free energy of a polynucleotide is correct. Meanwhile it is still De Voe and Tinoko who put forward the fundamental ideas and established theoretically the strong stacking interactions which stabilize the structure of single- and double-stranded polynucleotides; it seems that the hydrogen bonds of the Watson–Crick model are of the secondary importance. An analysis of the results of the calculations of the different authors leads to the conclusion that, regardless of quantitative differences between some of their results, the idea of De Voe and Tinoko that stacking interactions play the predominant roles in stabilizing the conformations of dinucleoside-phosphates, oligomers, and single- and double-stranded polynucleotides is qualitatively correct and supported by all the authors.

Let us consider some figures pertaining to the interactions of stacked bases [197]. For two adenine molecules, located one under the other, the electrostatic interaction energy is 0.98 kcal mole$^{-1}$, the attraction energy of nonbonded atoms is $-10.96$ kcal mole$^{-1}$ and the repulsion energy is 2.47 kcal mole$^{-1}$; for thymines the respective values are 0.62, $-7.00$, and 2.34 kcal mole$^{-1}$, etc. Thus, the greatest contribution is made by the dispersion interactions, which depend only slightly on the base sequence, and decrease rapidly with distance. (For bases separated by but one nucleotide the dispersion energy is 30–40-fold less than the interaction energy of adjacent bases.) The electrostatic energy in almost all versions has a positive sign, i.e., it tends to counteract the stacking of the bases.

Of course, such estimates are not totally valid since they do not account for the effect of the solvent; the data so obtained are true only for vacuum conditions. In aqueous solutions of polynucleotides the base stacking is additionally stabilized by hydrophobic interactions. Indeed, the stabilization provided by the nonbonded interactions (5–8 kcal mole$^{-1}$ per base) is insufficient to render the energy of the polynucleotide–solvent system minimal for the given conformation: In organic solvents (e.g., poly C in ethyleneglycol [198]) which are known to be effective denaturing agents, the free energy minimum corresponds to the disordered conformation of the polynucleotide. This conformation is due to the configurational entropy, which sharply increases when rotations around the single bonds of the ribose–phosphate backbone are suddenly allowed.

There is rich evidence in favor of the idea that strong stacking interactions are responsible for the ordered arrangements of purine and pyrimidine bases in aqueous solutions: NMR investigations [199–206]; optical rotatory dispersion [207–212]; circular dichroism [213–215]; and uv absorption [216, 217] spectral data, measurements of osmotic pressure [218, 219], etc. Finally, it is worthwhile to mention that the only existing structural data for adenosine-phosphateuridine [220] was obtained from an analysis of X-ray diffraction patterns. In this structure both bases are in the anti conformation which they

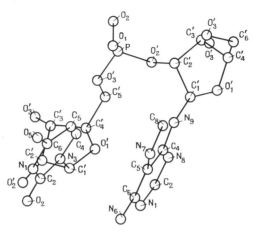

*Fig. 39.* Molecular structure of adenosine-phosphateuridine.

possess in nucleosides, both bases are approximately parallel (15° accuracy), and are located one under the other at a distance of about 3.4 Å (Fig. 39).

Polynucleotides and nucleic acids possess an interesting peculiarity, i.e., although various, rather strong, interactions are at work in them, they subtract from one another to produce a difference in free energy between the helical and disordered molecule close to zero. According to Brahms *et al.* [221, 222] the $\Delta H$ of the order–disorder transition in dinucleoside phosphates ranges from 6 to 8 kcal mole$^{-1}$; $\triangle S$, from 21 to 28 entropy units; and $\Delta F$ at 0°C is 0.2–0.7 kcal mole$^{-1}$. This small free energy difference (see also Table 4) accounts for the specific behavior of nucleic acids in small $pH$ and temperature ranges; it is for this very reason that double helices are capable of replicating when the conditions in the cell change.

## *Table 4*

THERMODYNAMIC PARAMETERS OF THERMAL DENATURATION
OF ADENYLIC OLIGOMERS AT NEUTRAL $pH$
(in 0.1 $M$ NaCl, $pH = 7.4$)

| Oligomer | $H$ (kcal mole$^{-1}$ per base) | $S$ (entropy units per mole of base) | $F$ (kcal mole$^{-1}$ per base at 0°C) |
|---|---|---|---|
| Dimer | 8.0 | 28 | 0.4 |
| Trimer | 8.0 | 28 | 0.4 |
| Pentamer | 8.1 | 28 | 0.5 |
| Heptamer | 8.1 | 27 | 0.6 |
| Dodecamer | 7.7 | 26 | 0.7 |
| Poly-A | 7.9 | 25 | 1.1 |

It is interesting that the stacking interactions in polynucleotides are essentially not cooperative. This means that the energy of interactions of, say, 10 bases stacked together is approximately nine-fold greater than the interaction energy of a single pair of bases in a dinucleosidephosphate. The absence of cooperativity in adenine oligomers is clearly shown in Table 4 which contains the thermodynamic parameters from paper [213]. Furthermore, Poland *et al.* [209] came to the conclusion that if the process of formation of ribonucleotides is absolutely noncooperative, then, in the case of deoxyribonucleotides it is to some extent anticooperative, i.e., the energy of formation of the ordered stacked structure of deoxyribonucleotides is even somewhat less than the total energy of the stacking interactions of the base pairs.

Of course, one could hardly expect that atom–atom potentials alone will be able to account for all details of the conformations of oligo- and polynucleotides in solutions. The most difficult problem will be, naturally, to take into account the hydrophobic interactions and the configurational entropy;

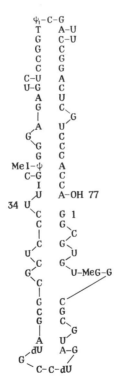

*Fig. 40.* One of the versions of the alanine *t*RNA secondary structure (two hairpins) established by means of the combinatorial method with the maximum pairing criterion. The complementary bases are opposite to each other.

therefore, sometimes, simpler models, for example, combinatorial, may be more tractable.

The above-said is largely true of predictions of the secondary structures of single-stranded polynucleotides in solutions. Several years ago it was believed that the equilibrium conformation of a single-stranded polynucleotide, like that of double-stranded native DNA, should necessarily involve base pairing in the same chain [223, 225], to form the so-called "hair-pin" structure.

These were the ideas underlying the combinatorial models intended for predicting the three-dimensional structures (or the secondary, "hair-pin," structures) of tRNA and mRNA. These models proceeded from the concept of maximal base-pairing in the hair-pin structures. Tumanyan [226] suggested a general algorithm for obtaining structures with maximum pairing with any base sequence in polynucleotides and applied it to calculate the optimal structure of alanine tRNA (Fig. 40).

Thus, it has become clear nowadays that it is not H bonds between the bases that determine the polynucleotide structure, as was postulated by Watson and Crick. Instead, it is the maximum length of a regularly complementary fragment with stacked bases which should be the true criterion for the secondary structures of tRNA and mRNA. The maximum base-pairing criterion with random sequences gives rather defective structures, as is shown in Fig. 40. It is still better to use these criteria together with their respective weights in such combinatorial calculations.

## c. Double Helices of Nucleic Acids

In the sixties alkaline metal salts of nucleic acids were intensively studied by X-ray crystallographic methods. Depending on the humidity and the content of metal salts DNA may exist in three crystalline modifications with different conformations of the polynucleotide chain. If the humidity exceeds 92% DNA acquires the B conformation [227] regardless of the salt. At a relative humidity of 75% DNA sodium salt acquires the A conformation [228], although the lithium salt still has the B conformation at 66% relative humidity. At 44% humidity the lithium salt has the C conformation [229]. Native DNA exists in the cell under high relative humidity conditions and, probably, has a B-like conformation though in this case no strict long-range order of bases is maintained.

Before discussing the similarities and differences among these crystalline structural modifications, designations characterizing the conformations of double helices should be agreed upon. Like crystalline stereoregular polymers, a nucleic acid crystal has an axis; X-ray diffraction patterns of its fibers furnish evidence about the identity period and the number of monomer units per period (in our case it is the number of Watson–Crick base pairs). These data are not, of course, sufficient to establish all atomic coordinates; as was indicated above, the conformation of a stereoregular polynucleotide is characterized by six independent rotation angles.

In structural investigations, polynucleotide conformations are usually described by the mutual positions of the bases. Arnott *et al.* [230] have suggested five parameters to describe the mutual positions of bases: $\theta_1$, the angle of rotation around the $y'$ axis; $\theta_2$, the angle of rotation around the $x'$ axis; $D$, the distance of the origin of the primed coordinate system from the axis; and two helix parameters, *viz*, $d$, the translation along the axis, and $\tau$, the angle of rotation around the axis of the helix. The last two parameters have the values $\tau = 2\pi/n$, where $n$ is the number of base pairs per period, and $d = c/n$, where $c$ is the identity period (see Fig. 41). The system of coordinates in Fig. 41 is chosen so that at $\theta_1 = \theta_2 = 0$ the $x'$ axis passes through the $C_4$ atom of a pyrimidine and is parallel to the line connecting the carbon atoms of the glycosidic bonds. The bases are assumed to be planar, and the Watson–Crick base pairs are assumed to have average parameters which seem not to change appreciably as one conformation turns into another; for example, the line connecting the carbon atoms of the glycosidic bonds is 10.85 Å long. Certainly, to determine the positions of all the atoms these parameters should be supplemented with at least one more (we assume, as Arnott [230] did, that the bonds and bond angles in polynucleotides are absolutely rigid). This parameter may be the $\chi$ or $\varphi_{CN}$ angle or, for example, one of the cylindrical coordinates of the phosphorus atom. Arnott gives all three of them.

DNA in the A form is a double helix with the 28.15 Å identity period and 11 Watson–Crick base pairs per period. According to Arnott *et al.* [230] who recently made accurate calculations of the DNA molecule structure based on the concept of constant bond lengths and bond angles in all forms of RNA

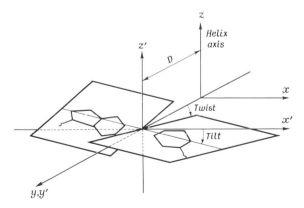

**Fig. 41.** Parameters determining the positions of bases: $\theta_1$, tilt; $\theta_2$, twist; and D, the distance between the axis of rotation and the helix axis.

and DNA double helices, the A form has the following $\theta_1$, $\theta_2$, $D$, $d$, and $\tau$ parameters: 20°, 8°, 4.25Å, 2.56Å and 32.7°; the $r$, $\theta$, $z$ cylindrical phosphorus atom parameters in the $xyz$ system of Fig. 41 are: 8.84Å, 67.2Å, −3.93°.

The B-conformation has the following base parameters: −2°, −5°, −0.63Å, 3.38Å and 36°; the coordinates of the phosphorus atom: 9.05Å, 94.8°, 2.04Å. Thus the A and B conformations are quite different: in the B form the bases are located close to the double helix axis and almost normal to this axis, whereas in the A form they are located at some distance from the axis and are inclined toward it. In addition, in the B conformation the furanose ring has the $C_{2'}$-endo conformation, and in the A form, $C_{3'}$-endo.†

The C form of DNA is close to the B form. It has the following base parameters: −6°, 5°, −2.13Å, 3.32Å and 38.6°; the position of the phosphorus atom is (9.05Å, 107.5Å, 2.95Å); the sugar conformation is $C_{2'}$-endo.

RNA molecules have several crystalline modifications which seem to contain double helices; Arnott [179] believes that all the investigated types of RNA have 11-fold helices (earlier some RNA types were suggested to have also 10-fold helices). The geometric parameters refined by Arnott are the following: 14°, 0.0°, 4.25Å, 32.7°, 2.73Å; the phosphorus coordinates are: (8.84Å, 68.5°, −3.62Å). Thus, the RNA molecule has a conformation similar to that of A-DNA.

Figure 42 shows the relative positions of the bases in A-DNA, B-DNA and RNA. It is seen that the planes of the bases are in contact with each other, which is one more proof of the significance of stacking interactions in poly-nucleotides. In A-DNA and RNA the bases overlap to a greater degree than in B-DNA; this seems to be the reason why DNA molecules, which in the cell have a conformation close to B, are less stable than RNA molecules—the former exists in the form of double helices in a narrower temperature range.

It is interesting to elucidate the question: is the base overlapping in DNA and RNA molecules optimal and does it determine the geometry of the double helices? Recent calculations [231] in which the potential function of base interaction was minimized over five parameters ($\theta_1$, $\theta_2$, $D$, $d$, and $\tau$) have shown that, first, the known structures of nucleic acids, though close to local minima in the base interaction energy, do not correspond to them; and secondly, there are rather deep minima far from the parameters typical of real polynucleotides: for example, with large D (of the order of 10–12Å) the base interaction energy may have a rather high absolute value. Thus, the geometry of nucleic acids is determined not only by the base interactions—a great role is played by limitations due to the riboso–phosphate backbone of

---

† Recent refinement of crystallographic data [230a] showed the preference to the $C_{3'}$-exo conformation of the sugar residue in B-DNA and confirmed the $C_{3'}$-endo conformation in A-DNA.

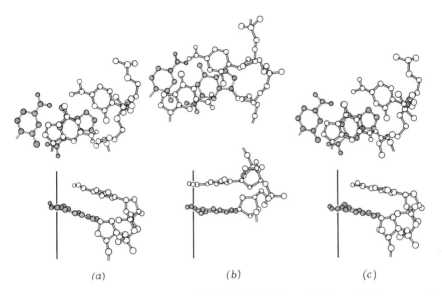

**Fig. 42.** Some molecular structures: a, A-DNA; b, B-DNA; c, RNA projected across and along the fiber axis; the point shows the position of the helix axis.

the double helix; that is why prediction of the structure of nucleic acids will require a search for an energy minimum in the interactions of all the atoms of a polynucleotide chain over all the most important geometric parameters, and primarily over $\varphi$, $\psi$, $\theta$, $\xi$, $\omega$, $\chi$.

Arnott and Hukins [180] distinguished some important features of the conformations of nucleotides, monomers and polymers, which are most valuable in theoretical conformational analysis. The conformational angles of monomers and polymers, unlike those of polypeptides and proteins, have small ranges of values. Moreover, the angles are almost identical in both monomers and polymers.

For $\varphi$, the rotation angle around the $P-O_1'$ bond (Fig. 35a), there are three ranges of values: $63° \pm 17°$, $196° \pm 25°$, and $294° \pm 14°$; in double-helical polymers there is only one range of values: $290° \pm 14°$, coinciding with the third range in monomers.

A similar situation is observed for $\psi$, the rotation angle around the $O_4-P$ bond: there are three ranges for monomers: $64° \pm 16°$, $182 \pm 5°$, and $306° \pm 11°$, and one range for polymers: $294° \pm 8°$ corresponding to the last conformation of monomers.

The angle of rotation around the $C_5'-O_4'$ bond, $\theta$, has two ranges of values for monomers: $262° \pm 14°$ and $181° \pm 22°$, the former range arising only in nucleosides. The polymer range, $188° \pm 4°$, corresponds to the latter range of values.

For rotation around the $C_4'–C_5'$ bond in monomers there is just one range of values $\xi = 48° \pm 7°$; almost similar values, $51° \pm 4°$, are typical of polymers.

It is only in the rotations around the $O_1'–C_3'$, $\omega$, that the angles in monomers and polymers do not correspond to each other; the former have the $252° \pm 16°$ range, and the latter, $196° \pm 10°$. It is possible that the second range of values has not been found in monomers because only three structures of nucleoside-3'-phosphates are known.

Finally, rotation around the glycosidic bond, $\chi$, (its dependence upon the angle $\varphi_{CN}$ was mentioned above) has the $299° \pm 31°$ range of values for the syn conformation in monomers and two ranges for the anti conformation: $\chi = 84° \pm 7°$ and $\chi = 110° \pm 10°$; the former range arises when the sugar ring is in the $C_3'$-endo and $C_3'$-exo conformations; the second range, in $C_2'$-endo-conformation. In polymers with the $C_3'$-endo conformation of the sugar ring (A-DNA, RNA and also in some synthetic polynucleotides), the $\chi$ angle has a very narrow range ($77° \pm 3°$) whereas in B-DNA and C-DNA, with their $C_2'$-endo-conformation of the sugar, $\chi$ is 173° and 139° respectively.

The sugar conformations which have the angles $(\sigma', \sigma'', \sigma''')$ (Fig. 35b) have the following ranges of values in monomers: $(204° \pm 2°, 37° \pm 4°, 97° \pm 6°)$ for the $C_{3'}$ endo conformation and $(266° \pm 4°, 323° \pm 4°, 152° \pm 3°)$ for the $C_{2'}$ endo conformation. In polymers there is almost no scattering in the values of these angles, i.e., $(\sigma', \sigma'', \sigma''') = (205°, 38°, 94°)$ for the $C_3$-endo sugar conformation and $(\sigma', \sigma'', \sigma''') = (266°, 323°, 158°)$ for the $C_2'$-endo conformations.

Thus, the conformational angles of monomers and polymers have rather small ranges of rotation angles (established experimentally) and the ranges are much smaller in polymers. It is remarkable that values of $(\varphi, \psi, \theta, \xi, \omega, \chi, \sigma', \sigma'', \sigma''')$ approximately equal to $(290°, 294°, 188°, 51°, 196°, 77°, 205°, 38°, 94°)$ give a rather adequate description of the conformations of all the polynucleotides studied by X-ray crystallography, with the exception of B- and C-DNA. In these two conformations only changes in the values of the angles $(\sigma', \sigma'', \sigma'''$ and $\chi)$ corresponding to discrete alterations in the sugar conformation, and the accompanying change in the value of $\chi$, are indispensible. Almost all the conformations so far investigated have mostly staggered bonds in the riboso–phosphate backbone; however, nothing is known as yet about the factors responsible for the range of values in each case.

The sum of all the above data constitutes comprehensive experimental evidence on which to base the theoretical conformational analysis of polynucleotides and nucleic acids. The following questions which can be clarified by the atom–atom potential method seem to be the most interesting: (1) Do the local and absolute energy minima of dinucleoside-phosphates and single-stranded polynucleotides correspond to the geometry of the double helices of nucleic acids? (2) What relative energies of the A and B forms of double helices and what calculated parameters give energy minima corresponding to these forms? (3) In what way does the conformational transition B → A occur, and is there a potential barrier to be overcome in the course of transition? In addition, the atom–atom potential method may be useful in obtaining some quantities pertaining to the properties of nucleic acids in solution, for example, the persistent length of DNA and RNA which is known to characterize the flexibility of this type of macromolecule.

REFERENCES

1. M. V. Volkenstein, "Configurational Statistics of Polymer Chains." Wiley (Interscience), New York, 1963.
2. P. J. Flory, "Statistical Mechanics of Chain Molscules." Wiley (Interscience), New York, 1969.
3. K. Ziegler, E. Holzkamp, H. Breil, and H. Martin, *Angew. Chem.* 67, 541 (1955).
4. G. Natta, P. Pino, P. Corradini, F. Danusso, E. Mantica, G. Mazzanti, and G. Moraglio, *J. Amer. Chem. Soc.* 77, 1708 (1955).
5. G. Natta, *Makromol. Chem.* 16, 213 (1955); *J. Polym. Sci.* 16, 143 (1955).
6. G. Natta, F. Danusso, and D. Sianesi, *Makromol. Chem.* 28, 253 (1958).
7. G. Natta and F. Danusso, eds., "Stereoregular and Stereospecific Polymerization," Vol. 1 and 2. Pergamon, Oxford, 1967.
8. M. L. Huggins, G. Natta, V. Desreux, and H. Mark, *Pure Appl. Chem.* 12, 645 (1966).
9. P. Corradini, *in* "The Stereochemistry of Macromolecules" (A. D. Ketley, ed.), p. 1. London and New York, 1968.
10. G. Natta and P. Corradini, *J. Polym. Sci.* 39, 29 (1959).
10a. G. Natta and P. Corradini, *Nuovo Cimento Suppl.* 15, 9 (1960).
11. W. Cochran, F. C. H. Crick, and V. Vand, *Acta Crystallogr.* 5, 581 (1952).
12. G. Natta, P. Corrandini, and I. Bassi, *J. Polym. Sci.* 51, 505 (1961).
13. T. M. Birstein and O. B. Ptitsin, "Conformations of Macromolecules." Wiley (Interscience), New York, 1966.
14. N. P. Borisova and T. M. Birstein, *Vysokomol. Soedin.* 5, 279 (1963); 6, 1234 (1964).
15. P. J. Flory and R. L. Jernigan, *J. Chem. Phys.* 42, 3509 (1965).
16. H. Eyring, *Phys. Rev.* 39, 746 (1936).
17. S. G. Galaktionov, *Inzh. Fiz. Zh.* 12, 765 (1967).
18. C. R. Eddy, *J. Chem. Phys.* 38, 1032 (1963).
19. T. Shimanouchi and S. Mizushima, *J. Chem. Phys.* 23, 707 (1958).
20. O. B. Ptitsin and Yu. A. Sharonov, *Zh. Tekh. Fiz.* 27, 2762 (1957).
21. R. Hughes and J. Lauer, *J. Chem. Phys.* 30, 1165 (1959).
22. T. Miyazawa, *J. Polym. Sci.* 55, 215 (1961).
23. K. Nagai and M. Kobayashi, *J. Chem. Phys.* 36, 1268 (1962).
24. H. Sugeta and T. Miyazawa, *Biopolymers* 5, 673 (1967); 6, 1387 (1968).
25. C. W. Bunn and D. Holmes, *Discuss. Faraday Soc.* 25, 95 (1958).
26. C. W. Bunn, *Trans. Faraday Soc.* 35, 482 (1939).
27. M. L. Huggins, *Pure Appl. Chem.* 15, 369 (1967).
28. S. Krimm, M. Tasumi, and C. G. Opascar, *J. Polym. Sci. Part B* 5, 105 (1967).
29. P. De Santis, E. Giglio, A. M. Liquori, and A. Ripamonti, *J. Polym. Sci. Part A* 1, 1383 (1963).
30. R. A. Scott and H. A. Scheraga, *J. Chem. Phys.* 44, 8 (1966).

31. V. G. Dashevsky and I. O. Murtasina, *Itogi. Nauki Khim. Nauki* **21**, 7 (1970).
32. C. W. Bunn and R. E. Howells, *Nature (London)* **174**, 549 (1954).
33. M. Iwasaki, *J. Polym. Sci. Part A* **1**, 1099 (1963).
34. V. Magnasco, G. Gay, and C. Nicora, *Nuovo Cimento* **34**, 1263 (1964).
35. R. L. McCullough and P. E. McMahon, *J. Phys. Chem.* **69**, 1747 (1965); *Trans. Faraday Soc.* **61**, 197, 201 (1965).
36. T. W. Bates, *Trans. Faraday Soc.* **63**, 1825 (1967).
37. A. Bondi, *J. Phys. Chem.* **68**, 44 (1964).
38. C. P. Smyth, "Dielectric Behaviour and Structure," p. 244. McGraw-Hill, New York, 1955.
39. E. L. Galperin, Yu. V. Strogalin, and M. P. Mlenik, *Vysokomol. Soedin,* **7**, 933 (1965).
40. J. B. Lando, H. G. Olf, and A. Peterlin, *J. Polym. Sci. Part A-1* **4**, 941 (1966).
41. R. O. Teulings, H. J. Dulleton, and R. L. Miller, *J. Polym. Sci. Part B* **6**, 441 (1968).
42. G. Cortili and G. Zerbi, *J. Mol. Spectrosc.* **23A**, 285 (1967).
43. R. C. Reinhardt, *Ind. Anal. Chem.* **35**, 422 (1943).
44. I. M. Gelfand, E. B. Wool, S. L. Ginsburg and Ju. C. Fedorov, "Valley method in the problems of X-ray analysis." Nauka, Moscow, 1966.
45. R. Brill and F. Halle, *Naturwissenschaften* **26**, 12 (1938).
46. C. W. Bunn, *J. Chem. Soc.* p. 297 (1947).
47. A. M. Liquori, *Acta Crystallogr.* **8**, 345 (1955).
48. G. Wasai, T. Saegura, and J. Furukawa, *J. Macromol. Chem.* **81**, 1 (1965).
48a. G. Allegra, E. Benedetti, C. Pedone, *Macromolecules* **3**, 727 (1970).
49. G. Natta, P. Corradini, and P. Ganis, *J. Polym. Sci.* **58**, 1191 (1962).
50. E. A. Mason and M. M. Kreevoy, *J. Amer. Chem. Soc.* **77**, 5808 (1955).
51. D. Badamy, *Polymer* **1**, 273 (1960).
52. G. Natta, P. Corradini, and I. Bassi, *Gazz. Chim. Ital.* **89**, 784 (1959).
53. G. Natta, P. Corradini, and I. Bassi, *Nuovo Cimento Suppl.* **15**, 68 (1960).
54. P. Corradini and P. Ganis, *Nuovo Cimento Suppl.* **15**, 96 (1960).
55. D. Sianesi, R. Serra, and F. Danusso, *Chim. Ind. (Milan)* **41**, 515 (1959).
56. S. Nurahashi, S. Nosakura, and H. Tadokoro, *Bull. Chem. Soc. Jap.* **32**, 534 (1959).
57. Y. Chatani, *J. Polym. Sci.* **47**, 491 (1960).
58. J. Hitta, Y. Chatani, and Y. Sakata, *Bull. Chem. Soc. Jap.* **33**, 125 (1960).
59. P. Corradini and P. Ganis, *Nuovo Cimento Suppl.* **15**, 104 (1960).
60. H. D. Noether, *J. Polym. Sci. Part C* **16**, 725 (1967).
61. C. Overberger, A. Bochert, and A. Katchman, *J. Polym. Sci.* **44**, 491 (1960).
62. G. Natta, *Makromol. Chem.* **35**, 94 (1960).
63. P. Burleigh, *J. Amer. Chem. Soc.* **82**, 749 (1960).
64. M. Asahina and K. Okuda, *Kobunshi Kagaku* **17**, 607 (1970).
65. R. Stefani, M. Chevreton, J. Terrier, and C. Eyrand, *C. R. Acad. Sci.* **248**, 2006 (1959).
66. T. Miyazawa and I. Ideguchi, *J. Polym. Sci. Part B* **1**, 389 (1963).
67. M. Goodman and J. S. Shulman, *J. Polym. Sci. Part C* **12**, 23 (1966).
68. G. Natta, M. Peraldo, M. Farina, and G. Bressan, *Makromol. Chem.* **55**, 139 (1962).
69. G. Natta, *J. Polym. Sci.* **48**, 219 (1960).
70. C. Hammer, T. Koch, and J. Witney, *J. Appl. Polym. Sci.* **1**, 169 (1959).
71. H. Tadokoro, T. Yasumoto, S. Murahashi, and I. Nitta, *J. Polym. Sci.* **44**, 866 (1960).
72. G. A. Carazzolo, *J. Polym. Sci. Part A* **1**, 1573 (1963).
73. C. Fuller, *Chem. Rev.* **26**, 143 (1940).
74. H. Tadokoro, Y. Chatani, I. Yoshihara, S. Tahara, and S. Murahashi, *Makromol. Chem.* **73**, 109 (1964).
75. H. Tadokoro, *J. Polym. Sci. Part C* **15**, 1 (1966).

76. K. Imada, T. Mijakawa, Y. Chatani, H. Tadokoro, and S. Murahashi, *Makromol. Chem.* **83**, 113 (1965).
77. M. Cesari, G. Peraldo, and A. Mazzei, *Makromol. Chem.* **83**, 196 (1965).
78. V. Sasisekharan, *Proc. Indian Acad. Sci. Sect. A* **53**, 296 (1961).
79. V. Sasisekharan, *in* "Collagen" (N. Ramanathan, ed.), p. 39. Wiley (Interscience), New York, 1962.
80. G. N. Ramachandran, C. Ramakrishnan, and V. Sasisekharan, *J. Mol. Biol.* **7**, 95 (1963).
81. G. N. Ramachandran, C. Ramakrishnan, and V. Sasisekharan, *in* "Aspects of Protein Structure" (G. N. Ramachandran, ed.), p. 121. Academic Press, New York, 1963.
82. P. De Santis, E. Giglio, A. M. Liquori, and A. Ripamonti, *Nature (London)* **206**, 456 (1965).
83. D. A. Brant and P. J. Flory, *J. Amer. Chem. Soc.* **87**, 663, 2791 (1965).
84. G. Nemethy and H. A. Scheraga, *Biopolymers* **3**, 155 (1965).
85. H. A. Scheraga, S. J. Leach, R. A. Scott, and D. Poland, *Fed. Proc. Fed. Amer. Soc. Exp. Biol.* **24**, 413 (1965).
86. H. A. Scheraga, S. J. Leach, R. A. Scott, and G. Nemethy, *Discuss. Faraday Soc.* **40**, 268 (1965).
87. S. J. Leach, G. Nemethy, and H. A. Scheraga, *Biopolymers* **4**, 369, 887 (1966).
88. G. Nemethy, S. J. Leach, and H. A. Scheraga, *J. Phys. Chem.* **70**, 998 (1966).
89. R. A. Scott and H. A. Scheraga, *J. Chem. Phys.* **45**, 2091 (1966).
90. T. Ooi, R. A. Scott, G. Vanderkooi, and R. F. Epand, *J. Amer. Chem. Soc.* **88**, 5680 (1966).
91. T. Ooi, R. A. Scott, G. Vanderkooi, and H. A. Scheraga, *J. Chem. Phys.* **46**, 4410 (1967).
92. G. Vanderkooi, S. J. Leach, G. Nemethy, R. A. Scott, and H. A. Scheraga, *Biochemistry* **5**, 2991 (1966).
93. H. A. Scheraga, R. A. Scott, G. Vanderkooi, S. J. Leach, K. B. Gibson, T. Ooi, and G. Nemethy, *in* "Conformation of Biopolymers" (G. N. Ramachandran, ed.), Vol. 1, p. 43. Academic Press, New York, 1967.
94. K. B. Gibson and H. A. Scheraga, *Proc. Nat. Acad. Sci. U.S.A.* **58**, 420 (1967).
95. G. N. Ramachandran and V. Sasisekharan, *Advan. Protein Chem.* **23**, 287 (1968).
96. H. A. Scheraga, *Advan. Phys. Org. Chem.* **6**, 103 (1968).
97. R. B. Corey and L. Pauling, *Proc. Roy. Soc. Ser. B* **141**, 10 (1953).
98. D. R. Davis, *Progr. Biophys. Mol. Biol.* **15**, 188 (1965).
99. R. E. Marsh and J. Donohue, *Advan. Protein Chem.* **22**, 234 (1967).
100. J. T. Edsall, P. J. Flory, J. C. Kendrew, A. M. Liquori, G. Nemethy, G. N. Ramachandran, and H. A. Scheraga, *J. Mol. Biol.* **15**, 399 (1966); *J. Biol. Chem.*, **241**, 1004 (1966); *Biopolymers* **4**, 121 (1966).
100a.—UPAC—UB Commission on Biochemical Nomenclature, *Biochemistry* **9**, 3471 (1970).
101. G. H. Huggis, D. Michie, A. R. Muir, K. B. Roberts, and P. M. B. Walker, "Introduction to Molecular Biology." Wiley, New York, 1955.
102. M. V. Volkenstein, "Molecules and Life," p. 101. Nauka, Moscow, 1965.
103. D. A. Green and R. F. Goldberger, "Molecular Insights into the Living Process." Academic Press, New York, 1967.
104. K. Ramakrishnan, *Proc. Indian Acad. Sci. Sect. A* **59**, 327 (1964).
105. G. N. Ramachandran, C. M. Venkatachalam, and S. Krimm, *Biophys. J.* **6**, 849 (1966).
106. G. Del Re, *J. Chem. Soc.* p. 403 (1958); *Theor. Chim. Acta* **1**, 188 (1963).
107. H. Berthod and H. Pullman, *J. Phys. Chim.* **62**, 942 (1965).
108. E. R. Lippincott and R. Schroeder, *J. Chem. Phys.* **23**, 1099 (1955).

109. R. Schroeder and E. R. Lippincott, *J. Phys. Chem.* **61**, 921 (1957).
110. E. M. Popov, V. G. Dashevsky, G. M. Lipkind, and S. F. Arkhipova, *Mol. Biol.* **2**, 612 (1968).
111. A. I. Kitaigorodsky, *Vysokomol. Soedin. Ser. A* **10**, 2669 (1968).
112. I. L. Katz and B. Post, *Acta Crystallogr.* **13**, 624 (1960).
113. M. Davies and D. R. Thomas, *J. Phys. Chem.* **60**, 763, 767 (1956).
114. C. D. Dickinson and J. Stewart, *Acta Crystallogr.* **21**, 663 (1966).
115. G. N. Ramachandran, *Biopolymers* **6**, 494 (1968).
116. C. M. Venkatachalam, *Biochem. Biophys. Acta* **168**, 397 (1968).
117. C. Ramakrishnan and K. P. Sarathy, *Biochim. Biophys. Acta* **168**, 402 (1968).
118. T. Miyazawa, T. Shimanouchi, and S. Mizushima, *J. Chem. Phys.* **28**, 611 (1958).
119. K. Fukushima, Y. Ideguchi, and T. Miyazawa, *Bull. Chem. Soc. Jap.* **36**, 1301 (1963).
120. K. D. Gibson and H. A. Scheraga, *Biopolymers* **4**, 709 (1966).
121. E. M. Popov, G. M. Lipkind, S. F. Arkhipova, and V. G. Dashevsky, *Mol. Biol.* **2**, 622 (1968).
122. O. B. Ptitsin, *Usp. Sovrem. Biol.* **63**, 3 (1967).
123. G. Nemethy and H. A. Scheraga, *J. Chem. Phys.* **66**, 1773 (1962).
124. H. A. Scheraga, *Ann. N.Y. Acad. Sci.* **125**, 273 (1965).
125. S. L. Portnova, V. F. Bistrov, V. I. Tsetlin, V. T. Ivanov, and Yu. A. Ovchinnikov, *Zh. Obshch. Khim.* **38**, 428 (1968).
126. P. J. Flory, D. A. Brant, and W. J. Miller, *J. Mol. Biol.* **23**, 47, 67 (1967).
127. P. J. Flory, *in* "Conformation of Biopolymers" (G. N. Ramachandran, ed.), p. 339. Academic Press, New York, 1967.
128. G. M. Lipkind, S. F. Arkhipova, and E. M. Popov, *Zh. Strukt. Khim.* **11**, 121 (1970).
129. I. M. Klotz and J. S. Fransen, *J. Amer. Chem. Soc.* **84**, 3461 (1962).
130. S. Krimm and J. E. Mark, *Proc. Nat. Acad. Sci. U.S.* **60**, 1122 (1968).
131. D. A. Brant and P. R. Schimmel, *Proc. Nat. Acad. Sci. U.S.* **58**, 52 (1967).
132. S. G. Galaktionov, *et al.*, *in* "Conformational Calculations of Complex Molecules," p. 71. House of Acad. Sci. of Bielorussian SSR, Minsk, 1970.
133. G. M. Lipkind, S. F. Arkhipova, and E. M. Popov, *Izv. Akad. Nauk SSSR Ser. Khim.* p. 315 (1970).
134. V. G. Dashevsky, *Itogi. Nauki Khim. Nauki* **21**, 93 (1970).
135. Yu. A. Ovchinnikov, *Vestn. Akad. Nauk. SSSR* **38**, No. 7 (1968).
136. Yu. A. Ovchinnikov, V. T. Ivanov, A. V. Evstratov, N. D. Abdullaev, E. M. Popov G. M. Lipkind, S. F. Arkhipova, E. S. Efremov, and M. M. Shemyakin, *Biochem. Biophys. Res. Commun.* **37**, 668 (1969).
137. G. M. J. Schmidt, D. C. Hodgkin, and B. M. Oughton, *Biochem. J.* **65**, 744 (1957).
138. A. M. Liquori, P. De Santis, A. L. Kovacs, and L. Mazzarella, *Nature (London)* **211**, 1039 (1966).
139. D. Balasubramanian, *J. Amer. Chem. Soc.* **89**, 5445 (1967).
140. I. M. Gelfand and M. L. Tsetlin, *Usp. Mat. Nauk* **17**, 103 (1962).
141. B. T. Polyak, *Ekon. Mat. Metodi* **3**, 881 (1967).
142. L. A. Rastrigin, *in* "Statistical Methods of Search." Nauka, Moscow, 1968.
143. C. B. Anfinsen, E. Haber, M. Sela, and F. H. White, *Proc. Nat. Acad. Sci. U.S.* **47**, 1309 (1961).
144. I. Isemura, *J. Chem. Soc. Jap.* **86**, 447 (1965).
145. A. G. Szent-Gyorgyi and C. Cohen, *Science* **126**, 697 (1957).
146. E. R. Blout, C. de Loze, S. M. Bloom, and G. D. Fasman, *J. Amer. Chem. Soc.* **82**, 3787 (1960).
147. A. V. Guzzo, *Biophys. J.* **5**, 809 (1965).
148. D. A. Cook, *J. Mol. Biol.* **29**, 167 (1967).

149. M. F. Perutz, *J. Mol. Biol.* **13**, 646 (1965).
150. O. B. Ptitsin, *Mol. Biol.* **3**, 627 (1969).
151. P. F. Periti, G. Quagliarotti, and A. M. Liquori, *J. Mol. Biol.* **24**, 313 (1967).
152. P. F. Periti, *Nature (London)* **215**, 509 (1967).
153. J. W. Prothero, *Biophys. J.* **6**, 367 (1966).
154. B. H. Havsteen, *J. Theor. Biol.* **10**, 1 (1966).
155. C. C. Bigelow, *J. Theor. Biol.* **16**, 187 (1967).
156. B. W. Low, F. M. Lovell, and A. D. Rudko, *Proc. Nat. Acad. Sci. U.S.* **60**, 1519 (1968).
157. D. Kotelchuck and H. A. Scheraga, *Proc. Nat. Acad. Sci. U.S.* **61**, 1163 (1968).
158. P. R. Schimmel and P. J. Flory, *J. Mol. Biol.* **34**, 105 (1968).
159. H. A. Scheraga, *in* "Molecular Architecture in Cell Physiology," p. 39. 1966; *J. Gen. Physiol.* **50**, 5 (1967).
160. S. V. Konev, V. M. Mozhul, and E. M. Chernitsky, *Dokl. Akad. Nauk Belorussk. SSR* **12**, 1122 (1968).
161. M. Brunory, *J. Mol. Biol.* **34**, 497 (1968).
162. P. Dunill, *Sci. Progr. (London)* **53**, 609 (1965).
163. O. B. Ptitsin, *Usp. Sovrem. Biol.* **69**, 26 (1970).
164. D. C. Phillips, *Proc. Nat. Acad. Sci. U.S.* **57**, 484 (1967).
165. R. Sinsheimer, *J. Mol. Biol.* **1**, 37 (1959).
166. R. E. Franklin, *Biochem. Biophys. Acta* **18**, 313 (1955).
167. M. Huber, *Acta Crystallogr.* **10**, 129 (1957).
168. M. Spencer, *Acta Crystallogr.* **12**, 59, 66 (1959).
169. A. A. Lugovskoy and V. G. Dashevsky, *Molek. Biol.* **6**, 440 (1972).
170. D. G. Watson, D. J. Sutor, and P. Tollin, *Acta Crystallogr.* **19**, 111 (1965).
171. P. Tollin, H. R. Wilson, and D. W. Young, *Nature (London)* **217**, 1148 (1968).
172. M. Sundaralingam, *J. Amer. Chem. Soc.* **87**, 599 (1965).
173. M. Sundaralingam and L. H. Jensen, *J. Mol. Biol.* **13**, 914 (1965).
174. M. Sundaralingam and L. H. Jensen, *J. Mol. Biol.* **13**, 930 (1965).
175. R. D. Harris and W. M. Macintyre, *Biophys. J.* **4**, 203 (1964).
176. J. Kraut and L. H. Jensen, *Acta Crystallogr.* **16**, 79 (1963).
177. K. N. Trueblood, P. Horn, and V. Luzzati, *Acta Crystallogr.* **14**, 965 (1965).
178. A. E. Kister and V. G. Dashevsky, *Itogi Nauki Khim. Nauki* **21**, 167 (1970).
179. S. Arnott, *Progr. Biophys. Mol. Biol.* **21**, 265 (1970).
180. S. Arnott and D. W. L. Hukins, *Nature (London)* **224**, 886 (1969).
181. J. Donohue and K. N. Trueblood, *J. Mol. Biol.* **2**, 363 (1960).
182. A. E. V. Haschemeyer and H. M. Sobell, *Acta Crystallogr.* **10**, 125 (1965).
183. T. R. Emerson, R. J. Swan, and T. L. V. Ullbricht, *Biochemistry* **6**, 843 (1967).
184. A. E. V. Haschemeyer and A. Rich, *J. Mol. Biol.* **27**, 369 (1967).
185. I. Tinoko, Jr., R. C. Davis, and S. R. Jaskunas, *in* "Molecular Associations in Biology" (B. Pullman, ed.), p. 77. Academic Press, New York, 1968.
186. A. V. Lakshminarayanan and V. Sasisekharan, *Biopolymers* **8**, 475 (1969).
186a. A. A. Lugovskoy, V. G. Dashevsky, and A. I. Kitaigorodsky, *Molec. Biol.* **6**, 449 614 (1972).
187. A. E. Kister, V. G. Dashevsky, and A. I. Kitaigorodsky, *Molek. Biol.* **5**, 232 (1971).
188. V. Sasisekharan, A. V. Lakshminarayanan, and G. N. Ramachandran, *in* "Conformation of Biopolymers" (G. N. Ramachandran, ed.), Vol. 2, p. 641. Academic Press, New York, 1967.
189. A. V. Lakshminarayanan and V. Sasisekharan, *Biopolymers* **8**, 489 (1969).
190. V. Sasisekharan and A. V. Lakshminarayanan, *Biopolymers* **8**, 505 (1969).
191. O. Sinanoğlu, and S. Abdulnur, *Fed. Proc. Fed. Amer. Soc. Exp. Biol. Part 2* **24**, 12 (1965).

192. H. De Voe and I. Tinoko, Jr., *J. Mol. Biol.* **4**, 500 (1962).
193. W. Kauzmann, *Advan. Protein Chem.* **14**, 1 (1959).
194. K. Pollak and R. Rein, *J. Chem. Phys.* **47**, 2045 (1967).
195. P. Claverie, B. Pullman, and J. Caillet, *J. Theor. Biol.* **12**, 419 (1966).
196. R. Rein, P. Claverie, and M. Pollak, *Int. J. Quantum Chem.* **2**, 129 (1968).
197. V. I. Poltev and B. I. Sukhorukov, *Biofizika* **12**, 763 (1967); **13**, 941 (1968).
198. G. D. Fasman, C. Lindblow, and L. Grossman, *Biochemistry* **3**, 1015 (1964).
199. R. C. Davis and I. Tinoko, Jr., *Biopolymers* **6**, 223 (1968).
200. S. I. Chan, M. P. Schweizer, P. O. P. T'so, and G. K. Helmkamp, *J. Amer. Chem. Soc.* **86**, 4182 (1964).
201. M. P. Schweizer, S. I. Chan, and P. O. P. T'so, *J. Amer. Chem. Soc.* **87**, 5241 (1965).
202. S. I. Chan, B. W. Bangerter, and H. H. Peter, *Proc. Nat. Acad. Sci. U.S.* **55**, 720 (1966).
203. C. C. McDonald, W. D. Phillips, and J. Lazar, *J. Amer. Chem. Soc.* **89**, 4166 (1967).
204. A. D. Broom, M. P. Schweizer, and P. O. P. T'so, *J. Amer. Chem. Soc.* **89**, 3612 (1967).
205. M. P. Schweizer, A. D. Broom, P. O. P. T'so, and D. P. Hollis, *J. Amer. Chem. Soc.* **90**, 1042 (1968).
206. I. Feldman, and R. P. Agarval, *J. Amer. Chem. Soc.* **90**, 7329 (1968).
207. D. N. Holcomb and I. Tinoko, Jr., *Biopolymers* **3**, 121 (1965).
208. M. M. Warshaw and I. Tinoko, Jr., *J. Mol. Biol.* **13**, 54 (1965).
209. D. Poland, J. N. Vournakis, and H. A. Scheraga, *Biopolmers* **4**, 223 (1966).
210. M. W. Warshaw, C. A. Bush, and I. Tinoko, Jr., *Biochem. Biophys. Res. Commun.* **18**, 633 (1965).
211. M. M. Warshaw and I. Tinoko, Jr., *J. Mol. Biol.* **20**, 29 (1966).
212. W. B. Gratzer and D. A. Cowbarn, *Nature* (*London*) **222**, 426 (1969).
213. J. Brahms, A. M. Michelson, and K. E. Van Holde, *J. Mol. Biol.* **15**, 467 (1966).
214. K. E. Van Holden, J. Brahms, and A. M. Michelson, *J. Mol. Biol.* **12**, 726 (1965).
215. C. A. Bush and J. Brahms, *J. Chem. Phys.* **46**, 79 (1967).
216. M. Leng and G. Felsenfeld, *J. Mol. Biol.* **15**, 455 (1966).
217. J. Applequist and V. Damle, *J. Amer. Chem. Soc.* **88**, 3895 (1966).
218. P. O. P. T'so, I. S. Melvin, and A. C. Olson, *J. Amer. Chem. Soc.* **85**, 1289 (1963).
219. P. O. P. T'so and S. I. Chan, *J. Amer. Chem. Soc.* **86**, 4176 (1964).
220. E. Shefter, M. Barlow, R. Sparks, and K. N. Trueblood, *J. Amer. Chem. Soc.* **86**, 1872 (1964).
221. J. Brahms, A. M. Michelson, and K. E. Van Holde, *J. Mol. Biol.* **15**, 467 (1966).
222. K. E. Van Holde, J. Brahms, and A. Michelson, *J. Mol. Biol.* **12**, 726 (1965).
223. J. R. Fresco, B. M. Alberts, and P. Doty, *Nature* (*London*) **188**, 98 (1960).
224. J. R. Fresco and B. M. Alberts, *Proc. Nat. Acad. Sci. U.S.* **46**, 311 (1960).
225. J. R. Fresco, L. C. Clotz, and E. J. Richards, *Cold Spring Harbor Symp. Quant. Biol.* **28**, 83 (1963).
226. V. G. Tumanyan, L. E. Sotnikova, and A. V. Kholopov, *Dokl. Akad. Nauk SSSR* **166**, 1465 (1966).
227. R. Langridge, D. A. Marvin, W. E. Seeds, H. R. Wilson, C. W. Hooper, M. H. F. Wilkins, and L. D. Hamilton, *J. Mol. Biol.* **2**, 19 (1960).
228. W. Fuller, M. H. F. Wilkins, H. R. Wilson, L. D. Hamilton, and S. Arnott, *J. Mol. Biol.* **12**, 60 (1965).
229. D. A. Marvin, M. Spencer, M. H. F. Wilkins, and L. D. Hamilton, *J. Mol. Biol.* **3**, 547 (1961).
230. S. Arnott, S. D. Dover, and A. J. Wonacott, *Acta Crystallogr. Sect. B* **25**, 2192 (1969).
230a. S. Arnott and D. W. L. Hukins, *Biochem. Biophys. Res. Commun.* **47**, 1504 (1972).
231. V. G. Dashevsky, and A. E. Kister, *in* "Conformation of Biopolymers in Solutions," p. 91. Nauka, Moscow, 1973.

# Author Index

Numbers in parentheses are reference numbers and indicate that an author's work is referred to although his name is not cited in the text. Numbers in italics show the page on which the complete reference is listed.

## A

Abdullaev, N. D., 500(136), *534*
Abdulnur, S., 521, *535*
Abrahams, S. C., 38(15), *131*, 178, *192*
Adrian, E. J., 385, 421(16), *446*
Afanassieva, G., 358, 379(15), *380*
Agarval, R. P., 523(206), *536*
Ahmed, F. R., 108(76), *132*, 419(128), *449*
Akopyan, Z. A., 14(8), *130*, 388(30), 421(30), 426(30, 139), *447, 449*
Alberts, B. M., 526(223}, *536*
Aleksanyan, V. T., 400(66), *447*
Alekseev, N. V., 402(88), *448*
Alexandrov, K. S., 308(3), 311(4), *336*
Ali, A. A., 416, *449*
Allegza. G., 471, *532*
Allen, G., 396(56), *447*
Allen, L. C., 395(49), *447*
Allen, N. C., *433*
Allen, T. L., 427, *449*
Allinger, N. L., 382(9), 394, 400, 402(9, 86),
407(68), 429(68), *446, 447, 448*
Almenningen, A., 385(15), *433, 446*
Amoros, J. L., 268(7), *293*
Amoros, M., 268(7), *293*
Andrew, E. R., 277, 278, *293*
Anfinsen, C. B., 503(143), *534*
Angyal, S. L., 382(9), 402(9), *446*
Applequist, J., 523(217), *536*
Arkhipova, S. F., 486, 487(110), 488(110, 121), 491(121), 492(121), 494(128), 495(128), 497(121), 499(133), 500(136), *534*
Arnott, S., 515, 516, 526(228), 527, 528, 529, *535, 536*
Arridge, R. G. S., 145(18), *191*
Asahina, M., 474(64), *532*
Aston, J. G., 401(82), *448*
Avakyan, V. G., 443, *450*
Avitabile, G., 182(55), *192*
Avoyan, R. L., 64(41), *131*, 382(10), 413(117), 423(134), 425(137,138), 426 (140), *446, 449*

537

Axilrod, B. M., 137, *191*

**B**

Babushkina, T. A., 88(59), *132*, 289, *293*
Bacon, G. E., 171, *191*
Badalyan, D. A., 93(62,62a), 118(62,87), *132*
Badamy, D., 473(51), *532*
Balasubramanian, D., 502, *534*
Balcon, Y., 89(60), *132*
Bangerter, B. W., 523(202), *536*
Barker, C., 437(173), *450*
Barlow, M., 523(220), *536*
Barnes, W. H., 108(76), *132*
Bartell, C. L., 399(64), *447*
Bartell, L. S., 183, *192*, 399, *429*, *433*, *447*
Bassi, I., 456(12), 473(12,52.53), *531*, *532*
Bastiansen, O., 385(15), 413(116), *433*, *446*, *449*
Bates, T. W., 465(36), 466, *532*
Bayer, H., 289, *293*
Bayle, G. G., 53(37), *131*
Bazhoulin, P. A., 222(9), *232*
Bekoe, D. A., 424(135), *449*
Belikova, G. S., 114, *132*
Belyaev, L. M., 114, *132*
Benedetti, E., 471(48a), *532*
Benson, S. W., 427(155), *450*
Bernstein, H. J., 427, 430, 434(168), 441(179), *449*, *450*
Berthod, H., 484, *533*
Bezzi, S., 73(50), *131*
Bigelow, C. C., 504(155), *535*
Binne, W. P., 81(55), *131*
Bird, R., 158(24), *191*
Birshtein, T. M., 394(43), *447*, 456(13), 474, 475, *531*
Bistrov, V. F., 494(125), *534*
Bixon, M., 393(38), 404, 429(38), *447*
Bloom, S. M., 504(146), *534*
Blout, E. R., 504(146), *534*
Bochert, A., 473(61), *532*
Boggus, J. D., 122(89), *132*
Bolotnikova. T. N., 115(84), *132*
Bolton, W., 10(9), 17(9), *130*
Bonamico, M., 129(95), *133*
Bondi, A., 17, *130*, 388, *447* 466(37), *532*
Bonham, R. A., 399(64), *429*, *433*, *447*
Bonnell, D. W., 429(162), *450*

Borisova, N. P., 410(105), *448*, 456(14), 474, 475, *531*
Born, M., 379(14), *380*
Bradley, C. A., 443, *450*
Brahms, J., 523(213, 214, 215), 524, (213), *536*
Brant, D. A., 388(29), 389(29), *446*, 479(83), 482(83), 483(83), 485, 487, 494(126), 496(131), 506, *533*, *534*
Brathovde, J. R., 50(28), *131*
Breil, H., 452(3), *531*
Brenner, S., 259(2), *293*
Bressan, G., 475(68), *532*
Bridgman, P. W., 330, *336*
Brier, P. N., 397(56), *447*
Brill, R., 470(45), *532*
Britts, K., 259(2), *293*, 401(73), *448*
Broom, A. D. 523(204,205), *536*
Brown, D. S., 122(90), *133*
Brunory, M., 509(161), *535*
Buckingham, R. A., 138, *191*
Bucourt, R., 411(109), *448*
Bunn, C. W., 51, *131*, 464, 465(32), 470, 471(25), 477(25), *531*, *532*
Burleigh, P., 474(63), *532*
Bush, C. A., 523(210,215), *536*
Buss, J. H., 427(155), *450*
Butcher, S. S., 412(112, 115), *448*, *449*

**C**

Caillet, J., 522(195), *536*
Campbell, E. S., 145, *191*
Cannon C. G., 145(18), *191*
Carazzolo, G. A., 477(72), *532*
Carle, J., 397(57), *447*
Caron, A., 65(42), *131*
Carrol, S., 429(162), *450*
Casalone, G. I., 190(63), *192*
Cesari, M., 478(77), *533*
Chan, S. I., 523(200, 201, 202, 219), *536*
Chatani, Y., 473(57, 58), 477(74), 478(74 76), *532*, *533*
Chernitsky, E. M., 509(160), *535*
Chevreton, J., 474(65), *532*
Chung, D. H., 325(12, 13), *336*
Clastre, J., 73(49), *131*
Claverie, P., 522(195, 196), *536*
Clementi, E., 394, *447*
Clotz, L. C., 526(225), *536*

Cochran, W., 454, *531*
Cohen, C., 504(145), *534*
Colin, G., *164*
Cook, D. A., 504, *534*
Corey, R. B., 480, *533*
Corner, J., 138, *191*
Corradini, F., 452(4), *531*
Corradini, P., 182, *192*, 452, 454, 456(12), 464, 472(49), 473(12, 52, 53, 54, 59), 474(49), *531, 532*
Cortili, G., 467(42), *532*
Cottrell, T. L., 426(142), *449*
Coulson, C. A., 414, 416, 417, 418, 419(126), 430, *449, 450*
Coulter, C. L., 424(136), *449*
Cowbarn. D. A., 523(212), *536*
Cox, D., *440*
Cox, E. G., 77(52), *131*, 171, 187(40), *191*
Craig, D. P., 145(17), 153, 160, 174, *191*
Craven, B., 288(23), *293*
Craven, B. M., 416(121), *449*
Crawford, B., *433*
Crick, F. C. H., 454(11), *531*
Cross, P. C., 442(182), 443(182), *450*
Cruickshank, D. W. J., 171(40), 181, 187(40), *191, 192*
Curry, N. A., 171(42), *191*
Curtiss, F., 158(24), *191*

**D**

Dailey, B P., 401(70), *448*
Dakhis, M. I., 443, *450*
Dallinga, G., *408*, 412(110, 113a), *448, 449*
Damiani, A., 396(52), *447*
Damle, V., 523(217), *536*
Danusso, F., 452(4. 6, 7), 473(55), *531, 532*
Dashevsky, V. G., 14(8), *130*, 189(61, 62), *192*, 382(10), 383, 387(23, 24), 388, 390, 393, 396(53), *405*, 407(100, 101), 409(104), 411(99, 104, 108), 413(104, 117), 420, 421, 426, 428, 429(160), 430(164), 436, 439(101), 441(177), 443, *446, 447, 448, 449, 450*, 464, 467(31), 468(31), 475(31), *486*, 487(110), 488(110, 121), 491(121), 492(121), 497(121), 500 (134), 507(134), 512(169), 514(178), 515(178), 518(187), 528(231), *532, 534, 535, 536*
Davies, M., 487(113), *534*
Davis, D. R., 394, *447*, 480(98), *533*

Davis, M., 402(89), *448*
Davis, R. C., 518(185), 523(199), *535, 536*
Dean, C., 288, *293*, 416(121), *449*
DeCoen, J. L., 396(52), *447*
Dehmelt, H., 282, *293*
de Loze, C., 504(146), *534*
Del Re, G., 484, *533*
Dereppe, J. M., 272, *293*
De Santis, P., 387(28), 388(28), *446, 463*, 464(29) 465(29), 466, 468(29), 470, 472, 477(29), 479(82), 502(138), *531, 533, 534*
Descius, J. C., 442(182), 443(182), *450*
Desreux. V., 452(8), *531*
De Voe, H., 522, *535*
Dewar, M. J. S., 430, 434(167), *450*
De Wette, F. W., 145, *191*
Dickinson, C. D., 487(114), *534*
Didenko, L. A., 426(140, 141), *449*
Donath, W. E., 401(84). *448*
Donohue, J., 22(14), 65(42), *131*, 480(99), 516, *533, 535*
Doty P., 526(223), *536*
Doty, P. M., 401(82), *448*
Dougill, M. W., 77(52), *131*
Douglass, D. C., 277, *293*
Dover, S. D., 527(230), *536*
Dulleton, H. J., 467(41), *532*
Duncan, N. E.. *440*
Dun-chai, L., 110(80), *132*
Dunill. P., 509(162), *535*
Dunitz, J. D., 62(40), *131*
Dunning, T. H., Jr., *395*

**E**

Eades, R., 277, *293*
Eddy, C. R., 459, *531*
Edsall, J. T., 480(100), *533*
Efremov, E. S., 500(136), *534*
Egelstaff, P. A., 268(8), *293*
Ehrlich, H. W. W., 188(60), *192*
Elefante, G., 396(52), *447*
Eliel, E. L., 382(9), 402(9), *446*
Elyashevich, M. A., 443(183), *450*
Emerson, T. R., 517(183), *535*
Emsley, J. W., 279, *293*
Epand, R. F., 479(90), *533*
Eveno, M., 89(60), *132*
Evstratov, A. V., 500(136), *534*
Ewald, P. P., 145, *191*

Ewens, R. N. G., 65(43), *131*
Eyrand, C., 474(65), *532*
Eyring, H., 457, *531*

**F**

Fajans, K., 426(143), *449*
Farina, M., 475(68), *532*
Fasman, G. D., 504(146), 523(198), *534, 536*
Favini, G., 421(131, 132), *449*
Fedin, E. I., 115(85) *132* 276, 280, 290, *293*
Fedorov, Ju. C., 468(44), *532*
Felchenfeld, H., 434(171), *450*
Feldman, I., 523(206), *536*
Felsenfeld, G., 523(216), *536*
Fink, H. L., 401(82), *448*
Fink, W. H., 395(49), *447*
Fisher, M. S., 401(79), *448*
Fletcher, R., 405(92, 95), *448*
Fletcher, W. H., *433*
Flitcroft, T., 439(175), *450*
Flory, P. J., 388, 389, *446*, 452(2), 456, 479, 480(100), 482(83), 483(83), 485, 487, 494, 506, *531, 533, 534, 535*
Fox, D., 89(61), *132*
Franklin, J. L., 427(154), 429(162), *450*
Franklin, R. E., 510(166), *535*
Fransen, J. S., 494(129), 495(129), *534*
Frasson, E., 73(50), *131*
Freiberg, L. A., 402(86), *448*
Fresco, J. R., 526(223, 225), *536*
Frish, M. A., 437(173), *450*
Frolova, A. A., 126, *133*
Fukushima, K., 488(119), *534*
Fuller, C., 477(73), *532*
Fuller, W., 526(228), *536*
Furberg, S. 402(85), *448*
Furukawa, J., 470(48), 471(48), *532*
Fyodorov, Yu. G., 260(5), *293*

**G**

Gafher, G., 421(133), *449*
Galaktionov, S. G., 459, 497, *531, 534*
Galashkevich, I. P., 325(14), *336*
Galperin, E. L., 467(39), *532*
Gamba, A., 421(132), *449*
Ganis, P., 182(55), *192*, 472(49), 473(54, 59), 474(49), *532*
Garbuglio, C., 73(50), *131*
Gay, G., 465(34), *532*
Geffrey, G. A., 416(121), *449*

Gelfand, I. M., 260(5), *293*, 468(44), 503(140), *532, 534*
Gibbions, C. S., *408*
Gibson, K. B., 479(93, 94), 489, *533*
Gibson, K. D., 405(94), *448*, 488, *534*
Giglio, E., 182, 184, *192*, 387(28), 388(28), *446*, *463*, 464(29), 465(29), 466(29), 468(29), 470(29), 472(29), 477(29), 479(82), *531, 533*
Ginsburg, S. L., 260(5), *293*, 468(44), *532*
Glockler, G., 434(170), *450*
Go, N., *405*
Goeppert-Mayer, M., 169, *191*
Goldberger, R. F., 481(103), *533*
Golebiewski, A., 419(126), *449*
Goodisman, J., 394, *447*
Goodman, M., 475(67), *532*
Goodman, T. H., 421(130), *449*
Gorskaya, N. V., 280, *293*
Grandal, W., 325(13), *336*
Gratzer, W. B., 523(212), *536*
Gray, T., 325(13), *336*
Green, D. A., 481(103), *533*
Greenberg, B., 401(78), *448*
Grosjean, D., 290(26), *293*
Grossman, L., 523(198), *536*
Gurskaya, G. V., 82(56), 83(56), *132*
Gutowsky, H. S., 274, 275, 276, *293*
Guzzo, A. V., 504(147), *534*
Gwinn, W. D., 401(69, 80), *447, 448*

**H**

Haber, E., 503(143), *534*
Haigh, C. W., 417, 418, *449*
Halle, F., 470(45), *532*
Hamilton, L. D., 526(227, 228, 229), *536*
Hammer, C., 477(70), *532*
Hamming, R., 271, 272(10), *293*
Hargreaves, A., 385(14), *446*
Harmony, M. D., 412(113), *449*
Harris, R. D., 513(175), *535*
Harsteen, B. H., 504(154), *535*
Haschemeyer, A. E. V., 517, 519(182), *535*
Hassel, O., 402(89), *448*
Haunaut, D., 411(109), *448*
Hearmon, R. F. S., 323(9) *336*
Heath, D. F., 444, *450*
Hedberg, K., 413(116), *449*
Hedberg, L., 413(116), *449*
Helmkamp, G. K., 523(200), *536*

Hendrickson. J. B., 387, 393, 401, 402, 405 (97), *446*, *448*
Herbstein F. H., 73(48). 123(91), *131*, *133*, 412, 421(133), *448*, *449*
Herraez, M. A., 414(119), *449*
Herzfeld, K. F., 169, *191*
Hill, R., 324, *336*
Hill, R. M., 158(22), 159, *191*
Hill, T. L., *164*, 382, 408, *446*
Hirota, E., 396(54), *447*
Hirschfelder, J. O., 158(24), *191*
Hirsh, J. A., 394(39), 400(68), 407(68), 429(68), *447*
Hitta, J., 473(58), *532*
Hoard, J. L., 401(71), *448*
Hodgkin, D. C., 502(137), *534*
Holcomb, D. N., 523(207), *536*
Hollis, D. P., 523(205), *536*
Hollowell C. D., *433*
Holzkamp E., 452(2), *531*
Hooper, C. W., 526(227), *536*
Horn, P., 513(177), *535*
Housty, J., 73(49), *131*
Hove, J., 167(33), *191*
Howells, R. E., 465(32) *532*
Hoyland J. R. 395, *447*
Huang, K., 379(14), *380*
Huber, M., 511(167), *535*
Huggins, M. L., 452(8), 464, *531*
Huggis, G. H., 481(101), *533*
Hughes, R., 460(21), *531*
Hukins, D. W. L., 516, 529, *535*
Huntington, H., *380*
Hyndman, D., 278, *293*

**I**

Ideguchi, I., 475(66), *532*
Ideguchi, Y., 488(119), *534*
Iitaka, Y., *81*, 249, *293*
Imada, K., 478(76), *533*
Ingham, L., 139, 167, *191*
Isemura, I., 503(144), *534*
Ismail-Zade, I. G., 70(46), *131*
Ito, M., 221(8), 222(8), *232*, 358(4), *380*
Ivanov, V. T., 494(125), 500(136), *534*
Iwasaki, M., 465(33), 466, *532*

**J**

Jansen, L., 137, *191*
Janz, G. J., *440*

Jaskunas, S. R., 518(185), *535*
Jeffrey, G., 288(23), *293*
Jeffrey, G. A., 77(52), 129, *131*, *133*
Jensen, L. H., 513, 514(173, 176), *535*
Jernigan, R. L., 456(15), *531*
Jordan, T., 129(93), *133*

**K**

Kabalkina, S. S., 62, *131*
Kaluski, Z. L., 64(41), *131*
Karl, I. L., 259(3), *293*
Karle, I. L., 401(72, 73), *448*
Karle, J., 401(72, 73), *448*
Katchman, A., 473(61), *532*
Katz, I. L., 486(112), *534*
Kauzmann, W., 522, *535*
Kendrew, J. C., 480(100), *533*
Ketelaar, J., 389(33), *447*
Khachaturyan, A. G., 93(62), 118(62, 87), *132*
Kholopov, A. V., 526(226), *536*
Khotsianova, T. L., 9(5), *130*, 88(59), *132*
Kilpatrick, J. E., 401(83), *448*
Kim, H., 401(69), *447*
Kirkwood, J., 136, 140, *191*
Kirkwood, J. G., 389, *447*
Kister, A. E., 514(178) 515(178) 518(187), 528(231), *535*, *536*
Kistiakowsky, G. B., 439(174), *450*
Kitaigorodsky, A. I., 1(1), 7(4), 9(5), 17(11), 33(1), 38(16), 50(29), 93(62), 95(63), 96(64), 97(67), 101(68), 103(71, 72, 73), 106(75), 110(77, 80), 114(81), 115(85), 116(86), 118(62), 122(75), 126 (92), *130*, *131* *132*, *133*, 144, 161(26), *164*, 170(39), 171, 173(43), 177(41), 181(51), 184(58), 189(61, 62), *191*, *192*, 212(2), 220(7), 222(10), *232*, 259(2), 260(4, 4a), 290, *293*, 351(1), 369(7), 373(9, 10), 374(11), 377(1), *380*, 382, 383, 387, 390(11, 12), 393, 397, 398, 402, 407, 413, 420 421(11, 12), 425(137, 138), 426, 436, *446*, *447*, *448*, *449*, 486, 518(187), *534*, *535*
Klimova, L. A., 115(84), *132*
Klinskikh, N. A., 322(8), *336*
Klotz, I. M., 494(129), 495(129), *534*
Kobayashi, M., 460(23), *531*
Koch, T., 477(70), *532*
Kogan, G. A., 434(172), *450*

Kohl, D. A., 399, *429*, *447*
Koide, T., 169(35), *191*
Kolosov, N. Ya., 110(78, 79), *132*
Konev, S. V., 509(160), *535*
Koreshkov, B. D., 352(2), 369(7), *380*
Koshin, V. M., 18(12), *130*
Kotelchuck, D., 504(157), 505, 508(157), *535*
Kovacs, A. L., 502(138), *534*
Kozhin, V. M., 171, 177(41), 179, 181(51, 53), *191*, *192*, 358(5), *380*
Kozlova, I. E., 170(38), 178(46), *191*, *192*
Kraut, J., 513(176), 514(176), *535*
Kreevoy, M. M., 394, *447*, 472, *532*
Krimm, S., 464(28), 482(105), *493*, 495, *531*, *533*, *534*
Krüger, H., 282, *293*
Krumhansl, J. A., 167(33), *191*
Kuchitsu, K., *408*, *433*
Kudchadker, A. P., 441(178), *450*
Kuziyanz, G. M., 400(66), *447*
Kybett, B. D., 429(162), *450*

### L

Laane, L., 411(106), *448*
Labes, M. M., 89(61), *132*
Laidler, K. J., 427, *450*
Lakshminarayanan, A. V., 518(186), 519 (186, 188), 520, 521, *535*
Landau, L. D., 162(28), *191*, 193(1), *232*, 334(17), *336*
Lando, J. B., 467(40), *532*
Langridge, R., 526(227), *536*
Lass, G., 412(113), *449*
Lauer, J., 460(21), *531*
Laurie, V. W., *433*
Lazar, J., 523(203), *536*
Leach, S. J., 479(85, 86, 87, 88. 92, 93), 485(85), 497(88), 502(92), *533*
Lehn, F. M., *395*
Lemaire, H. P., 401(74), *448*
Leng, M., 523(216) *536*
Lennard-Jones, J. E., 138, 139, 167, *191*
Leopoldo, S., *164*
Lepard, D. W., 396(55), *447*
Le Roy, L. V., *433*
Lide, D. R., *433*
Lifshitz, E. M., 162(28), *191*, 193(1), *232*, 334(17), *336*

Lifson, S., 393(38), 404, 429(38, 161), 445, *447*, *450*
Liguori, A. M., 387, 388, *446*
Lindblow, C., 523(198), *536*
Lingafelter, E. C., 50(28), *131*
Linnett, J. W., 444, *450*
Lipkind, G. M., *486*, 487(110), 488, 491 (121), 492(121), 494, 495(128), 497(121), 499(133), 500(136), 508, *534*
Lippincott, E. R., 485, *533*, *534*
Lipscomb, W. N., 394, *447*
Lipson, H., 262(6), *293*
Liquori, A. M., 182, 184, *192*, 396(52), *447*, *463*, 464, 465(29), 466(29), 468(29), 470, 472(29), 477(29), 479, 480(100), 502, *531*, *532*, *533*, *534*, *535*
Lisitsin, V. N., 426(140, 141), *449*
Livingston, R. L., 401(74), *448*
Loanem, J., 401(81), *448*
London, F., 135, *191*
Lord, C., 411(106), *448*
Lord, R. C., 401(81), *448*
Lovell. F. M., 504(156), *535*
Low, B. W., 504(156), *535*
Lugovskoy, A. A., *405*. 411(99), *448*, 512(170), *535*
Lukina, M. Yu., 400(66), *447*
Luntz, A. C., 401(75, 80), *448*
Luxmoore, A. R., 66(44), *131*
Luzzati, V., 513(177), *535*
Lwasaki, F., *164*

### M

McCall, D. W., 277, *293*
McCall, G., 271, 272(10), *293*
McCullough, J. P., 134, *190*, 328, 330, *336*
McCullough, R. L., 399, *447*, 465(35), 466(35), *532*
McDonald, C. C., 523(203), *536*
McGlynn, S. P., 122(89), *132*
Macintyre, W. M., 513(175), *535*
McLachlan, D., 68(45), *131*
McMahon, P. E. 399, *447*, 465(35), 466(35), *532*
McMullan, R. K., 129(93, 94, 95), *133*
Magnasco, V., 397, *447*, 465(34), *532*
Mak, T. C. W., 87(58), *132*
Mantica, E., 452(4), *531*
Margenau, H., 135, *191*

Margrave, J. L., 429(162), 437(173), *450*
Margulis, T. N., 401(76 77, 79), *448*
Mariani, C., 190(63), *192*
Marino, U., *408*
Mark, H., 452(8), *531*
Mark, J. E., 495, *534*
Markham, M., 325(11), *336*
Marsh, R. E., 480(99), *533*
Martin, H., 452(3), *531*
Martuscalli, E., 182(55), *192*
Marvin, D. A., 526(227, 229), *536*
Mason, E. A., 390(34), 394, *447*, 472, *532*
Mason, R., 153, 160(20), 174(20), 181, *191*, *192*
Mathieson, A. McL., 38(17), *131*, 181, *192*
Mayer, J. E., 382, *446*
Mazel, W. H., 53(37), *131*
Mazzanti, G., 452(4), *531*
Mazzarella, L., 502(138), *534*
Mazzei, A., 478(77), *533*
Meador, W. R., 439(176), *440*, *450*
Meiboom, S., 400(65), *447*
Meinnel, I., 89(60), *132*
Melvin, I. S., 523(218), *536*
Michelson, A. M., 523(213, 214), 524(221, 222), 525(213), *536*
Michie, D., 481(101), *533*
Mie, G., 138, *191*
Mijakawa, T., 478(76), *533*
Milazzo, G., *191*
Miller, M. A., 394(39), 400(68), 407(68), 429(68), *447*
Miller, R. L., 467(41), *532*
Miller, W. J., 388(29), 389(29), *446*, 494 (126), *534*
Mirskaya, K. V., 17(11), *130*, 144, *164*, 170(38), 178(46), *191*, *192*, 374(11), 377(12), 378(12), *380*
Miyazawa, T., 460(22, 24), 462, 475, 488 (118, 119), *531*, *532*, *534*
Mizushima, M., 158(23), *191*
Mizushima, S., 459, 488(118), *531*, *534*
Mlenik, M. P., 467(39), *532*
Mnyukh, Yu. V., 48(24), 74, 116(24, 86), *131*, *132*
Monfils, A., 290, *293*
Moraglio, G., 452(4), *531*
Morino, Y., 396(54), *447*
Morokuma, R., 395(50), *447*
Morrison, G. A., 382(9), 402(9), *446*

Mortimer, C. T., 427(157), 438(157), *450*
Mortimer, K., *440*
Morton-Blake, D. A., 421(130), *449*
Mozhul, V. M., 509(160), *535*
Müller, A., 48(23), 50, 52, 53, *131*, 162, *191*
Mugnoli, A., 190(63), *192*
Muir, A. R., 481(101), *533*
Mukhtarov, E. I., 220, 222(10), *232*, 372, 373(9, 10), *380*
Munn, D. E., *433*
Munn, R. J., 390(34), *447*
Murahashi, S., 477(71 74) 478(74, 76) *532*, *533*
Murtasina, I. O., 464, 467(31), 468(31), 475(31), *532*
Musgrave, M. J. P., 307(2), *336*
Myasnikova, R. M., 106(75), 110(77), 114(81), 122(75), *132*, 181, *192*

**N**

Nagai, K., 460(23), *531*
Nagai, O., 154, *191*
Nakamura, T., 154, *191*
Natalis, P., 429(162), *450*
Natta, G., 452, 454, 456, 464, 472, 473(12, 52, 53), 474, 475, *531*, *532*
Nauchitel, V. V., 17(11), *130*, *164*
Naumov, V. A., 407(100, 101), 409(104), 411(104, 108), 413(104), 429(160), 439 (101), *448*, *450*
Nemethy, G., 479(84, 86, 87, 88, 92), 480(100), 482(84), 488(123), 489(123), 497(88), 502(92), *533*, *534*
Nercesova. G. N., 115(84), *132*
Newman, M. S., 437(173), *450*
Nicora. C., 465(34), *532*
Nielsen, A. H., *433*
Nitta, I., 477(71), *532*
Noether, H. D., 473(60), 474(60), *532*
Nosakura, S., 473(56), *532*
Nowacki, W., 11(6), *130*
Nurahashi, S., 473(56), *532*

**O**

Okuda, K., 474(64), *532*
Olf, H. G., 467(40), *532*
Oliver, D. A., 220, *232*
Olson, A. C., 523(218), *536*

Onishi, S., 169(35), *191*
Ooi, T., 479(90, 91, 93), 482(91), 485, 498, *533*
Opascar, C. G.. 464(28), *531*
Orgel, L. E., 62(40), *131*
Oth, J. F. M., 430(163), *450*
Oughton, B. M., 502(137), *534*
Ovchinnikov, Yu. A., 500(135, 136), *534*
Ovchinnokiv, Yu. A., 494(125), *534*
Overberger, C., 473(61), *532*
Overend, J., *433*
Owen, T. B., 401(71), *448*

P

Pake, G. E., 274, *293*
Pake, G. S., 282, *293*
Palke, W. E., 395(51), *447*
Papulov, Yu. G., 441(180), *450*
Papulov, Yu. T., 427, *450*
Pauling, L., 158(25), *191*, 426(144), *449*, 480, *533*
Pauling. P., 153, 160(20), 174(20), *191*
Pawley, G. S.. 220, *232*, 268, *293*
Pedersen, D., 395(50), *447*
Pedley, J. B., 428, *450*
Pedone, C., 471(48a), *532*
Peraldo, G., 478(77), *533*
Peraldo, M., 475(68), *532*
Periti, P. F., 504(152), *535*
Perutz, M. F., 504(149), *534*
Peter, H. H., 523(202), *536*
Peterlin, A., 467(40), *532*
Phillips, D. C., 72(47), 108(76), *131*, *132*, 509, *535*
Phillips, W. D., 523(203), *536*
Pickett, H. M., 405(98), *448*
Pino, P., 452(4), *531*
Pinsker, Z. G., 50, *131*
Piper, S. H., 52, *131*
Pitzer, K. S., 49, *131*, 162, *191*, 401, *448*
Pitzer, R. M., 394, 395(51), *447*
Plyle, E. K., *433*
Poland, D., 479(85), 485(85), 523(209), 525, *533*, *536*
Pollak, K., 522(194, 196), *536*
Pollak, M., 288(23), *293*, 416(121), *449*
Poltev, V. I., 386, *446*, 522(197), 523(197), *536*
Polyak, B. T., 405(93), *448* 503(141), *534*

Popov, E. M., 434(172), *450*, *486*, 487, 488 (110, 121), 491(121), *492*, 494(128), 495(128), 497(121), 499(133), *500*, 508 (128), *534*
Portnova, S. L., 494(125), *534*
Post, B., 401(78), *448*, 486(112), *534*
Powell, H. M., 65(43), *131*
Powell, M. J. D., 405(95), *448*
Powles, G. J , 275, 276, *293*
Prothero, J. W., 504, 508(153), *535*
Ptitsyn, O. B., 394(43), *447*, 456(13), 460(20), 504, 509(163), *534*, *531*, *535*
Pullman, B., 522(195), *536*
Pullman, H., 484, *533*

Q

Quagliarotti, G., *535*

R

Rakhimov, A. A., 222(9), 223(11), *232*
Ramachandran, G. N.. 386(22), 387, 388, 389, *446*, 479, 480(100), 481, 482(105), 484, 485, 487, 488, 490, *491*, *493*, 500(95), 519, *533*, *534*, *535*
Ramakrishnan, C., 479(80, 81), 487(117). *533*, *534*
Ramakrishnan, K., 482, *533*
Rao, K. N., *433*
Rastrigin, L. A., 503(142), *534*
Rein, R., 522(194, 196), *536*
Reinhardt, R. C., 468(43), *532*
Remyga, S. A., 106(75), 122(75), *132*
Reuss, A. Z., 322, *336*
Rice, O. K., 134(1), 139, *190*, *191*
Rich, A., 62(40), *131*, 517, *535*
Richards, E. J., 526(225), *536*
Ried, C., 166, *191*
Riger, M., 382(4), *446*
Ripamonti, A., 387(28), 388(28), *446*, *463*, 464(29), 465(29), 466(29), 468(29). 470 (29), 472(29), 477(29), 479(82), *531*, *533*
Rizvi, S. H., 385(14), *446*
Robas, V. I., 88(59) *132*
Roberts, K. B., 481(101), *533*
Robertson, G. B., 385(13), *446*
Robertson, J. M., 38(15, 17), 40(18), 45(21), 46(22), 79(53, 54), 81(55), 86(57), *131*, *132*, 178(47), 181(49), *192*, 416(120), *449*

Rossini, F., 409(103), *448*
Rossini, F. D., 426(145), *449*
Roth, E. A., *433*
Rotshild W. G. 401(70), *448*
Ruchoff, J. R., 439(174), *450*
Rudko, A. D., 504(156), *535*
Ryzhenkov, A. P., 179, 181(53), *192*, 297, *336*, 358, *380*

S

Saegura, T., 470(48), 471(48), *532*
Sakata, Y., 473(58), *532*
Samaraskaya, V. D., 114(81), *132*
Santry, D. P., 153, 160(20), 174(20), *191*
Sarathy, K. P., 487(117), *534*
Sasisekharan, V., 387, 388, *446*, 479, 481, 484, 485, 488(95), 490, *491*, 500(95), 518(186), 519(186, 188), 520, 521, *533*, *535*
Schacher, G. E., 145, *191*
Schaerer, A. A., 53, *131*
Scharpen, H., *433*
Scheraga, H. A., 386, 388, 389, 390(18), 396(17), 399 405, *446*, *447*, *448*, 464, 479, 480(100), 481, 482(84, 91), 485, 487, 488, 489, 491(96), 493(96), 496(96), 497(88), 498(91), 500(96), 502, 504(157), 505, 508, 523(209), 525(209), *531*, *533*, *534*, *535*, *536*
Schimmel, P. R., 496(131), 506, *534*, *535*
Schmeising, H. N., 430, 434(167), *450*
Schmidt, G. M. J., 73(48), *131*, 502(137), *534*
Schoon, T., 57, *131*
Schroeder, R., 485, *533*, *534*
Schumann, S. L., 401(82), *448*
Schweizer, M. P., 523(200, 201, 204, 205), *536*
Scott, R. A., 386, 388, 389, 390(18), 396(17), 399, *446*, *447*, 464, 479(85, 86, 90, 91, 92, 93), 482(91), 485(85, 91), 498(91), *531*, *533*
Seeds, W. E., 526(227), *536*
Seki. S., 169(35) *191*
Sela, M., 503(143), *534*
Semin, G. K., 88(59), *132*
Senent, S., 414(118, 119), *449*
Serra, R., 473(55), *532*
Sharonov, Yu. A., 460(20), *531*
Shaw, D. E., 396(55), *447*

Shchedrin, B. M., 170(39), *191*
Shearer, H. M. M., 50(27), 86(57), *131*, *132*
Shefter, E., 523(220), *536*
Shemyakin, M. M., 500(136), *534*
Shimanouchi, T. 411(107), 443(184), 444 (189), *448*, *450*, 459, 488(118), *531*, *534*
Shoemaker, D. P., 229, *232*
Shpolskij, E. V., 115, *132*
Shulman, J. S., 475(67), *532*
Sianesi, D., 452(6), 473(55), *531*, *532*
Sim, G. A., 86(57), *132*
Simonetta, M., 190(63), *192*, 421(131), *449*
Sinanoğlu, O., 521, *535*
Sinclair, V. C., 38(17), *131*, 181(49), *192*
Sinsheimer, R., 510(165), *535*
Skinner, H. A., 427, 439(175), *440*, *449*, *450*
Slater, J., 136, 140, *191*
Slater, J. C., 389, *447*
Smith, A. E., 51, 53, *131*
Smith, E. A., 439(176), *440*, *450*
Smith, H. A., 439(174), *450*
Smith, J., 390(34), *447*
Smith, J. A., 279, *293*
Smith, J. A. S., 171(40), 187(40), *191*
Smith, W. V., 158(22), 159, *191*
Smyth, C. P., 466(38), 484(38), *532*
Snyder, L. C., 400(65), *447*
Snyman, I. A., 123(91), *133*
Sobell, H. M., 517(182), 519(182), *535*
Somayaijulu, G. P., 441(178), *450*
Sotnikova, L. E., 526(226), *536*
Spang, H. A., 405(91), *448*
Sparks, R., 523(220), *536*
Spencer, M., 511, 526(229), *535*, *536*
Spitzer, R. 401(83), *448*
Stefani, R., 474(65), *532*
Stepanov, B. I., 443(183), *450*
Stepanov, N. F., 426(147), *449*
Stewart, J., 487(114), *534*
Strauss, H. L., 405(98), *448*
Streltsova, I. N., 6(3), *130*
Strogalin, Yu. V., 467(39), *532*
Struchkov, Yu. T., 6(3), 9(5), 14(8), 64(41), *130*, *131*, 382(10), 388(30), 413(117), 421(30), 423(134), 425(137, 138), 426(30, 139), *446*, *447*, *449*
Sugeta, H., 460(24), 462, *531*
Sukhorukov, B. I., 386, *446*, 522(197), 523(197), *536*
Sumsion, H. T., 68(45), *131*

Sundaralingam, M., 513, 514(173), *535*
Sutor, D. J., 513(170), *535*
Suzuki, K., 169(35), *191*
Suzuki, M., 358(4), *380*
Suzuky, M., 221(8), 222(8), *232*
Swan, R. J., 517(183), *535*
Syvin, S., 419, *449*
Szent-Gyorgyi, A. G., 504(145), *534*

**T**

Tadokoro, H., 473(56), 477, 478(74, 76), *532*, *533*
Tahara, S., 477(74), 478(74), *532*
Takeniko, S., 400(67), *447*
Tasumi, M., 464(28), *531*
Tatevsky, V. M., 426, 427, 441(180), *449*, *450*
Taylor, C. A., 262(6), *293*
Taylor, W. J., 49, *131*
Teller, E., 137, *191*
Terrier, J., 474(65), *532*
Teslenko, V. F., 313, *336*
Teulings, R. O., 467(41), *532*
Thomas, D. R., 487(113), *534*
Timmermans, J., 122(88), *132*, 358(3), *380*
Tinoko, I., Jr., 518(185), 522, 523(199, 207, 208, 210, 211), *535*, *536*
Tollin, P., 513(170, 171), *535*
Toneman, L. H., *408*, 412(113a), *449*
Toneman, L. M., 412(110), *448*
Touillaux, R., 272(11), *293*
Tovbis, A. B., 170(39), *191*
Toyotoshi, V., 400(67), *447*
Traetteberg, M., 412(114), *433*, *449*
Trotter, J., 4(2), 40(18), 41(19), 43(20), 44(20), 87(58), *130*, *131*, *132*, 187(59), *192*, *408*, 419(127, 128), *449*
Trueblood, K. N., 229, *232*, 424(135, 136), *449*, 513(177), 516, 523(220), *535*, *536*
Truter, M. R., 66(44), *131*
Tsetlin, M. L., 503(140), *534*
Tsetlin, V. I. 494(125), *534*
T'so, P. O. P., 523(200, 201, 204, 205, 218, 219), *536*
Tumanyan, V. G., 526, *536*
Turner, R. B., 439(176), *440*, *450*
Tyminski, I. J., 400(68), 407(68), 429(68), *447*

**U**

Ueda, T., 411(107), *448*
Ullbricht, T. L. V., 517(183), *535*
Urey, H. C., 443, *450*
Utkina L. F., 115(84), *132*

**V**

Vainshtein, B. K., 50, *131*
Van-Catledge, F. A., 394(39), 400(68), 407(68), 429(68), *447*
Vand, V., 50(27), 51, *131*, 454(11), *531*
Vanderkooi, G., 479(90, 91, 92, 93), 482(91), 485(91), 498(91), 502(92), *533*
Van Holden, K. E., 523(213, 214), 524(221, 222), 525(213), *536*
Van Meersche, M., 272(11), *293*
Van Vleck, J. F., 270, *293*
Vaughan, P., 22(14), *131*
Vaughan, W. E., 439(174), *450*
Veillard, A., *395*
Venkatachalam, C. M., 386(22), 388(22), 389, *446*, 481(116), 482(105), *493*, *534*
Vittorio, C., *164*
Vitvitsky, A. I., 441(181), *450*
Voight W., 322, *336*
Volkenstein, M. V., 394(42), 443(183), *447*, *450*, 452(1), 481(102), *531*, *533*
Volkov, S. D., 322(8), *336*
Vostokova, E. I., 400(66), *447*
Vournakis, J. N., 523(209), 525(209), *536*
Vull, E. B., 260(5), *293*

**W**

Walker, P. M. B., 481(101), *533*
Walley, E., 169, *191*
Wallwork, S. C., 122(90), *133*
Walmsley, S. H., 145(17), *191*
Walsh A. D., 430, *450*
Warshaw, M. M., 523(208, 210, 211), *536*
Warshel, A., 429(161), 445, *450*
Wasai, G., 470(48) 471(48), *532*
Watson, D. G., 86(57), *132*, 513(170), *535*
Waugh, J., 276, *293*
Weghofer, H., *191*
Weissberger, A., 89(61), *132*
Wells, A. J. 409(102), *448*
Welsh, H. L., 396(55), *447*
Westheimer, F. H., 382, *446*
Westrum, E. E., 134, *190*, 328, 330(15), *336*

Weulersse, P., 220, *232*
White, F. H., 503(143) *534*
White, J. G., 38(15), 45(21), 46(22), *131*, 178(47), *192*
Whiting, M. C., 439(175), *450*
Wiberg, K., 393(37), 402, *447*
Wiener, K., 427(156), *450*
Wilkins, M. H. F., 526(227, 228, 229), *536*
Williams, D. E., 165, 182, *191*, *192*, 386, *446*
Willson, A., 122(90), *133*
Wilson, E. B., 395(48), 396(48), 409(102), 442(182), 443(182), *447*, *448*, *450*
Wilson, H. R., 513(171), 526(227, 228), *535*, *536*
Wilson, S. A., 171(42), *191*
Winter, N. W., *395*
Witney, J., 477(70), *532*
Wojtala, V. J., 137, *191*
Wolf, K., *191*
Wonacott, A. J., 527(230), *536*
Wool, E. B., 468(44), *532*

**Y**

Yaravoy, S. M., 426(147), *449*
Yasumoto T., 477(71), *532*
Yokozeki, A., *408*
Yoshihara, S., 477(74), 478(74), *532*
Young, D. W., 513(171), *535*
Young, J. E., *433*

**Z**

Zane, G., 397(56), *447*
Zaripov, N. M., 407(101), 409(104), 411(104, 108), 413(104), 429(160), 439(101), *448*, *450*
Zerbi, G., 467(42), *532*
Zhdanov, G. S., 70(46), *131*
Zhidkov, N. P., 170(39), *191*
Ziegler, K., 452, *531*
Zotova, S. V., 400(66), *447*
Zwolinski, B. J., 441(178), *450*

# Subject Index

## A

Acenaphthalene, 249
N-Acetylglycine methylamide, 491
O-Acetyl-D-L-oxyisovaleric acid
   dimethylamide, 501
N-Acetyl-L-valine-methylamide, 497
Acoustic frequencies, 205
Acridine, 72
Acridine–anthracene, 105
Activation energy of reorientational
   motion, 230
Adamantane, 11
Adenosine-phosphateuridine, 523
Adenylic acid, 513
Adiabatic
  approximation, 142
  calorimeter, 329
  compressibility, 340
  elasticity, 307, 341, 349
  tensor of elasticity, 341
Alanine dipeptide, 492, 496
Alloxan, 17
Anemonine, 401

Angular deformation energy, 383
Anharmonicity, 369
Anisotropy of velocities, 314
Anthracene, 38, 149, 373
Anthracene–phenanthrene, 105
Anthrone–anthraquinone, 96
Antisymmetric vibrations, 207
Arginase, 509
Asymmetry parameter, 284
Atomic charges, 234
Atomic scattering factor, 249
Atomization energy, 427
Average molecule, 85, 88
Average scattering, 263
Azulene, 86

## B

1,14-Benzbisanthrene, 44
Benzene, 187, 373
1,12-Benzperylene, 43
Bicycloheptane, 91
Biphenyl, 190
Bond lengths, 189, 249

Bond orientation energy, 396
Bragg's law 238
Bromocyclobutanes, 401
5'-Bromo-deoxythymidine, 511
2-Bromo-1,1-di-*p*-tolyl-ethylene, 190
*cis*-1,3-Butadiene, 434
*cis*-2-Butene, 409


## C

Calcium thymidylate, 513
Central forces, 143
Chair–boat conformations, 402
Characteristic temperature, 344
Chloranil, 291
*p*-Chlorobromobenzene, 86
Chlorocyclobutanes, 401
2-Chloro-2-nitropropane, 275
*cis*- and *trans*-Chlorovinyl esters, 475
Christoffel's equation, 308
Chrysene, 416, 419
Chymotrypsin, 505
Chymotrypsinogen, 509
Clahrate compounds, 126
Closest-packed layer, 25
Compliances, 358
Compliance moduli, 317
Compliance tensor, 317
Compressibility measurements, 330
Condition of stability, 184
Configuration statistics of polymer chains,
    398
Configurational entropy, 522
Conformers, 398
Contact specimen method, 106
Continuous series of solid solutions, 95
Coordination number, 15, 22, 23
Coronene, 46
Coupling coefficient, 195
Crystallizer, 107
Cubane, 429
Cyanuric triazide, 13
"Cyclization" potential, 405
Cyclodecane 402
Cycloheptane, 402
1,3,5-Cycloheptatriene, 412
1,4- and 1,3-Cyclohexadienes, 411
Cyclohexane, 402
Cyclohexene, 411

Cyclononane, 402
Cyclooctane, 402
Cyclooctatetraene, 409, 413
Cyclopentadiene, 410
Cyclopentane, 401
Cyclopentene, 410
Cytidylic acid, 513


## D

Debye approximation, 344
Degree of quasi harmonicity, 357, 363
Denaturation, 509
Deoxiribose, 402, 511, 512
Deoxyadenosine, 513
Deoxyribonuclease, 509
Determining contacts, 3, 10
1,3,5-Deuterobenzene, 277
*cis*-Deuteropropylene, 475
A-DNA, 528
B-DNA, 528
5:6-7:8 Dibenzoperylene, 416
3:4-5:6 Dibenzophenonthrene, 416
Dibenzyl, 313
Dibenzyl–stilbene, 105, 110
*p*-Dibromobenzene, 289
*p*-Dibromobenzene—*p*-chloronitrobenzene,
    106
1,2-Dibromo-1,2-dicarbomethoxycyclo-
    butane, 401
Dicarboxylic acids, 182
Dichloracenaphthene, 423
Dichlorodinitromethane, 291
*o*-Dichlorobenzene, 291
*p*-Dichlorobenzene, 73, 288, 292
1,2-Dichloroethane, 275
Dichlorodiphenylnaphthacene, 424
Diffuse scattering, 267
1,1-Difluorocyclobutane, 401
Dihalomethanes, 291
1,12-5,8-Dimethylbenzophenanthrene, 437
1,3-Dimethylenecyclobutane, 438
Dimethylhexahelicene, 437
2,6-Dimethylnaphthalene, 6, 260
2,7-Dimethylphenanthrene, 437
4,5-Dimethylphenanthrene, 437
2,2-Dimethylpropane, 275
4,4'-Dinitrodiphenyl–4-hydroxy-diphenyl,
    124

1,5-Dinitronaphthalene, 4
2,2-Dinitropropane, 275
Dinucleosidephosphates, 510
Dipeptides, 490
Diphenyl, 41
Diphenyl-α, α-dipyridyl, 95
Diphenyl mercury, 373
Dipole contribution, 151
Dipole interactions, 153
Di-*p*-xylylene, 35
Disordered crystals, 89
Dispersion "hypersurfaces", 195
Dispersion interaction, 135
Dispersion surfaces, 220, 267, 268, 343
Dovetail principle, 6


**E**


Effects of the crystalline field, 184
Eigenvectors, 207, 210, 220, 221
Elasticity tensor, 309, 313
Electron cloud, 242
Electron density, 234, 245
Electron density series, 258
Electrostatic contribution, 161
Electrostatic interactions, 141
Enantiotropic polymorphic change, 73
Energy surface, 174
Enniatine, 502
Entropy, 338
Equilibrium
    conformation, 384
    crystal structure, 169
    distance, 163
Ethylene, 373
Expansion
    coefficient, 340
    ellipsoid, 294
    patterns, 297
    tensor, 294


**F**


Ferrocene, 62
5-Fluoro-2'-deoxy-β-uridine, 513
Forbidden regions, 482, 490
Fourier method, 242
Fourth moment, 270

Framework structures, 128
Free energy, 337, 362
Furanose ring, 511


**G**


Gaseous-crystalline structure, 53
Gaseous hydrate type, 129
Geometrical analysis, 6, 8
Globular protein, 452
Glutamine, 481
Glycine, 82, 249, 490
Glycine dipeptide, 490
Glycosidic bond, 515, 518, 530
Gradient descent method, 260
Gramicidine, 502
Graphite, 12
Grüneisen
    coefficient, 369
    constant, 346
    tensors, 356, 373


**H**


Harmonic oscillations, 194
Heat capacities, 338, 340, 358
Heats of sublimation, 134
α-Helix, 493, 504
Helix-breaking, 506
Helix-making, 506
Helix parameters, 459
Heptane, 399
Hexachlorobenzene, 6
Hexaginase, 509
Hexamethylbenzene, 182, 184, 276
Hexamethylenetetramine, 22, 35
Hexamethylprismane, 429
Hexane, 399
Hexanitrobenzene, 14
High-frequency lattice vibrations, 224
Hooke's law, 309
Hydration energy, 489
Hydrogen bond energy, 166


**I**


Ideal bond angles, 382
Ideal packing, 3, 8, 18, 260
Interblock solubility, 114, 116

Intermolecular radii, 3
Intermolecular vibration frequencies, 220
Internal energy, 338, 362, 375
Internal parameters, 462
Internal rotation barrier, 394
Intramolecular frequencies, 377
Iodobenzene, 291
Iodoform, 9
Iron
    pentacarbonyl, 65
    tetracarbonyl, 66
Isobaric
    coefficient of volume expansion, 338
    process, 353
Ising statistical model, 119
Isomerization paths, 405
Isomorphous substitution, 261
Isotactic polymers, 453
Isothermal compressibility, 338, 340
Isothermal elasticity, 341
Isothermal tensor of elasticity, 349

K

$\beta$-Keratin, 493
Knudsen method, 334

L

Lattice component of free energy, 343
Lattice distortions, 115
Lattice sums, 145, 167
Layers of maximum density, 29
Layer stacking, 59
Least-squares method, 242
Libration frequencies, 222
Limiting frequencies, 220
Linear compressibility, 317
Low-frequency vibrations, 224
Lysozyme, 503, 509

M

Matrix of dynamic coefficients, 220, 226
Matrix lattice, 115
Meso-$\beta$, $\beta'$-dimethyladipic acid, 182
Method of steepest descent, 182

Methylchloroform, 275
1-Methyl-3 methylcyclobutene, 438
20-Methylcholanthrene, 416
Molecular
    coordination number, 205
    packing coefficient, 9, 19
    transforms, 261
Monobromopentachlorobenzene, 88
Morphotropic changes, 18
Multiple bond, 430
Myoglobin, 503

N

Naphthalene, 38, 249, 280, 313, 373, 379
Naphthlane-$\beta$-chloronaphthalene, 105
Naphthalene-$\beta$-naphthylamin, 105
p-Nitrochlorobenzene, 86
NMR line breadth, 274
NMR line widths, 231
NMR spectrometer, 281
Nonadditivity effect, 137
Nonbonded interactions, 382, 384
Noncentrality of the atom–atom potentials, 165
Nonhelical conformations, 507
Norcamphane, 407
Normal propylbenzene, 72
Norpinane, 407
Nortricyclene, 407
Nuclear
    amplitude, 254
    Scattering, 255
Nucleic acids, 452, 510, 515, 524
Nucleosides, 513
Nucleoside-3′-phosphate, 519
Nucleoside-5′-phosphate, 519
Nucleotides, 513

O

Octachlorocyclobutane, 401
Octachloronaphthalene, 421
Octafluorocyclobutane, 401
Optical
    branches, 224
    dispersion, 135
    masks, 261

Optimal
  conformations, 390, 509
  structure, 189
Orientation effect, 135
Ovalene, 47
Overcrowded molecules, 385
Oxalic acid, 77, 78
Oxalic acid dihydrate, 279


**P**


Packing of a bimolecular crystal, 121
Patterson series, 256
Paracyclophanes, 424
1,3-Pentadiene, 435, 438
1,4-Pentadiene, 438
Pentane, 399
Pepsin, 509
Perfluorethane, 275
Phenanthrene, 187, 419
Phenazine, 73
Plastic crystals, 89
Poisson equation, 253
Polyacrylonitryl, 474
Polybutadiene, 474
Polychlorvinyl, 474
Polyethylene, 399, 464
Polyethyleneoxide, 477
Polyisobutylene, 470, 472
Polymethyleneoxide, 476
Poly-2,5-methyl-styrene, 473
Poly-m-fluorostyrene, 473
Polymorphism in *n*-paraffins, 62
Polynucleotides, 511
Polynucleotide conformations, 516
Poly-*o*-fluorostyrene, 473
Poly-*o*-methyl-*p*-fluorostyrene, 473
Polyoxacyclobutane, 478
Polypeptide, 479
Polypeptide conformation, 480
Polypropylene, 472
Poly-p-trimethyl-silylstyrene, 473
Polystyrene, 454
Polytetrafluoroethylene, 464
Polyvinyl chloride, 453
Polyvinylidene
  chloride, 454, 468, 472
  fluoride, 467
Polyvinylcyclohexane, 473

Potential
  barrier, 276
  surface, 451
  well depth, 163
Propylene, 410
Pyrene, 45
Pyrimidine nucleosides, 517


**Q**


Quadratic mean displacement, 227
Quadrupole contribution, 160
Quadrupole interaction constant, 284
Quadrupole moments, 158, 282
Quantum chemistry, 234
Quasi-harmonic
  approximation, 219
  model, 346, 351
Quasi-longitudinal waves, 308
Quasi-shear waves, 308
Quaterphenyl, 42


**R**


Renaturation, 509
Reorientation of a molecule, 277
Repulsive forces, 137
Residual
  charge, 161
  electrical charges, 143
Resorcinol, 79
Reuss's method, 322
Ribose, 402, 512
Riboso-phosphate backbone, 515, 520
RNA, 528
Rotation barrier, 232
Rotation potentials, 519
Rotational csystalline state, 73


**S**


Scattering amplitude, 243
Second
  moment, 270
  virial coefficients, 138
Selection of conformations, 186
Shape of NMR lines, 270
Short-range order, 92, 118, 119
Shpolskij effect, 115

Silanecyclobutane, 401
Single-stranded polynucleotides, 526
Solid solutions of *n*-paraffins, 117
Solubility conditions, 98
Spectroscopic deformation constants, 384
Spin–lattice
    relaxation, 287
    relaxation times, 291
Stacking interactions in polynucleotides,
    525
Staggered conformation, 396
State equation, 338
Steric effects, 381
Stilbene, 313
*β*-Structures, 504
Structure
    amplitude, 241
    seeker, 7
Subcell, 51
Sublimation pressure, 332
Summation radius, 170
Symmetric vibrations, 207
Syndiotactic macromolecules, 453

T

Tensor of adiabatic elasticity, 358
Tensor of compliances, 342
Terphenyl, 42
Tetrabenzoperopyrene, 416
Tetrabromobenzene, 292
1,2,4,5-Tetrachlorobenzene, 288
Tetrachloro-*p*-dibromobenzene, 88
Tetracyanocyclobutane, 401
Tetraphenylcyclobutane, 401
Tetraphenylmethane, 68
Thermal
    diffuse X-ray scattering, 268
    expansion tensor, 358
    waves, 266
Thermodynamic potentials, 328
Thiourea, 279
Thymidine, 513
Tolan, 313
Tolane–diphenylmercury, 101, 114

Torsion energy, 399
*trans*-deuteropropylene, 475
*trans*-1,3-dicarboxicyclobutane, 401
*trans*-isobutyl-propenyl esters, 475
Translational vibrations, 205, 223
*trans*-emthyl, 475
1:12-5:6-7:8 Tribenzoperylene, 416
1,1,1-Trichloroethane, 275
Trimethylene oxide, 401
Trimethylene sulphide, 401
Trichloroacetamide, 291
Trichlorogermane, 291
Triphenylene, 419

U

Ultrasonic method, 311
Uniform compression, 319
Universal potential, 389
Urea, 278
Urease, 509

V

Van Vleck formula, 271, 278
Vibration
    amplitudes, 195
    frequencies, 206
Vibrational corrections, 377
Vinyl polymers, 464
Voigt's method, 322

W

Watson–Crick model, 523
Work of expansion of a lattice, 362

Z

Zeeman splitting, 288
Zero-point energy, 168, 343
Zero-point vibrations, 168, 378

# Physical Chemistry

## A Series of Monographs

**Ernest M. Loebl,** Editor

Department of Chemistry, Polytechnic Institute of

Brooklyn, Brooklyn, New York

1   W. Jost: Diffusion in Solids, Liquids, Gases, 1952
2   S. Mizushima: Structure of Molecules and Internal Rotation, 1954
3   H. H. G. Jellinek: Degradation of Vinyl Polymers, 1955
4   M. E. L. McBain and E. Hutchinson: Solubilization and Related Phenomena, 1955
5   C. H. Bamford, A. Elliott, and W. E. Hanby: Synthetic Polypeptides, 1956
6   George J. Janz: Thermodynamic Properties of Organic Compounds — Estimation Methods, Principles and Practice, Revised Edition, 1967
7   G. K. T. Conn and D. G. Avery: Infrared Methods, 1960
8   C. B. Monk: Electrolytic Dissociation, 1961
9   P. Leighton: Photochemistry of Air Pollution, 1961
10  P. J. Holmes: Electrochemistry of Semiconductors, 1962
11  H. Fujita: The Mathematical Theory of Sedimentation Analysis, 1962
12  K. Shinoda, T. Nakagawa, B. Tamamushi, and T. Isemura: Colloidal Surfactants, 1963
13  J. E. Wollrab: Rotational Spectra and Molecular Structure, 1967
14  A. Nelson Wright and C. A. Winkler: Active Nitrogen, 1968
15  R. B. Anderson: Experimental Methods in Catalytic Research, 1968
16  Milton Kerker: The Scattering of Light and Other Electromagnetic Radiation, 1969
17  Oleg V. Krylov: Catalysis by Nonmetals — Rules for Catalyst Selection, 1970
18  Alfred Clark: The Theory of Adsorption and Catalysis, 1970

# Physical Chemistry

A Series of Monographs

19  ARNOLD REISMAN : Phase Equilibria : Basic Principles, Applications, Experimental Techniques, 1970

20  J. J. BIKERMAN : Physical Surfaces, 1970

21  R. T. SANDERSON : Chemical Bonds and Bond Energy, 1970

22  S. PETRUCCI, ED. : Ionic Interactions : From Dilute Solutions to Fused Salts (In Two Volumes), 1971

23  A. B. F. DUNCAN : Rydberg Series in Atoms and Molecules, 1971

24  J. R. ANDERSON : Chemisorption and Reactions on Metallic Films, 1971

25  E. A. MOELWYN-HUGHES : Chemical Statics and Kinetics of Solution, 1971

26  IVAN DRAGANIC AND ZORICA DRAGANIC : The Radiation Chemistry of Water, 1971

27  M. B. HUGLIN : Light Scattering from Polymer Solutions, 1972

28  M. J. BLANDAMER : Introduction to Chemical Ultrasonics, 1973

29  A. I. KITAIGORODSKY : Molecular Crystals and Molecules, 1973

30  WENDELL FORST : Theory of Unimolecular Reactions, 1973

*In Preparation*

JERRY GOODISMAN : Diatomic Interaction Potential Theory. Volume 1, Fundamentals ; Volume 2, Applications

I. G. KAPLAN : Symmetry of Many-Electron Systems